T0295085

A Different Thermodynamics and Its True Heroes

A Different Thermodynamics and Its True Heroes

Evgeni B. Starikov

PAN STANFORD PUBLISHING

Published by

Pan Stanford Publishing Pte. Ltd.
Penthouse Level, Suntec Tower 3
8 Temasek Boulevard
Singapore 038988

Email: editorial@panstanford.com
Web: www.panstanford.com

British Library Cataloguing-in-Publication Data
A catalogue record for this book is available from the British Library.

Cover Image Copyright: Dr. George Augustus Linhart

ISBN 978-981-4774-91-8 (Hardcover)
ISBN 978-0-429-50650-5 (eBook)

To my dear mom, Lilia Pavlovna Starikova (1939–1973)
To my dear grandpa, Yosef Yakovlevich Volisson (1913–1987)
To my dear grandma, Marina Naumovna Gorer (1918–1991)

To my respected teacher, Vladimir Yakovlevich Maleev
(1930–2018)

To my dear friend, Amarin Chateau Sallarde,
the Oriental Shorthair Cat (2013–2018)

*"The fine arts are more captivating in their colours, it is true,
and general literature more attractive than science: But science,
embracing the whole circle of nature, is more important, more
various, and more dignified."*

—Charles Bucke (1781–1846)
*Amusements in Retirement; or, the Influence of Science, Literature,
and the Liberal Arts, on the Manners and Happiness of Private Life*
Henry Colburn, London, Great Britain, 1816

*"A noble human life does not end on earth with death. It continues in
the minds and the deeds of friends, as well as in the thoughts and the
activity of the nation."*

—*The Lost Manuscript: A Novel* by Gustav Freytag (1816–1895)
Authorised translation from the 16th German Ed. (in two volumes)
The Open Court Publishing Company, Chicago, USA, 1890

*"N'en déplaise à ces fous nommés sages de Grèce,
En ce monde il n'est point de parfaite sagesse;
Tous les hommes sont fous, et malgré tous leurs soins
Ne diffèrent entre eux que du plus ou du moins."*

—Nicolas Boileau-Despréaux (1636–1711)

*"**No offence to all Greek sages whichever** crazy things they told,
There's still never perfect wisdom anyhow in this world;
All the humans are so crazy, while in spite of all their care
They do differ from each other more or less, as we're aware."*

—Our response to Nicolas Boileau-Despréaux

"All great song has been sincere song."

—John Ruskin (1819–1900)

Contents

Preface

This book ought to be the very first of its kind as it represents a bold experiment, which, I hope, might be of some use to ascertain what should be in effect the very nature of scientific research work, whatever field it is embracing. With this in mind, I have sincerely illustrated here the zest of the chosen topic by introducing several bright examples in the field of thermodynamics.

While writing this book, I had formed somewhat accurate estimate of its probable effect: I flattered myself by the thought that young colleagues, who should be interested, will be pleased with it, fancy it, and would read it with much more than just a common pleasure. On the other hand, I believe that those who should dislike it would fall over it and read it with much more than common dislike.

Several of my friends, including my beloved relatives and spouse, are anxious for the actual success of the volume at hand, as they believe that the combination of a scientific monograph, historical investigations, philosophical deliberations plus a little bit of everyday poetry might turn out to be a mere wreck!

Now, at last, I leave the final judgement up to the readership.

Evgeni B. Starikov
Autumn 2018

Introduction

First and Second Law

The version of thermodynamics by Flanders & Swann (Years active: 1956–1967)

"The first law of thermodynamics
Heat is work and work is heat
Heat is work and work is heat
Very good

The second law of thermodynamics
Heat cannot of itself pass from one body to a hotter body
Heat cannot of itself pass from one body to a hotter body
Heat won't pass from a cooler to a hotter
Heat won't pass from a cooler to a hotter
You can try it if you like but you far better notter
You can try it if you like but you far better notter
'Cause the cold in the cooler will be hotter as a ruler
'Cause the cold in the cooler will be hotter as a ruler
Because the hotter body's heat will pass through the cooler

Heat is work and work is heat
And work is heat and heat is work
Heat will pass by conduction

A Different Thermodynamics and Its True Heroes
Evgeni B. Starikov
Copyright © 2019 Pan Stanford Publishing Pte. Ltd.
ISBN 978-981-4774-91-8 (Hardcover), 978-0-429-50650-5 (eBook)
www.panstanford.com

Heat will pass by conduction
And heat will pass by convection
And heat will pass by convection
And heat will pass by radiation
And heat will pass by radiation
And that's a physical law

Heat is work and work's a curse
And all the heat in the universe
It's gonna cool down as it can't increase
Then there'll be no more work
And they'll be perfect peace
Really?
Yeah, that's entropy, man!
And all because of the second law of thermodynamics, which lays down

That you can't pass heat from the cooler to the hotter
Try it if you like but you far better notter
'Cause the cold in the cooler will get hotter as a ruler
'Cause the hotter body's heat will pass through the cooler

Oh, you can't pass heat from the cooler to the hotter
You can try it if you like but you far better notter
'Cause the cold in the cooler will get hotter as a ruler
That's the physical law

Ooh, I'm hot!

What? That's because you've been working

Oh, Beatles? Nothing!

That's the first and second law of thermodynamics..."

Our present story begins with the above-cited jazzy setting of the basic laws of thermodynamics by the actor and singer Michael Flanders (1922–1975) and the composer, pianist and linguist Donald Swann (1923–1994). This formulation was indeed highly popular in the years 1956 to 1967, and it sounds like something fully applicable till nowadays. Definitely, the story as a whole does look like a funny light opera. Meanwhile, the funniest point here ought to be that the very gist of exactly this same story we might read in thousands of thermodynamics books in all the World's languages. We might hear exactly this same story in the university lecture halls and rooms worldwide—and—the last but not the least—the story lasts already more than hundred years and truly seems to be lasting till doomsday.

Howbeit, what could and can we hear from the recognized specialists in the field? Indeed, aside from the light operas we ought to have physics, chemistry, biology, etc. This is why so many questions do arise to the address of the leading specialists in these fields.

In general, and as a whole, thermodynamics turns out to be a truly multilevel knowledge area. Still, at the first glance the multilevel nature of the area seems to boil down to different fashions, flavors, tastes, and representations of thermodynamics having indeed practically nothing in common. An outstanding American physicist and thermodynamicist Prof. Dr. Mark Waldo Zemansky (1990–1981) could even distinguish among the three quite separate pieces of them: *an engineering thermodynamics*, *a chemical thermodynamics*, *a physical thermodynamics*—and clearly, he was not happy at all about his finding [1].

He concludes that *physical thermodynamics*, which we expect to serve as the solid conceptual basis for the other two branches just mentioned, is dealing mostly with the pure mathematics, instead of concentrating on the actual physics. Noteworthy, his conclusion has been drawn at the peak time of Flanders and Swann's light opera popularity. Still, to our sincere regret, this same assessment does remain valid till nowadays.

At first glance, Prof. Zemansky's claim sounds really strange, for we know about the tightest interrelationship between physics and chemistry, with the respective products being physical chemistry and chemical physics. Moreover, without any profound knowledge in physics/chemistry there is no way to success in the respective engineering branches. Therefore, how might thermodynamics, being possessed of no unique representation, serve as a true knowledge area? Isn't it then just a conglomeration of some information portions being not properly connected to each other?

The answer to the latter question is definitely negative. Sure, thermodynamics is long and well known to be fruitfully used for getting valid physical, chemical, and engineering results. Although the lack of (sometimes even just elementary!) logics in the thermodynamics' foundations has long been noticed, whereas the prevalent standpoint is being based upon the purely '*operational approach*' to the problem. A prominent American physicist and Nobel Prize winner Prof. Dr. Percy Williams Bridgman (1882–1961) did proclaim the latter already long ago. To paraphrase it in short terms: *We see that thermodynamics is somehow correct and is working, but we really don't know, why it is just like this, with the first part being the main point of our concern.*

Bearing this entirety in mind, one might immediately ask: How about mathematics? Mathematics is long and well known to be the pertinent basement for drawing logically perfect conclusions. There were a number of outstanding mathematicians who boldly accepted the challenge of trying to formulate the logical foundations of thermodynamics. The first and foremost colleagues in this raw were Prof. Dr. Jules Henri Poincaré (1854–1912) and Prof. Dr. Constantin Carathéodory (1873–1950). Prof. Poincaré, *inter alia*, was greatly helpful in properly understanding the ingenious contribution by the undoubted pioneer of thermodynamics, Nicolas Léonard Sadi Carnot (1796–1832). Prof. Carathéodory could manage building up the logical basement of thermodynamics, but apparently had dearly missed a physicist/chemist capable of putting his seminal inferences onto a firm physical/chemical basis.

Two prominent American mathematicians should also be mentioned here. They were Prof. Dr. Clifford Ambrose Truesdell III (1919–2000) and Prof. Dr. James B. Serrin (1926–2012). Their books on the theme published by Springer Verlag [2, 3] were, are, and ought to remain indispensable sources of ideas for young researchers, because these works could clearly show the proper directions of thoughts.

Meanwhile, it is back in 1957 that Prof. Zemansky has published his bitter words noteworthy in this connection [1]. As a kind of response to the above-mentioned statements by M. W. Zemansky, C. A. Truesdell had noted as follows [2]:

"Finally, I confess to a heartfelt hope—very slender but tough—that even some thermodynamicists of the old tribe will study this book, master the contents, and so share in my discovery: Thermodynamics need never have been the Dismal Swamp of Obscurity that from the first it was and that today in common instruction it is; in consequence, it need not so remain."

In addition, J. B. Serrin had added the following wise words [3]:

"*Since the earliest days of thermodynamical science, it has always been recognized that the conclusions of the subject were to be obtained deductively from general laws. At the same time, in contrast with the case of other sciences, these laws have not been expressed in any standard or usual mathematical formalism, thus making the deductions appear different from those in other branches of physics. Indeed,* **Buchdahl** *expressed the situation well when he wrote* **'There is no doubt that part of the difficulty of the classical arguments lies in the subtlety with which mathematical notions and ostensibly physical notions are almost inextricably interwoven'.** *Finally, since the early days of the subject there have been repeated calls for rigor in proofs. Thus, along with the discovery of an appropriate mathematical structure in which to carry out the deductions, modern research is also concerned with the struggle for precision. It is not that nothing correct has been available, but rather that rigor has been only occasional and confined to special circumstances. These facts themselves contributed to the general mysteries of the subject, since they caused further confusion between mathematical derivations and physical thinking.*"

Prof. Serrin mentions here Prof. Dr. Hans Adolf Buchdahl (1919–2010), an outstanding Australian physicist and thermodynamicist of German origin, whose thoughtful and instructive thermodynamics books [4, 5] definitely have not lost their significance till nowadays.

Howbeit, irrespective of the *dismal swamp of obscurity* produced by the *physical thermodynamics* mechanical and chemical engineers could properly find an interconnection between thermodynamics and energetics. As a result, they could manage fruitfully and successfully using the second basic law even without clarifying its actual sense. This is why, the motors and refrigerators are working, and the necessary chemical agents are duly synthesized.

On the other hand, physicists and advanced chemists were only producing hundreds of more or less equivalent definitions of the *two basic laws*, as well as further basic laws (the zeroth and third ones), which are in effect nothing more than just paraphrases of some purely mathematical constructions. Still, an unambiguous answer to the following four basic questions could not be found:

1. Equilibrium of **what**—or—**between what and what**—is described by the so-called *equilibrium thermodynamics*?
2. What is the true sense of the *entropy* notion?

3. Consequently, what is the true sense of the *second basic law* of thermodynamics?

4. How many basic laws at all should thermodynamics be possessed of?

The protagonists of the monograph at hand were devoting their lives to thinking over the correct answers to the above fundamental posers. Meanwhile, their seminal work was and is still remaining widely unknown. Instead, the *dismal swamp of obscurity* was successfully developing and presently constitutes the *actual mainstream* worldwide. This is why the present author has decided to try attracting the attention of the professional community to the latter fundamental inconsistency.

Bearing this in mind, the present volume is an attempt to try help interested youthful readership in recognizing that there was and still is another *part of the difficulty of the classical arguments*, which did and still does not lie solely in the mathematical subtleties, but had and still has a number of other quite objective reasons: inattentive/biased reading of the original sources; excessively aggressive marketing methods when striving for public recognition at any price—and—the last but not the least—*our human mortality*.

Except for the latter factor, which is anyway sheer unavoidable, we could connect the former both with the truly revolutionary developments in the physics during the end of the XIX-th and the beginning of the XX-th centuries. In this connection, the present author might recall a highly relevant 'hit' from the time of the former USSR—'*Нет у революции начала—Нет у революции конца!*'—'*There is no start of the revolution—and there is no end of the revolution!*'

Noteworthy, the monograph at hand is not a historical investigation of the actual modalities of the revolution in question; we would instead like to refer the readers to the serious investigations on this very interesting and important theme published in different languages [6–16]. Nonetheless, some important conclusions might be drawn after attentive reading of the sources cited. These conclusions ought to be published here. Moreover, in the chapters to follow we would therefore like to illustrate in detail the process of drawing them.

Along with this, it is of tremendous importance to introduce the colleagues mentioned above, who appear to look like 'sheer counter-revolutionary', but were practically devoting their lives to solving the fundamental problems listed above. A definite feeble point here is that these 'counter-revolutionaries' do remain widely unknown, unlike the 'revolutionaries'. So, why should the former become known at all? To try answering the latter poser in the well-founded way, the monograph at hand lists below a succinct summary of their seminal ideas, in the hope to excite the interest of the readership in reading the chapters to follow.

In fact, there is only one basic law of thermodynamics: It is *the law of energy conservation and transformation*. Thus, the latter *truly golden coin* should have *two sides*: *energy conservation* and *energy transformation*. Noteworthy, every child knows well that the *two sides* of the *golden coin* cannot be separated from each other.

Nonetheless, the true story of the physical revolution of the XIX-th and XX-th centuries seems to consist in passionate overcoming the basically false trend of *separating the two sides of the coin*. Indeed, two prominent theoretical physicists of that revolutionary time: Ludwig Boltzmann and Max Planck, when confronting the problem of the *two basic laws* have suggested the *probabilistic interpretation* of the *second basic law of thermodynamics*. Thus, they could manage killing the latter as a *separate physical reality*. Noteworthy, an outstanding experimental physicist of that time, Niels Bohr, has attempted to go this same step in regard to the *first basic law of thermodynamics*, but his attempt could have been proven to be unsuccessful. As a result, at least one aspect of the actual unique basic law of thermodynamics could remain conceptually intact.

Meanwhile, in speaking figuratively, the true image of the latter, namely the *golden coin*, somehow became fuzzy: As a result we definitely know that there is indeed something *truly golden*, but we cannot recognize the actual form of it. The great achievement of the *revolutionary physicists* headed by Max Planck at the beginning of the XX-th century was to demonstrate the way of how to build up a valid physical theory **without** any direct application to the basic law of thermodynamics. On the one hand, this could open the way to the meanwhile *old and good* quantum physics. On the other hand, it had driven thermodynamics apart from the over-all scientific progress. It is just this 'setting-apart' of an important branch of knowledge that has resulted in its conceptual fragmentation.

Should we then accuse the revolutionary physicists of "intentionally killing thermodynamics?" **Definitely not at all**!

Quantum physics has clearly proven its own usefulness to 100%. Consequently using the quantum physics could result in creating such powerful nuclear weapons as atomic and hydrogen bombs. To our sincere regret, these powerful deadly weapons could meanwhile come to the hands of unambiguous criminals, but this is definitely not due to the intrinsic failures of quantum physics and/or relativity theory.

On the other hand, this same course of scientific progress could definitely lead to an enormous development of energetics, pharmacology, molecular medicine, which could be considered clear successes for the humankind as a whole.

All the listed results could finally be viewed as those resulting from the giant efforts of the revolutionary physicists.

Howbeit, our main poser here ought to be about the destiny of thermodynamics as a branch of knowledge. *Whom might we make responsible for its gradual conceptual putrefaction?*

The only answer ought to be: *All of us*, who had found no better way to develop and then to stubbornly cultivate such truly sociopathic manias as *quantophrenia* and *numerology* based mainly upon the enormous achievements of the revolutionary scientists. Prof. Dr. Pitirim Alexandrovich Sorokin (1889–1968), an outstanding American sociologist of Russian origin, has introduced these terms [17]. Noteworthy, the author of the book at hand does not exclude himself from the crowds to be accused of the manias named.

What was the very 'turning point' of the continuous link between the scientific research and a definitely sociopathic development? To our mind, that was just Max Planck's bold hypothesis of the '*elementary disorder*', which helped him to arrive at the following seminal conclusion (below we would greatly appreciate summarizing the very zest of Max Planck's standpoint):

"**Any *macro*-state of matter is compatible with a huge numbers of its *micro*-states. Meanwhile, there are always such states among the latter whose *temporal evolution* ought to be *paradoxical* in that it does contradict the *second basic law of thermodynamics*. The hypothesis of the *elementary disorder* helps prove that the number of such paradoxical states ought to be infinitesimally** low. The formula connecting the entropy with the probability logarithm represents the unique proof of the Basic Hypothesis just mentioned."

This way we *kill* the second basic law in a virtuoso manner by forcing it to be just a ***probabilistic law*** (*That is, one always has to follow it, but some exclusions from it should still be possible, though very improbable*). The truly marvelous achievement here ought to be the perfect physical theories, which might be produced **without taking the second basic law into account seriously**. In other words, to produce valid physical results, we do not really need to have any spot of bother with the **physical sense of the probability**, which is formally connected to the entropy notion. Moreover, this same holds for the physical sense of the **basic second law**. It is just at this point of our logical inference that the **probability** as it is—*and nothing apart of it*—does finally become a **real physical paradigm**.

Noteworthy, this ought to be just the very first *sociopathic* step. Why? The correct way of the scientific research work would be trying to attack the problem of the actual rational sense of the probability involved. Otherwise, scientific research comes to its ultimate stop here.

At this very point attentive readers would definitely ask the author: *Well, but should the 'second basic law' be killed at all?* The proper answer is: *Sure, of course!* Its widespread formulation of the **revolutionary time** in question, due mainly to Rudolf Clausius, had to be abolished anyway.

The next question: *Might the 'probabilistic method' of abolishing the improper formulation be the only one to solve the problem?* The only answer is: *No, there must surely be different methods as well!* To put this in more detailed way: The Clausius' formulation of this law was not only definitely improper, but could not introduce some separate valid law in itself.

The next question: *Had the 'probabilistic method' helped solving the problem correctly?* The only answer: *Sure, of course!* The Boltzmann–Planck's logarithmic formula is absolutely correct and moreover possessed of the clear-cut physical sense. Now we know this for sure. In the book at hand, we would like to introduce the colleague who could even manage publishing some of his seminal work on this theme, but that was not the 'vogue' topic at the time of Boltzmann, Planck, and their immediate followers.

Remarkably, this was not interesting to Boltzmann and Planck's numerous followers, for the success achieved by the both peers just named was undoubtedly marvelous—and it is still remaining so!

Indeed, Max Planck's bold step is showing the straightforward and simple method of how to skillfully kill the notorious formulation of the second basic law by Rudolf Clausius. In addition to this definite success, Max Planck had but one more theoretical step to go: Along with the second basic law we would also have to kill any undesired *temporal* and *spatial* evolutions. This step has been gone by Albert Einstein, who suggested a total of two handy relativity theories. This opened the way for the further revolutionary physicists: Louis de Broglie and Werner Heisenberg had come up with their well-known seminal ideas.

It is this 'conquering a *conceptual mountain of microcosm*' that could allow Max Planck to introduce the valid theory of the blackbody radiation—and to further develop this success in bringing us to the very *quantum physics' realm*.

Without throwing any shadow onto the general qualities of the *revolutionary approach to the theoretical physics*, it is important to recognize that the marketing activities around the latter were—*and, to our sincere regret, are still remaining*—ultimately aggressive to the extent that sometimes we even tend to forget that the revolutionary physicists and chemists were **not** *deities*, but undoubtedly *outstanding and talented people*. Hence, nothing human was alien to them, including natural fallacies.

The next question: *Who had but cast the thermodynamics adrift?* The only answer is: *We, the faithful followers of the revolutionary scientists*.

The last but not the least important question: *How could we accomplish casting thermodynamics adrift?* The only answer is: *By declaring the revolutionary approach in physics to be the **only possible** theoretical approach, by blindly paradigmatizing it*. The present author guesses that we have been exceedingly glad to learn that "*The probability does govern our whole world*," and that was just the ultimate stop for the rational thermodynamics.

To sum up, it is by stubbornly refusing to look for the actual physical sense of the probability that we could have a unique opportunity to stop the research work on the thermodynamics' foundations.

What had but come instead of the latter was **physicalizing** the ingenious gadget by N. L. S. Carnot in declaring it '*a unique carrier of some deep physical sense*' and starting to fruitlessly look for the latter. Then, there was a powerful continuation of the trend initiated by R. Clausius, which is nicely reflected in inventing "*novel basic laws*" (third basic law—and even the zeroth basic law). The last but not the least important example ought to be the solemn and spectacular '*killing*' of the non-existent basic laws.

It is this trending entirety that could in fact be described by the handy terms *quantophrenia* and *numerology*, for this has nothing to do with the actual scientific research and turns out to be highly contagious.

Prof. Sorokin could detect the mentioned negative trend and thoroughly analyze it in full detail, see his monograph [17]. On the other hand, Prof. Dr. Bernard H. Lavenda in the Introduction part of his recent textbook entitled *A New Perspective on Thermodynamics* tends even to ascribe the difficulties of the thermodynamics as a scientific branch to some kind of *institutionalizing the scientific mafia*, by declaring Prof. Dr. Peter Guthrie Tait (1831–1901) to be the first and the foremost '*Capo di Tutti Capi*' of the latter [18].

In our opinion, both Prof. Sorokin and Prof. Lavenda are in fact diagnosing one and the same pathology, whereas Prof. Lavenda's verdict does sound like an emotional paraphrase of Prof. Sorokin's systematically weighted diagnosis. Should *emotional* be a negative note? By far not, for the details see the special monographs on this theme [19, 20].

With the above in mind, the truly important poser ought to be: **Was there but any kind of at least somewhat different thermodynamics**?

The correct answer is: **Sure it was—and sure it is!** In effect *there were indeed lots of colleagues*, who were actively and fruitfully working on the foundations and the applications of thermodynamics, effectively counteracting the actual conceptual putrefaction and gradual fragmentation of the latter. Just to present a volatile illustration of the actual sources for our standpoint, we would like to refer here to the book [21] and the references therein.

Here we would greatly appreciate introducing the lives and works of five unique colleagues, who managed to recognize and even skillfully employ the essential foundations of thermodynamics, irrespective of any *quantophrenia* or *numerology*. This is why here we would like to introduce them as the **true heroes** of the **different thermodynamics**. These colleagues are:

1. Peter Boas Freuchen (1866–1959) who lived all his life long in Denmark and was a courtier lecturer for the princes at the Danish Court—and thereafter just a schoolteacher. Peter Boas could carefully analyze the trends of thermodynamics' development and recognized their actual directions. He published two monographs in Danish, most probably on his own account.

2. Dr. Nils Engelbrektsson (1875–1963) who lived all his life long in Sweden having inferred the general thermodynamic equation of state. He had gotten his PhD degree when he was already 73 years old. Nonetheless, all his professional life long Nils was nothing more than just a schoolteacher. He has published a considerable number of his studies in Swedish, English and German, but most probably on his own account.

3. Karl Alexius Franzén (1882–1967) who was also a Swedish schoolteacher and a colleague of Nils, for he could manage to experimentally prove the correctness of Nils' theory and was his co-author.

4. Prof. Dr. Max Bernhard Weinstein (1856–1918), who was born in the Russian Empire (in city of Kaunas of today's Lithuania), and working at the University of Berlin, where he acted as a lecturer for physics and philosophy and actively published his results in prestigious German journals and in monographs, but had somehow remained in our memory as nothing more than a harsh critic of Albert Einstein's relativity theory.

5. Dr. Georg(e) Augustus Linhart (1884–1951) who was born in Austria, but as a lone small bub emigrated to the USA, where he first mastered English on the streets, then graduated the school, and finally successfully got his Master's and PhD degrees at not more and not less than the Yale University. Finally, after short, but intensive and instructive contacts with the US-American academic milieu, he was working as a schoolteacher in a small town in California all his life long. Howbeit, George could manage to clarify the actual physical sense of the meanwhile notorious Boltzmann–Planck formula for entropy. He could be viewed as an actual successor of Prof. Dr. Gibbs.

The main aim of the present volume is to analyze the achievements of the five colleagues named above and to show the possible fruitful directions of developing thermodynamics and kinetics.

In accordance with all the above, we would like to start with the presentation of Peter Boas Freuchen's life and work (Chapter 1), for his contribution was of truly fundamental significance. The same follows for Dr. Nils Engelbrektsson and Karl Alexius Franzén (Chapter 2), and for Prof. Dr. Weinstein (Chapter 3). Chapter 4 will tell about the life and work of Dr. Linhart.

Finally, Chapter 5 will critically summarize all the above points— and show the perspectives of developing the true foundations of thermodynamics—by taking into account all the past and modern trends as well. We have supplied a number of notes to all the chapters, where it was necessary to reveal the significance of the *different thermodynamics* in detail.

It is noteworthy that although our protagonists were practically never acquainted with each other and, except for Prof. Dr. Weinstein, had practically no intensive connection with the academic milieu, their life stories and their work allow seeing behind the curtains of the conventional thermodynamics. The pictures we thus capture are revealing many further colleagues whose contributions have been undeservedly forgotten. This is why, we tell the readership about these colleagues and their work as well—but solely to the extent allowed by the volume of the monograph at hand.

An attentive reader might cast grounded doubts on the fact that although thermodynamics and quantum physics are both fields of general physics, there could be no further apparent relationship between them and that the monograph's author might be trying to promote some of his own artificial logical constructions that may be of no relevance, interest, and use.

In fact, there is still much-much more to the story.

Indeed, Prof. Dr. Erwin Schrödinger (1887–1961) in his seminal paper [22] infers the now famous Schrödinger equation from the standard Hamiltonian formalism by recasting the function S in the standard Hamilton–Jacobi formulation in the form of $S = K \ln \psi$ by commenting his step as follows:

"*We now introduce a new unknown function ψ for S in such a way that ψ would appear as a product of intervening functions of the individual coordinates.*"

The actual physical sense of the Hamilton's principal function S ought to be the *mechanical action* (in German: *Wirkung*), whereas the physical sense of what every modern schoolchild knows to be the quantum-mechanical wavefunction ψ has remained largely unknown. In his paper Prof. Schrödinger himself has solely multiplied the logarithm with the constant K having the dimension of the mechanical action.

Remarkably, the thus recast mechanical action closely resembles the famous entropy formula ingeniously guessed by Prof. Schrödinger's outstanding compatriot Prof. Dr. Ludwig Boltzmann, while in Boltzmann's formula the place of ψ is occupied by the so-called 'thermodynamic probability' W having unclear physical sense as well. In addition, the Boltzmann's constant k serving as the coefficient for the logarithm is possessed of other physical dimension, namely that of entropy.

We know that Prof. Schrödinger was looking for the physical sense of the wavefunction all his life long. The physical grounds of the apparent parallel between the mechanical action and entropy have also been thoroughly analyzed by another grandfather of quantum physics, Prof. Dr. Louis de Broglie (1892–1987) [23].

Nonetheless, despite the Herculean efforts by many competent colleagues, it seems that any conclusive answer to this fundamental poser could not yet be found till nowadays (for more detailed discussion of this theme cf. the works [24–32] and references therein).

On the other hand there seems to be striking parallels between the great revolution in the natural sciences and the social revolutions known in history [33–35].

Bearing this entirety in mind, in what follows we shall touch a number of conceptual aspects inherent in quantum physics, which ought to be of relevance to thermodynamics.

References

1. M. W. Zemansky (1957): Fashions in thermodynamics, *Am. J. Phys.*, v. 25, p. 349.

2. C. A. Truesdell (1980): *The Tragicomic History of Thermodynamics, 1822–1854*. Springer-Verlag: Berlin, Heidelberg, New York, and Tokyo.

3. *New Perspectives in Thermodynamics* (1986): James B. Serrin (Ed.). Springer-Verlag: Berlin, Heidelberg, New York, and Tokyo.

4. H. A. Buchdahl (1966): *The Concepts of Classical Thermodynamics*. Cambridge University Press: Cambridge, U.K.

5. H. A. Buchdahl (1975): *Twenty Lectures on Thermodynamics*. Pergamon Press: Oxford, New York.

6. Henry Crew (1928): *The Rise of Modern Physics. A Popular Sketch*. The Williams and Wilkins Company: Baltimore, USA.

7. Hans Kangro (1970): *Vorgeschichte des Planck'schen Strahlungsgesetzes*. Franz Steiner Verlag: Wiesbaden, Germany.

8. Armin Hermann (1971): *The Genesis of Quantum Theory (1899–1913)*. MIT Press: Cambridge, Massachusetts, USA; London, Great Britain.

9. J. Mehra, H. Rechenberg (1982–2002): *The Historical Development of Quantum Theory*, Volumes 1–6. Springer-Verlag: Heidelberg, Berlin, New York.

10. J. Mehra (2001) *The Golden Age of Theoretical Physics*. Volumes 1 and 2. World Scientific Publishing: Singapore, River Edge, London.

11. Friedrich Hund (1984): *Geschichte der Quantentheorie*. Wissenschaftsverlag, Bibliographisches Institut: Mannheim, Wien, Zürich.

12. Th. S. Kuhn (1987): *Blackbody Theory and the Quantum Discontinuity: 1894–1912*. Revised Edition. University Of Chicago Press: Chicago, USA.

13. Christa Jungnickel, Russell McCormmach (1990): *Intellectual Mastery of Nature. Theoretical Physics from Ohm to Einstein*, Volume 1: *The Torch of Mathematics, 1800 to 1870*. University of Chicago Press: Chicago, USA.

14. Christa Jungnickel, Russell McCormmach (1990): *Intellectual Mastery of Nature. Theoretical Physics from Ohm to Einstein*, Volume 2: *The Now Mighty Theoretical Physics 1870–1925*. University of Chicago Press: Chicago, USA.

15. Res Jost (1995): *Das Märchen vom Elfenbeinernen Turm. Reden und Aufsätze*. Springer-Verlag: Berlin, Heidelberg, New York.

16. John G. Cramer (2016): *The Quantum Handshake: Entanglement, Nonlocality and Transactions*. Springer International Publishing: Cham, Heidelberg, New York, Dordrecht, London.

17. Pitirim A. Sorokin (1956): *Fads and Foibles in Modern Sociology and Related Sciences*. H. Regnery Co.: Chicago, USA.

18. Bernard H. Lavenda (2010): *A New Perspective on Thermodynamics*. Springer Science+Business Media LLC: New York, Dordrecht, Heidelberg, and London.

19. Alexander Bain (1876): *The Emotions and the Will*. D. Appleton & Company: New York, USA.

20. Katharina Lochner (2016): *Successful Emotions. How Emotions Drive Cognitive Performance*. Springer Fachmedien: Wiesbaden, Germany.

21. William Fay Luder (1967): *A Different Approach to Thermodynamics.* Reinhold Publishing Corporation, A subsidiary of Chapman-Reinhold, Inc.: New York, Amsterdam, London.

22. Erwin Schrödinger (1926): *Quantisierung als Eigenwertproblem. Annalen der Physik,* v. 384, pp. 361–376.

23. Louis de Broglie (1995): *Diverses questions de mécanique et de thermodynamique classiques et relativistes*; Georges Lochak, Michel Karatchentzeff, Daniel Fargue (Eds.). Springer-Verlag: Berlin, Heidelberg, Germany.

24. Bryce S. DeWitt, Neill Graham (Eds.) (1973): *The Many-Worlds Interpretation of Quantum Mechanics.* Princeton University Press: Princeton, New Jersey, USA.

25. Wheeler J. A., Zurek W. H. (Eds.) (1983), *Quantum Theory and Measurement.* Princeton University Press: Princeton, New Jersey, USA.

26. Gary Zukav (1984): *The Dancing Wu Li Masters. An Overview of the New Physics.* Bantham Books: New York, Toronto, London, Sydney, and Auckland.

27. Paul R. Gross, Norman Levitt, Martin W. Lewis (Eds.) (1996): *The Flight from Science and Reason.* New York Academy of Sciences: New York, USA.

28. Alan Sokal, Jean Bricmont (1999): *Fashionable Nonsense. Postmodern Intellectuals' Abuse of Science.* Picador: New York, USA.

29. Wojciech H. Zurek (Ed.) (1990): *Complexity, Entropy and the Physics of Information.* Addison-Wesley Publishing Company: Redwood City, Menlo Park, Reading, New York, Don Mills, Wokingham, Amsterdam, Bonn, Sidney, Singapore, Tokyo, Madrid, San Juan.

30. Ulrich Mohrhoff (2011): *The World According to Quantum Mechanics. Why the Laws of Physics Make Perfect Sense after All.* World Scientific: New Jersey, London, Singapore, Beijing, Shanghai, Hong Kong, Taipei, and Chennai.

31. Jean Bricmont (2016): *Making Sense of Quantum Mechanics.* Springer International Publishing: Cham, Switzerland.

32. Jean Bricmont (2017): *Quantum Sense and Nonsense.* Springer International Publishing: Cham, Switzerland.

33. Charles Tilly (1978): *From Mobilization to Revolution.* Random House: New York, USA.

34. Charles Tilly (1993): *European Revolutions, 1492–1992.* Blackwell: Oxford, U. K., and Cambridge, USA.

35. Theda Skocpol (2003): *States and Social Revolutions. A Comparative Analysis of France, Russia and China.* Cambridge University Press: Cambridge, New York, New Rochelle, Melbourne, Sydney.

Chapter 1

Peter Boas Freuchen (1866–1959): His Life and Work

I thought that my voyage had come to its end at the
Last limit of my power, that the path before me was
Closed, that provisions were exhausted and the time come
To take shelter in silent obscurity.

But I find that thy will knows no end in me; and
When old words die on the tongue, new melodies break
Forth from the heart; and where the old tracks are lost,
New country is revealed with its wonders.

Rabindranath Tagore (1861–1941).
From *Gitanjali*, Verse XXXVII.

This portrayal of Peter Boas Freuchen has been taken from the
'My Heritage' portal by giving his name, his date of birth and
his homeland. Its exact time is unknown.

A Different Thermodynamics and Its True Heroes
Evgeni B. Starikov
Copyright © 2019 Pan Stanford Publishing Pte. Ltd.
ISBN 978-981-4774-91-8 (Hardcover), 978-0-429-50650-5 (eBook)
www.panstanford.com

1.1   The Life of Peter Boas Freuchen

First, let us but try having a look at his CV. The story will not be very long, for there is practically no clear information about him. Our stubborn attempts to contact Danish Universities, Academy of Sciences and National Library to try fetching at least some traces of him in the Danish academic life were in vain. Either we had no answer at all (in most cases), or the answer was '*We do not know anything about him*'.

However, it was possible to determine that he was a professional natural scientist: mostly physicist and having insight into chemistry as well. He had published several works, including two monographs—a truly interesting puzzle—howbeit, we shall see that Peter's CV ought to be definitely devotional: physics was his true faith.

Peter Boas Freuchen was born into a family of traditional merchants from his father's side and a traditional farmer's family from his mother's side.

Peter's birthday and birthplace: 16 April 1866, Nakskov Kbst., Lollands Nørre h., Maribo a., Denmark.

His parents were

> *Father*: Ole Elfred August Freuchen (1821–1912)
>
> *Mother*: Pauline Johanne Boas (1831–1908)

This was not the first family for Peter's father. Earlier he had married (on 30 May 1851) Hanne Marie Børresen (1827–1861) who regretfully had to untimely depart. As a lone father of four children, he had decided to marry again. The wedding with Peter's mother took place on 28 December 1862.

This is how Peter had his elder sister, elder brother, as well as two elder half-brothers and two half-sisters. One of his half-brothers was Lorentz Benzon Freuchen (1859–1927), father of the outstanding Danish polar explorer, author, journalist, and anthropologist, the Grønland's father, Lorenz Peter Elfred Freuchen (1886–1957), who appears to be the only member of this family who could really manage gaining worldwide recognition.

1.1.1   P. B. Freuchen's School Study

Peter Boas started and finished his primary and secondary education at the Nykøbing College, currently known as 'Nykøbing F.—University College Sjælland'. The only information about Peter's activity during this time that could be found on the Internet was on the below-mentioned web site devoted to a lieutenant of the Danish army, J. P. Muller, a world-famous but nonetheless forgotten Danish sportsman:

http://www.learntomuller.com/j-p-muller-a-danish-sportsman-world-famous-and-forgotten/

"... In 1884, J. P. Muller graduated from the Nykøbing College. That year only three students graduated. Among his closest friends was someone who was later to become a lawyer Frederik Graae (died 1938), the founder of Marielyst, and P. B. Freuchen (the grandfather of Peter E. Freuchen, an arctic explorer). The latter was the only man J. P. Muller could not beat in a wrestling fight."

Thus, we know that Peter Boas was ready with his school education as he was about 18 years old. The only feeble point of the above-cited fragment is that Peter Boas was in effect a 'step-uncle' to the Grønland's father but definitely not his grandfather, as mentioned above.

1.1.2   P. B. Freuchen's University Studies

Howbeit, the three school friends have moved to København, and Peter went to the University to get the degree of *Candidatus Magisterii* (*cand. mag.*) after the four or five years of attending the university courses of his choice. Peter Boas was thus studying chemistry, physics and mathematics. To follow the university courses, he rented a flat in København and lived there during the time of his studies.

1.1.3   P. B. Freuchen: Professional Activity

Meanwhile, after this time the traces of Peter Boas are truly lost. The only information accessible in the Internet shows that Peter Boas was a teacher in the state gymnasium Schneekloths—from the year 1919 on.

The State College Schneekloths School was at that time a secondary school in Copenhagen. The active pedagogic work over there was but discontinued in 1992.

In 1854 Hans Schneekloths, who was an outstanding organizer of school education in Denmark, founded a school named 'Real School of Frederiksberg and Vesterbro', for it had moved to the København district of Vesterbro. In 1856 the school moved to Værnedamsvej (no. 13.A), also situated in the København city area, and adopted thereafter the name of 'Latin and Real School of Værnedamsvej'.

Later on, in 1883, the school again changed its name to 'Schneekloths Grammar School'. Then, in 1901, it was taken over by the 'United Latin and Real Schools', and, in 1904, was subsequently merged with the 'Hertz' Primary School' under the department names 'Schneekloths Grammar School' and 'Hertz' Primary School'.

The Danish state had taken over the gymnasium department in 1919 but not the primary school under the name of 'Svanholm Gymnasium'. However, in 1930, it finally switched its name to 'State College Schneekloths School'. The school had then acquired the necessary place for the handball club 'Schneekloths' (who had such prominent Danish players as John Bernth and the goalkeeper Bent Mortensen), as well as the cricket club 'Svanholm'.

The gymnasium was accepting only boy pupils, until it had moved to Brøndby Møllevej in 1969. In 1986, the København County took over this school under its original name 'Schneekloths Gymnasium', but 1992 a sudden termination of the school's activity had come.

Among those gymnasium graduates who might surely owe a significant part of their knowledge and skills to Peter Boas Freuchen are:

1. *Among the graduates of the year 1920*: Prof. Dr. Frederik Geert Fabricius-Bjerre (1903–1984), a professor of mathematics at the Danish Technical University, with its campus presently situated in Lyngby.

2. *Among the graduates of the year 1933*: Gunnar Andreasen (1914–1989), who was an outstanding Danish chemical engineer, and—from the year of 1938 on—the founder and director of the chemical company Cheminova, which is one of the leaders of the Danish chemical industry until now, known all over the world as a supplier of diverse pesticide agents.

Well, but what Peter Boas was busy with before he acquired the position of the gymnasium teacher in 1919? We might well estimate that he could have gotten his *Cand. Mag.* diploma in the years of 1888 and 1889.

To get more information about Peter and his everyday life, we could fortunately come across the website organized and supervised by Mrs. Bente Paulsen at the Internet portal of 'My Heritage', which is dedicated to Peter's memory and containing his photo, which we dare to present at the beginning of this chapter. I have applied to Mrs. Paulsen with a question:

> "... In the meantime, I could fetch Peter Boas' works, but have no information about his life. Could you please help me to reconstruct his actual Curriculum Vitae?"

Her answer was as follows:

> "... Unfortunately, I cannot help, but I have discussed your request with the mother of my son-in-law, for she is just from Freuchen family. She was young, as she had to nurse Peter Boas at his home, and in particular she had to read him aloud, so that she could remember that he was especially happy, when she was reading him the book entitled 'Atomer og andre småting' ('Atoms and Other Small Items').

[Please see Supplementary Note 1, Section 1.9, Pages 121–122]

> He was very happy at sailing. In addition, Greenland's Father, Lorenz Peter Elfred Freuchen was living at his home.

> Peter Boas was working as an Associate Professor with The Royal Family of Denmark. He was teaching Physics, Chemistry and Mathematics to Danish Princes Christian Alexander Robert, Count of Rosenborg (1887–1940), Axel Christian Georg, Prince of Denmark and Iceland (1888–1964), Erik Frederik Christian Alexander (1890–1950), Viggo Christian Adolf Georg (1893–1970) and to Princess Margrethe Françoise Louise Marie Helene of Denmark (1895–1992) ..."

Oh, that was but the hardest puzzle to solve!

On the one hand, Peter Boas was teaching for a very special audience, and on the other hand, he was publishing his works mostly in his mother tongue (there are no academic publications by him either in German, or in English, not to speak of some other language available upon the World).

1.1.4 Our Main Result Ought to Be the Following CV of Peter Boas Freuchen

An offspring of a wealthy family, Peter could get proper school education and prepare himself for university studies. He was a sporty young man, fond of wrestling, swimming and sailing—and, along with all this, he was utterly interested in mathematics, physics and chemistry.

After successfully completing his university studies, Peter at once got the position of an assistant professor at the Danish Royal Court to teach physics, chemistry and mathematics to all the five children of Prince Valdemar of Denmark (1858–1939) and his wife Princess Marie Amélie Françoise Hélène d'Orléans (1865–1909). In total, the overall period of Peter's service at the Danish Royal Court ought to be about 30 years and by then he was almost 53 years old. He moved to the position of a teacher at the Schneekloths Gymnasium in Copenhagen, where he duly worked till his retirement age of 65–70 years. As a pensioner, he lived alone (he had no family of his own) but as he started progressively losing his eyesight, his numerous relatives actively started taking care of him.

Concerning his professional activity, he was never officially frequenting and/or maintaining the university/academic sphere either in Denmark or elsewhere in the world. However, he was highly interested in and working on natural sciences such as physics and chemistry, while the special field of his devotion was thermodynamics.

Although he could perform and publish a very careful and thorough analysis of thermodynamics' foundations, there was no pertinent discussion of his work, which thus remained undeservedly unnoticed for a very long time. The reasons for this might be objective.

1.2 Why at All Peter Boas Freuchen Ought to Remain in Our Memories?

Frankly speaking, we would immediately expect such a question— even from Peter's compatriot colleagues. Moreover, particularly remarkable ought to be here the present author's own story about occasionally fetching from the Internet traces of Peter's publications, then gaining access to their copies and trying to investigate Peter's familial roots. This was not a trivial straightforward study for Peter was and is still obediently wearing his veil of oblivion.

The listing of his works is truly not very long. In effect, he seems to have mostly devoted his time to thermodynamics in general and in particular. However, if you would start studying them, you will be deeply impressed by the professionalism, attentiveness and skillfulness of this author—as well as by the absence of any conceptual bias in his writings!

As a rule, some definite bias is somehow stubbornly present in the conventional thermodynamic literature, quite irrespective of its language. This is why, here we would greatly appreciate taking the opportunity of opening the volumes published by Peter to try analyzing, where the bias in question does come from and why it is indeed so stubborn.

We start our analysis in the series of Supplementary Notes to this chapter placed in Section 1.9, from the Page 121 on.

While reading Peter's book carefully, we shall refer the readership to the pertinent Supplementary Note in the relevant stretches of the main text.

1.2.1 Peter's Publications (One Journal Paper and Then Two Books): Reactions to His Book

1. P. B. Freuchen: '*Om Nordlys*', Fysisk Tidskrift, v. 5, pp. 89–95, 1907.
   [Please see Supplementary Note 2, Section 1.9, Page 122]
2. P. B. Freuchen: '*Termodynamik: Grundtræk af termodynamikkens historie og de to hovedsætningers betydning*', København: Lehmann & Stages Forlag, 1915.
3. P. B. Freuchen: '*Kemisk potential og andre fysisk-kemiske emner*', P. Haase & Søns Forlag, København, 1936.

Meanwhile, the former book has encountered heavy criticism in Denmark. The official Danish engineering bulletin has published the referee's report just nearly one year after the publication, the corresponding reference is as follows: *Ingeniøren, Number 6: 19-01-1916, page 42.* Here we would like to present its English translation:

"As the book's title indicates, the author's intention could anyway have been to write a book on thermodynamics and thermodynamics' history. Noteworthy beforehand, any attempt to try reconciling two different tasks in the same work should as a rule lead to the only result that presenting the both tasks would fail, and the present case is a typical illustration of the latter viewpoint.

The author has designed layout of his book in such a way that each of the great names in the thermodynamics field gets his own chapter, containing a concise account of the works by the relevant protagonist. As it is, this does shear the general thread of narration every time, when readership arrives at the end of the respective section. Hence, especially when all the sections are numerous and short, it becomes truly difficult to find any correlation between the fragments in question. The author seems to mend such a deficiency of narration coherence in part by injecting on and off the sections, where specific chapters of Thermodynamics are treated, or when complementing a discussion of the works of some particular protagonist by including all the related contributions to the same topics by other colleagues. Sometimes, the author achieves a similar amendment by dividing the contributions of some particular protagonists into two parts and moving each part to its separate location in the book. Meanwhile, no listed measures are repealing the sturdy over-all feeling of dissension the book at hand excites, if considering the book's performance from the standpoint of thermodynamics. Nonetheless, the book discussed might possibly be worthwhile considering as a historical work to trigger evaluation of the contributions by individual Noblemen of Thermodynamics. Meanwhile, the aforementioned interference showing up in particular chapters among the concise summary of the works by the respective protagonist, and the lengthy discussions on the contributions by other colleagues does work disruptively also in this case. Furthermore, simply presenting all the individual formulas or statements is not helpful in getting a fruitful enough estimate of how important the respective protagonists, who could derive them, are. To sum up, the first part of the book looks like almost purely historical treatise, whereas its last part is indeed approaching the field of pure Thermodynamics by considering what should be the best way to deal with the newer Thermodynamics as a whole and the Chemical Thermodynamics in particular. Finally, it is in the Chemical Thermodynamics' section that the narration clarity is significantly increasing, especially when it comes to discussing some illuminating examples; but unfortunately, on the other hand, systematic realistic examples are mostly lacking in the book.

Among all the protagonists actually mentioned in the book, the absolute priority is definitely belonging to theorists to batten on experimenters;

specifically, among the true inventors of the heat equivalent, Mayer had received over four pages, Joule—solely 1.5, while the author had mentioned Colding's contribution just by the way. This is all the more remarkable, in view of the fully justified expectation that a Danish protagonist in a Danish book should not be driven to a clearly inferior place compared to the one he gets even in the foreign literature, for his forward-looking thoughts ought to fully justify his clear being to the fore. Contrariwise, the author mentions another Dane, much less known in effect, Hermann v. Kauffmann, but his work is anyway completely theoretical.

The book is of definite interest to anyone, who is already well versed in thermodynamics beforehand, and such a readership ought to gain some insight into the vicissitudes of life in getting the seminal research results. Nonetheless, the book is not properly reflecting the long and hard struggle by particular book's protagonists for their conceptions—for example, there is no information as to the hard and ruthless resistance Mayer encountered among other colleagues, including Helmholtz.

Howbeit, the book in question is definitely not a textbook, because of quite turbid form of its narration, if estimating in particular the quality of the first part of this treatise. Further, of striking technical significance ought to be almost complete absence of the practical examples and arbitrary skipping the possible uses of Thermodynamics for studying the substances and conditions of importance for engineering sciences. Thus, it is not quite clear to me whether the book's territory might fall within the physics' area or may perhaps rather within that of the chemical physics—or instead within the area of history of the both.

A. P. Hjortsø
M. Ing. F."

Most probably, Peter Boas was not very happy at encountering such a welcome of his work—for, obviously, the referee could not completely understand his actual goals—or, instead, could fully understand, but had furiously rejected them. Therefore, just a bit later, the same bulletin published traces of a short debate, which Peter Boas had triggered in connection with the above-mentioned report.

The correspondent reference is as follows: *Ingeniøren, Number 17: 26-02-1916, page 151* (below we would like to present the relevant English translation):

"*Reply to Mr. P. B. Freuchen*

Mr. Freuchen rebuts in connection to my review of his book "Thermodynamics" that he finds it unnecessary doing anything more than just mentioning Colding's name beside Mayer's and Joule's ones in his book, because Colding's work is sufficiently well known through other sources. To this point, I would only like to note that we truly know both Mayer and Joule's works just to the same extent as the work by Colding, and the same pertains to most of the protagonists in Mr. Freuchen's book. Moreover, both the book featured by Mr. Freuchen and other equally reputed, widely known sources do mention the entire set of works in question; therefore, Mr. Freuchen could definitely spare himself the efforts to write a separate book.

As for the relationship between Mayer and Helmholtz, I regret to give up documenting my opinion. My knowledge on the theme stems from an article in a periodical; but in which Journal exactly, I am now unable to recall. I have been looking in vain both for that article and for any comprehensive information elsewhere. In my memory, which should not deceive me, the actual point is that Helmholtz, who was in fact aware of Mayer's treatises, but had refused to read them, although the colleagues were asking him to do so. In addition, he had provided effective support to one of Mayer's worst enemies. I would guess the latter is sufficient to enable me using the terms 'hard' and 'ruthless' about Helmholtz's stance. If Mr. Freuchen would not be satisfied with this, I can only regret it. I really cannot spend more time by random searching for the evidence, since this particular topic as it is ought to be of somewhat secondary importance in this connection altogether.

In case Mr. Freuchen had any intention of writing about "The Basics of Thermodynamics, its History and the Meaning of the Two Basic Laws" – should this deal solely with the previous developments? – I think it is plainly wrong to entitle the book at hand this way: "The Thermodynamics"; this was definitely misleading for me and might probably cause further misunderstandings. Therefore, I would suggest viewing the book as a historic one, and this is just what I was most inclined to express in my review; but "Thermodynamics" ought to be something quite different, namely—Physics or Chemistry.

A. P. Hjortsø
M. Ing. F."

In our opinion, the both reports reveal a noticeable ill will of the referee concerning Peter Boas' work, or Peter Boas himself—or even both—but now it is difficult to say, which of these assumptions ought to be correct. Howbeit, this surely would not be of interest to the widest circles of our readership.

On summarizing the points raised by the referee, we get that

1. Freuchen's book at hand is surely not a standard handbook in the field—and, most probably, Peter Boas had no plan to create one such—in fact, he does hint to his very intentions in the preface to his book.

2. The actual purpose of Peter's book was to trigger a concise but thorough analysis of the past development and of the current trends in the thermodynamics' field as a whole. In addition, this is why, Peter Boas had chosen the mentioned layout form—moreover, this is why, he was focusing his and his intended readership's attention on the theorists—or, at least, on those experimenters, who had a tendency to think over their theoretical ideas themselves. This is why, the book is not addressing non-professionals or freshmen—indeed, it is aiming at the versed colleagues interested in trying to summarize the general trends.

3. With this in mind, it ought to be clear why Peter Boas was considering only a few practical examples: The latter should properly illustrate this or that interesting trend or idea, but ought to bear no deeper pedagogical significance.

4. Apparently, Peter Boas was sure that N. L. S. Carnot was the only thermodynamicist during a rather long period—and, with this in mind, duly trying to monitor the actual roles of all the Carnot's followers known at his time. Sure, such an approach *per se* was already a huge strike against the conventional academic science's Sancta Sanctorum of that time!

Contrariwise, truly remarkable in this connection should be the official reaction to Peter's book published in English in the journal *Nature, v. 94, Number 2417, p. 702, February 24, 1916.* Here we would like to cite that book review as a whole:

"Termodynamik. By P. B. Freuchen, p. 143 (København: Lehmann and Stages Forlag, 1915), No price.

The scope of this little book is best indicated by its sub-title: "An outline of the history of thermodynamics and the significance of the two chief laws." In the preface, the author declares his intention of tracing the development of thermodynamical ideas and their bearing on physics and chemistry. It is not a textbook, but rather a kind of thermodynamical "Who's Who"; successive short chapters deal with Carnot, Clapeyron, William and James Thomson, Robert Mayer, etc. One of these begins: "To read Planck's thermodynamical papers is to breathe pure, clear air."

The various parts of the subject are of unequal length; some topics dealt with in the larger textbooks of physics are entirely omitted. Julius Thomsen's and Horstmann's work is described more fully, but like many other histories, this does not concern itself greatly with Nernst's theorem, which occupies only half a page, and the quanta theory is referred to in a single sentence. Although unsuitable for beginners, the book should appeal to physicists, and particularly to chemists desirous of extending their outlook. Its publication in Danish speaks well for the scientific public of small countries, and we hope that by means of a translation it may become accessible to a larger number of readers.

G. B."

The above is but in truly striking contradiction with the reaction from Peter's compatriots! Thus, the immediate poser arises: '*So, what was the ultimate matter?*'

Most probably, Peter Boas had published his book on his own account, and no official state or private organization had ordered it, so there was no real chance that the publication would excite a pertinently wide interest. Furthermore, one more actual reason for the latter trend might be obvious: The First World War was already at its apogee during the very time of publication, although Denmark was never directly participating in it. Now, let us have a look at this possible connection in somewhat more detail.

"During the First World War, Denmark was neutral, a status achieved not without some difficulty and one which the country had to work hard to retain. This neutrality was not entirely unambiguous, since Foreign Minister Erik J. C. Scavenius declared that Denmark would 'show favorable neutrality' towards Germany, adding, however, 'as far as this is consistent with the notion of neutrality'.

It was the very sensitive and ambiguous relations with Denmark's southern neighbor, which created most of the difficulties. After the German–Danish wars of the nineteenth century and the still unresolved problem of sovereignty over Southern Jutland, it was imperative not to provoke mighty Germany and thus precipitate a favorable solution to the problem ..."

The above is a citation from the paper by Bent Blüdnikow: *'Denmark during the First World War', Journal of Contemporary History, Vol. 24, No. 4 (Oct. 1989), pp. 683–703.*

1.2.2   What Was Peter's Actual Driving Force?

Whatever the response of the Danish colleagues to Peter Boas' book, the latter is definitely Peter's purely personal presentation, but yet fully unbiased professional standpoint. Sure, one might in principle wonder, whether the following fragment from his book could contain some above-mentioned purely political taste.

In fact, what Peter writes about Max Planck, an active German thermodynamicist of that time, is nothing more than an attempt to analyze professionally and in full detail Planck's contribution by comparing the latter to the results of other colleagues.

There seems to be no politics in the background. So that, let us try checking the point in question using the English translation of the relevant chapter (*Planck*, pp. 114–115, *Termodynamik*, by P. B. Freuchen, 1915):

> *"107. Reading Planck's thermodynamic works ought to be like breathing the fresh clean air. The thermodynamic literature pieces are indeed very scattered and attract not so much attention, because of their content. This is why it is clear that the thermodynamics' laws were often penetrating much too slowly into the practical research. Indeed, some further 20–30 years were necessary to trigger at least mentioning the mass action law after the very emergence of the latter; similar was the destiny of several other laws. Finally, Planck with his deep insight could manage clarifying a considerable number of thermodynamic questions and thus opening new routes. His most important work in this field is definitely his book entitled 'Vorlesungen über Thermodynamik', and here we shall look at why it has become very important, especially as concerns his original studies of the Second Basic Law and the resulting functions deduced from the latter.*
>
> *Indeed, Planck notes:* The Second Basic Law is often formulated in such a way that the transformation of work into heat always takes place completely, whereas the transformation of heat to work is only an incomplete one. That is, every time considering a quantity of the heat transformable into work, there must necessarily be some other heat*

**Vorlesungen über Thermodynamik*, p. 74.

amount to serve as compensation, for example, by moving from a higher to a lower temperature. This is indeed true in some cases, but it does not constitute the core of what the following simple example ought to demonstrate. If a perfect gas expands isothermally while performing some work, then one can say that heat transferred from the heat source (for example, a large body of water) to the gas should be completely transformed to work, without any other energy conversion to take place. At the first glance, there are no objections concerning the latter statement. Meanwhile, it is likewise clear that we cannot completely convert unlimited amounts of heat to work.

To our mind, Peter is actually demonstrating here the fact that the thermodynamic work by Max Plank is reflecting the 'operational approach to thermodynamics'.

[For the detailed comments on this particular part of the text by Peter please see the Supplementary Note 3, Section 1.9, Pages 122–123]."

...After a certain expansion the air cannot work anymore; hence, to render it capable of working once again, primarily we must compress it.

Moreover, according to Planck's view, the Second Basic Law should find its principal expression in the difference between the reversible and irreversible processes.

The former ones ought to be ideal borderline cases possessed of a great theoretical interest. Nevertheless, in fact, any natural phenomenon is irreversible; namely, one can never completely avoid friction, heat conduction and/or other ubiquitous irreversible companions. It is therefore in principle not possible to reverse the process entirely—to ensure that we have brought all the bodies participating in the process back to their exact original states.

Finally, since the Second Basic Law is a definite result of experience, one has to place some experimentally verifiable rule as the very cornerstone to be capable of proving the desired law. Accordingly, Planck suggests the following phrase as the desired headline: "It is impossible to construct a periodically acting machine that has no effect other than lifting a burden and cooling of a heat reservoir." This statement establishes the well-known impossibility of a Perpetuum Mobile of the second kind. From here on Planck derives the Second Basic Law in its usual form: Any natural physical or chemical process requires that the sum of the entropies of all the bodies involved in the process is increasing, whereas in borderline cases, i.e., for reversible processes, the entropy sum remains unchanged.

To sum up, Planck does admit that his proof ought to have in fact the same basis as those adopted by Clausius, William Thomson or Maxwell; he prefers the above-sketched form only because of its prominent importance for the technique."

Indeed, Peter Boas helps us to reveal where Planck's logic fails to build up a closed circle. The highlighted parts in the above Peter's citation boil down to as follows:

First, **it is not necessary to prove mathematically what should already be evident from the common experience. The ultimate goal is nothing more than just** *'to call a spade a spade'.* Nonetheless, in their "thermodynamicists' appearances," Clapeyron, Clausius, Thomson brothers, Maxwell, Boltzmann, Planck and their numerous surroundings/followers/opponents were busy with trying to accomplish just the opposite. This way the actual physical backgrounds were/are remaining undercover.

Second, with this in mind all the possible linguistic-mathematical games around the Second Basic Law ought to represent nothing more and nothing less than imitating some furious activity. Therefore, the actual result of the latter was/is remaining the *physicalization* of the ingenious theoretic gadgets, like Carnot's cycle and the ideally perfect gas by Boyle and Mariotte—which both deliver just a tremendous theoretical help in rationalizing the realistic phenomena—but, **taken themselves, they bear absolutely no**— or even erroneous—**physical sense**. Indeed, whereas the ideal gas model does properly describe the properties of several important gases at non-extremal temperatures and pressures, by declaring the Carnot Cycle itself to be 'a realistic reversible process at the microscopic level' we produce a cyclic perpetuum mobile working without any physically sensible driving force. It is this way that conventional equilibrium thermodynamics allows us to prove that 'the perpetuum mobile of whatever kind is sheer impossible' by introducing a basic microscopic cyclic perpetuum mobile, which ought to be in equilibrium forever, thus escaping any sensible time evolution, that is, being truly timeless!

In fact, as Max Planck himself had pointed out many years later, *"... Unfortunately, however, as I was to learn only subsequently, the very same theorems were obtained before me, in fact partly in an even more universal form, by the great American theorist, Josiah Willard Gibbs, so that in this particular field no recognition was to be mine ..."* [1, 2].

Meanwhile, the true 'winners' in the field and at the time under study were indeed Hermann von Helmholtz and Josiah Willard Gibbs, whereas the negative aspect of the whole story was a gradual conceptual perishing of the thermodynamics.

[Please see Supplementary Note 4, Section 1.9, Pages 123–124]

To Peter Boas' mind, thermodynamics was not quite successfully developing before Max Planck. Sure, Max, who had started studying thermodynamics from his youthful days, was in fact one of the most promising researchers capable of grasping the true physical sense of thermodynamics—as far as one could judge from his publications of the then time. Remarkably, not only Peter was expecting the true stimuli for the proper development of thermodynamics to come from Max Planck—specifically, so did a 'widely unknown' Swedish thermodynamicist Nils Engelbrektsson independently of Peter, who was also dealing with Planck's thermodynamic results in detail (see the next chapter of the present book for further details). The trend *to expect a breakthrough in thermodynamics* from Max Planck's work was definitely worldwide.

Howbeit, irrespectively of Planck's tremendous achievements in the field of general physics, it is throughout clear that he could not properly deal with the second basic law. The colleague who could indeed manage finding the true essence of the Second Basic Law is also a protagonist of Peter's book. Later on, in this chapter, I shall revert in detail to the achievements of the colleague just mentioned.

Now, we revert to Peter Boas' book and immediately note that in the chapter next to that devoted to Max Planck Peter Boas extensively analyzes Planck's work, when introducing all the details of the thermodynamic potential notion. Interestingly, there is also a separate book published by Peter Boas and devoted exclusively to this important topic, namely the 3rd item in his publication list presented earlier in this subsection. The latter book ought to become Peter's very last publication. In that publication, he did take into account the comments by the referee of his first thermodynamic book, by rendering his second publication by far more plausible as a handbook. Nevertheless, the book did not seem to have excited any professional interest—for there had been no citations—not to speak of more referee's reports, to the best of our knowledge.

Well, but why it is sheer impossible to find any references to his work?

Interestingly, an outstanding Danish specialist in the field of physical chemistry, Prof. Dr. Johannes Nicolaus Brønsted (1879–

1947) was also working in detail on thermodynamics and its intrinsic interrelationship with the energetics, which was in fact nothing more than a desperate move against the 'cadaverous stank' emanating from the so-called conventional 'equilibrium thermodynamics'. In addition, Brønsted could carefully analyze this interrelationship and publish the results—especially in his truly renowned and (at earlier time) lively debated works [3–5].

In the marvelous treatises by Prof. Dr. Brønsted, the readership would encounter many interesting and important ideas, but not even a volatile commendation of the attempts undertaken by his compatriot in the same principal direction he was heading for himself (not to speak of just mentioning, just referring to Peter Boas' earlier work in the same field!)… Howbeit, the smile of Clio consists in sending Brønsted's seminal works in the same direction like Peter's efforts. Now we might try solving the resulting 'Danish puzzle Number 2', pondering on the endless theme of the 'significance of the worldwide catastrophes in our lives'—or looking for other plausible explanations.

Howbeit, to our mind, the both thermodynamic books by Peter Boas deserve a separate translation of them into English and their re-publication, for they are very different as compared to the conventional handbooks in the field. Still, this looks like a separate large-scaled project beyond the scope of the volume at hands.

Remarkably, both Peter Boas' and Prof. Brønsted's works turned out to be sheer inacceptable for the adepts of the above-mentioned *operational approach*—that is, for those who place the primary priority solely onto arriving at findings, at sentences, at verdicts, at awards (in whatever sense of the both latter terms!), whatever the conceptual highway they follow when riding.

Instead, both Peter Boas' and Prof. Brønsted's works might be of interest for those dealing with the Truth Cognizance Process, Getting the True Insights, Increasing the Knowledge Level, Widening the Area of Awareness. Meanwhile, where were/are such colleagues? An alternative poser ought to be "Is something rotten in the State of Denmark"?

Apparently, Peter Boas was trying to trigger a thorough analysis of the thermodynamics' appearance showing up at his time, and here we would just like to read and discuss several parts of the first book by P. B. Freuchen and on its basis start analyzing the foundations of thermodynamics.

This is why this chapter should pay him his fully deserved tribute. In that we shall now read Peter Boas' thermodynamics book together to try revealing the actual history and foundations of thermodynamics. Peter Boas' outlook is of extreme interest, for he was fully independent and unbiased, whereas thermodynamics in its development was experiencing a considerable parasitic smack still vivid and clearly perceptible even nowadays, to our sincere regret. It is with this in mind that I open Peter's thermodynamics book now and start reading it thoughtfully.

1.3   P. B. Freuchen's Research Work: In Trying to Find a Different Way

P. B. Freuchen (1915): *Termodynamik. Grundtræk af termodynamikkens historie og de to hovedsætningers betydning* (Thermodynamics. Basics of Its History and the Purport of Its Two Basic Laws)

Here we shall present an English translation of a selected number of chapters from this book by Peter Boas. We have selected the chapters for translation in looking at their actual relationship to the foundations of thermodynamics. Meanwhile, the value of his both books ought to be their fully unbiased way of presentation. With this in mind, it is sure that our chapters' choice is throughout subjective. Moreover, this urges us to suggest translating the both books by Peter Boas into English, just to re-publish them separately in getting a wider readership.

Preface

The two basic laws of thermodynamics belong to the major and the most comprehensive laws of physics. Whereas the first of these, the energy conservation law, is easily comprehensible to anyone who is at least familiar with the basic physics, the situation gets quite different as soon as we come to the second law. To become familiar with the latter, we have to examine its content from several standpoints. The most frequently encountered one is presenting it in its mathematical form, linked to the concept of entropy. However, understanding the latter requires several assumptions. Meanwhile, fortunately it is possible to express this law in several ways. The statements that not all the energy kinds are equally useful, or that any natural phenomenon ought to be irreversible, belong

---

to the two valid expressions of the Second Basic Law. Most recently, we have also learned that it is also possible to present the Second Basic Law as a probability statement, expressing thereby the idea that any natural phenomenon ought to reach its most probable state at last.

The main purpose of the book at hand is to give an account of the thermodynamic concepts in their historic development—and, alongside, to display the impact of the two basic laws on the physics and chemistry. Here we either fully skip, or just fleetingly discuss several topics fully treated in any major physics textbook. Meanwhile, some other topics that little known, like, for example, Horstmann's work to introduce the Second Basic Law into chemistry, have received a much more detailed attention here.

As for the works in which Helmholtz, Planck and Gibbs introduce some new thermodynamic functions I have decided just to give their short descriptions; whereas the practical use of these functions is illustrated by a number of examples. Thus, it is my sincere desire that the book could be able to guide those, who are interested in thermodynamic issues.

Later, in a separate publication I intend to tell about the use of the Gibbs' phase rule.

Copenhagen, September 1915,

P. B. Freuchen

[Please see Supplementary Note 5, Section 1.9, Page 124]

Table of contents of the original book by Peter:

The Standpoint of Physics Around 1800 AD (pp. 1–5)

1. The mechanical physics was really tending to reach its apogee at the beginning of the 19th century, owing to the work by such colleagues as Galileo, Newton, Johannes Bernoulli, Lagrange and others. Lagrange had shown that we might summarize the entire mechanics under one principle of virtual velocities, later denoted as the virtual shooting or the work principle. It is possible to express its content in simple terms as follows. Let us assume that you hold a system of bodies at rest and

suddenly let all of them go off. If they do start moving, they will get kinetic energy, and the acting forces must therefore perform work. It follows that the movement could not occur, if the sum of the work performed by any small movement is zero. The latter statement thus ought to be an equilibrium condition for the system. If the total work at a virtual displacement is equal to zero, one must distinguish between two cases. In one of the cases any displacement would correspond to the one in the opposite direction, then the machine does go straight, both back and forth; furthermore, the system ought to provide the positive work by shifting in one direction, and in this direction the system's real movement occurs by itself. In the second case there is an obstacle present, which only allows the movement in one direction; if now some work could be done resulting from a virtual shift, then it is positive, and the system itself will come to get moving. However, if the work is negative, then the system is at rest. An example of the latter is a ball that rests at the bottom of a funnel; any displacement corresponds here to a negative work, for the gravity's work is negative. We might think of all the natural motions, whose onset goes spontaneously, as providing some positive work or, to express it energetically, their potential energy is decreasing. For a system being already in motion, the opposite change might also take place due to the inertia, since the potential energy is increasing at the expense of the kinetic energy. This is observable, for example, in a pendulum movement, or in a planet's motion around the sun, or just in all the possible periodic movements, where the both changes are taking place alternately. Later on, we shall see, whether something similar could apply outside the pure mechanics as well.

2. Howbeit, other branches of physics were developing quite independently of each other. In thermodynamics field, there were significant works in the early 19th century, for example, that by Leslie about heat radiation, those by Dalton and Gay-Lussac about gas expansion, as well as Fourier's mathematical treatment of heat conduction. It was the general view that heat was a substance, although different voices uttering the contrary had risen in the course of time. Apart from the already mentioned authors, we have to note Huygens as well, for in his *Traits de la lumière,* published in 1690, he strongly

raised his voice in favor of the mechanical theories of heat and light. Lavoisier and Laplace claimed some similar belief about 100 years later. Still, they both were living in the period of imponderability, for in addition to representing heat and light in the form of weightless substances they had to introduce both magnetic and electric fluids as well. Meanwhile, there were already clear signs that some new conception might emerge. Rumford, Davy and Young were convinced that the heat was not a substance, but a kind of movement. Young attacked Newton's emission theory, as well as that by Huygens, so the time of Hooke and Euler seemed clearly to come to its end. Fresnel's famous studies of light had brought the emission theory to the ultimate stop. The wave theory had at last triumphed, for after Fresnel we had to view the light as transverse oscillations. Of interest is the work by Euler, who considered the light movement as longitudinal vibrations and expressed the opinion that light exerts a pressure on the illuminated body.*

Still, for the supporters of the substantial heat theory even more and more difficulties were arising. Specifically, several phenomena caused by an electric current—for example, the heat generation in an electric conductor—would not be properly allegeable in terms of the substantial theory of heat. Well, the adepts of the latter theory were still trying to save the situation by artificial means, but their position was gradually getting untenable.

Ørsted's discovery had shown that the electric current had magnetic effects. In addition, Ampère could manage removing the magnetic fluids from the imponderable series. Mellonis studies of thermal radiation could prove the identity of the light and heat rays. Further, Ørsted already knew about the connection between light and electricity, an idea that ought to become important later on. Meanwhile, the science was approaching a time when the idea of 'forces' transformation began to dawn, and that was something which could not be reconciled with the imponderable notion of some substance. Finally, a fully new task had appeared, which consisted in examining, whether or not 'forces' mutual transformation took place according to a certain quantitative relationship.

---

*P. Lebedew. *Die Druckkräfte des Lichtes. Ostwalds Klassiker*, No. 188, p. 5.

3. Regretfully, there were hitherto much too confusion around the notion of force.* Some colleagues were measuring a force by the momentum/movimentum, like Descartes, while others, like Leibnitz, by the product of mass and velocity squared. To the latter expression, Leibnitz had given the name *Vis Viva*, i.e., living/**livening** force, as opposed to the *Vis Mortua*, i.e., the dead force (e.g., a pressure of a resting body). Finally, d'Alembert had rightly considered this story 'a dispute about words'.

[Please see Supplementary Note 6, Section 1.9, Page 124]

Indeed, the main point is that they tended to define the force notion in two different ways. Specifically, if we understand the force as a moving body's ability to overcome some resistance/obstacles, then we do not think about the time required for the process, but, instead, chiefly about what we nowadays call 'the work', and thus we directly arrive at the *Vis Viva* notion by Leibnitz. But on the other hand, if we consider the force a unique cause behind the body's movement, then we take into account the concept of time by measuring the force developed at certain time points and denoting the result as 'the *momentum/movimentum*'—or, as we now mathematically express the latter—the product of the mass and acceleration. Moreover, it is immediately recognizable that the both expressions of force are possessed of quite different dimensions. Remarkably, the Descartes' standpoint was prevailing all the time and it was just Newton, who could finally manage to fix the situation in his studies.

The concept of *work* appears first in the early 19th century. It is true that in the 18th century they were often employing the product of force and path length, while giving it quite different names, but still not earlier than in 1826 Poncelet brings the notion of 'work' to the general attention, as he states correspondingly

"**Working, *livening* force cannot come out of nothing or fully disappear.**"

[Please see Supplementary Note 7, Section 1.9, Pages 124–125]

The concept of *energy*[†] was in fact introduced in 1800 by Young to describe the livening force, but both the *energy* and the *work*

---

*Rosenberger: *Die Geschichte der Physik II*; p. 252 ff.

[†]Already Aristotle had introduced the word and notion of *energy* in its physical sense; then, Galileo and Johannes Bernoulli were incidentally using it without assigning any particular importance thereto.

notions were that time used only provisionally by assigning them purely mechanical significance. The same applies to the notion of the Potential introduced somewhat later by Green, Gauss and Hamilton.

Only around the year of 1850, Rudolf Clausius and William Thomson have introduced the notions of *work* and *energy* into all the areas of physics.

4. One of the main achievements in the 19th century that physics could arrive at was the 'law of the livening forces conservation'. In a system of material points subjected to central forces, the livening force can only be dependent on the relative positions of these points. Let us now assume some change in the latter, i.e., in the system's potential energy. Then the change involved might be a measure of the work to be performable after properly transforming the potential into the kinetic energy, with the latter being the actual livening force. Moreover, if now the system would come back to its original position, then the total available amount of energy ought to remain the same as before.

On the other hand, it is already at the end of the 18th century that the impossibility of the Perpetuum Mobile became clear. Actually, the latter standpoint was solely applicable to purely mechanical effects, and it was quite comfortable to live with the idea that *perpetuum mobile* was impossible in general. However, we had still to grasp the true essence of this law.

Carnot (pp. 5–7*)*

5. The very first colleague, who could manage applying the law of the perpetual motion impossibility to studying non-mechanical phenomena, was Sadi Carnot.*

Indeed, he was capable of founding a fully novel theory of heat, as a result. After the vapor engine invention, it was necessary to suggest a handy theory capable of assessing the efficiency of such devices. Moreover, that was just Carnot, who could pioneer finding the true starting point of such a theory. Sure, using heat can in principle be turned into getting work, but for this to happen, it is in fact necessary to start with the temperature difference or, in other words, to have a *disturbed heat balance* from the beginning on.

_____

*Carnot: *Sur la puissance motrice du feu,* Paris 1824, *Ostwalds Klassiker*, Nr. 37.

The work might be performed "*not due to the heat consumption as it is, but owing to heat transfer from a hot body to a cold one, i.e., by regaining the heat balance disturbed by some chemical effects like, e.g., burning some fuel etc.*"

Therefore, "*it is not enough solely to generate the heat, one must anyway get some cold body beforehand; without this the sole heat ought to be useless.*"

"*Wherever the temperature differences ought to be present, it is throughout possible to produce the Livening Force. Conversely, wherever the latter force is applied, it is throughout possible to create temperature differences or to disturb the heat balance.*" The latter might be accomplished, for example, by a shock, friction, or by gas compression.

6. Further, Carnot suggests his famous cycle, in which a working substance absorbs heat from a heat reservoir, performs work and transfers heat to a cold body. He shows that this intrinsically circular procedure is capable of performing the maximum work as compared to any other cycle between the same temperature limits. Why it is so? This way we avoid any heat loss and pressure drop. Moreover, this maximum of work ought to be independent of the working substance's type. Hence, if the both latter points were not the case, we could easily build a machine with the perpetual motion. Remarkably, that such a process is in fact an ideal limiting case was perfectly clear to Carnot and is surely in no way to reduce the importance of his brilliant mind.

Carnot compares the driving force of heat with the power of falling water, with the both arriving at their maximum, irrespective of the working substance or working machine in use: "*When the water drops the moving force is exactly proportional to the height difference between the upper and lower containers. By employing a fuel, the moving force undoubtedly increases with the temperature difference between the hot and the cold bodies, but we do not know whether it is proportional to the latter difference.*"

[Please see Supplementary Note 8, Section 1.9, Page 125]

Further, as Carnot turns to study the problem of the driving force in detail, he encounters the notion of specific heat of the air species taken either at constant volume or at constant pressure. Still, he believes that there is of course no way to the real reason of why the former, and that the both should be different from each other. The conventional explanation of that time was that the volume increase ought to require more amount of heat, and it is this idea that could survive for a long time. It is with this in mind that Carnot concludes in his considerations: More amount of heat ought to be necessary to keep a certain amount of air at a temperature of 100° while doubling its volume, than to keep the same amount of air at a temperature of 1° and performing exactly the same extension.

> "*These unequal amounts of heat would be capable of producing equal amounts of driving force for the same wastage of heat substance at different temperatures; and this leads to the following conclusion: The same wastage of heat substance ought to produce more driving force at lower temperatures than at the higher ones.*" Remarkably, as we now know, the latter statement ought to be correct as it is, but the way to deduce it is not to 100% correct.

[Please see Supplementary Note 9, Section 1.9, Page 126]

> "*At our current position we are unable to determine the law according to which the driving force of heat would change upon the various degrees of the temperature scale. This law ought to be connected to changes in the specific heat of gases, but it's still not yet known with the sufficient accuracy*"

In a very interesting footnote* Carnot mentions that if one assumes the specific heat of a gas taken at constant volume to be constant, then you might come to the result that the driving force of the heat is exactly proportional to the heat substance wastage, that is, to the temperature difference.

7. In the final part of his work Carnot goes on to show that the nature of the working substance is not influencing the driving force of heat; he investigates the efficiency of the three different cycles, carried out with air, water vapor and alcohol vapor. Concerning the then known values for the vaporization heats of liquids he carried out thought experiments at different temperatures. In addition, Carnot finds the following values for the number of kilogram-meters won by one kilogram-calorie falls one degree in the three cases: 1,395, then 1,112 and 1,230. The latter results are indeed somewhat different, but we might not really compare processes at the same height of the thermometric scale, as Carnot rightfully notices.

Ostwalds Klassiker, Nr. 37, p. 42.

Meanwhile, these results have apparently brought Carnot to strong doubts as for his core results.

"The basis, on which the heat theory rests, does not seem to be of unshakable firmness. Only further tests might clarify the issue."

8. Carnot died in 1832 of cholera. It is with a warm, but to some extent whimsical, feeling that we are now reading his posthumous notes, which were first published only in 1878.* From these it is apparent that Carnot in his last years had finally dropped the idea of the invariable heat substance by viewing the heat as a motion, and that he already had a good knowledge of the heat's mechanical equivalent. Without even mentioning how he might get the result, he concludes that 1000 kg-meter are equivalent to 2.70 kg-calories, thus giving the heat's equivalent value of 370. Therefore, he arrives at the value in fact very close to that found by Mayer some dozen years after Carnot's death.

[Please see Supplementary Note 10, Section 1.9, Page 126]

Clapeyron (pp. 7–13)

9. Carnot's work is thought to have become known to wider professional circles by the excellent graphical and analytical representation, which B. Clapeyron[†] has given thereof. One might imagine some air amount, which passes through a Carnot cycle. If to choose the volume as the abscissa, then the ordinate ought to be the pressure (see Fig. 1.1 below).

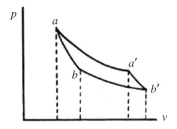

Figure 1.1

Comptes Rendus, 87, 967, 1878; *Mach: Principien der Wärmelehre*, p. 244.
[†]*Journal de l'école polytechnique*, 14, p. 170, 1834 ; *Pogg. Ann.*, 59, pp. 446 and 566, 1843.

The isothermal expansion of air is performed along the pathway a → a′, whereas its isothermal compression goes along b′ → b; and the adiabatic expansion and compression— along a′ → b′ and b → a, respectively. Clapeyron knew that a → a′ and b → b′ ought to be hyperbolas; but, on the other hand, he noted that the air pressure along a′ → b′ decreases by an unknown law; most probably, with this in mind, he was sure* that the Laplace's and Poisson's studies of the air dilation were unsafe.

Clapeyron stands quite firmly on Carnot's position and states that the actual result of the whole cyclic process ought to be the performed work—plus the correspondingly sunken heating degree—along the whole area a → a′ → b′ → b → a. He states explicitly that the process in question transfers the entire amount of heat obtained by the gas from the warmer body to the cooler one.

Of course, for the cycle with vapor the area adopts a quite different look (see Fig. 1.2).

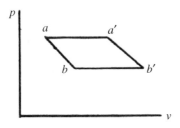

Figure 1.2

Hence, one has to carry out the isothermal compression or expansion achievable under the constant pressure in such a case.

10. To find the largest possible amount of work feasible by a certain wastage of the heat substance changing its temperature from a higher to a lower one, Clapeyron considers a cycle extended between the endless narrow limits. Just like Carnot, Clapeyron assumes that one has to inject the quantity of heat Q into a body, while its transition from one state into another ought to depend solely on the start and the end states, but not on the pathway; and therefore, Q ought to be a function of v and p.

Considering now a gas only, we recognize that the isothermal change occurs along the pathway a → b (Fig. 1.3).

Pogg. Ann., 59, p. 451, 1843.

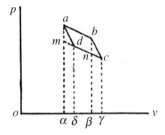

Figure 1.3

Then we can cast it as follows:

$$dQ = \frac{dQ}{dp}dp + \frac{dQ}{dv}dv,$$

or when *pv = const*,

$$vdp + pdv = 0$$

and then

$$dQ = \left(\frac{dQ}{dv} - \frac{p}{v}\frac{dQ}{dp} \right)dv .$$

This amount of heat is carried out at the temperature t in the working substance and dispensed (according to Carnot's law) unchanged at *t–dt* (isotherm *d → c*). The work won by the process parallelogram *abcd*, whose area is equal to *a · b · n · m* or *αβ · am*; setting *αβ = dv* and *am = dp*, the available work equals *dv*dp*. However, as soon as *vp = RT* is valid, the work becomes equal to $\frac{Rdvdt}{v}$.* Further, according to N. L. S. Carnot, the ratio between the work and the amount of heat is dependent only on the temperature and not on the material; hence we get

$$\frac{\text{Work}}{\text{Heat}} = \frac{RdT}{v\dfrac{dQ}{dv} - p\dfrac{dQ}{dp}} = \frac{dT}{C} .$$

Here *C*, the Carnot's function, depends only on the temperature. We might also recast this equation as follows

$$CR = v\frac{dQ}{dv} - p\frac{dQ}{dp} . \tag{1.1}$$

---

*As *T = t + 273*, then *dT = dt*.

[Please see Supplementary Note 11, Section 1.9, Page 126]

11. Then Clapeyron examines the relationship between work and heat during the cyclic process with vapor, in changing the temperature from t to $t-dt$. We guess that at some temperature $t°$ a vapor amount of the volume v might be produced, with the vapor density to be δ and that of the liquid to be ρ, so that $\dfrac{v\delta}{\rho}$ would be correspondent to the volume of the evaporated fluid. Hence, we might cast the volume increase due to the evaporation in the following way:

$$v - \frac{v\delta}{\rho} = v\left(1 - \frac{\delta}{\rho}\right).$$

The transferred amount of heat is kv, where k is the latent heat of the volume v at the temperature t. Then the work produced during the circuit process expressed by the area abcd (Fig. 1.4) ought to be equal to $dv\dfrac{dp}{dt}dt = v\left(1 - \dfrac{\delta}{\rho}\right)\dfrac{dp}{dt}dt$. After dividing the corresponding amount of work by the corresponding amount of heat we get

$$\frac{\left(1 - \dfrac{\delta}{\rho}\right)\dfrac{dp}{dt}}{k} = \frac{1}{C} \qquad (1.2)$$

If we consider the ratio $\dfrac{\delta}{\rho}$ negligible as compared to a unity, then we get $k = C\dfrac{dp}{dt}$.

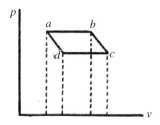

Figure 1.4

Since C at the same temperature has the same value for all the possible working substances, then the latent heat k at equal volumes of different vapors ought to be proportional to the derivative $\dfrac{dp}{dt}$.

Hence the latent heats for the liquids noticeably boiling at higher temperatures only ought to be very small, for example, mercury, for in such cases $\dfrac{dp}{dt}$ is small. Clapeyron also notes that the more common reformulation of the Eq. 1.2, namely $k = \left(1 - \dfrac{\delta}{\rho}\right) C \dfrac{dp}{dt}$, might better express the fact that k tends to zero if $\delta = \rho$, that is, if the vapor's density becomes equal to that of the liquid, when the pressure becomes so large and the temperature gets so high, since C and $\dfrac{dp}{dt}$ are not getting infinitely large.

12. Of the utmost importance ought to be Clapeyron's investigation* of a cyclic process for some arbitrary system. We start from the volume $0 - \alpha$ and pressure $\alpha - d$ (see Fig. 1.5) and consider heating the system from the temperature t to the $t + dt$ at the constant volume.

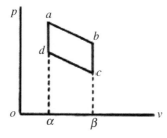

Figure 1.5

Then the pressure ought to increase from d to a. And now we let the system get the heat from a heat source A with the temperature $t + dt$ and isothermally expand from a to b. Thereafter the body is cooling down, so that the temperature should drop by dt degrees at the constant volume; along with this the pressure ought to decrease from b to c.

Finally, we let the system release the obtained heat to a container B with the temperature t as volume is decreasing by the amount $\alpha - \beta$, while the pressure is growing. The body is thus returning to its original state. This way the work has been performed during the cyclic process 'd-a-b-c' and the heat amount has decreased by dt degrees.

Pogg. Ann., 59, 569, 1843.

The work performed might then be calculated as

$$dvdp = dv\frac{dp}{dt}dt = \frac{dvdt}{\dfrac{dt}{dp}},$$

whereas the transferred heat amount can be calculated as

$$dQ = \frac{dQ}{dt}dv + \frac{dQ}{dp}dp.$$

But since extension of the body owing to the contact with the container A is isothermal, we arrive at the following relationship:

$$dt = \frac{dt}{dv}dv + \frac{dt}{dp}dp = 0,$$

so that $dp = -\dfrac{\dfrac{dt}{dv}}{\dfrac{dt}{dp}}dv$,

and we get

$$dQ = dv\left(\frac{dQ}{dv} - \frac{dQ}{dp}\frac{\dfrac{dt}{dv}}{\dfrac{dt}{dp}}\right) \tag{1.3}$$

By setting the relationship between the work done and the heat spent equal to $\dfrac{dt}{C}$, we cast the result as follows:

$$\frac{dt}{\dfrac{dQ}{dt}dv - \dfrac{dQ}{dp}dp} = \frac{dt}{C} \tag{1.4}$$

or

$$\frac{dQ}{dt}dv - \frac{dQ}{dp}dp = C$$

By combining Eqs. 1.3 and 1.4 we eliminate $\dfrac{dt}{dv}$ or $\dfrac{dt}{dp}$ to get

$$dQ = dv\frac{C}{\dfrac{dt}{dp}} = -dp\frac{C}{\dfrac{dt}{dv}} \tag{1.4a}$$

and after transforming the latter expression we arrive at

$$dQ = dv \frac{\dfrac{C}{dt}}{dp} = -dp \frac{\dfrac{C}{dt}}{dv} \tag{1.5}$$

where $\dfrac{dv}{dt}$ stands for the differential coefficient of v with respect to t at the constant pressure, that is, $\left(\dfrac{dv}{dt}\right)_p$. Thus we have arrived here at the famous Clapeyron's equation. It expresses the law applicable to all bodies experiencing isothermal changes, and amounts to that if the pressure on the various bodies captured at the same temperature increases with a small size, heat is generated in quantities proportional to their expansion coefficient.

Equation 1.4a can also be rewritten this way

$$dQ = dv\, C \frac{dp}{dt} \tag{1.6}$$

where we get $\dfrac{dp}{dt} = \left(\dfrac{dp}{dt}\right)_v$. Thus, it is also possible to formulate the above-mentioned law in such a way: If one decreases the volume by a small amount, there will be developed a quantity of heat, which is proportional to the pressure coefficient $\dfrac{dp}{dt}$. (We remember that negative dQ means heat deployment.)

For water both $\left(\dfrac{dv}{dt}\right)_p$ and $\left(\dfrac{dp}{dt}\right)_v$ are negative between the temperatures $0°$ and $4°$, and therefore the heat ought to be bonded by compression in this case. Later we shall see that an equation of the same appearance as Eq. 1.6 also apply to the change of state, as well as to several chemical processes.

13. Clapeyron highlights the importance of any information about the Carnot's function, and tries to calculate $\dfrac{1}{C}$ in two ways, both on the basis of Delaroche and Bérard's studies of the heat capacities of gases and also by starting from equation 1.2.

The calculation shows that $\dfrac{1}{C}$ decreases, when the temperature grows, that is, C grows along with the temperature. Using one method Clapeyron finds that $\dfrac{dC}{dt} = 0.002565$, and with the help of another one he arrives at the value of 0.00187; remarkably, the first value agrees quite well with the following approximate expression:

$$C = \frac{T}{425}$$ (as we shall see later) or with its consequence $$\frac{dC}{dt} = \frac{1}{425}$$

= 0.00235. Meanwhile, Clapeyron did not come beyond Carnot's standpoint.

William Thomson and James Thomson (pp. 13–15)

14. Carnot's ideas could find a very good soil in England, where William Thomson could further develop and deepen them.[*] It is in this time (1848) that Thomson had derived his first scale of the absolute temperature. According to Carnot and Clapeyron, the amount of work the heating unit could perform to lower its temperature by 1 degree, was different at different temperature levels. Meanwhile, Thomson had suggested to choose the temperature scale in such a way that the work necessary to decrease the heating unit's temperature by 1 degree would be the same at any temperature; then, the resulting scale should really be independent of any properties of the working substances chosen.

William Thomson was preliminarily remaining at Carnot's viewpoint. Of course, he knew that Joule in his experiments could have transformed mechanical work into heat, but he doubted that the reverse transformation could take place. Thomson had recognized the difficulty and emphasized the enigmatic point at posing such a fundamental question.

15. Moreover, his working with Carnot's theory could soon have led to the discovery of decreasing water's freezing point by applying external pressure.[†] Specifically, William Thomson could attract attention of his brother James Thomson to the fact that at 0° Celsius it is possible to transform water into ice by using a purely mechanical process without spending any work amount to achieve this result. In James Thomson's opinion, this fact should prove an impossibility that one might well enclose the freezing water in a container with piston and thereby carry out work for nothing.

---

[*]*Phil. Mag.*, 33, p. 313, 1848; W. Thomson: An Account of Carnot's theory of the motive power of heat. *Trans. Roy. Soc.*, Edinb., 16, p. 541, 1849.
[†]*Trans. Roy. Soc.*, Edinb., 16, p. 575, 1849.

[Please see Supplementary Note 12, Section 1.9, Pages 126–127]

The original train of thoughts was as follows. One should imagine a cylinder provided with a seamless stamp, with the both being possessed of a non-conductive material and a conductive basement; and now let the cylinder contain air at 0°. The bottom of the cylinder is placed in a lake having temperature 0°, so let us now compress the air slowly, thereby delivering heat to the water.*

Then let us place the cylinder's bottom into a limited body of water at 0° and allow air to expand to its original volume, in giving back the delivered heat quantity and the performed work involved, while the water freezes. Since there is no temperature difference, then, according to Carnot, there is no way to perform any work. However, the freezing water can in effect still carry out the work. Where does this work then come from? James Thomson could solve the poser by suggesting that a small body of water is contained in the reservoir with piston. As the water freezes, it is possible to carry out work by moving the piston. If it is now assumed that ice to be formed under pressure has a lower melting point and that the air expands thereby becoming colder than the air that was previously compressed—because the bottom of the cylinder is in contact with the colder water, so that everything gets in order, as concerns the applicability of Carnot's rule.

James Thomson could then devise a cyclic process in which the ice is brought to melt at 0°, and then the pressure is increased from 1 to 2 atmospheres, so that the pressed water freezes to ice, which is at last brought under one atmospheres pressure. By using Carnot's theory in its representation by Clapeyron, he finds that the ice's melting point goes down by 0.0075° per 1 atmosphere of the pressure. In the following year (1850) William Thomson could reproduce this result in his well-known experiments. James Thomson's cyclic process is in fact the first one in which air or vapor do not contribute.

16. Finally, William Thomson had calculated the values of Carnot's function for the absolute temperatures between 0° and 230° on the basis of Regnault's experiments with the pressure of saturated vapor, in that he made use of Eq. 1.2, but as long as he

---

*That any body ought to produce heat during its compression does not in itself contradict the substantial heat theory.

shared the Carnot's standpoint, he could not find any general expression for Carnot's function. However, this viewpoint got untenable, and William Thomson had left it aside in 1850, being influenced by the works of Clausius and Rankine, the both of whom had meanwhile found that saturated water vapor condenses by expansion.

Robert Mayer (pp. 15–19)

17. It was just the time around 1840s that various researchers discussed the issue of the heat's nature. Mohr (1837), Séguin (1839), and Grove (1842) were all familiar with the idea that heat and work were equivalent variables, but nobody of them managed to find the actual relationship between the both.

The true solution to the problem was found by Robert Mayer, whose basic work[*] was published in 1842.

In 1840–1841 Mayer had made a trip to the East Indies as a ship's doctor in Dutch service. This trip was of great importance. When performing bloodletting he noticed that the venous blood exhibited a striking pink color, as if it had been taken from an artery; German doctors on Java told him that this was a well-known phenomenon observed both for the natives and strangers. Mayer decided to put it into the context of Lavoisier's theory, according to which the heat produced by animals ought to be due to combustion processes. With this in mind, the lower heat loss to the environment results in less combustion in the organism, that is, in poorer work performance. Mayer was eagerly trying to think over this idea in all the necessary details, and he became for a long time so satisfied with his finding, that he could think almost of nothing else. With the genius' ability he envisaged an important context of all the things, which he could not even manage to properly express; and it was exactly this way that could lead him to a true enlightenment in the form of some basic law encompassing all of nature. In a letter he wrote in 1844 to Griesinger, he delivered the best depiction of his psychological state at the time. The letter was in Mayer's mother tongue German, a part of which has been translated to English and presented here:

---

[*]Mayer: *Bemerkungen über die Kräfte der unbelebten Natur.*, 1842.

"*...Therefore I had made an effort to grasp the physics and dwelt on the subject with such a fondness—somebody might wish to laugh at me—that I wasn't truly interested in the distant parts of the world, but rather preferred staying on board, where I was able to work incessantly, and where I sometimes felt such an inspiration that I could never experience previously or even later. Some flashes of thought that drove through me, and it was just on the road of Surabaya, were immediately pursued diligently and leading me back to new items. Those days are gone, but the calm examination of what then appeared in me, has taught me that it is the truth that might not only be subjectively felt, but also quite objectively proven; and whether this could be done by such an unqualified physicist as me myself, I had surely to leave undecided. Certainly, the day will come, that all these truths would become the common property of science; but who will cause it and when it will happen, who would be capable of predicting that?*"

18. In the beginning, Mayer was anyway not comfortable with the usual physical concepts—for example, he definitely confused the 'livening force' with the 'momentum'—but in his work* of 1845, he could finally reach the necessary full clarity.

Without carrying out experiments he was capable of demonstrating in his first publication of 1842 that previous experimental results for the gases were enough to calculate the thermal mechanical equivalent. And Mayer was the first colleague who could correctly grasp the physical sense of the outflow experiment by Gay-Lussac. In effect, already in 1807 Gay-Lussac could show that when an air mass flows from the one container into another evacuated vessel of the same dimensions then the cooling in the former is equal to the heating of the latter one. With this in mind Mayer could draw the correct conclusion that the actual expansion in itself does not require heat (this is inconsistent with the heat substance theory). It is equally certain, states Mayer, that an air mass expanding under pressure should be cooling down. Based upon the latter fact—and using the available to him contemporary information on the air's specific heat at constant pressure and constant volume—Mayer calculates the heat equivalent to be 365 work units (using the same method, but employing the most recent numerical data we find it to be equal to 423.8).

---

*Mayer: *Die organische Bewegung in ihrem Zusammenhang mit dem Stoffwechsel*, 1845.

19. Truly peculiar ought to be Mayer's use of the term force. He clearly recognized the importance of the various concepts, which Leibnitz and Newton had provided for the force notion; but both were possessed of different dimensions, and hence one of them had to be abandoned. Mayer joined Leibnitz' force concept (see § 3 above), because he considered this to be more fundamental than the other one. He would thereby like to highlight the contrast between the force and substance notions, as often happens even nowadays with the same terms, although these should be denoted energy and matter, respectively.

"One reason that raises a body," stated Mayer, "is a force. Its effect, the raised burden, is also a force. Any spatial difference between ponderable objects is a force; since this force causes the bodies to fall, we call it a gravity force. One might say that forces are motions or motions are forces, so that the both behave like cause and effect; they turn into each other, they are two different manifestations of the same thing. A burden resting on the ground is not force; it is not the cause of a movement or lifting another burden." It is clear that the effect here is similar to energy: either potential or kinetic. Mayer also says that heat is a power, chemical difference is a virtue, etc. "By all physical and chemical processes the given force is of a constant size." "There is in fact only a single force."

It is remarkable that Mayer does not consider heat as a motion. He states explicitly: heat, motion and the force of gravity let themselves turn to each other in specific proportions. And it is equally difficult to draw here the conclusion that the force of gravity and movement are identical, and to conclude that heat is motion. Instead, along with Mayer we rather ought to conclude that "to be converted into heat the movement must cease to be the movement."

In fact, following Mayer's deliberations it's possible to build up the entire heating theory without employing the notion that heat is of mechanical nature, for his two main statements say nothing about the latter. But it cannot be denied that the kinetic viewpoint has gained greater and greater value, particularly through the development of the gas theories. In his two papers Mayer had clearly stated what we now denote the Law of Conservation of Energy. Meanwhile, the term energy is used by Mayer occasionally, but rather undetermined.

20. In his next work* Mayer employs his theory to the cosmic relations. He gives an explanation of the solar heat maintenance (meteorites falling) and notes that the tides may hamper the rotation of the earth.

Finally, another couple of interesting citations by Mayer could be published here. "If one were to raise the earth's crust up on pillars, increasing this way the Earth's volume would be the highlight of this immense burden, thus requiring a tremendous amount of heat. Conversely, it is clear that properly reducing the Earth volume would release an equal amount of heat. But what applies to the entire earth's crust, would also be true for any fraction thereof. Indeed, a water drop causes a decline in soil volume; consequently, one can thus gain heat or some other force."

"A locomotive with its convoy could be comparable to a distillation device; a heat produced under the boiler transforms into the motion, and then is accumulating in the wheel axle again as a heat."

"It holds true both for a given mixture of chlorine, metal and oxygen, the small amount of which can be turned into chlorine acidic potash due to the formation of by-products, and for a given amount of heat as a whole to be transformed into a movement."

Remarkably, it's the latter couple of utterances that clearly sheds light in the direction of the 2nd law.

21. Finally, there is Mayer's small treatise on the "release"† where he jumps into the psychology.

The term "transformations of forces" is often connected to a triggering cause for the beginning to occur. Removal of an obstacle can thus bring a "tension" to come into operation; a small cause can thus trigger a major effect. The proper examples would include ignition, fermentation processes, etc. Mayer highlights the great importance that the releases do have in the living world. "All our lives represent a continuous release process." "A pleasant health sense stands for an undisturbed releasing device." The joy we feel when doing sports is then one outlier thereof. However, not only internal physiological actuations are the matter of well-being; Outer effects may also be connected to the human pleasure. For example, the desire to fire the guns, burning explosive substances and likewise.

---

*Mayer: *Beiträge zur Dynamik des Himmels*, 1848.
†Came out considerably later in the last publication by J. R. von Mayer: Die Toricellische Leere und über Auslösung (Stuttgart, 1876).

But unfortunately this desire to produce powerful effects can also lead to very bad results.

Indeed, such faults as assassinations, arson attacks are often urging violent cleardowns. "Yes, if our planet were designed in such a way that it might be possible for anyone to blow it like a dynamite-filled container, then probably there would always be people prepared to sacrifice their own lives to let our nice planet Earth explode into the Space."

All of Mayer's theses are published in a single volume—Mayer: *Mechanik der Wärme*, 1867.

Joule (pp. 19–21)

22. The experimental side of things was being developed in the early forties by Colding and Joule, later by Edlund and Hirn; Most of the related work is so well known that we can skip here any detailed consideration of them.

Already in 1840—i.e., independently of Mayer—Joule had started performing a series of experiments to find the laws that govern system's heating by an electric current. Joule found that the amount of heat was proportional to the resistance and square of the current's intensity and that the heat quantity thus developing in the electric circuit was equal to the heat amount produced by the chemical process taking place in the relevant battery.

Some years later Joule started studying thermal effects by induction currents. He put a copper wire wound over the iron core into a glass tube with water and turned it around between electromagnet poles; the power supply could be regulated by means of a commutator and was measured by a galvanometer.

Joule found that the heat generation was proportional to the square of the current's intensity. He had no doubt that these currents follow the same laws as other currents; but first of all Joule would like to learn about the heat to be really produced on site—whether this heat would be transferred from the one part of the apparatus to another, as it was just the case for the Peltier's heat.

The experiments were repeated by changing the iron core winding and the battery current sent to the galvanometer. By turning the system one or another way Joule was capable of generating induction current, which could either attenuate or reinforce the current from the battery in connection with the direction of rotation. In the first case the apparatus was acting as a motor and produced work, while in the other case—as an induction machine consuming work. Since it has thus turned out that it is possible to weaken or reinforce the power coming from battery at will, Joule could recognize the possibility of producing or destroying the heat by mechanical means, in using magneto-electric currents as intermediaries. "In magneto-electricity we have therefore an agent capable of destroying or generating heat by a simple mechanical action."*

23. Joule would then like to test whether there was a constant relationship between the lost or gained work and gained or lost heat. With this in mind he employed a gravity driven rotary shaker and was thus capable of measuring the desired equivalence. For the same purpose, he also carried out his well-known experiments on rubbing and pressing of water through narrow tubes as well as on the temperature changes[†] produced by compression and expansion of air. By these experiments Joule could find that it is possible to avoid binding or release of heat by changing the pertinent gas density,[‡] if the specific heat of this gas is independent of its volume. Therefore the gas energy ought to be of purely kinetic nature, and the heat regulation at the changes in the air volume ought to depend solely on the amount of the work done.

It is this way that Joule could unambiguously prove that the heat is not a substance, but a movement of the smallest parts of the systems under study. By his diverse studies Joule could definitely demonstrate that transformation of one energy type into another one ought to follow a certain quantitative regularity.

Helmholtz (pp. 21–27)

24. The above-mentioned new field of knowledge had gotten its firm scientific basis with the advent of Helmholtz' admirable treatise *Die Erhaltung der Kraft*, 1847. Sure, in the title of his

*Joule: Scientific Papers I, p. 146.
[†]Joule: On the changes of temperature produced by the rarefaction and condensation of air. *Phil. Mag.*, 26, 1845.
[‡]At that time Joule had still no information about Gay-Lussac's experiment on gaseous efflux.

work, as well as in a few other statements of him, Helmholtz demonstrates his acknowledgement of the Leibnitz force concept, whereas he is employing Newton's terminology elsewhere. While Helmholtz' move was clearly independent* of that by Mayer, he was still informed of the Joule's first works, and thus he tells in his introduction that one might either start from the proposition that it is impossible to produce work out of nothing, or from the assumption that all the natural effects would anyway boil down to the central forces. And just at the beginning of his work Helmholtz states that the latter two statements are in effect identical.

Helmholtz declares that his task ought to be extending the principle of "the impossibility of perpetuum mobile" to all the branches of physics. One might imagine a system of moving mass points with solely attractive and repulsive forces acting among them, with the latter to be dependent solely on the pairwise spacing between the points and to satisfy the (*Newton's*) law of action and reaction. Hence, in such a system the Law of Conservation of the livening forces ought to be applicable. Bearing this in mind, any increase in the livening force is therefore equal to the work performed by the Central Forces. Helmholtz uses the term 'tone/resilience/buoyancy/vigor' (Spannkräfte) to mention the forces, "as long as they have not yet produced any movement," as opposed to livening force, and expresses the following idea: The sum of the livening forces and clamping/tensioning forces ought to be constant. And this is just the law of the energy conservation. Nowadays we might willingly follow Rankine in using the terms of potential energy and kinetic energy instead of tension/clamping and livening force, but the main point, the law of energy conservation would still hold anyway.

25. As for the two assumptions, which Helmholtz considered to be ultimately valid, Planck had raised some comments.

Even assuming that a perpetuum mobile is impossible, one might however well imagine that energy could become nothing (as Clapeyron did it, for example[†]). For centuries they had had an inkling of that the work could not arise from nothing, without even thinking

---

*Ostwalds Classics No. 1, p. 56, wherein Helmholtz definitely spots Mayer as the true pioneer.
[†]Planck: *Erhaltung der Energie*, p. 16.

about whether the latter statement could equally be reversed;* indeed, there had hardly been a person, who were practically dealing with the task of destroying the work—instead, much more colleagues were rather trying to turn gold into lead. Nonetheless, the law of the energy conservation was understood completely, that is, assuming not only the impossibility of generating power from nothing, but also the impossibility of the energy destruction.

The second assumption that all the natural effects ought to be the consequences of the "central forces" has in fact played a major role. As soon as the substantial heat theory was clearly skipped, it had become immediately suggestive to consider heat and all other energy types in terms of kinetic or potential mechanical energy—and then, consequently, the Law of Energy Conservation might simply be cast using the equations of mechanics. Here are just the roots of the notion "Mechanical Theory of Heat," which had later to be superceded by the more comprehensive notion of "Thermodynamics." Apart from the hypothesis to consider all natural phenomena as the effects of central forces further assumptions were to be made and other difficulties were to be encountered that will be discussed in detail later on.

26. It is with great professional expertise that Helmholtz deals in "Erhaltung der Kraft" with various areas of physics. For example, in the section on "The Caloric equivalent of Force" he finds the expression for the Carnot's function. This could be done by comparing the Clapeyron formula (1) with the formula of Holtzmann.[†] The latter physicist could find that the amount of heat capable of increasing the temperature of 1 kg of water by 1° might also lift 374 kg by 1 meter, using the similar departure point and at about the same time as Mayer. Yet Holtzmann was essentially one of the adepts of the substantial heat theory,[‡] while arguing the "conservation of the heat substance."

---

[*]Ibid., p. 139.
[†]Holtzmann: *Über die Wärme und Elasticitat der Gase und Dämpfe*, 1845, *Pogg. Ann. Erg. II*, 183, 1848.
[‡]Holtzmann: *Über die bewegende Kraft der Wärme*, *Pogg. Ann.*, 82, 445, 1851. Remarkably, in his last treatise *Mechanische Wärmetheorie*, 1866, Holtzmann had skipped the substantial theory of heat. Rosenberger: *Die Geschichte der Physik III*, 373.

The Holtzmann's formula could in fact be cast as follows:

$$\frac{pv}{a} = v\frac{dQ}{dv} - p\frac{dQ}{dp}, \tag{1.7}$$

where a stands for the force equivalent of the heating unit (which was later on denoted using J). All other letters denote exactly the same as in the Clapeyron's formula

$$CR = v\frac{dQ}{dv} - p\frac{dQ}{dp}.$$

Helmholtz notes that the Holtzmann's formula is correct, because the heat capacity of a gas is independent of its volume "as it really seems to follow from the results of the above-mentioned Joule's experiment."

The both above-cited formulae ought to be correspondent to each other, if $CR = \frac{pv}{a}$ or when $pv = R(273 + t)$, then we immediately get

$$\frac{1}{C} = \frac{a}{273+t}. \tag{1.8}$$

Helmholtz notes that the C values calculated by Clapeyron are in very good accordance with the latter formula derived above.

27. In the section on 'Force Equivalent for Electrical Phenomena' Helmholtz refers firstly to the static electricity and finds a formula for heat generation by electrical discharge consistent with Riess' trials. In his discussion of "galvanism" Helmholtz deals with the two theories, which at his time were exhibiting a violent conflict, namely, the contact theory and the chemical theory. According to the genuine contact theory, claimed by Volta, the electricity ought to result exclusively from contacts between the metals, whereas the fluids ought to be only passive managers. A powerful contribution to the dispute had come in 1840 from Faraday, who had become an adept of the chemical theory, according to which the electric power has its origin in chemical processes. Faraday[*] writes in particular:

"The contact theory assumes that a force capable of overcoming a powerful resistance might arise from nothing. Practically this corresponds to the creation of a driving force out of nothing and

---

[*]Rosenberger: *Die Geschichte der Physik III*, p. 288.

thus appears to be quite different from any other natural force. Given diverse phenomena in which the expression form of the force might change in such a way that an apparent transformation of some kind of the force to another one takes place. In this way we can transform chemical forces to an electric current or vice versa. The beautiful experiments of Seebech and Peltier demonstrate the mutual transitions between heat and electricity, as well as other experiments conducted by Ørsted and me myself prove the electricity's and magnetism's mutual transformation abilities. But in no case, not even in such fishes as electric eels and skates, any generation of the force could be possible without the corresponding consumption of something else."

It appears that Faraday was quite familiar with the idea of the unity of the natural forces; meanwhile, his studies still could not lead him to finding a valid relationship between them.

28. Helmholtz considers the working equivalent of galvanic currents as a result of schisms going on in the conductors of the second class, and concludes that the contact theory ought to forcefully contradict the energy conservation law, provided there might be a single conductor of the second class (i.e., those not following the voltage row), which could not be cleaved by electric currents. However, if we assume that any conductor of the second class is an electrolyte, then there would be nothing wrong if we assume a valid contact force, for example, to be demonstrated by various metals capable of being electrical conductors with unequal capabilities, whereas the electrical current maintenance as it is might equally result from chemical processes in electrolytes. It is exactly this way that we might reconcile the conceptual gap between the two theories.

Helmholtz assumes that it is not remarkable, if the electromotive force in galvanic elements without polarization might be proportional to heat generated/consumed during the pertinent chemical process; and in effect, this was already a long time the general belief fitting into some particular cases (Daniell, Grove). But only many years later it had become apparent that this was Helmholtz himself, who could clearly solve this difficult poser.

Of much interest is surely the Helmholtz' treatment of induction current, where he could theoretically find the magnitude of the induced electromotive force. Still, the treatment of the mutual impact among different electrical currents ought to be considered incomplete, inasmuch as we lack a clear-cut expression for what we now call the electron kinetic energy. Sure that it is easy-peasy to point out some shortcomings from our today's point of view. But one must nonetheless admire the depth of Helmholtz' studies, for at his time he could manage to consider all the physical phenomena, for which the energy conservation law ought to be of importance.

29. Yet to mention in this connection ought to be the work by the Danish artillery officer Hermann v. Kauffmann: *'Die Arbeit der Wärme. Rückblick auf Vorträge, gehalten bei den Naturforscherversammlungen zu Nürnberg, Kopenhagen und Aachen'*, 1848. It turns out that Kauffmann had already in 1845 attended a naturalist meeting* in Nuremberg and announced there the new principle of 'the heat of action'. Moreover, he could also clearly show how it is possible to solve a number of tasks in heat theory by means of the latter principle. After reading his small book, where Kauffmann refers to Carnot, Clapeyron and Holtzmann, but does not mention Colding or Robert Mayer, one might immediately recognize that he had a definitely clear viewpoint as concerns the 'work transformation':

   "What disappears as livening force ought to re-appear as heat, light, electricity, depending on the circumstances." "Heat and mechanical work replace each other in turns, so that nothing of their sum ought to be lost."

Kauffmann applies his 'action principle' to various subjects such as temperature concept, properties of gases and vapors (He notices, i. a., that hydrogen is a gas with a negative cohesion). Based upon Suermann's experimental results he finds at one of his formulas that 'the heater's work amount is equal to 367 kilograms-meters'. Finally, he exactly calculates the work a certain amount of powder might perform.

---

*Amtlicher Bericht über die 23. Versammlung deutscher Naturforscher und Ärzte in Nürnberg, 1845, p. 90.

30. It is now clear from our above notes that the discovery of the Energy Conservation Law cannot be attributed to a single nation, let alone a single colleague. In various countries new ideas were growing out, but it still took several years before this basic principle could completely penetrate common knowledge. Most probably that was partly owing to the wide-spread 'wanton mood' of that time to deny anything that smacked of natural philosophy. Colding works remained unnoticed for many years; Mayer's theses triggered only little response at the beginning. Helmholtz was not aware of Mayer's work, when he wrote his *Die Erhaltung der Kraft*; Joule was also completely independent of Mayer. Indeed, Colding and Mayer were the pioneers, who managed to proclaim the law of general validity, while Joule managed to give the strongest experimental support to the law inquestion, and Helmholtz could finally manage to deliver its exact treatment, owing to his exceptional erudition and proficient outlook.

31. Similarly, Helmholtz' work could also not convey any significant impression in the first years after its publication. William Thomson* could become aware of it not before 1852. Still, in the late forties there were several researchers who had not made themselves familiar with the new doctrine. It is strange to read what William Thomson[†] writes 1849. Although he acknowledges the validity of Joule's experimental results on the heat of induction currents, he considers it to be not impossible that the heat might be taken from the inducing magnet and transferred in the conductor; although he is willing to admit that the work can be transformed into heat, whereas the opposite transformation ought to deserve a separate proof, according to him.

Neither Carnot's nor Joule's beliefs had anything to do with the energy conservation law, according to Thomson, "*Where disappears the heat transferred from a higher to a lower temperature just by conduction and without performing any work? What would you get then as a substitute for the work that might be won?*" These posers are still remaining eligible. In effect any transformation of energy

---

*Planck: *Erhaltung der Energie*, p. 48.
[†]Mach: *Prinzipien der Wärmelehre*, p. 270.

ought to be connected with a potential drop or a level decrease—depending on the energy type. Conversely, if a potential drop takes place, there is also a general transformation and therefore it results in a loss of the type of energy that falls from higher to lower potential. The latter does not seem to apply to heat. That, is, the heat can sink to a lower potential (temperature), without any loss of heat. It seems as if the heat has here a special position. Or may perhaps the chemical energy ought to be possessed of the same property?

Thomson continues, "*It might seem that any difficulty would disappear, if you skip the Carnot's fundamental axiom—an outlook which was strongly claimed by Joule. But if we do follow this claim, we could encounter other countless difficulties that might be impossible to overcome without further experimental studies, without a complete re-construction of the whole theory of heat.*"

[Please see Supplementary Note 13, Section 1.9, Page 127]

Clausius (pp. 27–33)

32. In our time one might wonder why it took so long before we managed to unite the Carnot principle of the heat drop with the doctrine of energy conservation; but we immediately see that even outstanding researchers had difficulties therewith. Holtzmann had tried it, but did not come through with it. Clausius was finally the one who brought the case to the absolute clarity. In his opinion, Carnot's conception of heat as an indestructible substance could not be reconciled with the law on energy conservation, for heat can carry out work, so that the latter consumes heat. On the other hand, Carnot was absolutely right that the heat drop is necessary to enable the heat's conversion into work, but it ought to be noticed that while a certain amount of heat is transformed into work, some other amount of heat ought to sink from a higher to a lower temperature.

We surely cannot be content with a mere statement of the Energy Conservation, that is, with the First Basic Law, as Clausius puts it, and would definitely need to add a new law based on a basic property of the heat. With this in mind, Clausius introduces the following principle: **Heat cannot be transferred by itself from a lower to a higher temperature**. The words 'by itself' mean that the transfer of heat from a colder to a warmer body can never be the only result of any process; something else must also happen here, for example, a transfer of heat from a warmer to a cooler body, a generation of heat during some work, diffusion of gases, etc. All the realistic processes might hence be divided into two groups; with the one group being called positive or natural, and the other being negative or unnatural. *The positive processes can take place by themselves, whereas the negative ones may not, for they do require some compensation.* Therefore, it means that any negative process cannot take place without a positive one taking place simultaneously. *The latter statement delivers in fact the essence of the Second Basic Law,* though later on we shall express it in a different way.

33. In his first treatise* Clausius had published a number of important relationships, some of which will briefly be reproduced here.

Let us assume that a body is in a state which is determined by two variables x and y, that is, for example, by the volume v and temperature t. If the body goes from the state $\{x, y\}$ to the state $\{x + dx, y + dy\}$, it will have to draw a quantity of heat dQ (positive or negative), which in any case can be expressed as follows: $dQ = Mdx + Ndy$, wherein M and N are some functions of x and y. One of the Clausius' greatest achievements is just the demonstration that Q is not a function of the instantaneous state, but depends on the way by which the change takes place, or, to express it in mathematical words, dQ is never a total differential itself.

In effect, it would be appropriate to introduce here some particular label (e.g., $\overline{dQ}$, according to C. Neumann), but we shall continue to employ the conventional notations up to the points, where misunderstandings would not be easily avoidable.

We imagine now that the amount of heat dQ was supplied to a body; so we can write then

$$dQ = dH + dI + dW,$$

where H stands for the body's heat content, I for the internal work, W for the external work. If we now put $U = H + I$, where U is called the internal energy, we get $dQ = dU + dW$. Now we shall choose the v and t as independent state variables and assume that the change

*Clausius: *Über die bewegende Kraft der Wärme.*, Pogg. Ann., 79, 378, and 500, 1850; *Ostwalds Klassiker*, No. 137.

takes place under some constant pressure. This way we immediately arrive at the following expression

$$dQ = dU + Apdv,$$

where $A = 1/J$ stands for the mechanical equivalent of heat. This might then be recast as follows:

$$dQ = \frac{dU}{dt}dt + \frac{dU}{dv}dv + Apdv .$$

Hence we have $dQ = Mdv + Ndt$, to finally get $M = \frac{dU}{dv}Ap, \quad N = \frac{dU}{dt},$

so that after differentiation:

$$\frac{dM}{dt} = \frac{d^2U}{dt\,dv} + A\frac{dp}{dt}, \quad \frac{dN}{dv} = \frac{d^2U}{dt\,dv} .$$

And we finally arrive at the following expression:

$$\frac{dM}{dt} = \frac{dN}{dv} + A\frac{dp}{dt} . \tag{1.9}$$

If, for example, we now assume that $pv = R\,(273 + t)$ is valid for a gas, then $\frac{dp}{dt} = \frac{R}{v}$, so we get

$$\frac{dM}{dt} - \frac{dN}{dv} = A\frac{R}{v} . \tag{1.10}$$

We immediately see that $\frac{dM}{dt}$ is never $\frac{dN}{dv}$, so that dQ can never be the total differential.

Clausius had drawn the same conclusion when considering the Clapeyron's cycle between the limits $\{v, t\}$ and $\{v + dv, t + dt\}$. And using the same cycle Clausius could arrive at the following estimate for the Carnot's function

$$C = A(273 + t), \tag{1.11}$$

which is identical to Eq. 1.8. By comparing its value of C with those by Clapeyron and William Thomson Clausius comes to the conclusion that he could find the true expression.

34. Clausius could also employ the mechanical theory of heat to study the saturated vapor. With this in mind, let us imagine a system consisting of fluid and vapor, whose state is determined by the temperature t, the quantity m of the vapor being in contact with the fluid quantity $l - m$, therefore a total amount

of stuff is equal to *l*. The liquid's specific heat is denoted by *c*, the heat of vaporization by *r*; being further on represented by *h*, that is, the saturated vapor's heat capacity, i. e., the amount of heat required to heat up 1 gram of saturated vapor by 1°, so that the vapor remains saturated. Let us imagine the system heated up by 1°, while m is kept constant. Each weight unit of the fluid would then require the amount of heat c, whereas each weight unit of the vapor—the amount of heat *h*; but for the vapor to remain saturated, the latter must be compressed, and thereby it releases the heat. The question is now, what is the actual situation with *h*. Here we ought to distinguish among the following 3 situations:

1. The vapor condenses in part by compression; but the liberated heat quantity is not sufficient to heat the vapor up by 1°; then this might be added to the heat quantity *h*. In such a case *h* is positive.

2. The very compression of the vapor is just sufficient to heat it up by 1°; then the *h* ought to be equal to zero.

3. The vapor compression releases so much heat that the vapor gets superheated; then one must snatch a certain amount of heat for the re-saturation of the vapor to become possible; Thus, *h* must be negative.

To sum up, the amount of heat required for any of the state changes can be cast as follows:

$$dQ = rdm + [(l - m)c + mh]dt. \qquad (1.12)$$

Here the heat amount required for evaporation ought to be equal to *rdm* only, whereas the heat necessary for heating-up would amount to $[(l - m)c + mh]dt$. By using Clapeyron's cyclic process Clausius derives the following three important equations

$$\frac{dr}{dt} + c - h = A(s - \sigma)\frac{dp}{dt} \qquad (1.13)$$

$$\frac{dr}{dt} + c - h = \frac{r}{T} \qquad (1.14)$$

$$r = AT(s - \sigma)\frac{dp}{dt}, * \qquad (1.15)$$

---

*By setting $s - \sigma = dv$, we see that Eq. 1.15 arrives at approximately the same appearance as Clapeyron's Eq. 1.6, but in terms of its contents it is rather correspondent to Eq. 1.2.

where s is the volume of 1 gram of the vapor, and σ stands for that of 1 gram of the fluid. By slightly re-arranging the designations we immediately see, that the same equations would apply to the solid–vapor and solid–liquid systems.

35. According to the old belief the dQ ought to be the total differential, so that $\dfrac{d}{dt}\left(\dfrac{dQ}{dm}\right)=\dfrac{d}{dm}\left(\dfrac{dQ}{dt}\right)$. But from Eq. 1.12 we see that $\dfrac{dQ}{dm}=r$, whereas $\dfrac{dQ}{dt}=(l-m)c+mh$, so that after substituting the latter expressions we get

$$\frac{dr}{dt}+c-h=0 \tag{1.16}$$

At the time of its inference this equation was regarded as the correct one; for to calculate h it is necessary to know $\dfrac{dr}{dt}$. James Watt had been measuring the water evaporation heat at different temperatures and had come to a result that, to put it in the shortest way, the sum of the free and latent heat was constant, that is, the sum of the two amounts of heat, where first you must heat up 1 gram of water from its freezing to $t°$ and then at this temperature it is transformed into vapor, is independent of temperature.

Hence, the Watts' law can be cast as follows:

$$r+\int_0^t cdt=\text{const}$$

or

$$\frac{dr}{dt}+c=0.$$

And if you compare this equation with Eq. 1.16, one gets $h = 0$. This was long considered to be true and they expressed it this way: when the vapor of maximum density in a heat impervious container is changing its volume, it continues to be in the same state.

But later on Regnault could arrive at the conclusion that

$$r+\int_0^t c\,dt=606.5+0.305t$$

so that $\dfrac{dr}{dt}+c=0.305$ or $h = 0.305$.

This value of h was introduced after Regnault's work into the vapor turbine theory. And they had concluded that the saturated vapor being compressed has to absorb heat from its surrounding—to experience the temperature rise just so high that the vapor continues to be saturated, and—vice versa—the saturated vapor at its expansion has to transfer heat to its surroundings—to experience such a cooling that the vapor continues to be saturated. Hence it was necessary to conclude that the saturated water vapor gets partially condensed by adiabatic compression, whereas at adiabatic expansion it might cease to be saturated.

Meanwhile, Eq. 1.14 teaches us something quite different, namely that the term $\dfrac{r}{T}$ ought to be substantial. For water Clausius found following values:

t	0°	100°	200°
h	−1.916	−1.133	−0.676

These results clearly show that the saturated water vapor as a result of adiabatic compression will get superheated, while at adiabatic expansion it will partly condense.

That h ought to be negative for water was simultaneously and independently found by Rankine. Furthermore, h is negative for carbon disulphide, acetonitrile, chloroform, benzene and chlorinated carbon, while it is positive for the ether. Water is the substance that has the highest numerical value of h, and it the latter fact that proves water being an unusual stuff. For any substance with the negative h value it is found that the numerical values of h decrease as the temperature grows; and therefore one has to assume that h can in principle reach 0 and then get positive. By setting $h = 0$ in Eq. 1.14, we get $\dfrac{dr}{dt} + c - \dfrac{r}{T} = 0$, and it is throughout possible to determine the temperature at which the $h = 0$. If we now apply Regnault's empirical formulas for c and r, one gets $h = 0$ for $t = 520°$. However, it should be noted in addition that these formulas should not be used very far beyond the limits within which Regnault performed his experiments, therefore temperatures not much higher than 200° ought to be considered. Furthermore, we are taking water critical temperature to be equal to 365°, thereby closing any discussion about water's saturated vapor. For other liquids, the temperature at which $h = 0$, ought to remain within the limits permitting the application of the formulas suggested by Regnault based upon his findings for different substances; his calculations give for benzene: 112°, for chloroform: 125° and for carbon chloride: 127°.

The experiments by Hirn and Cazin had fully supported the theory in several important cases.

36. At the end of his great treatise Clausius calculates the heat's mechanical equivalent. First of all, he uses a procedure similar to those by Mayer and Holtzmann, as he applies the well-known equation

$$c_p - c_v = AR \,, \tag{1.17}$$

which is applicable to all gases as c_p and c_v are heat capacities at constant pressure and constant volume, respectively, $A = 1/J$ and R the well known gas constant in the expression $vp = RT$. And then, he also uses Eq. 1.15 together with the expression $C = A\,(273 + t)$. Because of numerical uncertainty the results are somewhat different from each other; Clausius even highlights the value of 421 for J as a credible one. To sum up, Clausius' work of 1850 represents a turning point in the history of thermodynamics; for now the solid ground was laid on which the latter could be further developed. It is also worth noticing here that the exact numerical value of the heat's mechanical equivalent depends throughout on the thermometer, the gravitational acceleration and the heating unit being used. If we put the air thermometer to the ground, rendering the gravitational acceleration equal to 981 and calculating with the 15° calorie, we get using the Joule's experiments the value of 427, whereas, by calculating with zero-calorie we get it equal to 430.

William Thomson after 1850 (pp. 33–40)

37. In his treatise *On the Dynamical Theory of Heat** published in 1851, Thomson is leaving aside the Carnot standpoint. Thomson provides here an interesting and in many ways self-consistent picture of the new doctrine, although he willingly recognizes Clausius' priority. He is aware that the energy's actual value is dependent not only on its very magnitude,

---

*Edinb. Trans., 20, 261, 1851.

but also on its transformation capacity, and in this respect he stresses that the heat transformation can be quite impossible under certain circumstances. *It is impossible using inanimate things to win mechanical work by cooling a body under ambient lowest temperature.* This way Thomson expresses his basic idea. If this statement were incorrect, we could namely gain an unlimited amount of work by cooling the sea, land or air.

It is in a short and easily understandable manner that Thomson unites Carnot's principle with the first basic law. Let us assume that a body having the volume *v* and temperature *t* undergoes some two-fold state change, namely *dv* and *dt*. This would require the amount of heat equal to (*Mdv* + *Ndt*), where *M* and *N* are some functions of *v* and *t*. The mechanical equivalent of such a heat amount ought to be equal to *J*(*Mdv* + *Ndt*). If the state change in question is assumed to take place at constant pressure, then the work performed by this body is equal to *pdv*, so that the over-all impact on the surrounding becomes equal to

$$(p - JM)dv - JNdt$$

As Thomson avoids to designate the heat by a differential, he makes note that the integral of this impact may be equal to zero for one cyclic process, since the amount of the work done ought to be covered by the heat applied. Hence

$$\int \left[(p - JM)dv - JNdt \right] = 0 . \tag{1.18}$$

With this in mind we arrive at the following statements:

$$\frac{d(p - JM)}{dt} = \frac{d(-JN)}{dv}$$

so that

$$\frac{1}{J}\frac{dp}{dt} = \frac{dM}{dt} - \frac{dN}{dv} . \tag{1.19}$$

The above equation contains the first basic law and is in perfect accordance with Eq. 1.9 inferred by Clausius.

38. Similarly to Clapeyron Thomson puts $\dfrac{\text{work}}{\text{heat}} = \mu dt$, where $\mu = \dfrac{1}{C}$, known as the Carnot coefficient, which is dependent solely on the temperature. As the work of an infinitesimally

tiny cycle is $\frac{dp}{dt} = dtdv$, and the heat isothermally obtained from the working substance is equal to Mdv, so we immediately get $\frac{dp}{dt} = \mu M$, which is in effect nothing more than just an expression of the Second Basic Law. Thomson notes that only a part of the heat introduced during the process could be transformed into work, while 'the rest is irrevocably lost for the mankind, that is, it ought to be spilled, though not completely destroyed'.

With respect to the function μ Thomson, communicates* what Joule had suggested already in a 1848 letter to him, namely that the function μ may be inversely proportional to the absolute temperature and (with our notations) have the form $\frac{J}{273+t}$, which is the result Joule had arrived at through his experiments with gases. This appeared to be in discordance with Holtzmann's and Helmholtz' work (see Section 26 in the book by P. B.), but Thomson had still no information about this; on the other hand, he knew that Clausius could get the same value. Howbeit, Thomson could not readily be content with Joule's result; for the values Clapeyron and he himself had found for Carnot's function could not be in perfect accordance with Joule's expression.

In addition, Thomson had realized that the very condition for the formula's accuracy was the fact that the heat to be consumed or released during the gas isothermal expansion or contraction ought to be exactly the equivalent of the cast or applied work. Thomson had represented this as an arbitrary hypothesis of Mayer, however he was wrong, for Mayer had expressly relied on Gay-Lussac's experiments during his calculation of the thermal mechanical equivalent.

Meanwhile, the experiments by Gay-Lussac as well as Joule's experiments could just demonstrate that an air expansion process per se (i.e., without work) does not require any heat. Thomson did

---

*Indeed, Thomson writes (*Edinb. Trans.*, 20, 279, 1851), "It was suggested to me by Mr. Joule, in a letter dated December 9, 1848, that the true value of μ, may possibly be *inversely of the temperatures from zero*, and values for various temperatures calculated with the help of the following formula: $\mu = J\frac{E}{1+Et}$, were given for comparison with those I had calculated from data regarding steam." (Here E stands for the air expansion coefficient.)

not consider these studies as sufficient ones to settle the matter and devised a new method that he tried along with Joule.

Through a long tube, which is surrounded by water, we trigger by means of a pump a smooth current of air, which takes on the water temperature. In a well-isolated place in the tube the air current passes through a plug of cotton, and thus goes from the pressure p_1 to a lower pressure p_2, under which it flows more slowly. Just behind the plug a sensitive thermometer is placed. After the gas current was running some time, the over-all unit's state becomes stationary, so that each part of the unit gets a constant temperature. Thus, different processes might take place partly under the heat consumption and partly under the heat release. Sure, to overcome the friction of the plug, some heat is necessary, but due to the friction some heat should be developed as well. Moreover, right in front of and behind the plug certain changes in speed are taking place because of irregularities in the movement, and these changes would partly amplify each other, but if one uses only rather small speeds, then there is no need to take into account the kinetic energy. What remains ought to be the outer work and possibly the inner work.

Hence, the entire energy change becomes

$$Q = U_2 - U_1 + p_2 v_2 - p_1 v_1, \tag{1.20}$$

where the state of a certain air mass is going from volume v_1 and pressure p_1 to the volume v_2 and pressure p_2, whereas the temperature in front of the plug is T_1, corresponding to the internal energy U_1, and behind the plug the temperature is T_2 corresponding to U_2. A calculation the details of which should be neglected here, leads to the following conclusion

$$\frac{dT}{dp} = \frac{1}{c_p}\left(T\frac{\partial v}{\partial T} - v\right). \tag{1.21}$$

And if the gas is ideal, it obeys the relationship $pv = RT$, and then $\frac{\partial v}{\partial T} = \frac{R}{p} = \frac{v}{T}$, so finally we arrive at the condition $\frac{dT}{dp} = 0$.

In such a case, the pressure difference would therefore produce no change in temperature. But if the relationship $pv = RT$ does not apply, then the resulting heat quantity might change either due to the exterior work, $p_2 v_2 - p_1 v_1$, or owing to the internal work, $U_2 - U_1$.

39. Thomson and Joule have performed several series of experiments of this kind, for the first time in 1854* with dry atmospheric air, carbon dioxide and hydrogen, which all experienced cooling; for hydrogen's case but with inconclusive results, to our sincere regret.

Further studies[†] (1862) confirmed these results, with the exception of the hydrogen experiencing a slight heating; for other gases (oxygen, nitrogen, and carbon dioxide) the cooling is proportional to the pressure difference and inverse-square absolute temperature.

Therefore we might write

$$\frac{dT}{dp} = \frac{\alpha}{T^2} ,$$

(1.22)

where α stands for the specific gas constant. For the atmospheric air $\alpha = 0.267 \cdot 273^2$, while for CO_2 it is equal to $1.391 \cdot 273^2$, whereas the applied pressure is expressed in atmospheres.

From Eqs. 1.21 and 1.22, we get

$$T\frac{\partial v}{\partial T} - v = \frac{c_p \cdot \alpha}{T^2} ,$$

so that

$$\frac{\partial\left(\dfrac{v}{T}\right)}{\partial T} = \frac{c_p \cdot \alpha}{T^2} ,$$

And if we assume here that c_p is constant, the integration of the above expressions leads to the following result:

$$\frac{v}{T} = \frac{c_p \cdot \alpha}{3 \cdot T^3} + K ,$$

For $T = \infty$, together with $vp = RT$, or $\dfrac{v}{T} = \dfrac{R}{p}$, we arrive at

$$\frac{R}{p} = 0 + K ,$$

whereupon

$$\frac{v}{T} = -\frac{c_p \cdot \alpha}{3 \cdot T^3} + \frac{R}{p} ,$$

Phil. Trans., 1854, p. 321.
[†]*Phil. Trans.*, 1862. p. 579; Clausius: *Die mechanische Wärmetheorie*, 3. Aufl. 1. Bd. S. 228 ff.

or

$$v = \frac{RT}{p} - \frac{\beta}{T^2} , \qquad (1.23)$$

where β stands for some specific constant.

In effect the above result introduces a kind of equation of state, which is of interest insofar as it had been published much earlier than that by van der Waals, however it could not gain any particular attention.

[Please see Supplementary Note 14, Section 1.9, Page 128]

Remarkably, during Joule–Thomson's experiments the difference $p_2v_2 - p_1v_1$ was very small; The studies have thus demonstrated that for a real gas the internal work at a volume change is not equal to zero; When the gas expands, it uses the part of its energy to perform some work, that is, it is being cooled as a result, even if the external work is zero. Meanwhile, any real gas has not only kinetic, but also potential energy. As for the case of hydrogen, it might be conceivable that the molecules at some temperature, which is not very high, ought to repel each other, so that the internal forces generate heat when the volume grows.

40. Above we have already mentioned that Thomson had proposed the absolute temperature scale based upon Carnot's theory; and at the time he was unaware of Mayer's and Helmholtz' work, but he mentions Joule's attempts in a footnote. After Thomson* had recognized that $\mu = \dfrac{J}{273+t}$, he realized that the degrees on his scale would be unequally high in comparison with the degrees on the Celsius' thermometer, since μ decreases when the temperature grows, whereas μ is just the work we can get, when the temperature of our heating unit falls by 1 degree.

Thus, it was clear to Thomson that some other definition ought to be necessary to build up a basement under his absolute temperature scale.

With this in mind he imagines a reversible cyclic process in which some body takes up heat quantity Q_1 at the temperature T_1 and gives away the amount of heat Q_2 at the temperature T_2; both latter temperatures must be chosen to ensure that

---

*Phil. Trans., 144, 351, 1854.

$$\frac{T_1}{T_2} = \frac{Q_1}{Q_2} \tag{1.24}$$

To sum up, Thomson and Joule could have shown in the course of their gas outflow experiments, that the degrees of this novel temperature scale agree pretty well with the degrees of the conventional air thermometer.

41. A gas cooling at its flow from a higher to a lower pressure might be calculated by means of the van der Waals' equation

$$\left(p + \frac{a}{v^2}\right)(v - b) = RT. \tag{1.25}$$

Here $\dfrac{a}{v^2}$ stands for the internal pressure caused by the gas molecules' cohesion. *Thus, when a gas is expanding, it must perform some internal work to overcome the mutual molecular attractions. Such a work J equals to* $\displaystyle\int_{v_1}^{v_2} \frac{a}{v^2} dv = \frac{a}{v_1} - \frac{a}{v_2}$, *so that the volume v_1 grows to v_2; and the external work is equal to $p_2 v_2 - p_1 v_1$. Remarkably, the both above-mentioned work types ought to contribute to the cooling performance found* to be approximately equal to*

$$T_1 - T_2 = \frac{1}{J \cdot c_\mathrm{p}}\left(\frac{2a}{RT} - b\right)(p_1 - p_2), \tag{1.26}$$

where T_1 and T_2 are temperatures before and after the gas extension, respectively, whereas T stands for the corresponding average temperature.

[Please see Supplementary Note 15, Section 1.9, Page 128]

From the formula (1.26) it can be seen that gas cooling is directly proportional to the pressure difference and ought to decrease as the temperature grows. According to (1.22), the cooling should moreover be inversely proportional to the square of temperature; this could indicate that the parameter a is not constant, but ought to be dependent on the temperature. For atmospheric air (1.26) gives $T_1 - T_2 = 0.265 \cdot (p_1 - p_2)$, where p_1 and p_2 are expressed in atmospheres, which agrees well with Thomson and Joule's experimental results. For all gases except hydrogen the cooling at normal temperatures is positive, and might reach quite significant values, when large pressure differences are used. This principle was

---

*E. Riecke: *Lehrbuch der Physik.*, vol. 2, S. 533; Winkelmann: *Handbuch der Physik II b*, S. 480.

employed by Linde in his refrigerator machine, thus being capable of cooling down quite large quantities of air in relatively short time periods.

Formula (1.26) shows that $T_1 - T_2$ can come to zero with increasing temperature and then get negative; the temperature at which this difference changes sign is called the inversion temperature. As to the hydrogen, this temperature is located at $-80°$; Thus, hydrogen at $-80°$ ought to behave like other gases at ordinary temperatures. This has been experimentally confirmed by Olszewski.

By using the van der Waals' equation and other equations of state we would come to the conclusion that the inversion temperature is dependent upon the pressure, and that two different inversion temperatures correspond to each of the pressure points, although in such a way that there exists no inversion at high pressures. Meanwhile, different equations of state deliver quite different values, but we should not go in for the details here.

Rose-Innes has developed a formula that satisfactorily fits results obtained in Thomson and Joule's experiments; this expression reads as follows:

$$\beta = \frac{\alpha}{T} - \gamma , \qquad (1.27)$$

where β is cooling, whereas α and γ are constants for each gas, for example,

	α	γ
air	141.5	0.697
CO_2	2615	4.98
H_2	64.1	0.331

Joule and Thomson were working with the pressures up to 6 atmospheres. Newer works demonstrate* that gas cooling ought to be dependent not only on the pressure difference but also on the absolute value of the pressure. For various reasons, then, one must recast Eq. 1.22 in the following form:

$$\frac{dT}{dp} = \frac{\varphi(p)}{T^n} ,$$

where $\varphi(p)$ is a function of the pressure, whereas 'n' is dependent on the ratio between the heat capacities of the gas in question.

_____

*Physikal. Zeitschr. 15, 904, 1914.

The Second Basic Law (pp. 40–44)

42. As we have seen, it took a relatively long time before William Thomson could finally put aside the Carnot's standpoint to join the new doctrine, and once he had gone that step, he appeared soon in the forefront of thermodynamics' pioneers.

It is truly difficult to say who could have the greatest impact on the thermodynamics, either Thomson or Clausius. Partially independently from each other, they could arrive at the same results, as we can borrow from their excellent works published in the early fifties of the XIX-th century; they both could realize that a certain part of what is called energy, has in effect no practical value; whereas the mechanical energy might be completely transformed into heat, only a part of the heat energy might be rendered useful.

[Please see Supplementary Note 16, Section 1.9, Pages 128–129]

A lot of physical processes can on its own proceed only in one particular direction; but if we would like them to go in the opposite direction, then other things must also be done in parallel. The latter might be regarded as a kind of compensation, according to Clausius. The positive, natural processes are thus possessed of a clear preponderance with respect to the negative ones, he states. And, bearing this in mind, he finally comes to the conclusion that the world aims toward a state, where all the available energy has become the heat of a uniform temperature (heat death); that is, the total energy ought to become more and more spread out, its density ought to be reduced, the energy ought to be getting degraded.

[Please see Supplementary Note 17, Section 1.9, Page 129]

43. Thomson[*] and Clausius[†] find both an exceedingly important function, the entropy. As we have already mentioned in the previous paragraph, Thomson's final scale of absolute temperature was leading to the equation $\dfrac{Q_2}{T_2} = \dfrac{Q_1}{T_1}$, see Eq. 1.24. If we now assume that the heat obtained by the body is positive and that withdrawn from it is negative, then we immediately arrive at the relation $\dfrac{Q_2}{T_2} + \dfrac{Q_1}{T_1} = 0$.

---

[*]*Edinb. Trans. 2t*, 123, 1854.
[†]*Pogg. Ann.*, 93, 500, 1854.

This could easily be extended to the case of a number of reversible cyclic processes, so that we get $\dfrac{Q_1}{T_1} + \dfrac{Q_2}{T_2} + \cdots + \dfrac{Q_n}{T_n} = 0$.

or $\sum \dfrac{Q}{T} = 0$.

For any reversible cyclic process, which can be replaced by such a series of processes, the same must therefore apply.

During the same year (1854) Clausius arrives at a similar result. He delves into the idea he had already expressed in 1850, that the heat of itself does not pass from a colder to a hotter body. Based upon such a viewpoint Clausius shows that the efficacy of a reversible cycle does not depend on the nature of the substance, but only on temperature. Remarkably, the latter dependence might be inferred in a very convenient way, as Clausius* could demonstrate later on (1876).

Indeed, a gas circuit passes through the process a-b-c-d, which consists of two isotherms and a-b and c-d, as well as of two adiabates b-c and d-a. The amount of heat being recorded along the stretch a-b in Fig. 1.6 can be expressed as follows:

$$Q_1 = \frac{RT_1}{J} \log\left(\frac{V_1}{v_1}\right) \qquad (1.28)$$

And the amount of heat the gas loses along the stretch c-d ought to be

$$Q_2 = \frac{RT_2}{J} \log\left(\frac{V_2}{v_2}\right) \qquad (1.29)$$

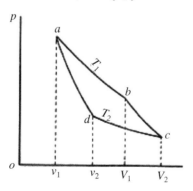

Figure 1.6

---

*Mach: *Principien der Wärmelehre. 2. Aufl.*, p. 293.

With the help of Poisson's equation and the equation of state $pv = RT$ we may arrive at the following relationships:

$$\frac{T_2}{T_1} = \left(\frac{V_2}{V_1}\right)^{k-1} \quad \text{and} \quad \frac{T_2}{T_1} = \left(\frac{v_2}{v_1}\right)^{k-1}$$

and, accordingly,

$$\frac{V_1}{V_2} = \frac{v_1}{v_2} \quad \text{or} \quad \frac{V_1}{v_1} = \frac{V_2}{v_2}$$

So, taking into account Eqs. 1.28 and 1.29 we get

$$\frac{Q_1}{Q_2} = \frac{T_1}{T_2}, \tag{1.30}$$

or

$$\frac{Q_1 - Q_2}{Q_1} = \frac{T_1 - T_2}{T_1}. \tag{1.31}$$

The amount of heat transformed into the work, $Q_1 - Q_2$, can then be expressed as follows $J(Q_1 - Q_2) = \dfrac{J}{T_1} Q_1(T_1 - T_2)$, in full accordance with the fact that the function of Carnot is equal to $\dfrac{T}{J}$.

The practical value of the heat quantity Q_1 captured at the temperature T_1 is therefore not exactly equal to $J \cdot Q_1$, but in fact to

$$J \cdot Q_1 \cdot \frac{T_1 - T_2}{T_1} \equiv J \cdot Q_1 \cdot \left(1 - \frac{T_2}{T_1}\right).$$ Hence, the smaller T_2 is compared to T_1, the close this variable approaches $J \cdot Q_1$.

44. Let us now imagine some reversible cyclic process followed by an arbitrary body (Fig. 1.7). The very cyclic curve can be replaced by the infinitely small pieces of isotherms and adiabates. The latter both might then be extended in such a way that the closed space is divided into the infinitely thin strips.

Each of these strips depicts a cycle in which Eq. 1.30 ought to be applicable. By going all the way around, and assuming that the supplied heat quantity is positive and the withdrawn is negative, we arrive at

$$\frac{dQ_1}{T_1} + \frac{dQ_2}{T_2} + \cdots + \frac{dQ_n}{T_n} = 0, \tag{1.32}$$

where $T_1, T_2, ..., T_n$, stand for the temperatures of the separate heat reservoirs, which the body successively comes into contact with. In general, it is possible to recast Eq. 1.32 in the shorter form

$$\int \frac{dQ}{T} = 0,$$ (1.33)

which is nothing more than just the 2nd basic law in its mathematical form, for the first time published by Clausius in 1854. The phrase can also be expressed so that $\dfrac{dQ}{T}$ is a total differential of some variable S, which therefore depends only on the body's state, but not on the pathway along which this state is achieved, or

$$\int_1^2 \frac{dQ}{T} = S_2 - S_1,$$ (1.34)

which describes a transition of the body from its state 1 to its state 2. And here the difference between the old and the new standpoints comes out very clearly. Indeed, according to Carnot, who thought that the heat be a substance, the cyclic process in question ought to deliver $\int dQ = 0$, but in fact, according to Clausius, it delivers $\int \dfrac{dQ}{T} = 0$; and only for isothermal cycles these two equations should coincide; meanwhile, such a process may, of course, produce no work, that is clearly shown by its graphical representation as well.

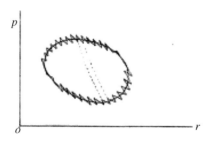

Figure 1.7

45. The variable S is called entropy. Meanwhile, this notion was introduced by Clausius somewhat later, in 1865.* He formed this term from the Greek word 'τροπη', meaning

*Pogg. Ann., 125, 390, 1865.

'transformation', and re-shaped it to mimic the word energy as much as possible. In fact Clausius was dealing with this variable since 1854, and that time he called it a 'metamorphosis content' or a 'transformation value'. But these both term choices cannot be considered satisfactory; as Wald* emphasizes, it ought to be throughout proper if we decide to call it a 'non-transformation value' as well.

If we now imagine a Carnot cycle arranged in such a way that the cooler has the absolute temperature 1, then we get in employing the usual notations

$$\frac{Q_1}{T_1} = \frac{Q_2}{T_2} = \frac{Q_2}{1} = S .$$

The entropy is thus equal to the amount of heat which cannot be transformed to work but continues to be heat as a result of the heat transfer to a body at absolute temperature 1. The right general definition of entropy as the amount of heat that cannot be transformed into work is therefore only correct when the minimum temperature is equal to 1.

To sum up, from the equation $\frac{Q_1}{T_1} = \frac{Q_2}{T_2} = S$ *one can see that the entropy is a variable, which after being multiplied by the lowest achievable temperature of T_2 in connection with the given process delivers the quantity of heat Q2, which cannot be transformed into work, but gets practically wasted.*[†]

[Please see Supplementary Note 18, Section 1.9, Page 129]

Reversible and Irreversible Processes (pp. 44–48)

46. For some process to be called reversible, it must be possible to completely reverse it, so that it might be applied in either direction, while every time passing the same intermediate states.

The result of the two reversible processes, where one of which is the reverse of the other one, ought to be equal to zero; hence, every body having participated in the process ought to be returned to their original states. It follows that, in a reversible process there should not be any final temperature difference between the body, which runs

---

*Wald: *Die Energie und ihre Entwertung* S., 73.
[†]See James Swinburne: Entropy, p. 6.

through the process and the container, which the body is currently in contact with, for otherwise heat conduction or radiation should immediately occur, and the latter both are irreversible phenomena. Nor must there be any final pressure difference between the working material and the surrounding area, for otherwise the kinetic energy generated by shock should be transformed into heat, and the latter process is irreversible.

It is therefore immediately clear that any reversible process defined in the above way becomes an ideal borderline case, and if it would proceed infinitely slowly, it consists entirely of equilibrium states. Still not diminished thereby is the accuracy of the statements applicable to such processes, likewise the deliberation that the pure mechanics' laws are less valid because a purely mechanical process does not occur in reality. Indeed, everywhere friction and other kinds of resistance occur, which render any realistic process under study an irreversible one.

That the electric conduction and radiations of any kind are irreversible ought to be an expression of the basic feature that the heat by itself cannot come from a system having lower temperature to a system having a higher one. Additionally, to irreversible processes belong all the diffusion phenomena, a gas outflow into the vacuum, or a gas expansion performed in such a way that there remain some final pressure differences.

Furthermore, irreversible seem to be all the processes initiated by what is called the 'trigger'. This means that such a process is initiated by an event exhibiting a rather low energy consumption, and this ought to be just the introduction to the actual process. Opening a tap to enable a gas or a fluid to flow, ignition, electric discharge, etc. might be viewed as the proper examples of irreversible events.

We shall immediately see that all the processes initiated by unstable equilibrium conditions may be considered irreversible as well.

47. Indeed, for any reversible circular process we have obtained the following relationship

$$\int \frac{dQ}{T} = 0 , \qquad (1.35)$$

where T stands for the temperature of the container contacting the body during the process under study, because if the process

---

*Planck: *Vorlesungen über Thermodynamik*, p. 45.

is reversible, then there shouldn't be any temperature difference between the body and container, hence T ought to be the temperature of the body as well.

With this in mind for any irreversible process we get

$$\int \frac{dQ}{T} < 0 , \tag{1.36}$$

where T stands for the temperature of the container, but not that of the body enclosed in the latter—because the body is now possessed of no definite temperature due to thermal currents, radiation, friction or impact.

It is not difficult to realize that the integral in the latter case must indeed be less than zero using the following examples.

1. If the hot container's temperature is the final value greater than the body's temperature, the positive elements under the integral should decrease.

2. If the chiller's temperature is the final value lower than the body's temperature, the negative elements under the integral should increase.

3. If the external pressure is lower than the body's internal one, then the body during its expansion would perform less work than that during a reversible process, and it therefore would withdraw smaller amount of heat from the hot container to perform work; hence, the positive parts of the expression under the integral should be reduced.

4. If the external pressure during the compression is greater than the body's internal one, the external forces would perform a larger amount of work than that during a reversible process, and the heat quantity to be obtained by the chiller will increase; therefore, the negative terms under the integral should be greater.

5. If friction or impact are present then an excess of heat will be released, which is transferred to the chiller, or the hot container would release less heat; then the resulting value of the integral must decrease.

48. Let us now assume that some body changes its state from A to B during some reversible process (Fig. 1.8). Then we get

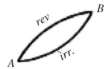

Figure 1.8

$$\int_A^B \frac{dQ}{T} = S_B - S_A \ ,$$

where S_A and S_B stand for the entropies of the respective states. And now let us assume that there exists an irreversible transition from A to B as well. With this in mind we may devise a cyclic process bringing the body irreversibly from A to B and reversibly from B back to A. Our previous considerations could then be summarized as follows:

$$\mathrm{irr}\int_A^B \frac{dQ}{T} + \mathrm{rev}\int_A^B \frac{dQ}{T} < 0$$

or

$$\mathrm{irr}\int_A^B \frac{dQ}{T} + S_A - S_B < 0$$

$$\mathrm{irr}\int_A^B \frac{dQ}{T} < S_B - S_A \ . \tag{1.37}$$

From here we see that during the irreversible process the entropy increase is higher than the value of the integral $\displaystyle\int_A^B \frac{dQ}{T}$.

If now we assume that the states A and B are in the nearest vicinity of each other, we get $S_A - S_B = dS$ and the integral sign disappears. We might then summarize our preceding discussion in the following manner:

For the reversible cyclic process: $\displaystyle\int_A^B \frac{dQ}{T} = 0$

For the irreversible cyclic process: $\displaystyle\int_A^B \frac{dQ}{T} < 0$

For the infinitely small reversible change: $dS = \dfrac{dQ}{T}$

For the infinitely small irreversible change: $dS > \dfrac{dQ}{T}$

For some isolated body we have $dQ = 0$; hence, for any change of the state $dS > 0$ must hold, i.e., the entropy must either remain constant or increase, but it might never diminish; further, **adiabatic** means in fact the same as **isentropic**; *however, these two concepts coincide only for the reversible processes.*

Besides, it is well known that energy might generally be divided into two factors, with the one of which expressing the intensity, while the other being the quantity or capacity. For thermodynamics, T would be intensive variable, while the entropy ought to be thought of as a quantity/capacity (extensive) variable.

Meanwhile, for irreversible processes we have $dQ < TdS$, so that the above-mentioned classification would not be pertinent in this case.

[Please see Supplementary Note 20, Section 1.9, Pages 129–133]

49. Now we shall consider a system of bodies;* the system under study must be conceivably isolated; the most diverse physical and chemical processes might take place in the system; still, it is assumed that absolutely no energy interchange takes place between the system and the surroundings.

Does the principle of entropy apply here as well? The correct answer ought to be that all experience has shown that the statement's validity could be extensible to the system mentioned here. The system's entropy can therefore never decrease; if reversible processes are taking place within the system, the entropy remains to be constant; and if the processes in the system are irreversible, the entropy grows. However, let us now assume that the system consists of two bodies with temperatures T_1 and T_2 ($T_1 > T_2$), and that the amount of heat dQ goes from the warmer to the colder body. Then the entropy decrease for the first of the bodies is equal to $\dfrac{dQ}{T_1}$, whereas the entropy increase for the second body would amount to $\dfrac{dQ}{T_2}$. To sum up, the total entropy change in the system would be $dS = \dfrac{dQ}{T_2} - \dfrac{dQ}{T_1} = dQ\left(\dfrac{1}{T_2} - \dfrac{1}{T_1}\right) = dQ\dfrac{T_1 - T_2}{T_1 \cdot T_2}$, so that it comes to be positive.

---

*Chwolson: *Lehrbuch der Physik Bd. III*, S. 514.

To sum up, if the system contains some elements that are not homogeneous or do not have the same temperature everywhere, we may imagine them to be divided into infinitely small parts; and entropy of each element is assumed to be the sum of the individual parts. If there are some chemical processes accompanied by the heat release, then the heat produced will spread out to other parts of the system; if endothermic processes are going on, then the heat will go the opposite way, namely by means of conduction or radiation. In either case, the entropy grows, but here it should be noted that the entropy of a chemical process might also be changed in other ways, for example, by volume changes, dissolution phenomena, diffusion, etc.

[Please see Supplementary Note 21, Section 1.9, Pages 133–146]

Denotation of the Two Basic Laws (pp. 55–57)

55. To describe the importance of the first basic law would be the same as travelling through large areas of physics, astronomy, meteorology and chemistry. Everywhere in natural sciences there ought to be at least one poser solvable only using the Energy Conservation Law. The latter Basic Law could be capable of bridging the gaps among the different sections of various natural sciences at the diverse points. *In addition to the concepts of space and time, the concept of energy is the one, which is common to all the areas of physics.*

That the second law is of paramount importance has become gradually clear as well. Meanwhile, some colleagues were continuing to feel that the first basic law would be enough, but this idea is incorrect. The content of the both basic laws is quite different from each other; moreover, it is impossible to deduce the one from the other one. It would not contradict to the first basic law if a part of the air in a room would become warmer by itself, while the rest of the air were cold, so that the total heat content was unchanged. This same holds, if we would suddenly and forcefully separate from each other two gases, which are initially mixed with each other. Furthermore, if we plan to create a device that could take heat from the seawater and transform it into work, our dream would not come true. Even in this case we could not dispense ourselves from fuel, although we seem to have the inexhaustible energy supply at hand. Such a machine would not create new energy anyway, but carrying out work without any expense would also be impossible.

According to W. Ostwald, the relevant name for the latter situation ought to be perpetuum mobile of the second kind, unlike perpetuum mobile of the first kind, which ought to be capable of providing work without any energy input. The perpetuum mobile of the second kind stands in the same relation to the second basic law, as the perpetuum mobile of the first kind to the first basic law. The latter both are fully impossible. The pertinent proof of the latter statement might come solely from the experience. Consequently, the both basic laws, just like the law of matter conservation, are fundamentally empirical— and therefore it is impossible to infer them somehow *a priori*.

The second basic law ought to teach us classifying all the phenomena into the reversible and irreversible ones. Specifically, the reversible phenomena ought to form the ideal borderline cases; any real natural phenomenon is to a greater or lesser extent irreversible and leaving its specific ineluctable trail everywhere.

[Please see Supplementary Note 22, Section 1.9, Pages 146–147]

The total energy remains constant forever, while its parts might at the same time change from the higher to the lower intensity; or, in other words, they degrade. Along with the latter processes, a certain function, the entropy, grows simultaneously. Nature, as a whole, progresses in some particular direction definable by using the second basic law, whereas the first basic law cannot foretell this effect.

[Please see Supplementary Note 23, Section 1.9, Pages 147]

56. Meanwhile, *there is a special case,** where the first basic law alone might prescribe a particular direction for a phenomenon. That ought to be true if the system under consideration is in such a state, where one of the various energy types is possessed of its absolute maximum or minimum. Then only a change might be possible that brings the mentioned energy type to decrease or increase, respectively. Now we shall illustrate this event using the conventional mechanics. If a system of several bodies remains at rest, then its kinetic energy is in absolute minimum, and, as a result, any possible

---

*Planck: *Vorlesungen über Thermodynamik*, S. 72.

change ought to increase the kinetic and reduce the potential energy, if no external effects take place.

Thus, here we may deduce a very important principle from the following fact: If a mechanical system starts moving by itself, the direction must be such that the potential energy decreases. Bearing this in mind, if both the kinetic and potential energy are at minimum, the former will not grow at the expense of the latter, and the system must hence remain at rest. The pertinent examples include a pendulum in its lower position and a fluid being at the same level within two connected pipes. As soon as we start from a position where the kinetic energy is not zero, we might still trigger a change in such a direction that the potential energy increases (periodic fluctuations). Meanwhile, there ought to be no states of minimum energy for the processes including the heat exchange. Hence, there is no prospect to derive rules for thermodynamic changes or equilibrium states that would apply to dormant systems only.

[Please see Supplementary Note 24, Section 1.9, Pages 147–148]

As to Max Planck's research activities concerning the basic laws of thermodynamics, we might immediately note that in fact he was duly following the approaches of Rudolf Clausius and William Thomson (Lord Kelvin).

These two trains of thoughts were but in principle not quite different from each other. The first colleague who could notice and had analyzed this was Tatyana Alexeyevna Ehrenfest-Afanassjewa (see her book [6]). Moreover, in her thermodynamics works Tatyana Alexeyevna has found and discussed in detail the clear logical inconsistencies in Max Planck's standpoint concerning general thermodynamics. Nonetheless, although her insightful publications have appeared much later than the book by Peter Boas, they were and are still not attracting the due attention of the colleagues. We shall revert to this topic later on in the book at hand, namely cf. the supplementary notes to Chapter 5 here for the detailed discussion of her actual contribution to the field.

At this very point, we shall punctuate for a while Peter Boas' original narration with gaining a closer look at the over-all topical structure of his book under study.

Indeed, after presenting all the main 'dramatis personae' in the thermodynamics field, as well as the details of all their works leading to the discoveries of the both Basic Laws of thermodynamics, Peter Boas informs us about the debates around the both. In effect, Peter devotes two chapters to the professional criticism concerning the second basic law.

Doubts about the Second Law (pp. 57–61)

In this paragraph, Peter Boas informs us about several immediate critical reactions to R. Clausius' suggestions. Still, Clausius could himself quite conclusively respond to the complaints mentioned, and Peter Boas is presenting the full list of the relevant references. However, nowadays we know very well that R. Clausius has stopped his research work and publishing in the early 70-ies of the XIX-th century. From the end of 1860-ies on, he was mostly involved into the political, as well as into the direct military activity in connection with the Prussian-French war (19 July 1870–10 May 1871). The latter story resulted in his angry wound, moreover at the same time he had severe familial problems. Due to this tragic entirety, Rudolf Clausius had practically to quit his research activities [7].

On the one hand, Peter's list of the critical voices is by far not complete. Indeed, there is much-much more to the story.

On the other hand, Peter Boas shows that there were physicists dealing with chemistry, who could ensure the proper development of thermodynamics. Peter attracts our attention to the colleague, Prof. Dr. August Friedrich Horstmann, who could manage to formulate the Second Basic Law correctly. Meanwhile, some important details remained untold by Peter. **Thus, our actual task here would be to complete solving Peter's puzzle!**

1.4 The Second Basic Law: What Is the Actual Physical Sense of It?

Thus, to our sincere regret, Rudolf Clausius himself had not enough time to duly react to the entire sum of criticism, although his competent contemporaries and colleagues did warn him that he should not get too carried away in mathematics at the expense of physics. Prof. Dr. Karl Friedrich Mohr (1802–1879) has described the details and presented a clear analysis of R. Clausius' actual standpoint in his book [8].

Who was Prof. Dr. Mohr?

Below comes the full list of his professional accomplishments:

– A doctor of philosophy and medicine
– Extraordinary professor at the University of Bonn
– Medical and pharmacological advisor for the Rhineland Medical Collegium of Koblenz
– A corresponding member of the Bavarian Academy of Science
– A regular member of pharmaceutical societies of Erlangen, Vienna, Antwerp, London, Brussels, St. Petersburg, Philadelphia
– A regular, a corresponding and an honorary member of the research society Pollichia (at Rhineland-Palatinate), of the numerous natural-scientific research societies and trading associations at Embden, Mainz, Aachen, Frankfurt/Main, Lahr, Darmstadt, Hamburg, etc.
– The last, but not the least: he was a fourth-class knight of the Red Eagle Order.

To sum up, he was a renowned professional physical chemist well known all over the world. In his book involved, Peter Boas is duly referring to his name in the row of those who was dealing with the first basic law of thermodynamics.

Being at work in the University of Bonn, Prof. Mohr could contact Rudolf Clausius personally. Thus, in his book [8] he describes his attending one of the seminars, where R. Clausius have reported in detail about his findings in the field of the 'mechanical theory of heat'. The topic was of immense interest for Prof. Dr. Mohr as well already since a considerable time—see F. Mohr's numerous works on the theme (we mention here his works in the field just to point out Prof. Mohr's immense competence in the field of the then report by R. Clausius) [9–11].

To our sincere regret, Peter Boas' book does not contain even a volatile reminiscence about Mohr's ideas concerning the second basic law. Sure, Peter was definitely aware of this important theme, but, most probably, had in effect no more opportunity to publish anything about it (let us recall the wild reaction to his book, we were discussing at the beginning of this chapter!).

Meanwhile, we may find a very nice review of Prof. Mohr's life and work published a dozen years later as compared to Peters' book in the paper [12].

This portrayal of Prof. Dr. Karl Friedrich Mohr (1802–1879) has been taken from
https://en.wikisource.org/wiki/Page:Popular_Science_Monthly_Volume_17.djvu/302

The seminar of R. Clausius attended by Prof. Dr. Mohr, as mentioned above, had taken place at the regional conference of German natural scientists and physicians on September 23, 1867 in Frankfurt/Main.

Prof. Mohr has devoted a separate chapter of his above-mentioned book to the thorough, careful and fully competent analysis of R. Clausius' report, the chapter is entitled: "*Entropie oder Rückkehr zum Gleichgewicht*" (*Entropy or Return to the Equilibrium*), see pp. 34–44 in his book. After carrying out a detailed philology and relevance analysis of the pertinent terminology (the notions of force, work and energy) Prof. Mohr had considered the very contents of R. Clausius' report and came to the following remarkable conclusions, which are really important for our present discussion (here we translate them into English and highlight the most relevant fragments):

"... The Work is in fact an impact of a movement, in which the former loses its nature, and enters the sum of an equal amount of other movements. The transformations of heat into a mass movement and vice versa represent only two particular examples of their kind, but the sense of the very notion might not be entirely exhausted in pointing out these both.

Instead, Clausius summarizes the above two laws as the First Basic Law of the mechanical theory of heat, but separates the latter from all other phenomena, where heat is being converted not into the conventional mass movements but into the molecular dynamics by summarizing the latter as the Second Basic Law of the mechanical theory of heat. For him the First Basic Law would be the Law of the Equivalence between Heat and Work, whereas the Second Basic Law he formulates accordingly as the Law of Transformations Equivalence.

Meanwhile, such a separation ought to be entirely unwarranted and in fact even rather confusing, for the implementation of mass movement into heat and vice versa is definitely a transformation, like melting of ice or evaporation of water. In other words, the mass movement underlying heat ceases to be the 'heat' as it is, but might now belong to the sum of the available mass movements. Still, we should be bearing in mind that the 'Livening Force' of this dynamics type ought to be of quite special nature and that such a movement type could be no longer easily recognizable.

[Please see Supplementary Note 25, Section 1.9, Pages 148]

Thus, there is absolutely no reason to exclude the contents of the First Basic Law from the generic class of transformations, so that these both Basic Laws would immediately collapse into a single unique law, which would then itself be nothing but an application of the energy conservation law.

If we would now adopt a more general viewpoint, we immediately recognize that there are in effect hundreds of transformations without any quantitative loss of the matter, which seemingly do not obey the both laws of the mechanical heat theory. If, for example, 1 part of hydrogen would chemically associate with 8 parts of oxygen, while 34462 thermal units of energy would be free, then we've got in this case no conventional mass movement, and hence there is seemingly no way to apply the first basic law. Further, the heat released is owing to some other (provisionally unknown) kind of motion, whereas the amount of heat involved does not disappear to perform some conventional 'useful work', hence there seems to be no way to apply the Clausius' Second Basic Law as well.

The actual one-sidedness of the Clausius' viewpoint consists just in that he was unaware of the chemical molecular dynamics as a quite different form of motion ...

... If we provide the mathematical expression of the work amount performed by some dynamic process with the negative sign, then the sum of all the energy transformations ought to be equal to zero. But by furnishing the work with the positive sign and also expressing it in equivalent quantities of the dynamical processes involved, we might be immediately entitled to conclude that the overall sum of all the movements before and after the fact of the energy conversion remains the same, and this again boils down to the General Energy Conservation Law. Through likewise elementary mathematical transformations, we definitely cannot win anything, but the overall clarity has been lost at once, because we might then easily draw the erroneous conclusion that we could thus have managed to proffer some novel important laws.

Clausius regrets that he could not mathematically prove these laws in front of the meeting. I cannot share this regret, for in fact we have to prove the theorems of such a kind not mathematically, but purely logically. The formula is nothing but the mathematical expression in the spirit of an already clearly recognized relationship of phenomena. ..."

From our modern viewpoint, Prof. Mohr's criticism is not only just apt. It does immensely help us to reveal *the severe methodological error*—sure, we should not be afraid of such a truly bold estimate—into which the followers of R. Clausius have trapped. We might recast their actual error into the following succinct phrase:

Introducing a conceptual wall separating the First and the Second Basic Laws of Thermodynamics, based upon playful mathematical exercises.

1.5 How Many Basic Laws Are There in Thermodynamics?

In fact, the actual basic law ought to be only one according to Prof. Mohr, namely the law of energy conservation and transformation. In such a case, to separate the conservation from the transformation seems to be dubious from the general standpoint.

Finally, truly *general* ought to be standpoint not restricted to just an ideal gas.

Meanwhile, this actually unsolicited introduction of more and even more novel basic laws was longer time remaining a 'cozy trend', as had clearly been demonstrated by the solemn announcement of the 'third basic law of thermodynamics'. See, for example, the original detailed survey of the very author of this 'law'. Until now, the latter publication is remaining a highly cited one [13].

Remarkably, Peter Boas was definitely aware of the story's actual conceptual background, so that he devoted just a couple of phrases to this alleged "basic law."

As we have already seen above in this chapter, the report on Peter's book delivered by the journal *Nature* expresses a sheer amazement concerning the truly tangential attention devoted by Peter's book to this quasi-basic law.

We guess the reasons for Peter's move are throughout clear. He is not discussing them explicitly, but attentive reading his book allows us to substantiate them. We shall do so, but first we would like to hear to what Peter would like to convey.

Nernst Theorem (pp. 136–137)

129. In 1906 Nernst has arrived at a theorem,* the gist of which might be expressed as follows: At the absolute zero temperature the entropy value of any chemically homogeneous solid or liquid substance is independent of the physical state and chemical modification of the substance. One can go a step further together with Planck to add the constant to the entropy expression, for the latter is equal to zero.

Mathematically, the theorem sounds: At the absolute zero temperature, the entropy of any chemically homogeneous solid or liquid substance is equal to zero.

This theorem leads even to more striking consequences. For example, the specific heat at a constant pressure, c_p, for any substance of the mentioned kind converges towards zero as the temperature

---

*Planck: *Vorlesungen über Thermodynamik*, 3. Aufl. p. 266.
Planck: *Über neuere thermodynamische Theorien*.
Pollitzer: *Die Berechnung Chemischer Affinitäten nach dem Nernst'schen Wärmetheorem*.

approaches its absolute zero. Moreover, the expansion coefficient of such a substance also converges to zero under the same conditions. Comparative measurements of both specific heats and expansion coefficients at very low temperatures demonstrate their heading for the same direction vs. temperature.

Nernst theorem stands in close relationship to the quantum theory put forward in 1900 by Planck and extended by Einstein and others. In following this novel theoretical branch, we might expect interesting results.

To sum up: *The theorem as it is ought to be due to Max Plank. This had helped him to build up the conceptual basement for the quantum theory. On the other hand, Walther Nernst and his coworkers could experimentally demonstrate that it is sheer impossible to reach the zero of the absolute temperature (i.e., –273ºC).*

The only real reason for 'forcing-through' such a law was the absence of any detailed, thorough mathematical inference of the Boltzmann–Planck formula $S = k*\ln(W)$. Sure, by the way, of extreme interest in this connection was the reaction of Prof. Dr. Walther Nernst (1864–1941) himself, we read this just in the preface of his above-cited book (our English translation):

> " ... We know that Planck is incidentally the first theorist, who has dealt extensively with my thermodynamics law—in giving an excellent representation of the latter in the last editions of his textbook of thermodynamics. The view put forward here ought to be a somewhat broader representation than just the law valid for gases of finite density, for, naturally, we might a fortiori assume its validity for solutions. ...

Walther Nernst, December 1917

In no way is the above quick detour an attempt to negate the marvelous systematic experimental work by Prof. Dr. Nernst himself, as well by his students and followers all over the world. Indeed, all of them could clearly demonstrate that there is some specific physical rule in connection with the zero value of the absolute (Hon. Lord Kelvin's) temperature scale. Instead, the story about forcing-through some additional basic law is rather a failure of a number of theoreticians to explain this ingenious experimental result in the correct way.

Sure, the rule involved ought to be of undoubtedly broad validity, and the *a fortiori* assumption put forward by W. Nernst is in effect his and his collaborators' tremendous achievement. Meanwhile, it is anyway **not** the basic law. Accordingly, the 'intrinsic interrelationship between the notions of entropy and probability' is **not** the very expression of the second basic law. **As they are**, these both **are not at all** expressing any basic law. To sum up, the actual basic law is immediately recognizable behind these both.

The most important points to be stated here: First of all, the notorious Boltzmann–Planck formula $S = k*\ln(W)$ might easily be derived using the formal school mathematics. Dr. G. A. Linhart could convincingly demonstrate this. Secondly, this same holds for the temperature dependence inherent to the formula in question, showing how mathematically easily may entropy S reach its zero value, when the absolute temperature arrives at its zero as well— see Chapter 4 of the present book, where we discuss this interesting and important topic in much more detail.

Meanwhile, already when reading Peter's book we do get a feeling of some global inconsistency in the field of thermodynamics at least at the time of the book's publication... A consistent analysis of the then theoretical results does leave a definitely sour taste ... so let us now revert to what Peter thinks about this matter.

Our Conception of Nature (pp. 137–141)

130. As soon as the law of energy conservation was at last widely recognized to be a basic doctrine, it has become obvious to assume that all the natural phenomena do consist of movements. Meanwhile, the specifics of the latter are unobservable directly, since these ought to happen at the molecular and/or atomic level of consideration. Anyway, the basis of the entire physics should therefore be attributable to the mechanics, for it became necessary to trace the diverse movements and express them in the relevant mathematical equations. This mechanical view could lead in many cases to marvelous results, for example, in the kinetic theory of gases.

However, in other respects, such an approach occurs to be inadequate. As soon as we penetrate the area of liquid and solid bodies, any use of atomistic conceptions becomes rather difficult or even impossible, when trying to formulate general rules without mathematical intermediations. Contrariwise, by properly using thermodynamic methods, one might master productive consideration of all the three aggregate states to find important laws for melting, evaporation, osmotic pressure, dissociation, and so on, so forth. Hence, it appears that one might come forward even without always resorting to mechanical hypotheses. Furthermore, an awareness of the energy conservation law tells in fact nothing about the very nature of energy—a stance that Mayer had initially adopted. This had naturally triggered the doubts, whether we should trust the old conceptual basis at all, and would not it be better to consider at least possibilities of changing it.

131. At the end of the last century Ostwald* was seeking to replace the atomistic conception by the energetic one stating that energy ought to be the only natural thing possessed of the real existence, being the actual physical sense of the vacuum. Each time as we go through our senses we get an impression from the outside world because of some energy utterance. Without the temperature difference one gets no feeling of heat, without perceiving the pressure difference one cannot recognize the air pressure, without the sense of touch, one cannot perceive the presence of an object, and so on. Any realistic process requires an energetic phenomenon, namely the transfer of energy from its higher to its lower intensity, and only this way it would produce our feeling of it. Looking for some independent core reasons behind observable matter properties is not helpful. The energy corresponds to the ultimate reality, whereas the matter is a complex of properties dictated by the collection of energy species interwoven in the given location at the given moment.

The new doctrine sounded largely in the above-sketched way. In accordance with it, one should not dare to regard matter as a carrier of energy, as it was usually considered, but let the energy stand on its own feet. Ostwald believed that such a gradual change of mind would prove to be a relief. A number of papers were published in trying to derive the well-known laws, for example, the stoichiometric rule, in a

---

*Ostwald: *Die Überwindung des wissenschaftlichen Materialismus. Zeitschrift für Physikalische Chemie*, 18, 305, 1895.

purely energetic way. However, in fact, those works as a whole, were carrying a very artificial touch, and soon it had become clear that we could not skip the fully tangible/fully material image implied by the atomistic picture of the matter. On the other hand, the energetic view was that time consequently meeting a strong opposition coming in particular from the side of Ludwig Boltzmann* and Max Planck, at that time—the brightest proponents of the atomistic picture.

132. Meanwhile, a serious objection to the mechanical view is based upon the fact that natural phenomena are irreversible (Zermelo, Ostwald, and many others). If all the natural phenomena were of a mechanical nature, they had to be reversible, since any purely mechanical process is reversible. It is also apparent that in the differential equations describing reversible processes the time differential is always possessed of an even power, with the result that time direction can be reversed without changing the equation (*in mathematical expressions: the negative-to-positive sign change, and vice versa*). This ought to be applicable, for example, to the mechanical pendulum and electrical oscillations. The processes very well known in thermodynamics, whose trajectories ought to consist entirely of equilibrium states, the processes, which proceed infinitely slowly, are also of the same nature. In the latter phenomena, the time plays absolutely no role or, as one can also mathematically put it, the time enters here being possessed of the zero power.

An arbitrary system of material points, which are influenced by forces dependent only on the positions of those points in space, must eventually be able to return to its initial state. One could thus expect the nature to initiate processes, which would contradict the second law. Consequently, irreversible processes in a purely mechanical world would really be impossible, unless the molecules could be able to spread out indefinitely.

133. By introducing the probability concept Boltzmann could reconcile the atomism with the second law. Thus, he had brought the discussion described above to a certain conclusion.

Although it is possible at whim to change the mathematical sign of the time, it does not immediately follow as a result that the given

---

Wied. Ann., 57, 39, 1896.

dynamic process might run in the one direction equally well as in the other. The point is that any triggered dynamic phenomenon is determined not only by the pertinent differential equations, but also by the relevant initial conditions. Therefore, even if we know that a particular mass of a kind of gas was some time ago in some definite initial state, it is not absolutely impossible for the former to exactly return to the latter, but the likelihood of such an event ought to be inconceivably small.

If we are dealing with a system consisting of a few molecules, or with visible movements of some large objects, the situation is quite simple, and we can easily imagine the system's return to a certain state.

Contrariwise, as soon as there is a very large number of molecules, we must wait for the change from a less probable to a more probable state, that is, from a more ordered to a less ordered state. Say, in a mixture of hydrogen and oxygen atoms, it is definitely not impossible that the diversity of the latter both at a given moment could have such speeds that they would dismantle or transform themselves somehow; Still, the likelihood of such a result is negligible, unless the molecules' number is quite small. The most probable mode is that in which the disorder is highest, i.e., where the two gases form a homogeneous mixture.

The story is definitely independent of the exact nature of the system under study. We do encounter the probability of the lasting change virtually everywhere.

As one can recognize, in addition to the energy of the movements of a visible body as a whole there is a far greater amount of energy hidden in the body's tiniest parts: molecules, atoms and electrons. Obviously, the latter amount of energy continues to grow at the expense of the former. Consequently, this means that the entropy is still growing.* This way Boltzmann has concluded that it is advisable to put the system's entropy S proportional to the logarithm of the probability W of the state in which the system is likely to stay. To sum this up mathematically:

$$S = k \log W, \text{ where } k \text{ is a constant.}$$

*The time ought to have the same property, if we start measure the latter from some given moment. If the Nature would stand still, and therefore the entropy had reached its maximum, we should not notice any progress of time; hence, it follows: Assuming that Nature is at a standstill, time must also stand still (H. Hort: *Der Entropiesatz*, pp. 8–14 [Please see Supplementary Note 26, Section 1.9, Pages 148–149]).

Owing to such a connection with the probability concept, the second law has assumed the snugly form where formidable items can be rationalized from the statistical viewpoints; in the recent times this has become a very important trend.[*]

Boltzmann's prediction that the atomism ought to be indispensable had come true this way. Most recently, the electron theory added a new applicability area to the atomistic conception and thus ensured its further development. Based only upon the kinetic atomism, it is possible to grasp the due sense of ionization and radioactivity phenomena. As a result, the contradiction between the mechanics and electrodynamics, between the physics of matter and the physics of ether might seem to be diminishing.

Newer Objections to the Second Basic Law (pp. 141–143)

134. Recently, one had put forward the idea that entropy should be able to decrease in the interstellar dust clouds. Because of the small influence of gravity in the vicinity of some dust clouds the molecules with the high speeds should escape from the latter and forma hotter mass cluster, while the remaining part of the cloud would retain molecules with the lower speeds and will thus cool down. Hence, the entropy of the system ought to become lower.

Schwartzschild[†] has addressed this issue and come to a different conclusion. Since molecules with speeds of about h and corresponding to the temperature T are leaving the cloud, the presence of gravity would anyway decrease their speeds. Let us denote the temperature of the escaping mass T' and the final temperature of the remaining mass T''. Schwartzschild finds that if the attraction is relatively small, then T' would be greater than T''; if the attraction is relatively large, then the T' be less than T''. But in any case, T' would be less than T, the temperature of the intact cloud; i.e., the escaped mass would therefore be cooling down, as well as the part that is left.

The final state of the above system should also gain a certain amount of potential energy, for the cloud is divided into two parts. Consequently, in such a state the cloud's resulting temperature should not be homogeneous any more, at the expense of the initial

---

[*]Planck: *Acht Vorlesungen über theoretische Physik.*
[†]*Comptes Rendus* 157, 101, 1913.

kinetic energy. Howbeit, the cloud would still not become strictly divided into two parts, with the one of them being definitely warmer than the other one.

135. Laue* has identified an interference phenomenon, which might appear to contravene the second law. By Fresnel and by Michelson's experiment, it is established that two beams with the same radiation intensity at the interference can form two others whose intensity is different, thereby entailing the temperature difference. This would cause the entropy decrease without any sensible change or compensation having taken place.

Laue comes to such a conclusion in the following way. In solving thermodynamic problems, one has always to assume that the entropy is an additive variable, i.e., the total system's entropy should always be equal to the sum of the partial entropies. If the latter is true, then the interference phenomenon involved should undeniably deliver a deviation from the second law.

Now, the question is whether we should abandon the law of entropy growth in this case, or, instead, the entropy's additivity principle. The conventional thermodynamics cannot be helpful in finding the right solution to the above-sketched problem. It is possible to consider the whole story from the Boltzmann's viewpoint. Bearing this in mind, one might express the entropy S for any system as follows:

$$S = k \log W.$$

Here W is the probability of the system's state, k is a constant.

The system consists of two parts, so that, we might express the respective entropies of the latter both using the similar designations:

$$S1 = k \log W1 \text{ and } S2 = k \log W2.$$

Thus, the entropy's additivity principle ensures that $S = S1 + S2$. However, the latter applies only when $W = W1 \bullet W2$, for only then $\log W = \log W1 + \log W2$ is true. To sum up, the condition introduced by the expression $W = W1 \bullet W2$ means that the system's two parts are statistically independent of each other.

The next question is how to describe the actual situation physically correctly in such a case?

---

* *Annalen der Physik*, 20, 365, 1906.

Two beams, which can be brought to interference, will not be independent of each other anymore. They are stemming from the same light source and therefore should now be coherent, since oscillations inherent in one beam are closely connected with fluctuations in the other one. Hence, the addition theorem should no longer hold true, and this way the phenomenon is not in conflict with the second law.

Indeed, if we would place two small magnets inside each of the two dices,* respectively, we can no more expect that any number of dice rolls would now give the same result, as if the magnets were absent. The point is that the dynamics of the both dices is now coherent; in fact, the magnetic power lines interweave now them both.

*Classen: Das Entropie-Gesetz, p. 27.

[Please see Supplementary Note 27, Section 1.9, Pages 149–151]

Still, taking care of the consistency in our calculation, we find that for the both light beams and cubes the probability laws do continue to apply; one has but simply to transform the results in the correct way.

Taking this into account, one might together with Plank* imagine the possibility that some remote coherent system would correspond to the system of celestial bodies, and that the two systems ought to relate quite normally to each other, as long as they are separated from each other. Meanwhile, by starting some reciprocal action could then cause apparent exceptions to the entropy law. In this way, the danger of the so-called Heat Death may be averted without any gross violation of the second law.

[Please see Supplementary Note 28, Section 1.9, Pages 151–161]

To sum up: *Peter Boas presents a comprehensive description of the then conceptual situation. He has told us about the revolution in physics, which was consisting in the big war between two main paradigms.*

Noteworthy, the basics of thermodynamics as a unique branch of natural sciences seem to be strongly endangered during the then revolution. Meanwhile, Peter Boas does not draw such a conclusion explicitly, but his book does lead us at least to calling it in question. A clear evidence for our latter impression ought to be the fact that he devotes so many pages to considering in full detail the 'chemical'

*Planck: *Die Einheit des physikalischen Weltbildes*, p. 2.

contribution to thermodynamics, as well as to physics as a whole (see pp. 66–143 in his book we are discussing).

Remarkably, this is but exactly the direction shown by Prof. Dr. Mohr, as we have already seen above! Still, in Peter's book we find only a casual reference to Mohr's contribution—in duly underlining his work on the first basic law, but without anyhow mentioning—*not to put: stressing*!—Prof. Mohr's groundbreaking ideas as for the second basic law.

Meanwhile, Peter does still bring to light the 'Mohr's shadow' in his book, when he describes the work by Prof. Dr. August Friedrich Horstmann (1842–1929). Here we would greatly appreciate to tell about this colleague even much more than what we might find about him and his work in Peter's book.

The main point is that to the best of our knowledge Prof. Dr. Horstmann could pioneer revealing the true physical sense of the second basic law (entropy). There is indeed much more to the story, even as compared to the most detailed treatise by Peter: Namely, the relationship of Prof. Horstmann to Prof. Mohr on the one hand and Prof. Gibbs on the other hand.

We shall but first revert to our discussion about the main paradigms in the natural sciences.

1.6 The Discord Between Two Main Paradigms

Meanwhile, at this point Prof. Dr. Mohr calls once more the tune [8]:

"... What one might deduce from the formulae should already be contained in them and is hence not a discovery of the mathematician. If the primary assumptions are wrong, then the mathematically consequent conclusions will be wrong as well. ...

... It is therefore anyway an overestimation of mathematics, when Clausius believes that through the application of his Second Law, which is the Law of the Equivalence of Transformations, he might obtain a number of important results. Among the latter, he lists the determination of the volume of saturated vapors, of the amount of vapor, which precipitates when saturated steam expands in an impenetrable shell for heat, the transformation of the steam engines' doctrine & so on.

We know all the mentioned phenomena from detailed experiments, and as concerns the actual relationship between vapor pressure and temperature there is for the present no detailed explanation, but only thorough tables determined by experiment, hence engineers might answer all questions about density and saturation of vapors using these tables. Even the steam engine doctrine has obtained no useful practical indications from the scientific research. We have already started to apply high tensions long before we have learned that less heat is included in the exiting vapor of the truly working machine, as compared to that of a machine working on empty.

In fact, Carnot was the first mathematician who thoroughly dealt with this matter and had as a result formulated the rule that the escaping heat is the same in the both cases just mentioned above. Meanwhile, that was only Mayer who could clearly notice: It is one heat equivalent corresponding to the energy of the work being performed that should be missing in the vapor escaping from the working steam engine. Although all of us know and believe this, it is tremendously difficult to demonstrate experimentally the mentioned deficit for any realistic steam engine and we can disclose this mathematically.

The Energy Conservation Law, which includes the mechanical theory of heat as well as the Equivalence of the Transformations, the mechanical theory of affinity, and in general the physics as a whole, is expressing nothing more than just the rule of correspondence between the causes and the effects. The latter is just the law of the common sense, and therefore it might be straightforwardly proven like the well-known rule that 3 plus 1 is equal to 4.

Bearing in mind such simplicity of the basic concept, there is absolutely no way for the higher mathematics to provide us with some discoveries; the relationships are so simple that it is always possible to express them by a combination of addition and subtraction operations. If we cannot convert the whole dynamics available into some new way of motion, then it is enough to state that the Rest + the transformed Part = the Original Amount of Motion. Hence, all the application of the higher mathematical analysis proves to be unnecessary and fruitless. Therewith, there is also no point to discover, no point to derive using deliberations other than those based upon the Cause-and-Effect Relationship Law.

Today we are still building steam engines according to the Watt's system, and their advantages concerning the older ones consist merely in the more precise way of working and in better fire jigs and tools, but not on the usage of some laws that we have allegedly derived because of mathematical transformations.

There is still one important circumstance not highlighted at all by Clausius in this sense.

Indeed, the transformation of every movement into heat ought to be natural and complete. On the other hand, the reverse is always intrinsically incomplete. A mass movement might completely go over into heat during the shock of inelastic bodies, and the dynamics disappears as a result. However, if we want to implement the reverse situation, i.e., convert the heat into the mass movement, then in the steam engine we must be happy with getting only 2.05 percent of the initial heat amount, whereas 97.05 percent heat would escape in remaining unchanged and without having made any work.

Light disappears at last to re-appear as a heat.

[Please see Supplementary Note 29, Section 1.9, Page 161]

The electric current at any moment after its creation does perfectly go over into heat; but, on the other hand, we get only a minimum amount of light from a lot of heat, and in the thermopile, just a minimum amount of electricity might arise, but the latter both do instantly go over back into heat as a result.

*It comes out but slightly cheaper in the chemical processes. Indeed, during the combustion a large portion of dynamics is existing as the chemical affinity in the intact components (educts) and being converted into heat, while only a smaller part of the motion remains in the combustion product as **molecular dynamics**—which is in fact a chemical property, but **not just an unserviceable** heat.*

Naturally, every motion goes over into heat, that is, each kind of motion might be convertible into heat. Meanwhile, restoring the mass movement at the expense of heat or restoring the chemical difference at the expense of light, takes place upon Earth through the evaporation of water and, respectively, through the decomposition of carbonic acid in the plants.

In fact, it is definitely striking that Clausius, dealing so much with the heat as a driving force, was not discussing in detail the condition, under which the heat might at all occur as a kind of motion. This condition consists in the extension of bodies by heat, which is just nothing more and nothing less than a volume change as it is, and hence is a kind of movement in itself. In addition, after inserting the body into a flask with movable walls, this would appear to be a conventional mass movement. Such an explanation ought to be very simple and obvious, once we have uttered it, but it does not exist prior to its utterance.

Now, we might take into account that all the movement types upon Earth should at last boil down to heat. Furthermore, that the natural processes, like the expansion of the air, evaporation of water and the growth of plants, do convert the heat again to the mass movement, or, respectively, preserve it as a chemical difference in the plants. Then it is throughout possible to use the former directly as a wind and running water, while the latter might at last act as a heat source to produce a mass movement with the help of the steam engine—again with a great loss of heat. To sum up, there seems to be absolutely no reason for taking into account some 'Entropy of the World', or a 'Gradual Equilibration of all the possible movements upon Earth to finally render the Heat of Equal Temperature', as feared by Clausius. Contrariwise, we see that we do not take advantage of all the existing resources, and that in effect all the mass movements involved evince an immense abundance in storms, hail, floods, glaciers, earthquakes and volcanoes.

Although Clausius is quite right to view the total sum of the movements present in the world as a Constant, he would not like to go so far in assuming that the whole state of the universe is immutable and represents a kind of eternal circuitry. Definitely, according to Clausius, the Second Basic Law of thermodynamics ought to contradict the latter ideas. Our total experience does show that converting diverse movements into heat is always much more complete and occurs much more frequently than the vice versa converting heat into other forms of movement. As soon as this is the case, Clausius is afraid that the work the Natural Forces perform, the work contained in the existing movements of the heavenly bodies, should gradually be turning into heat. Further, the resulting heat ought to offset more and more to arrive finally at a state of the Total Equilibrium, Total Rest, Dead Inertia and Torpor, which he called the Entropy and represented as an alleged output of all the scientific research to await. To sum up, he expresses all this in a final general proposition:

'The entropy of the world tends to its maximum'.

Nevertheless, there is no reason for ending up with such a result. Rather, one might state with certainty that we have reached the maximum entropy already since an infinite time. In the World as a whole, we see but no phenomenon, whereat any mass movement (and we do not know anything more intrinsically dynamical in the Whole Universe except for heat and light) would be converted into heat, so we have to go back to our Earth. In addition, the Latter is already in equilibrium with the Sun, in terms of heat and light: the ice does forever block the rigid poles, *whereas the tropical belt is permanently located under the vertical rays of the Sun.*

... Therefore, the question of whether it might and ought to shine forever we have to address to the Sun.

... When the sum of the heat located on the Sun ought to decrease, then we must somehow be able to extend this principle to other celestial bodies.

... To sum up, it is clear that the assumption of the existence of the world without any kick-off ought to completely exclude any fear for entropy increase or return to the state of the over-all equilibrium.

Howbeit, Clausius believes that the current state of the world is still very far away from this boundary condition and that approaching the latter would be happening so slowly that all the historical periods ought to be only margins compared to the immense periods necessary for the World to perform just picayune re-organizations. To his mind, he could at least manage to announce an important event, namely that a law of nature was discovered, which let's conclude with certainty that the World does not exhibit circulation, but that it ought to be continually changing in the sense that it is aspiring to The Border State.

*In effect, nobody could find such a law of nature. Presently, the actual conceptual basis of what is available to us presently is solely an overhasty estimate of the relationship between the conversions of different movement types into heat and vice versa. **However, if there is already the World for all eternity, so nothing ought to proceed with it in the course of time, nothing that could be solely due to the course of time, for all what is already available could survive for every conceivable amount of time.** As the ultimately global entropy maximization has not yet occurred since eternity, it should not take place in any conceivable future."*

In his above-cited statement, Prof. Dr. Mohr is clearly demonstrating the apparent weaknesses of the Paradigm triggered by R. Clausius.

To sum up, its main weakness is the prevalence of some kind of a global emotion, instead of a sober standpoint of a researcher.

Interesting might then be the following immediate posers: If Prof. Mohr was such a truly discerning colleague, why was he 'The Lone Bird'? Where were other serious researchers hiding at that time?

The immediate answer we shall get: '*ah, do not fuss! That was a time of great revolution in all the fields of natural science!*'

Sure, of course! We have nothing against the great revolution, we have nothing against its heroes, and we have nothing against their great achievements. However, it is rather easy to kill the roots of scientific research (*in fact, it is to 100% true for any kind of human activities!*) in declaring somebody 'the genius of all the times and the entire manhood'. Well, hopefully, least said, soonest mended.

To our sincere regret, the work of Prof. Mohr had somehow trapped into oblivion—see, for example, the work by Ralph E. Oesper [12], although Dr. Oesper's conclusion was not in fact to 100% correct. In the following, we shall see why we are right.

1.7 Prof. Horstmann: Embodying Ideas by Prof. Mohr, Transferring Them to Prof. Gibbs

Indeed, at least one colleague—namely, an eminent German physical chemist at the University of Heidelberg, Prof. Dr. August Friedrich Horstmann (1842–1929) was in fact following carefully and systematically the guidelines suggested by Prof. Dr. Mohr. He could duly take into account his criticism. Prof. Horstmann but seemed to be carrying out his work quite completely independently of other colleagues at his time.

This authentic portrayal of Prof. Dr. Horstmann has been taken from http://www.family-horstmann.net/ho_texte/ho_tx020.html

Thus, we definitely arrive at the very point, where we should owe truly extreme gratitude to Peter Boas Freuchen for his books, because he looks like the very first colleague all over the world, who did perform a detailed analysis of Prof. Horstmann's seminal contribution to the general thermodynamics and had tried to attract the colleagues' attention to Horstmann's finding. Still, in effect, Prof. Dr. Horstmann was and is still remaining mostly in shadow. This urges us to present here his biography first of all.

1.7.1 The Life of August Friedrich Horstmann

Indeed, the only noticeable reaction concerning Horstmann's work has appeared in 1903, as a part of the series *Ostwalds Klassiker, 137*, being a small volume re-publishing some of his papers in different periodicals [14]. In this volume, we might find at least two seminal contributions due to Prof. Horstmann, which deserve our concentrated attention.

At the end of the above volume, its editor, Prof. Dr. Jacobus Henricus van't Hoff (1852–1911) had published the following biographical sketch. Here we present its English translation:

August Friedrich Horstmann was born on November 20, 1842 in Mannheim, as the eldest son of a prominent local merchant. His definite inclination toward science had surfaced rather early. Having graduated from the local gymnasium, Horstmann received particular encouragements from the then gymnasium Director H. Schröder already in his junior high school years.

Meanwhile, due to his organic myopia having later developed into a dramatic eye condition, already from his childhood, serious concerns had arisen as for the suitability of any scientific or technical career for him. Therefore, Horstmann had entered into his father's business as a merchant and spent three years in such a quality, even without absolving gymnasium at all. Later on, he regrets very often and vividly the loss of his precious time.

Only as late as in 1862, Horstmann was able at last to follow his scientific impulse and started attending the University of Heidelberg. First, he wanted to hear lectures aiming at his general education, but without specific destination, and the growing opulence of his father in the then flowering Mannheim could financially enable his following this way.

However, very soon the scope of his actual interests had been clearly concentrated on the chemistry and related sciences. Emil Erlenmeyer, in whose laboratory Horstmann started working, had immediately suggested theoretical chemistry to be the field in which Horstmann could most likely achieve something with his abilities. Indeed, he considered Erlenmeyer and soon Hans Landolt his actual teachers.

In the spring of 1865, Horstmann could successfully obtain his doctorate degree already after just attending the lectures of Bunsen, Kopp, Kirchhoff and others, for at that time there were no overly great difficulties in Heidelberg to accomplish this even without the Higher School Certificate.

Then he moved to Zürich, to attend the lectures of Wislicenius as well as those of Clausius on the mechanical theory of heat, which ought to have formed the factual basis of his work re-published in the current collection.

[Please see Supplementary Note 30, Section 1.9, Page 161]

Afterwards he went to Bonn, where he started his work on variable vapor densities in the laboratory of Landolt. He continued this work after returning to Heidelberg and completed it under the leadership of Bunsen. That was just the topic of his Habilitationsschrift.*

[Please see Supplementary Note 31, Section 1.9, Page 161]

In winter 1867, the University of Heidelberg has credentialed Horstmann as a Docent after the ordinary disputation, where Victor Meyer was his opponent, whereas Horstmann's thesis was devoted to the then active dispute as for the vapor density and molecular weight of ammonium chloride. It is in that time that he published a number of preparatory works that chronologically came prior to Horstmann's major paper appeared about the year 1873 entitled "The Theory of Dissociation". The latter works have been included here for the sake of historical importance; namely, those entitled "vapor tension and vaporization heat of ammonium chloride" (1869), as well as "About the second law of thermodynamics and its application to some decomposition processes" (1870–1872).

In 1872, Horstmann started acting as an Associate Professor, while in 1889 as an honorary professor. However, in the early years he was rather playing the role of a student than of a Docent. Indeed, he was striving to extend his mathematical and physical knowledge, and the lectures held in Heidelberg at that time by Kirchhoff, Helmholtz and Königsberger were offering an excellent opportunity to achieve such a goal. Still, henceforth he decided to confine his teaching activity.

**Berichte der Deutschen chemischen Gesellschaft*, 1868.

Then, especially after his marriage in 1873, he was acting mostly as a private tutor. Horstmann carried out his experimental work in the small private laboratory, which he had set up according to his needs with limited resources. In this period, he carried out the work on carbamin-ammonia salts, which we re-publish on page 42 of the present collection. Finally, he decided to stop his laboratory works gradually, as well as his literary activities, due to the severe limitations caused by his deteriorating health state. For the last time he published his report in 1881, which is re-published on the page 56 of this collection.

Horstmann's last literary activity consisted in editing the papers on physical chemistry for the Journal of the American Chemical Society, but finally he had to stop this as well, because of increasing blindness. We sincerely hope that the present re-publication of his works would at least be a kind of consolation for our long-suffering colleague.

Frankly speaking, the above-cited statement causes a truly split feeling. On the one hand, it is definitely a kind of 'encomium' for a highly esteemed, but a clearly unlucky colleague. On the other hand, this demonstrates a clear attempt to diminish the contributions by the latter. Indeed, after reading the above fragment, it looks like that the main Horstmann's achievements ought to consist solely in skillfully investigating some particular chemical phenomena. Consequently, his thermodynamic results of actual general significance ought to be nothing more than just a kind of prelude to his theoretical–chemical considerations.

In reality, Horstmann has published his '*Habilitationsschrift*' not in 1868, but just one year earlier—and indeed in Heidelberg [15].

In effect, the work [15] is a general physical–chemical treatise concerning the relationship between molecular weights and specific weights of elastic-liquid systems.

Meanwhile, the exact reference to Horstmann's work introduced above as that in the '*Berichte der Deutschen chemischen Gesellschaft. 1868*' sounds other way (see Ref. [16]).

The latter paper contains no reference to his '*Habilitationsschrift*', but nonetheless, one of its main points consists in the detailed discussion of the basic hypothesis introduced by Clausius, Maxwell et al. and based upon the probability theory.

Horstmann presents a valid and constructive criticism of the hypotheses constituting the basis of the then kinetic gas theories by

Clausius, Maxwell, Boltzmann and others. Horstmann concludes that such a conventional approach ought to be dispensable, whereas some other ideas might be of use. In reading this paper by Horstmann, a poser arises:

> *Was the mentioned criticism the true reason for not re-publishing Horstmann's 1868 paper in 1903?*

Moreover, of extreme interest for our current discussion ought to be the obituary published by the German Chemical Society, Berlin, right after A. F. Horstmann's ultimate departure (below we shall present the English translation of its most relevant parts) [17]:

> *"Physical chemistry has recently lost one of its pioneers, one of its Great Fathers, who could recognize in thermodynamics a powerful weapon to penetrate the jungle of chemical facts, to strike the first streets in it, where the researchers are still wandering nowadays. Having deficient eyesight from his adolescence time, he was anyway following the inner light of his strong spirit. Now he passed away in extreme old age. Being almost 87-year-old, he closed his tired eyes that already 40 years ago failed to serve him.*

> *He has practically spent his entire lifetime at his homeland country, the Palatinate, namely in Mannheim and Heidelberg. However, the male line of his family was in fact not at home there. As in most families of the West Germany, or at least those of the Southern-Western Germany, we may trace their destinies back up to the 30-years war. ...*

> *...The only information about the earliest ancestors of Horstmann reveals that they came originally from the Paderborn region, where even nowadays there should be a 'Horstmann's yard'. Still, we are not sure about any further details of their familial story. ...*

> *...Unlike our colleague Horstmann, one of his greatest contemporaries, unforgettable molecular physicist Boltzmann had truly unpleasant experiences. Remarkably, they both were fervently fond of music as well as of scientific truth. They both were interested in rather similar problems, although they have chosen quite different ways to arrive at their solutions. Meanwhile, the both colleagues were losing their eyesight. Whereas Horstmann experienced this rather slowly, from his childhood on, Boltzmann*

had but entered this corridor much faster, being already much elder, moreover, having bitter disappointments and pursuing harsh battles at that time. Most probably, the entirety of the latter factors has caused Boltzmann to feel the over-all Senselessness in an extremely hard way . . .

[Please see Supplementary Note 32, Section 1.9, Page 161–162]

. . . That was Boltzmann's fate, a fate of a delicately sensitive person who valued Beethoven's music above all other things. Instead, Horstmann's soul was living in the sounds of the great Thomas-Kantor Bach and probably he would say the following about himself:

> *"I know, demons, you would hardly vanish.*
> *The rigorous ties with you are truly wedding,*
> *Your power, oh, Worry, might be colossal,*
> *Still, I shall never-never recognize this!"*

With this in mind, we would now like to bid farewell to A. F. Horstmann, who always ought to have the nerve to be strong-minded enough in striving for the knowledge against any kind of stream:

> *The night might penetrate the deepest depths,*
> *Still, solely Inside the brightest light does burn.*
> *FAUST, J. W. Goethe.*

> *M. Trautz,*
> *Physikalisch-Chemisches Institut der Universität Heidelberg"*

The author of the above very instructive story was an outstanding German physical chemist Prof. Dr. Max Theodor Trautz (1880–1960), who was productively working on finding the true physical sense of the "*Activation Energy*" notion employed in the chemical kinetics. We shall revert to Prof. Trautz, his works and results in the Supplementary Note 35 to this chapter (see Section 1.9, Pages 163–177). The notion of '*Activation Energy*' ought to have the tight interconnection with thermodynamics, and in the present volume we shall meet another colleague, who was looking for the true interrelationship between the both fields—Prof. Dr. Max Bernhard Weinstein, see Chapter 3 of the volume at hand.

The above story by Prof. Trautz is clearly underlining the difference between the undoubtedly talented scientific research workers, who were *either truly capable or sheer incapable of overcoming the everyday vanity…* Definitely, Ludwig Boltzmann belongs to the latter team, whereas August Friedrich Horstmann, Peter Boas Freuchen and all the protagonists of the present volume— to the former cohort.

Finally, Prof. Dr. Trautz' report clearly shows a rather tight professional connection between Profs van't Hoff and Horstmann. Meanwhile, the whole World knows the former colleague as one of the Holy Fathers of the modern Physical Chemistry, whereas the latter (together with his inspired and talented venerator, Prof. Trautz, a propos!) deserves solely some occasional sluggish comments, if any. Well, the conclusion is immediately clear, but somehow sorrowful:

> *If you consistently strive for veneration, you would surely get it finally. The latter might be either a truly hard-earned one, or even more or less just so. Indeed, the Manhood, until it is populating our Earth, would always need something or somebody to hold in veneration.*

If we are striving for finding the ultimate truth, there are solely two variants: either we do find it, or do miss it—but the both sole variants would be all you might really count on.

There is nothing to modify on the above, but the author of the volume at hand would like to mention here only one serious point:

It is not OK to fully forget anybody and/or anything, so the sole fair stance ought to be: *Nobody is forgotten and nothing is forgotten—* which might also be a rather difficult task, for every realistic phenomenon (the memory is definitely among those) is possessed of its natural limits.

Then, the only solution to the problem seems to be a thorough and detailed selection of the events/phenomena/persons to estimate their actual values, aiming at the correct decision about their exclusive right to remain in our memories. However, is it possible to arrive at the point where *everybody* upon earth is a fine professional, a skillful specialist in **all the possible** areas of the human activity? Looks like a gloomy horizon anyhow…

1.7.2 The Work of August Friedrich Horstmann

Meanwhile, for the present it is only in German that we may find a detailed historical analysis of Prof. Horstmann's life and work, that is, in Prof. Horstmann's mother tongue. An outstanding Soviet historian of science, Dr. Alexander Yakovlevich Kipnis, has undertaken a detailed analysis of Prof. Dr. Horstmann's life and work and published a book about this [18].

[Please see Supplementary Note 33, Section 1.9, Page 162–163]

Here we would just like to cite what Dr. Kipnis could conclude as for the contribution by Prof. Dr. Horstmann [18] (below we present the pertinent English translation):

> *"... Most important were the theoretical foundations that A. F. Horstmann was developing in connection with his experimental research. This was the emergence of a new discipline, namely that of chemical thermodynamics. First he tried to describe the chemical equilibria using the Clapeyron-Clausius' equation developed for evaporation phenomena— that was already quite successful for a separate class of chemical systems—but only in 1873 he had managed to find the general solution. Namely, A. F. Horstmann could re-design the famous Clausius' aphorism – "The entropy of the world strives towards a maximum" – into the fundamental scientific principle. The actual physical sense of the latter aphorism is that all the changes in a closed system might be complete. Along with this, the equilibrium state occurs, when the entropy of the system has become as high as possible for the process in question. Next, he could demonstrate how to apply this principle in chemistry: It has become possible for him to thermodynamically infer the law of mass action and confirm this inference later on with a number of his own experiments.*
>
> *Certainly, A. F. Horstmann ought to deserve a place of honor in the history of science as the true founder of chemical thermodynamics. Noteworthy, thermodynamics as it was not autotelic for him, but merely a tool for analyzing diverse chemical processes. Indeed, he was thinking over a lot as for the mechanisms of chemical reactions, [sic] and he was one of the first chemists, who had introduced the latter term. His most interesting achievements in this field relate to catalysis, reactivity and phase formation; in particular he could manage to provide us with the first authentic and general description of the chemical process' acceleration through its products (autocatalysis)."*

We would just dare to point out here that the notorious collocation 'closed system' is another stark example of the numerous linguistic misnomers in the field of the conventional 'equilibrium thermodynamics', which are cementing the notorious (still lasting nowadays and inexplicable) splitting of a unique knowledge domain into three hardly compatible branches [19]. The mentioned paper by an outstanding thermodynamicist Prof. Dr. Mark Waldo Zemansky (1900–1981) is demonstrating us in fact not a unique branch of the natural sciences, but a kind of 'conceptual cadaver' burst into three independent parts.

In effect, the actual roots of all the misnomers mentioned in [19] and the likewise ones lie in the ubiquitous over-interpretation of the ingenious widget suggested by N. L. S. Carnot. Indeed, the 'closed systems' are throughout akin to such notorious misnomers, as the 'reversible' and 'adiabatic' processes, etc. we have considered in this chapter with the help of Peter Boas.

Here we shall consider in detail Prof. Horstmann's seminal contribution in Chapter 5 of the present volume. Still, it is truly important to note already at this point that in his studies on the theory of thermodynamics he was taking into account Prof. Mohr's indications [20]. (We present here the pertinent English translation.):

"98. In one of his letters, obviously written immediately after F. Mohr's publication Horstmann expresses the following:

'...I tend to consider Mohr an insolent chatterbox, at least in the areas in question, and would greatly appreciate to disparage him, to wit, in a quite coarse manner. On the other hand, owing to my known lack of self-confidence, I am afraid of his sharp pen, because I could conclude from our personal communication in Bonn that he knows well, how to use every mistake. Sure, mistakes happen to me often enough, so I am working the whole time on the correction to my last essay, where indeed not less than three blunders have happened to me, which but certainly do not affect the quality of the entire essay' [Letter number 69].

99. Here we should still mention that the angry remarks by F. Mohr did contain a kernel of truth. Mohr was categorically denying the possible heat death of the world. If one accepts this view, it comes to the following statement: '... if in the course of time any approach to a limit state were possible, this would have definitely taken place in the already elapsed

eternity, or, otherwise we have to grant the whole world a beginning in time, which is impossible:

F. Mohr: "Theory of Dissociation or Thermolysis" Liebigs Annalen of Chemistry, 1874, v. 171, pp. 361-379.

Well, now it is quite clear that the philosophical culmination in the formulation of thermodynamics triggered by Clausius is of no help in the thermodynamics' application to finite systems accessible to experimental research.

Howbeit, both Horstmann and Gibbs, who even decided to render Clausius' postulate the unique epigraph to his own most important work on the thermodynamics, were in fact employing not the philosophical but, instead, the very physical formulation of the both Basic Laws in their investigations.

[Please see Supplementary Note 34, Section 1.9, Page 163]

Contrariwise, it is ultimately choosing the Clausius' "philosophical" form that either forced/forces or, may perhaps, made/makes it possible to tie up the real application of thermodynamics with the ultimate truthfulness of the Clausius' postulates. But this alleged truthfulness is by far not uncontroversial, and F. Mohr was actually one of the first among the colleagues to clearly express this fact."

Interestingly, the historic investigations by Dr. Kipnis have brought even much more dramatic and throughout unexpected result [21]:

"... Although it would be unfair to state that Gibbs made no contacts with chemical experience[45], these contacts were not easy to see. ...

45. It seems to be a Clio's smile that Gibbs took lessons on theoretical chemistry, namely on thermochemistry, by Horstmann in Heidelberg: University Archive Heidelberg:

Akademische Quästur A. Horstmanns (Rep. 27, Nr. 621, Blatt 3)."

Remarkably, J. W. Gibbs was in direct personal contact with Prof. Dr. Horstmann and could even attend his lectures in Heidelberg. In addition, the library of the Heidelberg University has a special Internet site devoted to J. W. Gibbs' time spent in Germany [22]. We can get the following information in German from over there (We present the English translation below):

"From the summer semester of 1868 until the spring of 1869 Josiah W. Gibbs spent a time with his sister Anna for studying in Heidelberg; he was not enrolled at the University. During the period of the summer semester of 1868, he attended the course by Jakob Lüroth "Theory of Algebraic Forms" and during the WS 1868/69 probably also the course by Friedrich Eisenlohr "The Theoretical Optics." Indeed, Josiah Willard Gibbs was later sending Jakob Lüroth reprints of all his mathematical work, while to the address of Eisenlohr his work on optics only. During this time, Gibbs might also attend the courses by such already famous colleagues as Bunsen, Helmholtz and Kirchhoff.

At the same time he could also attend the courses by August Horstmann, namely 'The Physical Theoretical Chemistry' and 'Thermodynamics in Regard to the Mechanical Theory of Heat'. Meanwhile, Horstmann was in fact the colleague who pioneered using the entropy notion in considering chemical equilibria.

Prof. Horstmann was even mentioning an American student of him in his private correspondence.

(A note by Prof. Dr. Herbert Wenzel, University of Erlangen)"

All the above stories deliver in fact the most probable clue to understanding what ought to be happening with the field of thermodynamics at that time. The constructive criticism of Clausius' working direction was not effective, most probably partly due to the personal and professional demurs to the address of Friedrich Mohr—as well as one of his very active proponents, Peter Guthrie Tait [23].

Meanwhile, the ideas of Prof. Tait, although in somewhat modified form, could still find their audience and productive followers, whereas the legacy of Friedrich Mohr was undeservedly thrown away to the historical garbage, most probably because he was in fact a very competent and consequent proponent of the Energetics, true conceptual basis for thermodynamics.

As to Prof. Tait's legacy, we shall revert to it in the Supplementary Note 21 to this chapter (see Section 1.9, Page 143–146).

Well, howbeit, the development of the thermodynamics' foundations was most probably going along the following direction (we consider the last step in Chapter 4 here):

R. Clausius $\to$ K. F. Mohr $\to$ A. F. Horstmann $\to$ J. W. Gibbs $\to$ G. A. Linhart.

Not to forget as well ought to be the British branch at that time, owing to the work of Prof. Dr. George Downing Liveing (1827–1924) of Cambridge University, especially his book bearing the stimulating title *Chemical Equilibrium, the Result of the Dissipation of Energy*, published in 1885 [24]. The ideas of this book do parallel the ideas by A. F. Horstmann, but till nowadays we were somehow undeservedly overlooking them.

W. Thomson (Lord Kelvin) $\to$ George Downing Liveing $\to \ldots$?

As for the latter theme, see for example, the relatively recent stimulating discussion between William B. Jensen and Frank L. Lambert, as published in the *J. Chem. Educ.*, v. 81, pp. 639–640, 2004 and the references therein.

Importantly in this regard, that was Peter Boas Freuchen, who could at the very least pioneer the detailed analysis of Prof. Horstmann's seminal contribution.

1.8 Conclusion

The books by Peter Boas Freuchen have helped us consider in full detail the situation around the thermodynamics from the middle of the XIX-th till the beginning of the XX-th century.

After a careful analysis of all of the then happenings, we could get an adequate picture of what exactly was the achievement of the well-recognized founders of thermodynamics and what were the feeble points of their results.

The main drawback ought to consist in too much emotion, too many human relationships connected with the research progress in the natural sciences experiencing the true revolution in physics. Revolutions definitely bring new achievements and insights, they accelerate the research progress; we should not to 100% deny the significance of the revolutionary achievements. Sure, the contributions of all the particular revolutionists deserve the due honor. Still, at the same time it is necessary to retain the unbiased look at the revolutionary situation—whatever revolution we are looking at.

Furthermore, the consistent look at the XIX-th – XX-th centuries' revolution in the natural sciences leads us to the conclusion that along with all the undoubted achievements at least the work in the field of thermodynamics and statistical physics was not correctly finalized. Moreover, a number of seminal contributions have been fully undeservedly dropped and forgotten—the failures the present monograph is intending to compensate, at least partially.

In Chapter 2, we remain in Scandinavia to meet Dr. Nils Engelbrektsson and Carl Alexius Franzén—two Swedish gymnasium teachers, with the former having managed theoretical working out the truly general form of the thermodynamic equation of state, and with the latter being capable of experimental verification of Engelbrektsson's seminal result.

In Chapter 3, we shall learn about the life and the professional contribution by Prof. Dr. Max Bernhard Weinstein, an outstanding specialist in the fields of the natural sciences and natural philosophy, who started building up the conceptually unbiased thermodynamics and kinetics.

Chapter 4 gives the presentation of the professional contribution by Dr. Georg(e) Augustus Linhart, a physical chemist of Austrian origin working his entire professionally active lifetime in the USA, who could formally infer the famous Boltzmann–Planck formula and also help estimating the actual sense of the third basic (Nernst) law correctly.

In Chapter 5 we would like to summarize everything we discussed in the previous chapters to try drawing the picture of the different thermodynamics and its proper mathematical tool—in following the work directions suggested by the protagonists of our present book.

1.9 Supplementary Notes

[Note 1]: Most probably, Peter Boas had completely lost or was progressively losing his eyesight at the time, whereas the popular book Mrs. Paulsen mentioned is devoted to the microscopic physics and was published by Prof. Dr. Christian Møller (22.12.1904–14.1.1980), a Danish theoretical physicist. Christian Møller was throughout his career working at Niels Bohr Institute in Copenhagen,

1943–75 as a professor. Being an expert in quantum mechanics and relativity, he was in 1932 employing these theories to the study of electron collisions.

[Note 2]: This is an interesting, but rather short paper about the physical nature of the polar light. The German periodical '*Astronomischer Jahresbericht, 1899*' lists several of his reviews of the pertinent literature, as well as his earlier report about the polar light, but in fact—there was nothing more at the time.

[Note 3]: Here it is an important point to interrupt Peter's story for a while. Indeed, Max Planck formulates his actual standpoint on the page 74 of his treatise entitled '*Lectures about Thermodynamics*' in the following way (here is our English translation):

> "... *This law is not provable a priori and it counts for no definition. Instead, it contains a specific allegation testable against the relevant facts in each separate particular case by carrying out experiments in the relevant direction, to see whether it is right or wrong. ...*"

That is, in trying to translate the latter fragment into the 'normal human language', we ought to arrive at something like as follows:

> Actually, it is not completely clear what exactly this particular law is teaching us. Nonetheless, it is every time possible to formulate some plausible statement in connection to the current topic under study— and even to verify this statement by the pertinent experiments.

Putting it in other, somewhat more optimistic words: It is throughout possible to successfully make use of this unclear law— or, to formulate this in a more succinct and formal way. Then, our true scientific method ought to be a purely '*operational approach*' in all the difficult cases. Here we use the word combination '*operational approach*' after Prof. Dr. Percy Williams Bridgman's suggestion [3.1]. Whatever this kind of approach should introduce, Max Planck could as a result build up his theories without taking the 2nd basic law into account. Still, the success of Max Planck does not mean that the 2nd basic law ought to be conceptually killed! Meanwhile, the latter ought to be the roughest interpretation of the operational approach we are discussing here.

The following is just to describe the essence of the **operational approach**. If we try solving some particular problem, the construction of the relevant theory must not be to 100% logically perfect, if it is still possible to solve the problem posed using the current version of the theory. It is just solving the problem that truly ought to have the primary priority. Instead, furnishing the logical structure of the theory at hand should have less time-critical priority, for the latter activity has no direct, no immediate relevance to the actual particular problems we have to solve now and then.

Of course, we should never consider the above-sketched operational concept an erroneous or misleading motivation—still, which guideline to follow during the research activity is a matter of personal choice. With this in mind, we are returning to Peter's book.

Reference to the Note 3

3.1 Percy Williams Bridgman (**1927**): *'The Logic of Modern Physics'.* MacMillan, New York, USA.

[**Note 4**]: With this entire train of thoughts in mind, one of the essential topics of our present book should be to analyze the details of the latter process. This way we sincerely hope to trigger the due continuation of the work initiated by all the protagonists of the book at hand.

Apparently, this ought to be the one of the true zests of the Peter Boas' book in question, which he had never formulated explicitly. Nevertheless, it is exactly this point that could heavily strike the Danish referee of Peter's book. Now we dare to summarize it as follows:

The **actual thermodynamicists** after N. L. S. Carnot seem to be rather **rare**, for all of his immediate followers were mainly busy with thoroughly studying different important physical–chemical problems of *no direct thermodynamic relevance*—as they were sure—for thermodynamics as it is being in effect totally unknown to them.

Indeed, in many cases quite independently of each other, they were but accidentally lucked out in revealing that Carnot's result might be of tremendous importance for prompting their own studies.

Finally, as soon as the applicability of Carnot's result could prove to be of rather general nature, the true story of actual thermodynamics begins.

Sure, this is definitely not to dethrone Clapeyron, Clausius, and Thomson brothers, who have definitely been the true pioneers of (a) finding, (b) reading and (c) analyzing the N. L. S. Carnot's seminal work. Meanwhile, there were also many other colleagues taking part in the activities b) and c). Moreover, the last but not the least, it is not all right to single out any one of the actual crowds... Remarkably, none of them was a true Deity, whereas it is surely long known that "Errare humanum est, sed in errare perseverare diabolicum." (Hieronymus; Seneca, Epistulae morales VI,57,12; Cicero, Orationes Philippicae 12,2).'Errare Humanum Est'.

Hence the main point is:

> We definitely have to learn from errors—from our own ones and/or those of other human beings. In fact, this ought to be the most consequent position, but such a stance was, is and most probably will be truly rarely detectable upon Earth.

[Note 5]: In full accordance with Peter's original wish, the chapter aims at analyzing the foundations of thermodynamics, and the framework introduced by Peter in his book allows straightforward fulfilling this task.

[Note 6]: *Still, our suggestion for the reasonable translation:* **Livening**—*while the truly conventional designation ought to be* **Driving**.

[Note 7]: If we now mathematically take an integral of the Descartes/Newton's force expression over time, we get the expression for the impulse, whose physical dimensions are correspondent to 'Force-times Time.' The everyday meaning of the term 'Impulse' is in effect to 100% correspondent to the Leibnitz' idea of Vis Viva, the livening force, that is, the Impulse ought to represent 'the Force to liven the movement'. With the latter exposition we would simply like to underline the principal absence of any difference between the Descartes/Newton's and Leibnitz' standpoints. Furthermore, here it is important to mention that it is in effect possible to assess the actual progress of the livening/driving actions by introducing the notion of power/performance/proficiency/merit etc., which is mathematically expressed as the "Work fulfilled during some time," with the dimensions of 'Energy-divided-by-Time.' We dare to interrupt here the narration by P. B. Freuchen, for we would greatly appreciate to underline his main idea:

In no way the so-called classical, conventional thermodynamics ought to be 'a thermostatics', as it might indeed seem after casting a careful look at the conventional 'Equilibrium Thermodynamics'.

See, for example, a book by an outstanding American thermodynamicist, Prof. Dr. Myron Tribus (October 30, 1921–August 31, 2016), as well as much later accounts on the theme (see the works [7.1–7.3] and the references therein). Remarkably, there is a stubborn trend to replace the Time by the Probability in describing the foundations of thermodynamics. One of the brightest examples of the trend involved ought to be forcing the 'basic interrelationship' between the entropy and information notions. We shall continue this important discussion in Chapters 3–5 here.

Howbeit, the thermodynamics undoubtedly deserves its name! The point is that the time notion is in effect implicitly present in all the deliberations by the founders of thermodynamics, so the problem to solve was—and still remains to be—*how to rightly disclose* the role of this tremendously important variable.

References to Note 7

7.1 Myron Tribus (**1961**): *Thermostatics and Thermodynamics: An Introduction to Energy, Information and States of Matter, with Engineering Applications.* D. van Nostrand Company Inc., New York, USA.

7.2 Roger Balian (**2003**): 'Entropy, a Protean Concept'. *Séminaire Poincaré*, v. 2, pp. 13–27.

7.3 Roger Balian (**2005**): 'Information in statistical physics'. *Stud. Hist. Phil. Sci., Part B, Stud. Hist. Phil. Mod. Phys.* v. 36, pp. 323–353.

[Note 8]: It is exactly this, (indeed, surely rather poetical, than purely engineering) analogy spawned by N. L. S. Carnot that causes notorious posthumous blackmailing of Carnot as a "True Adept of the Substantial Heat Theory," which had and has nothing to do with the actual state of affairs. Peter Boas seemingly 'joins the row', to reflect the actual blackmailing, but along with this duly clarifies the actual standpoint of Carnot, see below.

[Note 9]: Many sincere thanks to Peter Boas for clarifying just what was in fact the **actual** deficiency of Carnot's train of thoughts, due solely to his time being! Down with the notorious representation of him as the **true adept of the substantial heat theory**!

[Note 10]: Noteworthy, the first adequate translation of Carnot's memoir into Russian came out in 1923, and the two outstanding Russian/Soviet physicists had commented it very profoundly and insightfully: Виктор Робертович Бурсиан (1886–1945) (Prof. Dr. Victor Robertovich Bursian) and Юрий Александрович Крутков (1890–1952) Prof. Dr. Yuri Alexandrovich Krutkov, who were among the numerous victims of the Stalinist regime. The both colleagues had expressly noted that if Carnot could somehow stay alive, he would definitely arrive himself at the correct theory of thermodynamics.

[Note 11]: Equation 1.1 clearly demonstrates that the 'Carnot's function' is a mathematical function of the temperature only. This very fact is very important to realize which exactly feature of the Carnot's cycle renders it a purely theoretical gadget. Indeed, the only physically realistic processes are *isothermal*, but not the *adiabatic* ones!

Whereas the former ought to be a strong physical idealization, for to fix the temperature at its given value requires non-trivial engineering efforts, as we well know now—the latter are in fact 'isentropic', that is 'free of entropy', which is impossible in physically realistic systems!

Peter Boas hints at this by demonstrating that B. Clapeyron's theoretical consideration boils down to adopting the 'adiabatic stretches' to be in fact of infinitesimal length.

Interestingly, the latter is but nothing more and nothing less than just a mathematical gimmick intrinsically related to the axiomatic introduced by Constantin Carathéodory (1873–1950).

Peter Boas is discussing neither the work by Carathéodory (plus the response to it), nor similar efforts by other colleagues, which did remain practically unnoticed. Howbeit, we shall revert to this extremely important topic in Chapter 5 of the monograph at hand.

[Note 12]: Here is another important point to notice: Indeed, while re-iterating the actually senseless collocation 'Carnot's standpoint', Peter Boas still stresses the true initial aim of Thomson brothers, namely once more proving the Energy Conservation Law. Peter Boas was going this way throughout his whole book.

[Note 13]: Our comment ought to be in order here, and we dare again to interrupt Peter Boas Freuchen in trying to estimate his own achievement from our today's standpoint.

The above fragments definitely demonstrate P. B. Freuchen's very attentive, unbiased and critically analytic reading of all the relevant literature, that is, not only of the works by the commonly recognized founders of thermodynamics plus interpretations by M. Planck, but also of the seminal works by E. Mach and G. F. Helm, whose 'energetic' views are commonly seen as those 'overthrown by the Genius of L. Boltzmann, M. Planck and their school'.

In the latter part of his thesis P. B. Freuchen clearly re-formulates the doubts originally expressed by Hon. W. Thomson, Lord Kelvin, as to the standpoint delivered by N. L. S. Carnot. Indeed, in skipping the latter we clearly seem to undeservedly throw out the baby together with the bathwater. To this point, P. B. Freuchen aptly highlights **"the wide-spread wanton mood of that time to deny anything that smacked of natural philosophy."** Interestingly, this might indeed be the actual reason for why a **clearly dialectic** suggestion by Carnot about 'hot and cold reservoirs' couldn't trigger the adequate and professional reaction of his descendants. Meanwhile, Carnot was in fact also by far not alien to the general idea of the Energy Conservation, as we well know now, although the most of the conventional thermodynamics handbooks are still ascribing to Carnot an alleged "full ignorance in the field of the First Basic Law."

Moreover, Prof. Dr. H. von Helmholtz definitely attempts to employ the full mechanical analogy in considering the heat processes. And this is duly underlined in the book by P. B. Freuchen, up to mentioning the **clearly dialectic Third Basic Law of Mechanics**, as formulated by Hon. Sir Isaac Newton—concerning the **action and reaction.** Remarkably, the latter important point hasn't been dwelt on in the conventional—**Equilibrium**—thermodynamics till nowadays, therefore it is still remaining fully unclear—the **Equilibrium of What, or between What and What** is 'so duly described' by the latter scientific discipline.

A very interesting topic, isn't it? But now let us first continue to follow P. B. Freuchen in his description of the salient work by Rudolf Clausius.

[Note 14]: Here Peter Boas touches an extremely interesting and important point: The availability of an equation of state for the true realistic systems beyond the ideal gas gadget.

That both Rudolf Clausius and Lord Kelvin were busy with deriving such an equation of state is well known. That the both prominent colleagues could never bring this research project to a successful result is very well known as well. Every student properly attending the physical chemistry course knows well that another outstanding colleague, J. D. van der Waals was also busy with trying to derive the equation of state for the realistic gas and that he could achieve significant progress in this research direction. More attentive students know well that in effect there are lots of different forms of the realistic equation of state [14.1].

Hence the true poser would sound as follows: Is it possible to derive a fully general equation of state equally applicable to all the aggregate states of the matter? The correct answer was/is/will be throughout positive, but it was not known to Peter Boas and until this day remains unknown, as we shall learn from the next chapter of this book, where we shall get acquainted with the life and work by a contemporary, colleague and close neighbor of Peter Boas—a Swedish school teacher, Nils Engelbrektsson (see Chapter 2).

Reference to Note 14

14.1 J. Richard Elliott, Carl T. Lira (**2012**): *Introductory Chemical Engineering Thermodynamics.* Prentice Hall, Upper Saddle River, N.J., USA.

[Note 15]: Highlighed above is one of the most important parts of the Peter Boas' book at our hands. Here he illustrates the basic principle of any realistic natural/artificial process: The pertinent driving forces should not only ensure the achievement of the desired aim, but—first and foremost of all—help overcome the ubiquitous hindrances/obstacles on the way to the desired aim, and this is exactly what all of us know very well from our everyday lives.

[Note 16]: The 'revolution in physics' during the end of the XIX-th and the beginning of the XX-th century, while rightfully introducing the idea of the energy quantum, had in effect mixed up everything on the field of thermodynamics. The book of Peter, attentive, competent, and at the same time fully unbiassed contemporary and witness of this 'revolution', allows to analyze the very roots and mechanisms having lead to the thermodynamics' illness. To perform our analysis we would need to extend Peter's book by a number of external sources, in part simply unknown to Peter. But first, we would like to continue following Peter's fascinating narration.

[Note 17]: This conclusion of Clausius was competently and fruitfully criticized, and even corrective efforts have been undertaken. But presently all this activity remains largely unknown, which urges us to consider it in detail here, in the current chapter.

[Note 18]: Peter Boas rightly mentions here the passionate confession by James Swinburne, as we have highlighted above. Remarkably, it is **much too emotional**—just like those of the most of his predecessors, allies and opponents in the internationally wide "nasty squabble" around the entropy notion at the beginning of the XX-th century [18.1].

In effect, the 'energy loss', which is the only true physical basis of the entropy notion, is not quite equivalent to the 'energy waste'— being indeed not the 'direct payment' for achieving 'the desired aim', but 'paying the actual expenditure' for 'overcoming all the ubiquitous hindrances/obstacles' on the way to 'the desired aim'.

We shall continue this important discussion in Chapter 5 of this volume.

References to the Note 18

18.1 Paul J. Nahin (**2002**): *Oliver Heaviside: The Life, Work, and Times of an Electrical Genius of the Victorian Age.* Johns Hopkins University Press, Baltimore, Maryland, USA.

[Note 19]: *Here our sincere gratitude must be due to Peter Boas for his unambiguous emphasizing one of the **most unlogical** aspects of the so-called "equilibrium thermodynamics," as well as pointing to the original source of the latter! We shall now lay Peter's book aside for a while to discuss a very important point. This is just a succinct summary of what Peter would like to communicate his readership.*

Concerning the deficiency of the conventional 'equilibrium' thermodynamics

Indeed, the conventional train of thoughts behind the "equilibrium thermodynamics" ought to look like as follows:

First and foremost, we take the cyclic process suggested by N. L. S. Carnot and carefully think it over by trying to find its intrinsic physical sense. Although it should seemingly be clear that the Carnot's cycle is nothing more than an ideal gadget, we are looking for some basic physical principle(s) underlying all the realistic processes—and therefore not immediately/directly visible/perceivable.

Well done! But what should the above-mentioned basic principle(s) be looking like?

The answer is immediate and throughout clear:

The True Basement of Everything ought to be An Equilibrium!

Indeed, it is never immediately perceivable, but it is definitely the **heat death**, which everything is heading to, as ingeniously predicted by Rudolf Clausius!

And how Clausius could be capable of arriving at such a truly basic conclusion? By carefully analyzing the modalities of the Carnot's cycle! Long live the Carnot's Cycle and its hidden physical sense!

But at least at the time of suggesting his gadget Carnot was definitely sharing the standpoint of the so-called "substantial heat theory," ain't it? Yes, sure—then, down with this obsolete theory! The ingenious works of the Carnot's outstanding followers could properly reconcile all the conceptual gaps!

Well-well, the followers of Carnot were definitely experiencing serious mathematical problems with introducing and explaining the true physical sense of the new concept—of the entropy, ain't it? Sure, they have experienced them up to the moment that Ludwig Boltzmann had ingeniously suggested that the notions of entropy and probability are intrinsically connected to each other—and, moreover, he could even guess the true functional interrelationship between the both—long live the logarithm—long live the exponent—its inverse function—these both functions ought to govern all the realistic processes!

But what about the way of reaching the basic equilibrium, what about the factors being basically equilibrated? Equilibrium between What and what should this be? The basic equilibrium is in fact extremely difficult to achieve under the conventional conditions. The only possible way coming to our minds ought to be performing the process under study in an infinitely slow way!

OK, well, but what about the notion of time? In effect there ought to be no *real basic time* as well as no *real basic space*, see the relativity theory by Albert Einstein! We shall continue the detailed discussion of this very important theme in Chapters 3, 4, and 5 of the present monograph.

Does the above story look like physics? To our mind—of course, not—this is something like extremely emotional bargaining-higgling at some eastern bazaar. And in this book we shall come back to the serious problem of interrelationship between entropy and probability, to the serious problem of interrelationship between time and thermodynamics.

Hence, what we would like to stress already here should be the obvious fact that reaching any kind of *desired aim* requires first of all enough driving force. Secondly, there are always *pertinent hindrances/obstacles* on the way to the *desired aim*, and the former both ought to be neutralized, before the latter one could be achieved. Interestingly, the very neutralization of the obstacles/hindrances should be nothing more and nothing less than just arriving at the equilibrium—yes-yes, the equilibrium between our very striving and all the available opposition to our very striving. Reaching this basic equilibrium means that we could have successfully achieved our *desired aim*, this is just the very unique result of any kind of the realistic processes.

What is remaining ought to be the correct estimate of the energy balance. We know that the total energy is always conserved, while there are different kinds of energy capable of transformations among each other.

This energetic story is immediately explicable in terms of money amounts we have on our bank accounts, but surely in the non-accounting sense only:

We have debits and credits, but nothing comes here out of nothing, for we get credits—conventionally from our salaries—and produce debits—conventionally to ensure our sustainment. And we cannot spend more than we are possessed of at the time of spending (we are herewith leaving behind such purely banking/financial phenomena as bank loans and overdrafts, Internet banking, etc.).

Assume we need to buy a fridge. We consult our bank and realize that we do have money to pay its price. Very nice! Can we immediately buy it? Obviously, **not**! But why? Sure, for we first of all have to arrive at the store where they sell fridges, buy it, pick it up and finally bring our purchase home.

With this in mind, we have to surmount some external (and possibly even internal) obstacles. What are these? Surely, first we need to somehow get to the store which is located not in the nearest vicinity to our living place. Thus, we have to spend some money to pay for the public traffic or to buy gasoline for our own car, otherwise we cannot get to the store and pay the price of the fridge.

Second, even if we could surmount the latter obstacle and pay the price to cashier, we realize that the fridge is too awkward/heavy to bring it home with our car or on our backs. This is why, we have to pay for the delivery to our address, for the fridge we have just gotten hold of. Thus, we get an additional hindrance to surmount.

To sum up, all the above are clearly the **external** hindrances to be surmounted, but there might be internal problems as well—for example, we know about all the external obstacles, say from the Internet—so we perform the preliminary calculation of the total debit to be perpetrated and realize that our present asset doesn't allow such an immediate debit. But we need the fridge urgently, so that we have to solve the problem by somehow loaning the required sum. Hence, all the latter difficulties ought to result in our **internal** obstacles to be overcome.

To sum up, this same story ought to happen in all the realistic natural/artificial processes—debits/credits could be translated as transforming the available energy to create some adequate driving force. 'Adequate' would mean here 'enough to overcome all the hindrances (let us denote these 'entropy') and achieve the desired result'. Further, the driving force to be 'adequate' ought to be correspondent with available asset. The 'asset' would physically mean the amount of 'potential energy', which should be enough to be transformed into the proper amount of the 'kinetic energy'—or the 'livening force', if we put it in the archaic terms.

With all this in mind, the actual story about the 'reversible' and 'irreversible' processes ought to be nothing more than just a skillfully constructed speculation having rather few interconnections with actual natural science. Exactly this same holds for the so-called 'adiabatic processes', which are in effect to be proceeding without any entropy change, which is nothing more than a brusque, abrasive idealization, resulting from the difficulties with grasping the actual sense of the entropy notion.

Noteworthy, the above-mentioned speculations/idealizations had engendered comprehensive professional efforts to build up such fields of knowledge as 'non-equilibrium thermodynamics', 'thermodynamics of irreversible processes', etc. and the resulting fields have already been transformed into the well established, useful, rather widely applicable theories. Is there really any serious scope of some further work?

Our immediate answer: Sure, there is!!!

Peter Boas had considered the foundations of thermodynamics in a very succinct, but nonetheless truly coherent, revealing and stimulating way, by considering the work by colleagues that all their lifetime were skillfully working on the theme, but had still remained in the historical shadow.

Indeed, the numerous doubts and objections concerning the 2nd basic law, the notorious production of the further basic laws, like the Nernst's law, the names of such true experts as the Danish physical chemist Julius Thomsen and August Friedrich Horstmann, the analysis of Helmholtz's and Gibbs' contributions to building-up a rational thermodynamics.

We have decided to attach here a number of fragments that, to our mind, ought to widen the scope suggested by the Peter Boas' seminal book.

But, first of all, let us continue to follow Peter Boas' gripping narration.

[Note 20]: Many thanks to Peter Boas for his clearly pointing out the conceptual blind corner produced by prestidigitating with the '(ir)reversibility' and 'adiabaticity' misnomers!

[Note 21]: At this very point, our sincere gratitude must again be due to Peter Boas for his citing the German translation of the renowned course book by Орест Даниилович Хвольсон (1852–1934) (Prof. Dr. Orest Daniilovich Chwolson), an outstanding Russian/Soviet physicist. After graduating the St. Petersburg university he started his career as an ordinary school teacher, like Peter Boas, but elsewhere, namely in St. Petersburg, in Russia, and then had yet moved to the position of the lecturer at his Alma Mater.

Remarkably, Prof. Chwolson's course book in physics was— and definitely still remains—a kind of encyclopedia, where the 3rd volume is to 100% devoted to the foundations and applications of thermodynamics. Prof. Chwolson was in fact a skillful and passionate attorney of the, at the time, novel, revolutionary stream in physics, represented by meanwhile famous colleagues, like Ludwig Boltzmann, Max Planck, Albert Einstein, and others. Meanwhile, along with this his book allows to reconstruct all the actual details of the then trains of thoughts.

When reading the book by Prof. Chwolson it is of interest and use to follow the true logics of building up the conventional 'equilibrium thermodynamics', the detailed interpretation of findings by Carnot, Clapeyron, Clausius, Thomson, et al.

Specifically, it is of definite interest to compare the books by Chwolson and those by Peter Boas. To sum up, Peter Boas' book at hand is much more sincere, objective and unbiased, whereas some very important, even throughout key details still lack in it. Prof. Chwolson's book is truly a physical encyclopedia of the then time, a work looking for the very fundamental logics of physics.

We shall encounter Prof. Chwolson once more in Chapter 3 of the volume at hand for a more detailed analysis of his standpoint.

Here we would just like to note that all the story about the reversible, irreversible and adiabatic processes, as discussed by Peter Boas above, was in effect triggered by purely mathematical difficulties experienced for the first time by Rudolf Clausius himself.

21.1 Reversibility–irreversibility prestidigitation: is it solely about how to successfully bite one's own tail?

The point is that R. Clausius employed the ingenious cyclic gadget by N. L. S. Carnot to derive the handy mathematical expression for the Second Basic Law and introduced the novel variable, that is the function of entropy. The next step ought to be finding the true physical sense of the latter—and thus of the Second Basic Law. Meanwhile, to mathematically extend this inference to realistic processes ought to be a non-trivial problem. Prof. Chwolson describes the actual situation as it was looking like at his time in the following way (we present here the translation of his original Russian text):

"For reversible circular processes the integral $\int \frac{dQ}{T}$ definitely ought to be equal to zero, but there is no general and strict proof that for irreversible processes this integral always ought to be negative. There are many literature references on this theme, from which we would like to mention here an interesting work by Carvallo (1899) [21.1]. He could strictly prove that the values of the integral $\int \frac{dQ}{T}$ may never be greater than zero. In the numerous treatises on the topic one might even fetch variants of the proof that this integral must have values less than zero, but all these proves might not be considered persuasive. In fact, we are dealing here with a particular case of a much more general principle, according to which in all the thermodynamic formulas, where a choice is to be made between the definite equality and inequality signs, one must choose either the equality sign, when speaking of reversible processes, or the inequality sign, if mentioning the irreversible processes.

Above we have already seen that during irreversible processes the parameters describing the body's states become indefinite. If, e.g., the temperatures of the heater or chiller are different from the temperature of the body experiencing a circular process, then during heating or cooling the latter body its temperature does not have a definite value. This is why, T in the integral expression involved might never adequately play the role of the working body's temperature, as one might erroneously think, and instead—T ought to be quite the temperature of the heat sources, which we assume to be inexhaustible. It is easy to show using a number of examples that for the irreversible processes the value of the integral in question is always less than that for the reversible ones."

Finally, the five points listed above by Peter Boas follow, as given by Prof. Chwolson.

The honest work by Prof. Dr. Chwolson helps us immensely. Indeed, the latter report clearly shows no reasonable physical grounds for the conventional deliberations on the theme 'reversible' versus 'irreversible' processes. Meanwhile, there are strictly speaking no perfectly reversible processes all over the world—because virtually all the natural/artificial processes are intrinsically irreversible. Moreover—based upon the latter fact—'the course of time' all of us perceive during our lives is definitely unidirectional.

21.2 The 'timelessness' of the conventional thermodynamics: outline of the conventional equilibrium thermodynamics

This way, our predecessors have encountered a serious poser. Specifically, on the one hand, everything given to us in our sensations is unidirectional in time. The relevant processes are therefore intrinsically irreversible. On the other hand, we face tremendous conceptual difficulties mentioned above. What would be hence the best way to describe this?

Two variants come immediately to our mind: (A) First of all let us carefully try to think over virtually everything we have already at hand, then the solution would come as the result —this is but looking like a much too lengthy story. (B) Otherwise, the folk wisdom conveys the following method: 'A smart guy would never climb mountains; the smartest way would readily bypass the latter'.

Now, in considering our plans we still have to set up, what do we have already at our hands?

On the one hand, we have the never-known-to-fail irreversibility of all the realistic processes. Still, on the other hand, all the realistic matter is consistent of tiny permanently moving mechanical particles, each of which is obeying the Hamiltonian equation of motion. We know well that the latter is symmetric with respect to the optional signing of the time involved into such equations.

Time is thus our enemy! We have to kill, to dismiss it. Consequently, everything pretending to be a fundamental phenomenon ought to be **timeless** in fact. The fundamental phenomena should therefore obey the relativity rules. Vice versa: It is the Relativity that renders all the pertinent events truly **timeless**.

In coming back to thermodynamics, we immediately note that there is an intrinsic parallel between the time and entropy: The latter both are **always increasing** in the realistic processes.

Then, a truly bright idea would be to reveal the entropy as an intrinsic source of the time's directionality. That is, the entropy constitutes the time's fundamental roots. *This ought to be the fundamental explanation of the irreversibility!*

Therefore, the entire fundamental physics is already contained in the Carnot's cycle. It is clear that such a cyclic process is finally arriving at the same point, from which it has started. *To sum up, the* **fundamental** *significance of such a cycle is that there is equilibrium at its every point.*

Do not ask us, please, **between what and what** *ought to be those equilibria; they are just* **equilibria as they are**. To ensure the system's eternal persistence at such an eternal equilibrium, the process of Carnot ought to be **quasistatic,** that is, it must be a perfect approximation of **a truly timeless process**. *It is just this way that it forms the conceptual* **basis of the entropy.**

The Clausius expression $\delta S = \delta Q/T$ is therefore true only for the Carnot cycle as a whole. Meanwhile, the very fact that the entropy changes permanently does not allow the determination of its absolute value, but only of its differences: $\delta S \geq \delta Q/T$. This is just the general expression of the entropy growth, that is, the intrinsic sense of the second basic law.

However, how to deal with **entropy** in the actual theoretical inferences? What should be the correct and **practical mathematical expression** for it? Here comes the ingenious guess of L. Boltzmann: $S = k \ln W$. Max Planck has just taken over this **true gem** as it is. Nobody could manage a mathematically formal inference of the latter, with considering in detail the actual physical sense of the functions participating in that formula. This problem was remaining unsolved up to the advent of Dr. Georg(e) Augustus Linhart (see Chapter 4 of the present monograph).

Have all of Linhart's predecessors been unskillful theorists to such an extent? Of course, not at all! They were absolutely sure that there is no need for any inferences, for the entropy–probability link is a fundamental one. What was it then? A self-hypnosis? We shall revert to this important theme in Chapters 3–5 of the monograph at hand.

To sum up, the actual achievement of L. Boltzmann, M. Planck, their comrade-in-arms and followers in the field of thermodynamics was in fact to skillfully 'kill' the very **time notion**—to render the conventional thermodynamics 'timeless'.

Sure, *timeless* might solely be masterpieces, the actual values of which we ought to check very carefully in the course of time [*sic*].

However, no serious problem ought to arise therefrom for the respective 'broachers'!

Indeed: if you cannot grasp, say, **why** Michelangelo's masterpieces are beautiful, then it is definitely your own problem—but in no way the problem of Michelangelo and/or his masterpieces. This seems to be just the true zest behind the known quote by Albert Einstein, appearing in his autobiographical notes:

"A theory ought to be the more impressive the greater the simplicity of its premises is, the more different kinds of things it relates, and the more extended is its area of applicability. Therefrom stems the deep impression that classical thermodynamics made upon me. It is the only physical theory of universal content concerning which I am convinced that within the framework of the applicability of its basic concepts, it will never be overthrown."

(Paul Arthur Schilpp (**2001**): *Albert Einstein: Philosopher-Scientist.*
MJF Books: New York, USA. Page 49)

The above story presents the quintessence of the resulting 'Thermodynamics' obtainable from the hundreds (if not thousands!) of standard thermodynamics textbook published during the last 150 years in all the languages available upon Earth. A good poser: Is the above-sketched equilibrium thermodynamics really a physical theory, or, instead, a conceptual corpse? Whatever the answer, the next poser would come as follows: Is the thermodynamics as it is applicable to analyzing realistic processes at all?

21.3 Thermodynamics of realistic irreversible processes: how to formulate it?

To sum up, our achievements for the present consist in revealing the purely subjective nature of the time notion, its intrinsic dependence on such fundamental variable as entropy and—the last but not the least—on the probability, being largely the source of literally everything upon the world. This diminishes the very time notion to being just a useful parameter during our mathematical exercises in solving the differential equations of motion.

Still, before solving any equation, we have to cast it in a physically correct form. To perform the latter task, we first have to clarify the driving forces of the process under study.

A good poser arises after all our above deliberations: How should we choose the plausible descriptors? What ought to be the actual logics of our inferences?

Howbeit, in this connection it is truly important to recall the statement on this theme by an outstanding German theoretical physicist Richard Becker (1887–1955) in his marvelous book [21.2].

Below we present our authorized translation of the original German text.

21.4   Explanations by Prof. Dr. Richard Becker: their significance

"*Chapter 7. Thermodynamics of Irreversible Processes*

87. Increase of entropy by irreversible processes

Classical thermodynamics deals only with reversible changes. Above in this book we have demonstrated the meaning of this restriction on the example of a simple Carnot engine. Reversible processes have to run "infinitely slowly." Meanwhile, any real process occurs with a finite velocity and, therefore, is necessarily irreversible.

For instance, an exchange of heat between two bodies A and B is possible only if A is warmer than B or a piston between two gas containers moves only if the pressure in the two containers differs. In both cases, the actual process is associated with an increase of entropy. Thermodynamics deals only with reversible processes, and this is quite strange. Indeed, reversible processes conserve the entropy of a closed system. Meanwhile, the entropy increases in all the actual realistic processes.

Howbeit, for the subject dealt with in the following section we have to emphasize a new and characteristic point of view. We have seen that both the increase of entropy and irreversibility always occur simultaneously. Therefore, the sole statement that one of the phenomena causes the other one would bring nothing physically new. Nevertheless, such a statement does sometimes introduce **a more lively and convincing formulation of the basic laws**.

In our case, for instance, we might say, either "the entropy ought to increase because an irreversible process is taking place" or, vice versa, "the irreversible process ought to occur because it is associated with an increase of entropy." In the earlier days, the first formulation was more popular. Nowadays the second point of view is widespread. It is the tendency of entropy to increase that ought to cause the irreversible process.

Sometimes one even speaks of a "force" due to increase of entropy, which pushes the irreversible change. Assuming this kind of picture, we expect that the irreversible change proceeds faster if the associated increase of entropy is larger. This leads to the suggestion that the velocity of the irreversible process is in direct connection with the corresponding change of entropy. ..."

Herewith Prof. Dr. Becker had helped us to reconstruct the actual logic of the relevant train of thoughts and could thus clearly point out, where this logic is incomplete, where it even clearly comes out to be deficient.

Indeed, playing with the relationship between the notions of entropy and irreversibility it is easy to lose the true logical thread by *'putting upside down' the physical sense of the entropy notion.* The approach is straightforward. We ought to use the tried-and-true cause-and-effect relationship—in that we just state:

After adopting that the entropy increase causes the irreversible process, one might even invoke the notion of the 'entropic forces'.

Furthermore, as the irreversible processes are actually the only realistic ones, the 'entropic forces' ought to be throughout realistic as well. Meanwhile, this is nothing more than just the old idea by Prof. Dr. Peter Guthrie Tait (1831–1901), who was quite busy with trying to refine the ideas of Rudolf Clausius.

Meanwhile, what ought to be the actual physical sense of these 'entropic forces'? Interestingly, James Clerk Maxwell (1831–1879) has enthusiastically adopted and then clearly omitted the suggestion by Prof. Dr. Tait [21.3]. The interested readership might follow the most recent detailed discussion on the latter theme elsewhere [21.4].

Meanwhile, Prof. Dr. Becker points out as well, that such a train of thoughts does not introduce any new physical idea. Instead, this ought to be practically nothing more than just facilitating mathematical description of some realistic events. As to the "***more lively and convincing formulation of the basic laws***": As a result, for the present we are possessed of some several hundreds of these formulations, but such a *liveliness* is somehow not quite convincing [21.5].

To our mind, it ought to be much more productive for the whole thermodynamics field if we would carefully re-consider the logics of the conventional trains of thoughts and try to look at the practical strengths/weaknesses of thermodynamics, for example, as the most recent authors do [21.6, 21.7].

Interestingly, it is already long ago that the renowned Norwegian physicist Lars Onsager (1903–1976) has re-furbished the 'entropy-irreversibility clinch' by declaring the entropy to be in effect the 'actual potential energy'. Remarkably, this is just in line with the logics described by Prof. Becker: Sure, if there might in principle be 'entropic forces', then it is throughout straightforward to speak of the entropy as the 'actual potential energy'.

Well, how could he come to this marvelous result? 15 years had passed after Peter's book; Peter's appeal went unnoticed even in Scandinavia. Sure, the story is of clear relevance to our main theme here!

First, the actual irreversibility ought to result from the action of physically definite forces originating from here upon Earth, causing/promoting the pertinent observable dynamical processes/flows. So, what are these driving forces? Lars Onsager, before deriving the general form of his famous reciprocal relations [21.8, 21.9], gives a clear and definite answer (We cite this below, while the highlighting is ours):

> "*Among the relations to be derived many have been proposed before, but some will be new. An important group among these relations can be summarized in a variation-principle, which is nothing but an extension of Lord Rayleigh's "principle of the least dissipation of energy"; we shall retain the name for the extended principle. According to this theorem* **the rate of increase of the entropy plays the role of a potential**."

In other words, the driving force for the irreversible processes under study ought to be of entropic nature. Lars Onsager seems to have formulated the latter idea by himself—and for the first time ever. A careful analysis of all the works cited in his original papers supports drawing such a conclusion.

To our regret, in his papers Lars Onsager is not describing in detail, how he could come to exactly such an idea of all others possible. In this regard, he refers us to the work by Hon. John William Strutt, Baron Rayleigh.

Meanwhile, in his preliminary report, Baron Rayleigh clearly states [21.10] that his theory is about rather small deviations from some mechanical equilibrium, and then he uses the results of William Thomson (Lord Kelvin) and Peter Guthrie Tait on the "*slightly disturbed equilibrium*" [21.11].

Then, Baron Rayleigh writes the relevant mathematical expressions for the kinetic and potential energies, thus assuming nothing more than just the conservative mechanical systems. However, he goes a huge step forward in his famous book on the theory of sound.

Indeed, he introduces the dissipation function, F, aside from the kinetic energy, T, and the potential energy, V, with the function F being physically correspondent (we cite, [21.12, 21.13]):

"*... to another group of forces, whose existence it is often advantageous to recognize specially, namely those arising from friction or viscosity. ...*

Although in an important class of cases the effects of viscosity are represented by the function F, the question remains open, whether such a method of representation is applicable in all cases. I think it probable that it is so; but it is evident that we cannot expect to prove any general property of viscous forces in the absence of a strict definition, which will enable us to determine with certainty, what forces are viscous and what are not. In some cases, considerations of symmetry are sufficient to shew that the retarding forces may be represented as derived from a dissipation function. At any rate, whenever the retarding forces are proportional to the absolute or relative velocities of the parts of the system, we shall have equations of motion on the form ...

... We may now introduce the condition, that the motion takes place in the immediate neighborhood of a configuration of thoroughly stable equilibrium. Then, T and F ought to be homogeneous quadratic functions of the velocities with coefficients, which are to be treated as constant, and V is a similar function of the coordinates themselves, provided that (as we supposed to be the case) the origin of each coordinate is taken to correspond with the configuration of equilibrium. ..."

This is exactly the point where the development of Lars Onsager's theory starts! Meanwhile, of considerable interest for our present discussion ought to be the exact information about what the actual reasons led Lars Onsager to start developing his theory were. He reported all the reasons in question to a conference of the Scandinavian Society Natural Scientists in 1929, in his lecture entitled: "*Simultaneous Irreversible Processes.*" The Norwegian-to-English translation of the introduction to that lecture has been re-published relatively recently [21.14]. Indeed, that time Lars Onsager has clearly stated, that (we cite):

(a) *Several authors (including such prominent authors as Lord Kelvin and Hermann von Helmholtz) have derived certain relations among two or more irreversible transport processes taking place simultaneously (e.g., heat conduction, electrical conduction and diffusion) using apparently thermodynamic reasoning. A number of other authors have used somewhat different approaches (W. Nernst, E. D. Eastman).*

(b) *Common for all the earlier theories is a consideration of a cyclic process we cannot carry out reversibly. It is therefore impossible to base such theories on the two laws of thermodynamics. On this ground, Ludwig Boltzmann has pointed out that by purely thermodynamic reasoning one cannot deduce more than certain inequalities.*

(c) *The proper rigorous derivation of the reciprocal relations could be possible by adopting the statistical standpoint like that developed in Einstein's theory of fluctuations. And it is necessary to make only one additional assumption, viz. the past and the future are on the same footing, in the sense, that any dynamically possible trajectory might also be traversed in the opposite direction.*"

[sic] **With all this in mind, Lars Onsager took the results by Lord Rayleigh and boldly threw away the potential energy by substituting it with the "rate of entropy increase."**

As we have already seen above, it was the prominent German theoretical physicist Richard Becker, who had presented the only detailed analysis of Lars Onsager's logical train of thought. Furthermore, Lars Onsager was trying to solve the problem within the frame of the revolutionary ideas by L. Boltzmann, M. Planck and their comrades-in-arms. *Was that the only valid/possible train of thoughts?*

It is a truly good poser, but now we would like to revert to it in Chapter 5 here.

Howbeit, when using the ingenious ideas by Hon. Baron Rayleigh, we are still in the infinitesimal vicinity of equilibrium, and it is sheer impossible to study both the physical reasons and the actual stages of approaching the equilibrium in such a way. This is why it surely seems to be much too early to speak about any clear-cut and final formulation of the "non-equilibrium thermodynamics," when sticking to L. Onsager's seminal results only.

Indeed, the mathematical work in this direction was/is going on quite steadily and successfully [21.15, 21.16]. There is but definitely much more to the story, so we shall revert to our discussion on this important theme in Chapter 5 here.

Obviously, the huge benefit provided by Lars Onsager's groundbreaking work consists anyway in clearly demonstrating the insufficient formulation of the conventional, so-called "equilibrium thermodynamics" and successfully invoking stimuli to look for its proper physical re-formulation. Anyway, one of such stimuli ought to come from the conventional logical train of thoughts, as clearly shown by Richard Becker.

Bearing the entire story in mind, we recognize that Lars Onsager made a skillful use of statistical thermodynamics. Using the latter enables us to dive into the microscopic (initially hidden) factors underlying the macroscopic observations. Therefore, it ought to be of crucial importance to re-analyze in detail the relevance and effectiveness of statistical thermodynamics as our working instrument, as well as the proper ways of its usage. We revert to this immensely important topic in Chapter 5 here.

Meanwhile, there is another problem of crucial importance to be pointed out before dealing with the statistical thermodynamics, namely: it is the **timelessness** of the conventional thermodynamics. On the one hand, Prof. Dr. Onsager is discussing the '*Rate of Entropy Increase*', and therefore incorporating the notion of time into his theoretical development, although only implicitly. On the other hand, he is '*adopting the statistical standpoint like that developed in Einstein's theory of fluctuations*', i.e., his deliberations are blindly based upon the famous Boltzmann–Planck formula. We employ the term '*blindly*', for Albert Einstein was never analyzing the possible conceptual sources of this formula in detail. That was Dr. George Augustus Linhart, who seems to be the only colleague undertaken and successfully accomplished the proper analysis (see Chapter 4 here for the pertinent details). Finally, to Chapter 3 here we shall postpone discussing the actual contributions by Albert Einstein to all the fields of our current interest.

To sum up here, we would just like to note the two crucial deficiency points of the conventional "*equilibrium thermodynamics*": (A) its apparent timelessness accompanied by some mystic interconnection between the time and the entropy, and (B) the notorious interrelationship between The entropy and the probability. We shall come back to the detailed discussion about all these fundamental points in the present volume.

Meanwhile, here comes the very point where we would like to turn our attention back to the book by Peter Boas Freuchen. Indeed, the Entropy concept remains anyway in the tightest relationship to the basic laws of thermodynamics, especially to its Second Basic Law. Peter Boas was carefully considering this crucial topic, so let us continue following his narration now.

References to the Note 21

21.1   E. Carvallo (**1899**) : Sur les cycles irréversibles et le théorème de Clausius. *Journal de Physique Théorique Appliquée*, v. 8, pp. 161–165.

21.2   Richard Becker (**1966**): *Wärmetheorie*. Springer-Verlag, Berlin, Heidelberg, New York; Its English translation: *Theory of Heat*. Springer-Verlag: Berlin, Heidelberg, New York, 1967.

21.3   J. C. Maxwell (**2001**): *Theory of Heat.* Dover Books: Mineola, New York USA.

21.4   W. B. Jensen (**2013**): George Downing Liveing and the early history of chemical thermodynamics. *Bull. Hist. Chem.*, v. 38, pp. 37–51.

21.5   D. Daudrich (**2002**): "*Der Zweite Hauptsatz der Thermodynamik und seine Macht.*" First-Minute-Taschenbuch Verlag, Emsdetten, Germany, ISBN 3-932-805-33-X: http://www.storyal.de/story2005/hauptsatz.htm.

21.6   Iosif Lvov (**2014**): "*Thermodynamics: Logical Analysis. What is Energy?*" LAP Lambert Academic Publishing, Saarbrücken, Germany.

21.7   Yuri Kornyushin (**2015**): *"Thermodynamic Approach in the Theory of Materials."* LAP Lambert Academic Publishing, Saarbrücken, Germany.

21.8   L. Onsager: Reciprocal relations in irreversible processes. I. *Phys. Rev.* v. 37, pp. 405-424, 1931.

21.9   L. Onsager: Reciprocal relations in irreversible processes. II. *Phys. Rev.* v. 38, pp. 2265–2279, 1931.

21.10 John William Strutt (Baron Rayleigh), 'Some general theorems relating to vibrations' , *Proc. Math. Soc. London*, v. 4, pp. 357–368, 1873.

21.11 Sir William Thomson (Lord Kelvin), Tait P. G., *'Treatise on Natural Philosophy'*. Volume I, Cambridge University Press, Cambridge, UK. 1867 (§§ 337–338).

21.12 John William Strutt (Baron Rayleigh): *The Theory of Sound. Volume I.* McMillan & Co., London, UK, 1877 (§§ 80–89).

21.13 John William Strutt (Baron Rayleigh): *The Theory of Sound. Volume I.* 2nd Edition. Revised and Enlarged. McMillan & Co., London, New York, 1894 (§§ 80–89).

21.14 *'The Collected Works of Lars Onsager (with Commentary)'*, edited by Hemmer P. C., Holder H., Ratkje S. K., World Scientific, Singapore, New Jersey, London, Hong Kong, 1996.

21.15 H. Haken: *Synergetik. Eine Einführung*, Springer-Verlag, Berlin, Heidelberg, New York, London, Paris, Tokyo, Hong Kong, 1990.

21.16 G. Jeschke: *Mathematik der Selbstorganisation*, Verlag Harri Deutsch, Frankfurt am Main, Germany, 2009.

[Note 22]: As we have already discussed earlier in this chapter, *it is definitely not* the Second Basic Law itself that teaches us to introduce such a kind of classification. Quite contrariwise, these are the mathematical difficulties at trying to rationalize how to apply the Second Basic Law to the realistic natural/artificial processes. It is very important to point out here once more that only the ingenious gadget of N. L. S. Carnot ought to be the actual basis for the concept of 'reversible processes', there is no realistic natural-scientific background behind the latter. Indeed, the only realistic physical basis under Carnot cycle's 'reversibility' is its cyclic nature. Indeed, Carnot's cycle does help rationalize the existence, the importance and the implications of the Second Basic Law, but there is **no more fundamental physics** behind it to look for.

Therefore, the mysterious *intrinsically reversible* processes ought to be obsolete in our discussions, for they have played their roles to the end, and they definitely belong to the legacy of natural sciences (see Supplementary Note 21).

Most probably, Peter felt impelled to tell us about the conventional concepts. Anyway, this is definitely passable, for he was by far not a herald of the latter, but instead, their attentive critical referee. This is why, reading Peter's work rather stimulates re-evaluation than immediate acceptance of the conventional thermodynamics.

[Note 23]: Here Peter Boas is just reiterating the emotional outcry by Rudolf Clausius.

Both the outcry involved and its whole pre-history definitely belong to the respectable legacy of the natural sciences. Remarkably, while carrying out his everyday research work Rudolf Clausius could fully unexpectedly arrive at some sudden discovery of the over-all importance and significance, namely, at the Second Basic Law. Peter had nicely depicted his working situation, as we could see above. Still, to our sincere regret, Clausius himself had not enough time to put his study of the Second Basic Law to the logical end.

Well, the possible explanation for the survival of Clausius' anyway quite awkward conclusion about the Energy and Entropy of the whole world might be that the immediate followers of R. Clausius, especially his contemporary colleagues, were not quite critical as concerns his actual bequest.

Meanwhile, the main point here ought to be the perfect applicability of the Second Basic Law to ALL the realistic natural/artificial processes. Interestingly, R. Clausius could himself launch the correct train of thoughts to rationalize his important discovery, in that he had introduced the notions of '*positive*' and '*negative*' processes together with their '*compensation*'. Fortunately, one of the Clausius' apprentices could still manage bringing this train of thoughts to its true logical station. In his book, Peter had told us about the further work on the Second Basic Law, so let us first continue to follow Peter's narration.

[Note 24]: The above fragment of Peter Boas' book clearly demonstrates that M. Planck was actively thinking over in detail the foundations of thermodynamics. However, the result of those efforts was rather strange indeed, as one might see.

Apparently, while pointing out some 'blind-corner' situations, he was not busy with analyzing the modalities of energetics. Instead, M. Planck has clearly formulated his standpoint in his publications—we shall revert to this important topic in Chapter 5 here.

Meanwhile, the adepts of the so-to-speak 'dissident energetics' (dissident was the latter solely with respect to Boltzmann–Planck's standpoint!) were trying to rationalize the basic laws of thermodynamics, like L. Boltzmann and M. Planck, but using different conceptual premises.

Peter Boas has devoted a significant portion of his book to considering the actual slant on the Second Basic Law at the then time. Indeed, his definite task was not to deliver a comprehensive review on the theme. Meanwhile, his text does present a clear hint to the truly great diversity of the professional stances. It is even possible to classify the latter into two general paradigms leading to no immediate consensus, at least at Peter's time. We shall revert to looking at the both paradigms in detail at the end of this chapter.

It is with this in mind, that we have decided to carry out the detailed review in the field—to pay the singularly clear and intelligible tribute to Peter's unique effort. It is clear that Max Planck was one of the central authorities in the thermodynamic field at the time of Peter's book publication. Hence, we feel obliged to re-analyze the actual value of Planck's contribution to the general thermodynamics and try to draw a more detailed picture of the field. We postpone this discussion to Chapter 5 here.

[Note 25]: [*sic*] Prof. Mohr's deliberations touch here the actual nature of **molecular dynamics**.

[Note 26]: Peter refers here to an extremely interesting book published by a (presently widely unknown) German engineer, Dr. H. Hort [26.1]. This publication was definitely aiming at reconciling the 'nasty squabble' concerning the actual physical sense of the entropy notion (see Note 18 to the present chapter, Ref. [18.1]). The book by H. Hort is truly commendable, for a pragmatist has written it in looking for the answer to the poser: *If they are so stubbornly advertising something, so what is then the actual practical value of what they are touting?* Generally, such a standpoint ought to be very useful and especially helpful, when trying to realize, so what is this 'damned' Entropy, after all. Moreover, the book by Dr. H. Hort [26.1] contains a clear-cut analysis of the entire standpoint available at the time of its publication.

Reference to the Note 26

26.1 Dr. Phil. H. Hort, Diplom-Ingenieur in Dortmund (1910): *Der Entropiesatz, oder der zweite Hauptsatz der mechanischen Wärmetheorie.* Springer-Verlag, Berlin, Germany.

[Note 27]: Peter refers here to a remarkable publication by an experimental physicist, Prof. Dr. Johannes Wilhelm Classen (1864– 1928). He was studying mathematics and natural sciences at the Universities of Breslau and Jena. In 1889, in Jena, he has defended his PhD thesis devoted to the specific heat of liquid sulfur. Prof. Dr. Classen had started his professional activity at the State Physical Laboratory in Hamburg, and his working task was connected with organizing the Office for Electrical Testing within the frame of this State Laboratory. He was then both heading that Office and conducting an intensive pedagogical work: He was holding lectures and publishing handbooks, mainly about the mathematical optics, as well as on the theories of electricity and magnetism.

Besides, Prof. Dr. Classen was a sincerely religious person actively engaged in the Evangelical Lutheran church community in Hamburg. This way he could directly act out his interest in reconciling the conceptual philosophic bridges between the Religion and the Natural Sciences. He was actively holding the relevant seminars and publishing his ideas. His book cited by Peter belongs just to the latter series. This is why, it is anyway of separate interest for us.

Specifically, in his book [27.1] he states from the beginning on the following problem (below we present our authorized English translation).

"Nowadays the energy conservation law ought to be well known to everybody, who is interested in scientific research, and this law is also being understood correctly.*

---

*In any closed system, the total sum of the energies in connection with any process remains unchanged.

That the physics does reveal the second law of the equal generality and the same comprehensive validity, in addition to this well-recognized first basic law of thermodynamics, is much less known. This second law is not just the derivative or amendment of that first law, but it conveys an entirely new matter of fact and leads to a large number of inferences that cannot be immediately derivable based upon the first law alone.

Many colleagues could have heard of this second law, they might express it and know how to apply it correctly. Still, there is a truly stubborn sensation that this second law is surrounded by a certain darkness; that it is very hard to duly grasp its general justification, and in particular, to overview the significance of his conclusions clearly enough.

Actually, the modern physics is ascribing an exceedingly general, and, one may even say, an absolute validity to the second law, hence we would like to discuss its origin and its justification in somewhat more detail below."

Remarkably, Prof. Classen immediately reveals the three (!!!) groups of the second law formulation. He states that in this small row every next formulation is '*clarifying and deepening the previous one*'.

The very first formulation in the row by Prof. Classen is due to R. Clausius and introduces the notorious 'Heat Death' of the 'whole Universe'. Nowadays we know already very well that Clausius has in fact uttered a very strange statement, which does contain the rational fundamental nucleus, but its 'wrapping-up' has nothing to do with the actual physics. Prof. Classen concludes: '*This formulation is quite mystical and hence sheer unsatisfactory*'.

The second formulation detected by Prof. Classen is the mathematical introduction to the notion of entropy. R. Clausius had pioneered this, but there were several other colleagues as well, who could come to the like conclusion independently. Meanwhile, Prof. Classen has noted here that '*This way, the physicists are hiding themselves behind the mathematical dens and rocks of the overcomplicated mathematical signs, which are never accessible to everybody*'. We might only add that here the mathematical wraps could perfectly hide the basic physics behind, to recall the relevant criticism expressed by Prof. Mohr.

The third formulation is due to L. Boltzmann: '*in any process, the final state is much more probable/likely/plausible than the initial one'*. Prof. Classen is very enthusiastic concerning this statement. He notes that *'if it might be possible to estimate the probability of the physical state in the pertinent manner, this would pave the way to the new levels of understanding'*. Here we agree with Prof. Classen to 100%! Ludwig Boltzmann had no more time to accomplish solving this problem, to our sincere regret. This problem could be solved by Dr. G. A. Linhart (see Chapter 4 for the details), but the solution had come in about a dozen of years after the publication by Prof. Classen.

Meanwhile, it is clear that 'paving the way' 'in the pertinent manner' requires much more work to accomplish. To our regret, we must note here that, in connection to this fully and generally valid third formulation, the negative trend could take over the initiative.

Indeed, the 'dens and rocks' of the sophisticated probability theory could successfully hide and even strangulate any attempts to analyze the actual physical sense of this probabilistic statement in detail.

We sincerely hope that our present monograph at hand might be helpful in trying to clarify the then general situation.

Reference to the Note 27

27.1   Johannes Wilhelm Classen (1910): *Das Entropiegesetz, oder der Zweite Hauptsatz der Wärmelehre.* Naturwissenschaftlichen Verlag, Abteilung des Keplerbundes, Godesberg bei Bonn. Germany.

[Note 28]: *Of extreme interest and importance for our present discussion is the tightest relationship of M. Planck's ideas to the standpoint of R. Clausius, as pointed out by Peter.*

On the one hand, it is nowadays clear that M. Planck was acting based upon the R. Clausius' maxim of the 'heat death'. On the other hand, Ludwig Boltzmann was trying to 'avoid the heat death' by introducing 'its low probability'. The 'real achievement' was to 'realize' that 'the **heat death** is not 'stringently required', it is 'solely probable'. This is just boiling down to the belief that 'the heat death' might somehow be avoided, 'the perpetuum mobile of the 2nd kind' might still be possible somehow. In other words, the second basic law is universal, but 'probably' there are still some hidden corners upon the world, where we might encounter violations of this law. This is just the very zest of the paradigm so forcefully introduced— and at last even successfully pushed through—by L. Boltzmann, M. Planck, together with their numerous comrades-in-arms and followers. In fact, both M. Planck and L. Boltzmann could find the way of *how to avoid using the basic law*, introduced by R. Clausius, when formulating valid physical theories. As it is, that was undoubtedly a truly positive achievement. Meanwhile, to absolutize the named approach, to decidedly skip looking for the actual physical sense of this law—the direction the Boltzmann–Planck followers were and still are heading—means the death of thermodynamics. We shall revert to the detailed discussion of this important theme in Chapter 5 here.

To this end, Prof. Dr. Bernard H. Lavenda, the author of many serious monographs, in particular in the field of thermodynamics and statistical physics, has posed an interesting question in our discussion on the theme [28.1]:

> "*Why only one philosophy becomes so ingrained in the literature: Namely that of Clausius? Would this be the actual root of confusion around Tait's inverting the physical sense of the Clausius formulation of the Second Basic Law inequality, as well as around the final thermodynamic state— be it the minimum dissipation of energy by W. Thomson (Lord Kelvin)' —or Clausius' 'heat death'?*"

The possible answer comes to mind when reading in detail Peter's book and the references therein. As we see, the revolutionary physicists (the circles around L. Boltzmann and M. Planck) were deeply impressed by the Clausius' paradigm; it looks like that they had drastically overestimated its 'basic philosophical'/'fundamental' appearance. They were vigorously trying to render R. Clausius' standpoint as widespread as possible. A plausible explanation of such a kind of 'marketing' might anyway be the best correspondence between the foggy paradigm by Clausius and their own innovating ideas.

However, to our sincere regret, nowadays nobody of the 'dramatis personae' could any more be reachable by us for clarifying this extremely important point.

Of course, success is never blamed, and the victors do not have to justify themselves, the only point here is how to correctly estimate all those who have remained defeated.

28.1 Peter Guthrie Tait: his actual role in formulating thermodynamics

As to Prof. Dr. Peter Guthrie Tait (1831–1901), he was a brilliant Scottish mathematical physicist best known for his works in the field of theoretical physics. Moreover, he was (and is) well known as a fierce castigator of R. Clausius work, especially as concerns the entropy notion.

We have already started the detailed analysis of his ideas here (see Peter's book notes regarding his contribution to the field, as well as [Note 21] to this chapter plus the references therein). A very important point is that Prof. Tait was stubbornly promoting the work by Prof. Dr. Mohr among the British colleagues (Prof. Tait was translating Prof. Mohr's publications into English and commented them).

We might even guess this part of Tait's professional activity to have stimulated (at least partially?) the seminal thermodynamic studies by the outstanding British physical chemist Prof. Dr. George Downing Liveing (1827–1924). In fact, Prof. Liveing is one of the widely unknown thermodynamicists; see Ref. 21.4 in [Note 21] to this chapter.

To this end, of interest and importance for the correct analysis of Prof. Tait's actual contribution to thermodynamics is also a most recent Internet publication by Prof. Dr. Jeremy Dunning-Davies [28.2].

28.2 The bitter struggle between the two paradigms

Howbeit, most probably, it ought to be misleading if we would demand from Prof. Tait some handy and seminal conclusions in the thermodynamics field, for in effect he was a brilliant physicist, but never a chemist. Indeed, in this chapter we have already learned that Prof. Mohr was objecting to the following two crucial points of R. Clausius' paradigm, namely:

(a) 'Much too much involvement into the mathematics at the expense of physics'.

(b) The sheer absence of serious attention to diverse chemical processes. Sure, that was an exaggeration due to Prof. Mohr, for R. Clausius—likewise P. G. Tait—were outstanding physicists. Then, R. Clausius was, unlike P. G. Tait, a renowned experimentalist. Meanwhile, neither R. Clausius, nor P. G. Tait were chemists! This is why grammatically considering chemical processes ought to be a kind of 'mission impossible' for the both renowned colleagues.

Remarkably, Max Planck, an outstanding theoretical physicist, but never a chemist, was nonetheless tending to consider chemical processes. Indeed, he was publishing papers in the pertinent periodicals and finally published a remarkable book [28.3]. The preface to this book is of extreme interest for our present discussion, and below we would like to insert our authorized English translation thereof.

"In fact, the present work is the offprint of the article entitled 'Thermochemistry' in the Handbook/Dictionary of the Whole Chemical Science by A. Ladenburg. First, I have tried to do my very best in particularizing as clearly as possible the notions and laws of thermochemistry based on the facts underlying them, but regardless of some more specific ideas, e.g., atomism. Meanwhile, here I am far from expressing the opinion that the latter ideas might yet be worthless or even dispensable. Certainly, it is not the logic, but the imagination that ignites the first flashes of new knowledge in the souls of researchers advancing into dark areas. In the course of its progress, the scientific research surely requires certain hypotheses, which are to combine clearly visible series of individual facts into a single intuitive image composed of the well-known figures. Moreover, what becomes important here, over and over again, ought to be the task to separate as sharp as possible the immutable facts as they are from the ever-changing subjective ideas, which the facts gave rise to. Indeed, the history has shown repeatedly that the best hypotheses, once they have already rendered its favor, do prove to grow the most dangerous enemy of the scientific progress with the latter leading over and above the former. It is just this way that one might overcome severe crises arising in science now and then.

Remarkably, overcoming such crises would anyway require the lesser sacrificing the treasures already won the more carefully we criticize the latter prior to sacrificing them.

Actually, I had completed writing this paper in the summer of 1892, while there was some delay in its actual publication by some external circumstances only. Meanwhile, the publication of the two excellent detailed works by W. Ostwald (Chemical Energy), by W. Nernst (Theoretical Chemistry), and of the smaller treatise by J. J. van Laar (Thermodynamics in Chemistry) took place beforehand. This is why, to my sincere regret, I could not anymore incorporate the named three treatises timely. I guess or at least I may hope that my present publication might at least provide the readership with some benefit in addition to the works mentioned above.

To facilitate overviewing the content of the second law of thermodynamics for the chemists less experienced in physics, I am affixing here a short essay entitled: "The Core of the Second Law of Thermodynamics." The latter is in fact correspondent to my invited contribution published a short time ago in the Journal of the Physical and Chemical Education edited by F. Poske, Verlag Julius Springer (vol. 6, p. 217. 1893). The last, but not the least: Herewith I would like to extend my sincere gratitude to Dr. Eugen Röber for his assistance during the preparation of the alphabetical subject index.

Grundnerhof, September 1893
The author."

Apart of suggesting a very constructive working plan formulated above, Max Planck does present in his book a rather broad review of the up-to-date physical–chemical results having to do with the application of thermodynamic laws to chemical phenomena. Among the results by other numerous authors the work by August Friedrich Horstmann is duly mentioned, but never pointed out and/or analyzed in detail. Contrariwise, Peter could duly carry out this work in his book. Unlike Max Planck, Peter Boas Freuchen is implicitly adhering to the standpoint of Prof. Mohr, who is trying to persuade his readership that there would never be any clear picture of thermodynamics, without any detailed analysis of the entire variety of chemical phenomena. In reading attentively Max Planck's report [28.3] we get but a stubborn feeling that R. Clausius had performed all the necessary preliminary work, whereas all other colleagues, especially the chemists, had to and could just carry out solely a number of 'cosmetic repairs'.

The announced Appendix dealing with the second basic law is first going in for the details of different realistic natural/artificial processes and ensures us in conclusion that the basic law in question is **'based upon the fundamental existence of irreversible processes'**. Then we learn that deciding whether some particular process is reversible or irreversible requires looking at the entropy change during the process in question. If the final entropy value is greater than the initial one, the process is irreversible. If the both entropy values are equal to each other, then we are dealing with the reversible process. To sum up, the top-most mathematical expression for the entropy notion should boil down to the fundamental inequality. The latter becomes the strict equality for the reversible processes only.

After sounding all the above-listed preparatory statements, Max Planck points out the particular significance of the circular processes for the handy theory of thermodynamics. Indeed, such processes do uniquely combine the initial and the final states of the system in question, and this way we arrive at the enormous simplification of the desired theory.

It is exactly this way that R. Clausius could prove his famous inequality for the entropy; the explicit expression for the latter follows immediately together with a short summary. Moreover, Max Planck notes that it is in principle possible to express entropy as a particular function of the system's state. Hence, any particular state might be correspondent to a particular value of this function, up to an additive constant. In conclusion, Max Planck mentions that the actual particular form of the function involved is being intensively studied, and the present it is well known solely for ideal gases and non-concentrated solutions. The appendix involved skips but the actual mathematical expression.

Meanwhile, the interested readership could nonetheless find the special thermodynamics handbook by Max Planck [28.4], which is until nowadays a truly famous publication translated into all the possible languages. In the latter book, Max Planck is unambiguously connecting the Entropy with the Probability of the system's state the Entropy is describing. Further, it is also possible to find the additive constant mentioned above in a straightforward manner, namely owing to the third basic law of thermodynamics, discovered by W. Nernst and his school in Berlin. Finally, we learn that the latter Basic Law is helpful when solving the fundamental chemical problems like *a priori* estimating the chemical affinity, and we get the true reference to the important source: the book [28.5]. Meanwhile, Max Planck's book does not deliver any eye-opening information as for the physical sense of the second law and its entropy notion.

However, the above is not true, would be the immediate response of the atomism's adepts, for Prof. Dr. Clausius along with such outstanding colleagues as James Clerk Maxwell and Ludwig Boltzmann could still manage to build up the true basis of thermodynamics. The latter consists in the complicated dynamics of myriads of atoms/molecules/electrons etc. To sum up, in effect, there is no more basic research to perform, perhaps except for some cosmetic refinements. This is why *statistics*, *probability* remain to be the *General Roots of Everything*. As a result, there is a stubborn feeling of a certain concealment we get after reading the encomia to the address of R. Clausius by the atomism's adepts [28.6–28.7].

This feeling does not escape even when reading the most modern treatises on thermodynamics [28.8–28.11], not to speak about hundreds/thousands elder ones.

What we might clearly recognize is the True War between two different paradigms: the atomism and the energetics, whereas the latter was to 100% defeated by the former. Max Planck was one of the brightest representatives of the triumphant atomism. Meanwhile, we do know his conclusion [1, 2]:

> "*A new scientific truth does not triumph by convincing its opponents and making them see the light, but rather because its opponents eventually die, and a new generation grows up that is familiar with it.*"

Well, this is definitely true, but to our ears, this sounds cynical, forsooth. Meanwhile, a further truly cynic idea but comes to mind immediately.

Indeed, when considering the then war between the energetics (anti-atomism) and atomism (anti-energetics) there was another extremely important factor: A trustful financial support from such very well known, serious and successful businesspersons as Ernest Gaston Joseph Solvay (1838–1922) and Alfred Bernhard Nobel (1833–1896).

Howbeit, now we know very well that the "anti-atomists" were successfully defeated by the "atomists," whereas among the latter were such outstanding physicists as Ludwig Boltzmann and Max Planck. So, what is about chemists?

Interestingly, the **big war** still does not seem to come to its natural end even nowadays. Indeed, in the most recent, very interesting and very stimulating report by Klaus Ruthenberg [28.12] we read:

> "*During the XIX-th century, atomism became a central model in the chemical sciences. However, the particular sort of atomism, which is sometimes called 'chemical atomism', was by no means founded on convincing empirical data, and developed no significant explanatory power. In fact, it has been a metaphysical concept, a heuristic speculation, put forward with a view to explaining certain law-like statements of general or theoretical chemistry. Among these law-like statements were the basic stoichiometric principles of chemistry (e.g., the proportion laws). Atoms were postulated as tiny material balls with purely mechanical functions. Later, ad-hoc attempts to attach additional properties to the atom concept were made.*
>
> *Only with the rise of quantum mechanics and quantum chemistry in the third decade of the 20-th century did the atomistic picture change dramatically, departing entirely from the classical realist interpretation of small particles, although most chemists kept and still cling to their traditional 'naive-realistic'—some may prefer 'pragmatic'—attitude.*"

Interested readership might find full professional accounts as for the history and philosophy of chemistry in the books [28.13–28.16].

In effect both 'atomists' (anti-energetists) and 'energetists' (anti-atomists) are not completely wrong. Thus, the both must remain to be our intellectual property. The main reason for the above-mentioned Big War between the both paradigms looks like to be down to the human relationships, mostly. The actual main point consists in the finiteness of our lives. None from the row of such great theorists as J. C. Maxwell, L. Boltzmann, and J. W. Gibbs could finalize their seminal studies. This is why their seminal efforts do require enthusiastic followers, who would duly pursue the Eternal Hunt for the Truth, which we shall never-never-never manage to finalize.

With this in mind, of interest would be to compare the dynamics of the two processes, namely, the historical development of the chemical philosophy/chemical theory vs. the historical development of the probabilistic reasoning [28.17–28.18]. This way, one might succeed in rationalizing the roots of concealment/conceptual disorder still present even in the most recent thermodynamics handbooks/treatises [28.8–28.11].

This same holds for facilitating the necessary amendments to be done (see [28.19–28.21] and the references therein).

We shall revert to this extremely important theme in Chapters 3–5 of the present monograph.

References to the Note 28

28.1   Bernard H. Lavenda (2010): *A New Perspective on Thermodynamics.* Springer Science+Business Media LLC: New York, Dordrecht, Heidelberg, and London.

28.2   Jeremy Dunning Davies: *Tait, Force and Entropy.* http://www.noeticadvancedstudies.us/JeremyX4.pdf

28.3   Max Planck (1893): *Grundriss der allgemeinen Thermochimie. Mit einem Anhang: Der Kern des zweiten Hauptsatzes der Wärmetheorie.* Verlag von Eduard Trewendt, Breslau, Germany.

28.4   Dr. M. Planck, Professor der theoretischen Physik an der Universität Berlin (1922): *Vorlesungen über Thermodynamik,* Siebte Auflage. Verlag von Walter de Gruyter & Co., Leipzig, Germany.

28.5   F. Pollitzer (1912): *Die Berechnung chemischer Affinitäten nach dem Nernst'schen Wärmetheorem,* Verlag von F. Enke, Stuttgart, Germany.

28.6   Walther Nernst (1922): *Rudolf Clausius, geb. 2. Januar 1822, gest. 24. August 1888, 1869–1888 Professor der Physik an der Universität Bonn*: Rede, gehalten am 24. Juni 1922. Röhrscheid, Bonn, Germany.

28.7   Max von Laue (1957): *Clausius, Rudolf Julius Emanuel.* In: *Neue Deutsche Biographie (NDB).* Band 3, Duncker & Humblot: Berlin, Germany.

28.8   Jeremy Dunning-Davies (2010): *Concise Thermodynamics: Principles and Applications in Physical Science and Engineering.* Woodhead Publishing, Oxford, Cambridge, Philadelphia, New Delhi.

28.9 H. J. Kreuzer and Isaac Tamblyn (2010): *Thermodynamics*. World Scientific Publishing, Singapore, Hackensack, London.

28.10 И. А. Леенсон (2010): *Как и почему происходят химические реакции? Элементы химической термодинамики и кинетики.* Издательский Дом Интеллект, Долгопрудный, Russia.

28.11 Christoph Strunk (2015): *Moderne Thermodynamik*. Walter de Gruyter GmbH, Berlin, München, Boston.

28.12 Klaus Ruthenberg (2012): *František Wald (1861–1930)*, in: *Philosophy of Chemistry*, Andrea I. Woody, Robert Findlay Hendry, Paul Needham, Editors; Elsevier, Amsterdam, The Netherlands; pp. 125–131.

28.13 Mary Jo Nye (1993): *From Chemical Philosophy to Theoretical Chemistry: Dynamics of Matter and Dynamics of Disciplines (1800–1950)*. California University Press, Berkeley, Los Angeles, London.

28.14 Davis Baird, Eric R. Scerri, and Lee McIntyre (2006): *Philosophy of Chemistry: Synthesis of a New Discipline*. Springer-Verlag: Dordrecht, the Netherlands.

28.15 Eric R. Scerri (2008): *Collected Papers on Philosophy of Chemistry*. World Scientific Publishing, Singapore, Hackensack, London.

28.16 Andreas Karachalios (2010): *Erich Hückel (1896-1980): From Physics to Quantum Chemistry*, Springer-Verlag, Dordrecht, Heidelberg, London, New York.

28.17 Ian Hacking (2001): *An Introduction to Probability and Inductive Logic.* Cambridge University Press, Cambridge, New York, Melbourne, Madrid, Cape Town, Singapore, São Paulo, Delhi, Dubai, Tokyo, Mexico City.

28.18 Ian Hacking (2007): *The Emergence of Probability: A Philosophical Study of Early Ideas about Probability, Induction and Statistical Inference.* Cambridge University Press, Cambridge, New York, Melbourne, Madrid, Cape Town, Singapore, São Paulo, Delhi, Dubai, Tokyo, Mexico City.

28.19 И. П. Базаров (2003): *Заблуждения и ошибки в термодинамике.* Едиториал УРСС, Москва, Russia. (I. P. Bazarov: *Errors and Mistakes in Thermodynamics*).

28.20   J. Dunning-Davies and D. Sands (2010): *Confusions in Thermodynamics.* Arxiv: 1103.4360.

28.21   D. Sands and J. Dunning-Davies (2011): *How Applicable is Maxwell–Boltzmann Statistics.* Apeiron, v. 18, pp. 10–14.

[Note 29]: Sure, for nowadays, every diligent schoolchild knows: Both heat and light ought to be different kinds of wave motions!

[Note 30]: Noteworthy, in Zürich he had also a possibility to attend the lectures of Prof. Dr. Gustav Anton Zeuner (1828–1907), an outstanding German specialist in the field of engineering thermodynamics, who was also teaching in Zürich that time. In his papers, Horstmann refers to Zeuner's works.

[Note 31]: Habilitation is in Germany, Austria and Switzerland correspondent to the second PhD (i.e., postdoctoral) thesis.

[Note 32]: What follows is a kind of poem about Ludwig Boltzmann and his place in our memories. To sum up, the author tried to poetically represent the intrinsic drastic personal conflict of Boltzmann, who was fancifully combining an outstanding talent of a scientific research worker with a flagrant endeavor toward the highest publicity (not less outstanding!). The acute conflict between these both powerful stimuli, and his briskly deteriorating health condition, could most probably be the actual causes for his abrupt doing away with himself. We just cite here the original German version of the poem by Max Theodor Trautz' and then present its approximate English translation):

> *Wer hat dem Geist den Unfug eingegeben, den Erdenkloß mit seinem Stoff zu tränken,*
> *In diese Leiber seinen Strahl zu senken. sie zu entfrieden mit bewusstem Leben?*
> *War's Eitelkeit, daß reicher er zu glänzen geschäftig war in hunderttausend Spiegeln.*
> *Gedachte er sich freier zu entsiegeln in unseres Denkens lichtgeschützten Tänzen?*
> *Er schmückte nur sich selbst mit seinen Gaben, die wie zum Hohne er ob uns verhängte*
> *Die Armut noch beraubend trügend schenkte; nie wird es uns, was jene Kleinsten haben:*

Auf Sonnenstrahlen schwerelos sich wiegen, die florgewobnen blauen Schwingen regen.
Die Liebesfühler ohne Schmerz bewegen, die tiefe Eintagslust der Eintagsfliegen!
Was gab er uns? Todahnendes Verlangen! Im Wissen jedes Leiden zwiefach fühlen,
Zerquältes Ringen nach versagten Zielen und Bangen nach des Abgrunds dunklen Hängen.

Who did endow the Spirit a mischief of penetrating into Clods of Dirt, by plunging Shafts of Light into the Bodies, depriving them of their peace by turning their existence to a conscious living?
Was it his vanity that he was rather like a busy bee that gleams reflecting in the hundred thousands silver mirrors. Well, may perhaps, he would unseal himself this way more freely in the light-protected dances of our own thoughts?
He was bedighting solely himself with his talents, by administering them as if to add insult to us, by demonstrating our alleged voidance; indeed, we would be never possessed of what those little ones do have: To sway ourselves in Sunrays weightlessly by moving their blue bloom-woven vans.
To move our love sensors without any pain, the daytime zestfulness of mayflies!
What is he leaving us? His appetence in smoking Death! Re-feeling doubly every known ailment, hagridden struggle for elusive goals and yearning for abysm's somber hillsides.

[Note 33]: Moreover, Dr. Kipnis was studying historical circumstances of the lives and works of the two outstanding Dutch physical chemists: Prof. Dr. Johannes Diederik van der Waals (1837–1923) and Jacobus Henricus van't Hoff (1852–1911). The both studies of Dr. Kipnis are of clear and direct relevance to our present discussion, so we shall refer to the pertinent results throughout the narration to follow.

By the way, in the same book series we might find the biography of Prof. Dr. Max Theodor Trautz published by Dr. Kipnis in German [33.1]. Remarkably, Prof. Dr. Trautz seems to be the only noticeable colleague stubbornly and successfully pursuing the research direction envisaged and started by Prof. Dr. Horstmann at his time. This is why, here we would greatly appreciate presenting the full

English translation of Dr. Kipnis' story about Prof. Trautz and his work in a separate supplementary note, see **Note 35**.

Reference to Note 33

33.1 Dr. Alexander Yakovlevich Kipnis: "*Max Theodor Trautz.*" In: "*Badische Biographien*", W. Kohlhammer Verlag Stuttgart, Germany, 1996.

[Note 34]: A direct apprentice of Clausius in Zürich

[Note 35]: Max Theodor Trautz, chemist

This authentic portrayal of Prof. Dr. Trautz has been taken from
http://america.pink/maxtrautz_2928937.html

35.1 A concise CV of Max Trautz

Born on 19.3.1880, Karlsruhe, evangelical confession; died on 19.8.1960, Karlsruhe.

Father: Julius Theodor (1845–1897), Oberkirchenrat (a member of the High Consistory).

Mother: Marie Luise Laura Johanna, born Hauer (1857–1941).

Siblings: Friedrich Max (1877–1952); Luise Agnes Marie, in her marriage: Krumm (1878–1920).

Marriage: 1912 (Pembury / Kent, England) Mona Janet, born Drysdale (1885–1971).

Children: Dieter Max Fritz Ferdinand (1914–1941); Fritz Alexander Theodor (1917–2001).

*****

1887–1889: Elementary school in Karlsruhe

1889–1898: High school in Karlsruhe

October 1898–July 1900: Study of chemistry at the TH Karlsruhe

October 1900–March 1903: Study of Chemistry at the University of Leipzig

August 3, 1903: Doctoral studies graduation (getting Dr. Phil. degree *summa cum laude*) University of Leipzig

October 1903: Assistant at the Chemical Laboratory at the University of Freiburg

February 1905: *Habilitation* (The second doctoral studies graduation and getting teaching license) at the University of Freiburg; Sample Lecture: "*The Principles of Chemical Kinetics*"

February 7, 1910: Extraordinary professor of physical chemistry at the University of Freiburg

April 1, 1910: Professor for physical chemistry at the University of Heidelberg

December 1921: Extraordinary member of the Academy of Sciences at Heidelberg

January 1927: Full professor and director of the Institute of Physical Chemistry at the University of Heidelberg

July 1928: Full member of the Academy of Sciences at Heidelberg

May 1934–September 1936: Full professor and director of the Chemistry Institute at the University of Rostock

October 1936: Full professor and director of the Chemistry Institute at the University of Münster

October 1943: Bombing of the Institute

September 1944: Bombing of the apartment, relocation to Bomlitz powder factory, owned by Gebrüder Wolf

October 1945: Professor emeritus

1948: Return to Karlsruhe

November 1952: Dr. Rer. Nat. h. c. TH Karlsruhe

*****

Honors: Member of the Heidelberg Academy of Sciences (1921, 1928); Dr. Rer. Nat. Honoris Causa, TH Karlsruhe (1952).

35.2 The life of Max Trautz

In his early childhood, Max suffered a severe illness, which led to lifelong consequences; His health was always weak, and thus he had avoided a military service draft. Meanwhile, Max was able to overcome his weakness by tireless working and to experience many professionally fertile years. He obtained a good education, first in his parents' house. His family, and then the humanist high school, had taught him "*striving for spiritual goods,*" as he estimated this himself, which was remaining typical for him throughout his whole life.

After graduation from high school with the "good" grade, Max studied chemistry at the Karlsruhe University of Technology. Over there he could get the best marks from the final exams, and moved to Leipzig, where he studied physical chemistry with Wilhelm Ostwald and could get his PhD with the *summa cum laude* note. Thereafter he obtained a position as a research assistant at the Chemistry Laboratory of the University of Freiburg under the leadership of the organic chemist Ludwig Gattermann and at the same time could begin with his own researches. His first research field was photochemistry, to which he had devoted his habilitation thesis (the 2nd PhD in Germany). From the winter semester 1905/06 until the summer semester 1910, he held courses on the theoretical chemistry, and especially on the photochemistry.

The years in Freiburg were decisive for the development of Max as a self-initiated scientist: It is over there that the most important research directions of his work arose. He was the first to elaborate the idea of the chemical activation of the molecules by the light theoretically and experimentally. In doing so, he combined the latest Planck discoveries on the laws of radiation with chemical results on the activation of the reactions. At the same time, he turned to thermal activation of chemical reactions.

In 1910, he could get a position of an ordinary professor in physical chemistry at the University of Heidelberg. Over there, he held not only a general cycle of lectures entitled *The Principles of Physical Chemistry*, but also the lectures on a number of chosen chapters of the latter, like *Physical Chemistry of Solutions* or *Activation and Catalysis*.

Along with the extensive pedagogic activity, he was permanently pursuing his scientific research. In effect, his working conditions over there were very unfavorable; indeed, the laboratory was crammed, whereas Max enjoyed a rather restricted organizational and operational vastness. Nevertheless, from morning until evening, he was able to carry out important experimental work on the chemical kinetics of gas reactions, on the physical properties of gases and binary gas mixtures. Some of the experimental regularities he found for gas mixtures were in use by other colleagues for many decades. In addition, he wrote an internship book and a textbook on general chemistry. After Prof. Dr. Karl Johann Freudenberg became the new Chair of Chemistry Department at the University of Heidelberg in 1926, Max had obtained the position of an ordinary professor and director of the new Institute for Physical Chemistry. The Institute had even awarded Max with his own house (that was the own house of the German chemist Robert Wilhelm Bunsen (1811–1899) and then of the German literary theorist, Ernst Robert Curtius (1886–1956), who moved to Bonn). Prof. Freudenberg supported Max. Later, he recommended Max as a successor to Prof. Dr. Paul Walden, director of the chemical institute at the University of Rostock. Max could take over this truly honorable place in 1934. However, after serving there for five semesters, he had moved to the approximately equal position at the University of Münster.

In Rostock and Münster, Max was chiefly studying various inorganic reactions, in part for their usage for practical synthetic purposes. During the time of the Third Reich, Max was loyal to the regime—his attitude, as he estimated it himself, was "always in the national sense." Formally, he was accepting the "rules of the game," but never being a member of the NSDAP. Remarkably, he had neither a rector nor a dean as a position until the year 1933, when the Nazi functionaries had dismissed the then Jewish dean.

The military air raids resulted in destroying first his institute and then his apartment, so that Max was forced to abandon his scientific activity and move to the town of Bomlitz, where he could find employment at the local powder factory, in agreement with the university administration. Max spent the last years of his life in his hometown of Karlsruhe and was still able to enjoy the recognition of his achievements by his contemporaries. Those who knew him personally could especially appreciate his "absolutely upright personality," according to Prof. Freudenberg.

35.3 The work of Max Trautz

The full publication list of Max contains 190 works covering a very wide range of topics. A tireless worker, he could practically embrace all the fields of chemistry at that time. His textbook on general chemistry (more than 140 printed sheets) is particularly interesting as the most recent attempt to represent in a single book all the then chemistry as a unique integral field of knowledge, both theoretical and experimental, both descriptive and conceptual, both organic and inorganic. Even nowadays, the conception of that work is exemplary. Max's own research includes contributions to electrochemistry, photochemistry, preparative inorganic chemistry, and especially general physical chemistry and chemical kinetics. From 1904 on, Max was developing the so-called *radiation hypothesis*, according to which infrared radiation of the reaction vessel walls might activate the molecules reacting in the latter. Definitely, his hypothesis had appeared to be fruitful for the over-all development of chemical kinetics and could find numerous proponents and opponents, during two decades after its formulation. Although there were ultimate rebuttals to this hypothesis, it could nevertheless help arriving at some realistic results that were of great importance.

On the other hand, among the 10 different formulas for the temperature coefficient of the reaction rate Max had chosen that by Svante Arrhenius, although it was not the widespread one at that time. Then, in 1910, he could introduce the concept of the activation energy as a new, not substance-specific, but reaction-specific energy quantity. At an advanced age, Max used to call the activation energy notion "*my dearest scientific child.*"

Moreover, both Max and the British scientist William Lewis (1885–1956) are the founders of collision theory. While Trautz published his work in 1916, Lewis published it in 1918. However, they were unaware of each other's work due to the World War I. Following this way, the both colleagues could determine the pre-exponential factor (frequency factor) in the Arrhenius formula. Howbeit, it was only in the 1930s that the collision theory could finally boil down to the notion of the transitional complex conventional nowadays. Importantly, Max Trautz was in fact the pioneer to investigate the molecular activation energies by connecting systematic results of chemical experiments with Max Planck's theoretical findings concerning radiation physics, which at that time was still a novel trend.

Another important contribution to chemical kinetics was the discovery and exploration of the first realistic monomolecular reaction (1922), namely that of the cyclopropane isomerization into propylene. To sum up, Prof. Dr. Trautz has gained his definite place of honor in the history of chemistry, especially thanks to his detailed research on chemical kinetics.

35.4 The contribution of Max Trautz to a possible reconcilement of thermodynamics' conceptual gap

Meanwhile, of special interest for our present discussion ought to be the unique standpoint of Max Trautz as a research worker. Considering the latter is of immense importance for trying to work out the necessary and effective immunity against the notorious and stubborn *quantophrenia* and *numerology* disorders mentioned in the introduction to the book at hand.

Indeed, Max has definitely not lost himself in the flood of *enormous achievements of the revolutionary physics* of his time, but could immediately recognize the rational nuclei introduced by the then emerging quantum physics. This is just what we might readily borrow from his fundamental chemistry handbook, especially when reading attentively all of its three volumes [35.1].

In his chemistry book, Max is reviewing very carefully and thoroughly the quality of correspondence between the novel and conventional concepts. He is checking how the latter both are related to the valid experimental data. The last but not the least, practically each chapter of his book contains a careful review of the pertinent mathematical toolboxes.

On the other hand, he is complaining clearly and in a conclusively justified way about *an over-all loss of clarity* thermodynamics suffered due to the incompleteness of the then degree of docking the then still novel microscopic statistical–mechanical ideas to the modalities of the macroscopic thermodynamics.

Most importantly, from Max's book we learn that the *bitter struggle of the paradigms* taken place during the revolution in physics at the beginning of the XX-th century, as we were describing in the Supplementary Note 28 to this chapter ought to be a pure fiction in fact. Indeed, Prof. Trautz clearly demonstrates that the both approaches are throughout valid and might even complement each other, if they both are considered seriously.

As we could borrow from Max's obituary to August Friedrich Horstmann, the English translation of which is published above in this chapter, he recognized the true significance of Horstmann's seminal ideas. In reading his chemistry textbook, we notice that Max was connecting the actual modalities of the macroscopic thermodynamics with Horstmann's ideas.

Moreover, in his thorough work on the foundations of the chemical kinetics he was looking for plausible and fruitful theoretical ideas and concepts. In reading attentively Max's chemistry handbook together with his earlier and later review papers, we might recognize his definite striving for rationalization of chemical kinetics on the thermodynamic basis.

Of special interest ought to be Max's review paper on chemical kinetics and catalysis he published in 1912 in the renowned German professional journal *Zeitschrift für Elektrochemie* [35.2]. In this paper, he was analyzing the works published between the years of 1909 and 1912. Remarkably, it was just the time of truly furious struggles between the revolutionary physicists just introducing what we now know as the *quantum physics*, and their opponents, the 'counter-revolutionary' *energetists*.

Meanwhile, Max's story carries absolutely no sigh of the latter struggles. Although the topic he is discussing has the direct relationship to physical chemistry—and therefore to the core physics, the physical revolution all around did not seem to deter him from his scientific research. Himself he was apparently thinking about Prof. Horstmann's seminal contribution to thermodynamics—and how to make use of the latter in working out the valid and well-grounded concepts of chemical kinetics.

For example, Max has discussed in detail the relevant publications by Prof. Dr. Hans von Halban, the senior (1877–1947), an outstanding Austrian-Swiss physical chemist. At the beginning of the XX-th century, they have encountered each other at the University of Leipzig in the laboratory of Wilhelm Ostwald. Hans was lively interested in chemical kinetics and studying in detail the modalities of the solvent's type influence on reaction rates. This was the theme of his second PhD thesis, accepted at the University of Würzburg in 1909 (for more information about him in see this work [35.2]). However, he could not duly pursue his studies, first due to his serious health problems, then owing to the working place problems. Meanwhile, his son, Hans Halban, the junior (1908–1964), had initially started following the father's trail, but finally changed to the field of nuclear physics, where he could make a name for himself with joining the famous Manhattan Project.

This authentic portrayal of Prof. Dr. Hans von Halban,
the senior, has been taken from
https://de.wikipedia.org/wiki/Hans_von_Halban_(Chemiker)

Interestingly, Max could still fetch a colleague who was working along the seminal direction shown by Hans, the senior.

We would like to present here the English translation of the relevant fragment of the first part of Max's review paper [35.3].

"...The conclusion by von Halban that a greater solubility of the reaction educts in some pertinent solvent should anyway lead to an increase in the reaction rate of these substances is shared by Zawriew (the Reference), who describes this effect as a catalytic one, and understands every catalytic effect as overcoming of some passive obstacles. Remarkably, the use of the word 'catalytic' in this case is nothing more than just a manifest, likewise using the 'passive obstacles' concept. Indeed, we are employing the both notions far too often in proportion to their imprecise meaning. ..."

This authentic portrayal of Prof. Dr. David Christopher Zawriew/ Zavryan
has been taken from
http://am.hayazg.info/%D4%B6%D5%A1%D5%BE%D6%80%D5%AB%D6%87_
%D4%B4%D5%A1%D5%BE%D5%AB%D5%A9_%D5%94%D6%80%D5%AB%D5%
BD%D5%BF%D5%A1%D6%83%D5%B8%D6%80%D5%AB

Here Max is referring to and starting to discuss the work by Prof.
Dr. David Christopher Zawriew/Zavryan; Давид Христофорович
Завриев/Заврян (1889–1957); Զավրիև Դավիթ Քրիստափորի, a
physical chemist from Armenia (at the time of cited work he was a
citizen of Russian Empire). First, it is of interest to tell his story here
in short.

David Zavryan was born in Tiflis (now, Tbilisi the capital of
Georgia) in a wealthy Armenian family. He had graduated the primary
and secondary school in the same city. Thereafter he could continue
his studies at the Chemical Department in the Imperial-Royal
University of Petersburg, where his teacher and mentor was Prof.
Dr. Dmitry Ivanovich Mendeleev (1834–1907) a prominent Russian
chemist. Already during his University studies, David Zavryan started
travelling abroad, in particular to France, where he could carry out
scientific research under the direction of the competent professors
in Paris. As a postgraduate, he could get a position of the laboratory
assistant at the State Technological Institute of St. Petersburg and
could pursue his research studies in the field of physical chemistry. In
1909, he had published the detailed report on his results in Russian
[35.4]. It was just the paper attracted the attention of Max Trautz,
although Max could read only the short abstract of it published in a
German periodical.

Still, the seminal work by D. Zavryan could find no continuation,
to our sincere regret.

As it was conventional for the young educated people in the Russian Empire at the beginning of XX-th century, D. Zavryan was tightly involved into the illegal anti-state activity in the years of 1909–1913. Thereafter, in 1914–1917 he was taking part in the First World War and the Russian revolutions.

Only in 1920s, he started his pedagogical activity by holding lectures in chemistry at different higher educational organizations in Transcaucasia (Georgia and Armenia). Specifically, in 1919, Yuri Ghambarian (the first rector of the Yerevan University) and David Zavryan initiated the opening of the University in Yerevan. In January 1921, David had moved from Tbilisi to Yerevan. At the University over there, he had obtained the Dean post at the Faculty of Engineering. Then, for the short time from October 1921 to February 1922, he could represent the Rector of the Yerevan University. From 1928 until 1930, he was holding the director position at the Transcaucasia's Chamber of Weights and Measures in Tbilisi and afterwards had obtained the professorship at the Technical University of Tbilisi. At that time, he could re-start his research work and had mail correspondence with Prof. Dr. Vladimir Ivanovich Vernadsky (1863–1945), an outstanding Russian natural scientist and philosopher. Finally, in 1938 the Stalinists had suddenly taken David into custody, among numerous other people. Somehow, he could at last avoid the repression, and even get the honor of the *"Outstanding Research Worker of Armenian Soviet Socialist Republic"* in 1939, but since that time, he was concentrating his entire effort on the pedagogic work and had completely quit the research community.

Howbeit, of special interest for our present discussion ought to be the deliberations about representation of *every catalytic effect as overcoming of some passive obstacles*. Max Trautz had pinpointed this aspect, for this is just in full accordance with the formulation of the Second Basic Law by August Horstmann. Indeed, Prof. Dr. Horstmann has suggested connecting the entropy notion with the sum of obstacles/hindrances/resistance arising when a process driven by some *livening force* (*Vis Viva*) starts its evolution in time. Before the ultimate start of the process, there ought to be no apparent resistance to the latter. Hence, in this sense, we might view such a resistance to be *theoretically passive*.

To our sincere regret, Prof. Horstmann himself had no opportunity—and finally no more time—to drive his train of thoughts through the complete line of its pertinent logical stations.

Prof. Trautz was definitely trying to do his best for keeping Professor Horstmann's train of thoughts in its proper movement, but that was not trendy at his time; moreover, the over-all political situation was not the supportive one.

We would greatly appreciate concluding this supplementary note with an appeal to pay much more attention to Prof. Trautz works. A re-publication of a commented English translation of his chemistry handbook ought to be in order for rendering it widely accessible. The last, but not the least: We are about to discuss here the seminal ideas of Prof. Horstmann in Chapter 5 of the book at hand.

35.5 Documentary sources concerning the life and work of Prof. Dr. Trautz

Archives of the University of Freiburg: information;

Archives of the University of Heidelberg: PA 6113, PA 6114, Rep. 14-5, and HAW 477;

University Library Heidelberg: Hs 3695E;

General Archives of Baden-Württemberg County, Department of Karlsruhe: 235/2598;

Archive of the University of Rostock: Phil. Faculty, personal files No. 272;

Archive of the University of Münster: information;

Archives of the University of Karlsruhe: 0/1/52, 0/1/810.

35.6 Selected publications by Prof. Trautz, which are most representative of his scientific interests

Zur physikalischen Chemie des Bleikammerprozesses (Doktorarbeit), Zeitschrift für physikalische Chemie 1904, 47, 513–610.

Studien über Chemilumineszenz (Habilitationsschrift), ebenda, 1905, 53, 1–111.

Der Temperaturkoeffizient chemischer Reaktionsgeschwindigkeiten, 1908, ebenda, 64, 53–88, 1909, 66, 496–511, 67, 93–104, 1910, 68, 295–315, 1911, 76, 129–144.

Beitrag zur chemischen Kinetik, Zeitschrift für Elektrochemie 1909, 15, 692–695.

Geschwindigkeit von Gasreaktionen, ebenda 1912, 18, 513–520.

Das Gesetz der Reaktionsgeschwindigkeit und der Gleichgewichte in Gasen, Zeitschrift für anorganische und allgemeine Chemie, 96, 1916, 1–28.

Praktische Einführung in die Allgemeine Chemie: Anleitung zu physikalisch-chemischem Praktikum und selbständiger Arbeit, 1917.

Verlauf der chemischen Vorgänge im Dunkeln und im Licht. Sitzungsberichte der Akademie der Wissenschaften zu Heidelberg, 1917, 8, A14, 1–36.

Das Gesetz der thermochemischen Vorgänge und das der photochemischen Vorgänge, Zeitschrift für anorganische und allgemeine Chemie, 1918, 102, 81–129.

Galvanische Elemente, im *Handbuch der Elektrizität und des Magnetismus, herausgegeben von L. Graetz*, Band l, 1918, 421–698.

(mit K. Winkler) *Die Reindarstellung des Trimethylens*, Journal für praktische Chemie 104, 1922, 37–52.

(mit K. Winkler) *Die Geschwindigkeit von Ringsprengungen in Gasen. Trimethylenisomerisation*, ebenda, 53–79

Lehrbuch der Chemie. Zu eigenem Studium und zum Gebrauch bei Vorlesungen. Erster Band: *Stoffe.* Zweiter Band: *Zustände.* Dritter (Schluß-) Band: *Umwandlungen.* 1922–1924.

August Friedrich Horstmann, Nachruf, Berichte der Deutschen Chemiker-Gesellschaft, 1930, 63 A, 61–86.

Abbrandsreaktionen (mit J. D. Holtz), Journal für praktische Chemie. 1937, 148, 225–265.

Vorstellung von chemischen Reaktionsereignissen: einiges über Ursprung, Leistung und Grenzen davon, ebenda, 1943, 162, 121–147.

The English translations of the above headlines follow:

On the Physical Chemistry of the Lead Chamber Process (PhD thesis), *Zeitschrift für physikalische Chemie* 1904, 47, 513-610.

Studies on Chemiluminescence (*Habilitationsschrift*, 2nd PhD thesis), *ibid*, 1905, 53, 1–111.

The Temperature Coefficient of Chemical Reaction Rates, 1908, *ibid*, 64, 53–88, 1909, 66, 496–511, 67, 93–104, 1910, 68, 295–315, 1911, 76, 129–144.

Contribution to Chemical Kinetics, *Zeitschrift fur Elektrochemie*, 1909, 15, 692–695.

Rates of gas reactions, *ibid.*, 1912, 18, 513–520.

The Law of Reaction Rate and Equilibrium in Gases, *Zeitschrift für anorganische und allgemeine Chemie*, 96, 1916, 1–28.

Practical Introduction to General Chemistry: Instruction on Physics and Chemistry for Practical and Own Studies, 1917.

Chemical processes in the dark and light, a report to the Academy of Sciences, Heidelberg, 1917, 8, A14, 1–36.

The Law of Thermochemical Processes and That of Photochemical Processes, *Zeitschrift für anorganische und allgemeine Chemie*, 1918, 102, 81–129.

Galvanic Elements, Manual of Electricity and Magnetism, edited by L. Graetz, Vol. 1, 1918, 421–698.

(With K. Winkler) Getting Trimethylene in Its Pure State, *Journal for Practical Chemistry* 104, 1922, 37–52.

(With K. Winkler) The Rate of Ring Explosions in Gases. Trimethylene Isomerization, ibid. 53–79.

Textbook of Chemistry. For own studies and for use in lectures. First volume: *Substances.* Second volume: *Conditions.* Third (final) volume: *Transformations.* 1922–1924.

August Friedrich Horstmann, Obituary, *Reports of the German Chemists Society*, 1930, 63 A, 61–86.

Burning Reactions (with J. D. Holtz), *Journal of Practical Chemistry.* 1937, 148, 225–265.

Introduction of Chemical Reactions: Some of the Origin, Power, and Limitations of it, *ibid.* 1943, 162, 121–147).

References to Note 35

35.1 Dr. Phil. Max Trautz; Professor für physikalische Chemie und Elektrochemie an der Universität Heidelberg (1922–1924): *Lehrbuch der Chemie. Zu eigenem Studium und zum Gebrauch bei Vorlesungen.* Band I: *Stoffe*; Band II: *Zustände*; Band III: *Umwandlungen.* Walter de Gruyter: Berlin, Germany.

35.2 M. Kofler (1948): Hans v. Halban, 1877–1947. *Helvetica Chimica Acta*, v. 31, pp. 120–128.

35.3 Max Trautz (1912): Reaktionsgeschwindigkeit und Katalyse. Januar 1909 bis 1912: Sammelreferat, Teil I.; Teil II. *Zeitschrift für Elektrochemie*, v. 18, pp. 908–919; v. 19, pp. 133–151.

35.4 Д. Завриев (1909): Опытное изслѣдованіе диссоціаціи углекислаго кальція. *Журнал Русского Физико-Химического Общества*, v. 41, pp. 34–56. (*Experimental Investigation of Calcium Carbonate Dissociation*).

1.10   References

1. Max Planck (1948): *Wissenschaftliche Selbstbiographie*, Barth: Leipzig, Germany.

2. Max Planck (1949): *Scientific Autobiography*, Philosophic Library: New York, USA.

3. Johannes Nicolaus Brønsted (1940): '*The Fundamental Principles of Energetics*', Philosophical Magazine Seventh Ser. Vol. 29, pp. 449–470.

4. Johannes Nicolaus Brønsted (1941): '*On the Concept of Heat*', Kongelige Danske Videnskabernes Selskabs Skrifter, Vol. 19, no. 8, 79 ff.

5. Johannes Nicolaus Brønsted (1946): '*Principper og problemer i energetikken*', København; and its English translation: '*Principles and problems in energetics*': Wiley Interscience 1955.

6. T. Ehrenfest-Afanassjewa (1956): *Grundlagen der Thermodynamik.* E. J. Brill, Leiden, the Netherlands.

7. Karl von Meÿenn (Herausgeber) (1997): *Die großen Physiker*—in 2 Volumes. Vol. 1: Von Aristoteles bis Kelvin. Vol. 2: *Von Maxwell bis Gell-Mann*. Verlag C. H. Beck oHG, München, Germany.

8. Karl Friedrich Mohr (1869): *Allgemeine Theorie der Bewegung und Kraft als Grundlage der Physik und Chemie. Ein Nachtrag zur mechanischen Theorie der chemischen Affinität.* Verlag Friedrich Vieweg & Sohn, Braunschweig, Germany.

9. Karl Friedrich Mohr (1837): *"Über die Natur der Wärme."*—in *Zeitschrift für Physik und verwandte Wissenschaften*; Dr. A. Baumgartner & Dr. J. Ritter von Holger, Eds., Volume 5, pp. 419-445, Verlag J. G. Heubner, Vienna, Austria.

10. K. F. Mohr (1837): *"Ansichten über die Natur der Wärme."*—in: *Ann. der Pharm.*, v. 24, pp. 141–147.

11. K. F. Mohr (1876): (*translated by Prof. Dr. P. G. Tait*) *"Views of the Nature of Heat."* – in: *Philosophical Magazine*, v. 2, pp. 110–114.

12. Ralph E. Oesper (1927): *"Karl Friedrich Mohr." The Journal of Chemical Education*, v. 4, pp. 1357–1363.

13. Walther Nernst (1918): *Die theoretischen und experimentellen Grundlagen des neuen Wärmesatzes.* Verlag Wilhelm Knapp, Halle an der Saale, Germany.

14. *Abhandlungen zur Thermodynamik chemischer Vorgänge von August Horstmann.* Herausgegeben von J. H. van't Hoff. Verlag von Wilhelm Engelmann, Leipzig, Germany 1903.

15. August Friedrich Horstmann (1867): *Über die Beziehungen zwischen dem Moleculargewicht und specifischen Gewicht elastisch-flüssiger Körper. Habilitationsschrift von Dr. August Horstmann.* Buchdruckerei von G. Mohr, Heidelberg, Germany.

16. August Friedrich Horstmann (1868): *Zur Theorie der Dissociationserscheinungen.* Berichte der Deutschen chemischen Gesellschaft zu Berlin, v. 1, pp. 210–215.

17. M. Trautz (1930): *August Friedrich Horstmann. Nachruf.* Berichte der Deutschen Chemischen Gesellschaft, Abteilung A, Vereinsnachrichten, v. 63, pp. 61–84.

18. Dr. Alexander Yakovlevich Kipnis (1996): *"August Friedrich Horstmann".* In: *"Badische Biographien"*, W. Kohlhammer Verlag Stuttgart, Germany.

19. M. W. Zemansky (1957): *Fashions in Thermodynamics.* Am. J. Phys., vol. 25, pp. 349–351.

20. Dr. Alexander Kipnis (1997): *August Friedrich Horstmann und die physikalische Chemie*, in der Reihe *"Berliner Beiträge zur Geschichte der Naturwissenschaft und Technik,"* ERS Verlag Berlin, Germany, pp. 176–177.

21. Alexander Y. Kipnis (2001): *"Early Chemical Thermodynamics: Its Duality Embodied in van't Hoff and Gibbs"*, in: *"Van't Hoff and the Emergence of Chemical Thermodynamics. Centennial of the First Nobel Prize in Chemistry, 1901-2001"*, Willem J. Hornix and S. H. W. M. Mannaerts, Editors, pp. 212–242.

22. http://www.ub.uni-heidelberg.de/helios/fachinfo/www/math/homo-heid/gibbs.htm

23. G. W. Rachel (1880): "*The Priority of the Late Friedrich Mohr in Regard to the Principle of the Conservation of Energy*," Science, vol. 1, pp. 203–205.

24. George Downing Liveing (1885): *Chemical Equilibrium. The Result of the Dissipation of Energy.* Cambridge: Deighton, Bell & Co.; London: George Bell and Sons, Great Britain.

Chapter 2

The Life and Work of Nils Engelbrektsson (1875–1963) and Karl Alexius Franzén (1882–1967)

ENGELBREKT:
Ja, jag har haft en dröm;
Men jag skall endast säga den åt dig...
Jag var på ett skepp, det var storm...
Jag stod i fören och såg utåt...
Vattnet...
Men vattnet blommade just då,
Så att vi färdades som genom en gata av guld...
Då kom ett litet flickebarn på vattnet gående;
Och när hon fick se mig, pekade hon och sade:
Se kronan!

MÅNS BENGTSSON, Natt och Dag:
Det har vuxit upp som ett snår ur händelsernas brokiga sådd;
Därför tar jag yxan och röjer.
Då skall du se en enda tistel längst under buskarna,
Och för att den icke skall fröa av sig,
Så rycker jag upp den med roten...
Det må tydas: Engelbrekt reste trälar mot herrar;
Vi äro herrar, alltså: död åt trälarna och deras man!

A Different Thermodynamics and Its True Heroes
Evgeni B. Starikov
Copyright © 2019 Pan Stanford Publishing Pte. Ltd.
ISBN 978-981-4774-91-8 (Hardcover), 978-0-429-50650-5 (eBook)
www.panstanford.com

ENGELBREKT:
Oh, yes, I had a dream,
But only you will hear about it from me...
I was on ship, the weather – stormy...
While I stood on the bow to watch surroundings...
The unfrequented waters...
Suddenly I saw: The water was in bloom,
As if we travel through a street of gold...
I noticed little girl just walking on the water,
She saw me too; she pointed at me and told me:
Behold the Crown!

MÅNS BENGTSSON, Night and Day:
All the events remind me kaleidoscopic plants,
Then I should take my ax and clear away them.
Lo! Here's a single thistle behind the bushes,
Sure, that to prevent it's seeding,
I must eradicate it...
What this should mean:
That Engelbrekt drives Slaves against Masters,
We are the Masters: Down with the Slaves and their chief!

Engelbrekt: Skådespel i fyra akter (Drama in four acts), by August
Strindberg (1849–1912)

Wie nur dem Kopf nicht alle Hoffnung schwindet,
Der immerfort an schmalem Zeuge klebt,
Mit gierger Hand nach Schätzen gräbt
Und froh ist, wenn er Regenwürmer findet.
Some minds don't lose their hope and find leisure,
For evermore adherence to fine thorns,
They duly grab with eager hands for treasure,
Rejoicing when they find just earthworms.

Faust: Der Tragödie Erster Teil (The first part of the tragedy), by Johann
Wolfgang von Goethe (1749–1832), our authorized translation.

2.1 The Life of Nils Engelbrektsson

The exact date of this Nils Engelbrektsson's photo is not quite clear.
The Internet portal Ancestry could have fetched the photo, after we

had typed-in Nils' name, homeland country, as well as his birth and death years as input.

Nils Engelbrektsson

Nils Engelbrektsson was born to a noble Swedish family on June 2, 1875 in the village of Stavnäs, Swedish province Värmland. His father was first a landlord and farmer in that village, and then the family moved consequently to the village of Alster and to the provincial capital, i.e., the city of Karlstad, where his father has founded a butchery. The family was relatively big, so that Nils had seven siblings, an elder sister Anna, as well as two younger brothers (Engelbrekt and Anders Gunnar) and also four younger sisters (Elin Maria, Emma Hildur, Agnes Maria, and Dagmar Augusta).

2.1.1 Nils' School Education

His actual education might in fact have started at home from his early years on. Finally, on August 28, 1884 Nils had very successfully passed his entrance exams at the '*Högre Allmänna Läroverket i Karlstad*' (Higher Educational Institution of Karlstad) and started his school studies.

His school time was continuing till 1892, when he passed the graduation exam after getting the full education corresponding to the classic Latin gymnasium. Below we place the English translation of his maturity rating.

Maturity Rating

Handed over to the adolescent named Nils Engelbrektsson, born on June 2, 1875 in the community of Stavnäs, county of Värmland, who was introduced into our grammar school in the autumn semester of 1884, and in the highest class of which he had spent two semesters, which he had cast off and ended at this day, in accordance with the Provisions of the Gracious Charter of April 11, 1862, by receiving as a result the leaving certificate with the following credentials:

In the written examination*:*

The essay in the mother tongue: approved
For translation into the Latin language: with distinction
For translation into the French language: approved

In the oral examination*:*

For insights into Christianity: with distinction
>*Latin language: with distinction*
>*Greek language: with distinction*
>*Hebrew language: commendably*
>*German: with distinction*
>*French: with distinction*
>*English: with distinction*
>*Mathematics and physics: with distinction*
>*Natural history: approved*
>*History and Geography: commendably*
>*Philosophical propaedeutic: with distinction*

Moreover, during the time he stayed in our grammar school, the aforementioned Nils Engelbrektsson had evinced praiseworthy diligence and very good behavior.

Due to the above, in regard to the maturity of Nils Engelbrektsson, who managed to complete the elementary education required for the latter, it has been found that over-all he ought to earn the grade with distinction.

Karlstad, June 14, 1894

Signed: (handwritten signature)

Rector of Karlstad Higher Public Grammar School

All this information, including the written proofs of Nils' school days in Karlstad, we have obtained from the Swedish archive Värmland, which is the Swedish county, where the village of Stavnäs and the city of Karlstad are situated.

2.1.2 Nils' Higher Education and Working Experience

The following formal CV of Nils might be reconstructed from the wealth of the Swedish archive data.

The family of Nils has moved from Karlstad to the Masthuggs community of Göteborg (Gothenburg), the county of Göteborg and Bohus, on August 17, 1894. The main aim of that move was most probably to ensure that Nils might enter and duly graduate his studies in physics and mathematics at the Gothenburg's University College.

Well, but why not in the city of Karlstad? Indeed, Karlstad had that time the so-called 'Public School Teacher Seminar' founded in 1843. After the great fire in Karlstad in 1865 the seminar had moved for a short period of time to Nysäter, county of Värmland, but very soon returned to Karlstad and was remaining over there. Meanwhile, it was a period in the 1860s, when the seminar was on the eve of its closure, but nonetheless, it could continue its work after the Bishop of Karlstad, Anton Niklas Sundberg, had sent the relevant enquiry to the Swedish parliament.

It is clear that such a story was not inspiring for the parents of Nils, and moreover, himself he was from his school days on definitely interested in the research work, but not in the purely pedagogic activity. Howbeit, they had most probably decided to send Nils to Göteborg, for just at that time this ought to be the nearest university campus, with respect to the Värmland county: Indeed, just several years have elapsed since the University College of Gothenburg was founded (namely, in 1891).

We could find no documentary proof of Nils' studies in Göteborg—but most probably it has taken place just over there. As a result, Nils could thus get his *Filosofie licentiat* (the license to serve as a school teacher). The time necessary for this ought to be some 4 to 5 years, so that we might estimate that Nils' university graduation might take place at some time in the years of 1898–1899.

Further on, Nils' first working place was most probably the grammar school at the town of Ystad, county of Skåne—the province in the southern–western corner of Sweden.

Nils was living several years in Ystad and teaching at the school over there, but according to the archive records, he moved to Högbergsgatan, Maria community of Stockholm on January 14, 1907.

Our tentative explanation of that move could be that from the beginning of his career Nils did not appear to be satisfied with his teacher's position and was definitely flirting with the hope for joining some academic organization of higher level than a grammar school or gymnasium. Definitely, he had already started his theoretical studies on the foundations of thermodynamics during that time.

Still, his hopes were most probably in vain, for his first publication in 1909 entitled *Meddelande till matematiska kongressen i Stockholm (22–25 Sept. 1909)* (*A Report to the Mathematic Congress in Stockholm*) had appeared in Lund, and Nils' own notes show that at that time he was already a teacher at a Latin gymnasium in Göteborg.

In the foreword to his 1909 publication he explains that he was invited to attend the Scandinavian Mathematical Symposium and could have many discussions with different colleagues, e.g., with such noticeable mathematicians as Profs. I. Bendixson, M. G. Mittag-Leffler, E. Phragmén, and M. Wiman, among others. Nils notes further on that, though he could reach a rather high estimate for his train of thoughts, there was no positive decision to go on with publishing his work in some corresponding official medium. The colleagues noted that the result he could arrive at ought to be not just pure mathematics, but a rather peculiar mixture between the latter and philosophy. This is why he finally decided to publish his results on his own initiative and costs, for he had no doubts in the correctness of his work.

With such an outcome in his pocket Nils returns to Gothenburg, he is duly pursuing his gymnasium-teaching career and is alongside busy with working on and consequently publishing of his above-mentioned results.

Meanwhile, the school education system of the then Sweden was not rendering him enthusiastic, for already in 1911 he publishes (now in Gothenburg) a lampoon entitled *Kungliga Läroverksöfverstyrelsen och Lärarepersonligheten* (*The Royal Administration of Grammar Schools and the Personalities of Teachers*).

This was a truly withering criticism, which had also been followed by the not less withering criticism of the Royal Society for Science and Letters in Gothenburg, also published in 1911. Here we would like to consider the latter critical lampoon in more detail, by giving its English translation.

Still, before getting further, we would like to mention that Nils' criticism could definitely expel him from Göteborg, for, most probably, he was viewed by his surroundings as a kind of 'cheap grievance-monger'. Consequently, according to his archive records, he had to move to the city of Kungsbacka, the county of Halland, that is, in the nearest vicinity to Göteborg. Meanwhile, his constructive criticism had seemingly been forgotten practically immediately, for we could not fetch any notice about Nils' lampoons or about Nils himself in the works on the theme, see the following reference:

Årsböcker i Svensk Undervisningshistoria [1941] Bokserie med Understöd av Föreningen för Svensk undervisningshistoria. [Vol. 63 i serie] [Vol. III i årg.] Goda Lärare. Minnesbilder av f. d. lärjungar o. a. Samlade och Utgivna av B. Rud. Hall.

(Yearbooks in Swedish Teaching History [1941] Book series with the support of the Swedish Association for Teaching History. [Vol. 63 of the Series] [Vol. III of that year] *Good Teachers: Memoirs of the Former Disciples and Other Stories*. Collected and Published by B. Rud. Hall).

This truly looks like a very sad story about a young, active, proactive pedagogue, and researcher, who was encountering a really enormous resistance of his surroundings, whereas he was struggling, struggling and—once more—struggling, although everything seemed to be in vain.

Meanwhile, the World had gotten crazy—in the August 1914 the First World War began. Remarkably, the history of Nils' homeland was following a very complicated trajectory as well, apart from all the well-known worldwide cataclysms.

It must be mentioned here that from the second half of the XIX-th century till the first two decades of the XX-th century there were several peaks of emigration from Sweden—mostly to the USA. Especially the time between the years 1903 and 1917 was characterized by several high emigration waves—so that the story started to finally come to its end in the mid-30s of the XX-th century, when the Swedish government could finally cope with the critical situation and build up the well-known Swedish social state.

The relevant historical details are thoroughly described and analyzed in the following book:

H. Arnold Barton (**1994**): *A Folk Divided. Homeland Swedes and Swedish Americans, 1840–1940*. Southern Illinois University Press, Carbondale and Edwardsville, USA.

Still, Nils continues his research and diverse publishing activity, aside from the pedagogic service. In the meantime, the following works by him appear:

Krigskulten. Uddevalla, 1915. (*The Cult of War*)

This is just a brilliant journalistic report about Nils' impressions concerning the First World War; in fact, this publication has nothing to do with thermodynamics or any other areas of physics.

En skandinavs tankeliv. Göteborg, 1916. (*A Spiritual Life of a Scandinavian*)

Nils' latter publication is a very interesting digression into the world of Scandinavian philosophic tradition. This particular story mentions philosophic searches of Nils himself and is trying to find the pertinent place of the latter ones within the Scandinavian tradition as a whole. This work contains no direct and explicit connections to Nils' theoretical thermodynamics studies.

Entropieekvationens dS = dZ/T integrationsteorem och kontaktsprincip. Lund, 1916. (*The Integration Theorem for the Entropy Equation dS = dZ/T and a contact principle*)

The latter booklet should just be Nils' first attempt to work and point out the proper mathematical formalism applicable in the thermodynamics' field. After this Swedish publication, Nils had also published a more detailed report in German, in the year of 1917 and also in Lund. The latter consists of the two parts entitled:

I. *Über die Bestimmtheit des zweiten Hauptsatzes der Thermodynamik. Lösung mittels des zweiten Hauptsatzes von Problemen, bei denen der Entropiesatz keine bestimmte Lösung gibt. (Anwendung: Herleitung der Temperaturfunktion des Dampfdruckes. Analytische Definition des kritischen Zustandes bei Substanzen).*

II. Zusammenhang zwischen Wärmefunktionen bei einer Substanz, die einem thermischen Prozess unterworfen ist, und dem unveränderlichen Gewicht der Substanz. (Anwendung: Herleitung des Gesetzes von Dulong-Petit und der Regel von Trouton).

(I. About the Determinacy of the Second Basic Law of Thermo-dynamics. Using the Latter when Solving the Problems, where the Entropy Law delivers no Definite Solution. Application: Inference of the Vapour Pressure as a Function of Temperature; Analytical Definition of the Critical States of Substances.

II. Interrelationship Between the Heat Functions of a Substance Undergoing Some Thermal Process and the Constant Weight of the Substance Involved. Application: Inference of the Dulong-Petit Law and of the Trouton Rule)

This publication is showing in detail the trajectory of Nils' theoretical train of thoughts in the field of the conventional macroscopic thermodynamics and its physical–chemical applications.

In general, it was clear that the extremely unfavorable local and global socio-political situation on the one hand, and the striking (from the then conventional standpoint!) innovations suggested by Nils on the other hand were strongly opposing the dissemination of Nils' truly seminal ideas.

Meanwhile, according to the archive data, Nils had never his own family, his own children, so the true poser might arise: So were there at least some lucky glimmers in the above-mentioned row of failures?

The answer is definite: *Yes, sure*—Nils could finally encounter his true colleague—Karl Alexius Franzén.

2.2   The Life of Karl Alexius Franzén

Whereas Nils had no family of his own, that is, according to the archive records, he was never married and had no children; Karl did have not only siblings, but also his own family, that is—his wife and children. Karl has married in Stockholm, on October 18, 1913.

Bearing this in mind, our investigation had then to be concentrated not on digging the Swedish state archives, but on trying to fetch the direct ancestors of Karl, who might be capable of providing me with the historical documents from the familial archives in regard to Karl's CV. On the other hand there is no conventional academic track of publications by Karl. The only reference to his work is connected solely with the research activities of Nils Engelbrektsson.

Karl Alexius Franzén

Meanwhile, in trying to fetch and contact at least some of Karl's direct ancestors, it was fortunately possible for us to reach solely his great-nephew, Dr. Anders Franzén, the chief antiquarian at the County Museum of Jönköping, historically corresponding to the Swedish county of Småland. Anders is dealing with the reconstruction of historical buildings.

He has kindly handed over to me the above photo of Karl in his youthful days (Karl is about 20–30 years old in the picture above here), as well as his photo as a retiree, which we would also like to place here below.

Sincere gratitude to Dr. A. Franzén for providing this photograph.

Karl was born on June 16, 1882 to a merchant family in the city of Växjö, the county of Kronoberg (also in Småland). The peak of his education: He had graduated from the University of Lund by obtaining the degree of Fil. Mag. (Magister of Philosophy) over there in the year of 1911.

2.2.1 Karl's Working Activity

Bearing such an educational luggage, Karl had immediately entered the pedagogical activity and was a school teacher in Karlskrona, county of Blekinge, 1911–1913, then in Gävle, county of Gävleborg, 1913–1914, then in Stockholm, 1914, and finally in Hudiksvall, also the Gävleborg county (though, historically, this is the Swedish county of Hälsingland), from 1915 on. This is just where Nils Engelbrektsson could fetch him.

As mentioned above, the publication together with Nils was Karl's only research work publication. All along his professional life he was prevalently involved into the pedagogic activity in different schools in Sweden.

After his retirement he settled in the Alingsås, county of Västra Götaland, where he had completed his earthly journey in the year of 1967.

2.2.2 Nils' and Karl's Collaboration

Likewise Nils, Karl was also never mentioned in the periodical about the history of the school education in Sweden cited—but this is definitely not the main point, for his teacher's work was definitely and duly distinguished by the Swedish state, as we might see from his above photo.

Of truly international significance and value ought to be the fact that Karl has demonstrated the perfect abilities of a world-class experimentalist in the field of physical chemistry in mastering the verification of Nils' theoretical inferences—and, to our mind, this ought to be his principal professional achievement!

After the common publication in 1920 by Nils and Karl, Karl had probably no more leisure and/or mood to follow the thermodynamic research, whereas Nils hasn't given up everything—and was continuing to work and publish. Hence, up to the Second World War, Nils' alone had mastered six publications more. The last of them came out in the year of 1935 and was written in German. That was a polemic summary of his thermodynamic works.

Why Nils was keeping a pause from 1935 on? He was already 60 years old at the time, while the Second World War was already on its eve. Although Sweden was neutral as ever, the atmosphere was very complicated and wasn't just simply 'letting-on'. How intellectual Swedes were accepting and evaluating that time might be clearly seen from the diaries of the famous author Astrid Lindgren (1907–2002):

Astrid Lindgren: *Die Menschheit hat den Verstand verloren. Tagebücher von 1939–1945.* Ullstein, Berlin, Germany, 2015. (The Manhood Loses its Nous. Diaries in 1939–1945)

Howbeit, Nils had once more returned to Fridhemsgatan 14 in the St. Görans community in Stockholm from Kungsbacka on October 22, 1940, where he was still teaching, and finally to Vasa community in Gothenburg, on December 7, 1944. Practically, he was already 70 years old by that time point, so that in Gothenburg he would like to spend his retirement period together with his siblings and their families. By that time the only sister of him, who was still living in the city, was his younger sister Dagmar Augusta.

But Nils would not like to simply get rid of the topic of his life. So that he decided to spend his retirement leisure by defending his PhD at the Uppsala University. The disputation had taken place in the year of 1946, while his PhD thesis was published in 1948 in two languages. That was his very last story. At that time Nils was already 71 years old [*sic*]. Here we would like to present the English part of his PhD thesis, especially because it is discussing in detail all the foregoing publications by Nils—and it really allows reconstructing practically everything in regard to Nils and Karl's work, including the professional fates of them both.

Moreover, the Nils' 1948 publication allows seeing everything up to the finest details as concerns the actual thermodynamics, Nils' own seminal suggestions in the field etc.

With this in mind, our plan for the following story in this chapter would be as follows:

1. We would like to provide the readership with a concise presentation of the essential work done by Nils and Karl;
2. We present the translation and republication of the critical lampoons by Nils.

2.3 Who Are Nils and Karl, and Why They Should Be of Interest to Us?

First, we would like to sum up all the hints scattered above.

Dr. Nils Engelbrektsson was an outstanding Swedish theoretical physicist, who could be able to mathematically derive the truly universal thermodynamic equation of state.

Nils Engelbrektsson's equation of state is not only mathematically rigorous; no drastic physical–chemical approximations underlie this equation as well.

Nils' equation ought to be truly universal, because it is strictly valid not only for the perfect ideal gas, but also for the real gases, liquids and solids.

Karl Alexius Franzén was a colleague of Nils, who could manage to experimentally check the equation by Nils in full detail—and thus to unambiguously demonstrate its universal character.

2.3.1   What Was Nils' Driving Force?

In 1920 both Nils and Karl have published a monograph—most probably on their own costs. We have learned about this publication only occasionally, by finding its main part, meanwhile without the supplement, at one of the antiquarian bookshops in Stockholm. The book is decorated with the *Ex Libris* by Dagmar Augusta Engelbrektsson, which represents just the family coat of arms of Engelbrekt Engelbrektsson (1390–1436). Two possibilities immediately come to mind—a wife of Nils, or his sister? The Internet search gives the correct answer: She was Nils' younger sister.

Nils Engelbrektsson/Karl Franzén (**1920**): '*Termodynamikens grunddrag*' ('Outline   of   Thermodynamics')/*Verification   of Engelbrektsson's equation.*

[Please see Supplementary Note 1, Section 2.8, Page 296]

And now, our English translation of the two parts of the Nils Engelbrektsson's above treatise in Swedish is to follow, and namely: 'Preface', as well as of its 'Chapter I' contents.

Noteworthy, this treatise—as well as the entirety of the publications by Nils Engelbrektsson—definitely deserves an English translation, a separate professional re-working and the following re-publication.

Most probably Nils was publishing his works at his own costs, and his writings were thus never properly distributed. Except for his ***Termodynamikens grunddrag***, the full electronic copy of which (the main book plus its supplement) might be downloaded in electronic form via the Internet portal '*HathiTrust*', only a couple of his publications have been available in different Swedish libraries. For instance, initially we could fetch only five publications of him: Of them the two copies were available to the Swedish National Library (SNL), four copies at the Library of the Chalmers University of Technology/University of Gothenburg. Our efforts have resulted in fetching of the entire set of Nils' works, at least in the SNL. The present author is extremely grateful to Mr. Mårten Asp from

SNL—for his immense librarian activity, for the production of the electronic copies of many of the Nils' publications and placing them to the Internet under the following address:

http://libris.kb.se/hitlist?q=zper%3A%22%5EEngelbrektsson+Nils%5E%22&r=&f =browse&t=v&s=r&g=&m=50

Of extreme interest for our present discussion ought to be the Preface to Nils and Karl's 1920 book, so we would like to present here its English translation.

Preface

The present work introduces the empirical approach to thermodynamics.

The mathematics is considered here from the empirical point of view (in striving for reaching the valid correspondence of the general scientific standpoint to the Nature being the World of Phenomena). With this in mind, all the inferences here have been carried out to a point, where they might generally convey the solutions to the thermodynamics problems. The present work does not intend to give a broad representation of the topic as a whole. Instead, it ought to be focused on the general principles by oozing all the pettifogging topics.

I have asked the Swedish government to provide me with a possibility to teach the topics I am discussing in this book. But my offer has been left without response. The Appendix (The verification of the general equation of state) accompanying the present work is written in such a way that it can be read independently of the main treatise.

Mr. Karl Franzén from Hudiksvall, Sweden, has written this appendix. To my mind his publication delivers a sturdy proof that my ideas, which the Swedish academic world had taken for a savage alien heresy, are nonetheless valid, hence, I have now no need to lose more words about this theme. Mr. Franzén's results speak for themselves.

The Author

What can we read from the above fragment? Indeed, Nils has carried out some, to his mind, very important theoretical work; Karl has successfully helped him to verify the correctness of his inferences. Still, the Swedish academic community has not accepted the results thus obtained. The application to the Swedish government could not help. The questions immediately arise: Why? What has but happened at all?

To try an objective evaluation of the whole story, let us first have a look at what has been accomplished by Nils and Karl.

2.3.2   What Is the Essence of the Contribution by Nils?

The first question would be: What is Engelbrektsson's equation of state? Karl Franzén introduces it as follows—we just cite here his English description of the problem and the tool:

The term 'coexistent phases', as referring immediately to different phenomena and generally to a notion, might be expressed analytically by Engelbrektsson's equation of state.

Let us take two rectangular axes: the axis of T representing the temperature and the axis of p representing the pressure (Fig. 2.1).

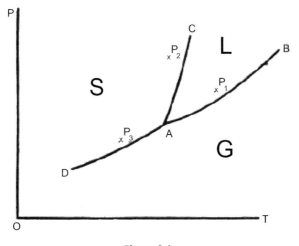

Figure 2.1

The portion of the plane, G, being to the right of the line BAD, may represent the gaseous state of the substance considered. The portion of the plane, S, being to the left of the line CAD, may represent the solid state. The portion of the plane, L, being between the lines AB and AC, represents the liquid state. The point A is then the 'triple point'. The substance, even if solid, is regarded to be isotropic.

Let us denote any two of the phases G, L, S by I and II, respectively, liquid and gas, solid and liquid, solid and gas. The mass of the substance considered is 1 gram, (1 – μ) of which is in the phase I, and μ in the phase II.

[Please see Supplementary Note 2, Section 2.8, Page 296]

(α) *Let us take any point P_1, whatever, on the left of the curve AB and in the vicinity of the curve. At the point P, liquid exists with a value for volume and thermodynamic potential. At the point P, also gas is capable of existing with a value for volume and thermodynamic potential; gas, when existing at P_1, is supersaturated vapor. Denote by v^I, z^I respectively the specific volume and the thermodynamic potential of a gram at the point P_1 for the liquid. Further denote by v^{II}, z^{II}, respectively the specific volume and the thermodynamic potential of a gram at the point P. for the gas. According to Engelbrektsson, at the point P_1 the quantities T, p, v^I, z^I, v^{II}, z^{II}, μ are subject to the following relation:*

$$-T\,\frac{\dfrac{\partial v^I}{\partial p} + \eta\left(\dfrac{\Delta z}{T}\right)\dfrac{\partial \Delta v}{\partial p}}{\Delta v} = \vartheta\left(\frac{\Delta z}{T}\right) \tag{1}$$

$$\mu \equiv \eta\left(\frac{\Delta z}{T}\right) \tag{1a}$$

$$\eta\left(\frac{\Delta z}{T}\right) \equiv \eta_0 + \eta_1 \frac{\Delta z}{T} \tag{1b}$$

$$\vartheta\left(\frac{\Delta z}{T}\right) \equiv \vartheta_0 + \vartheta_1 \frac{\Delta z}{T} \tag{1c}$$

$$\Delta z \equiv z^{II} - z^I \tag{1d}$$

$$\Delta v \equiv v^{II} - v^I \tag{1e}$$

Equation (1) is a partial differential equation having temperature, T and pressure, p as independent variables.
The order of the constants (h_0, h_1) and (q_0, q_1) is that of decreasing magnitude.

(β) *Instead of taking AB, P_1, L, G let us take respectively AC, P_2, S, and L. On introducing v^I, z^I, v^{II}, z^{II} let I refer to S and II to L. At the point P_2 the quantities T, p, v^I, z^I, v^{II}, z^{II}, μ are subject to the following relation*

$$-T\frac{\dfrac{\partial v^I}{\partial p}+\eta\left(\dfrac{\Delta z}{T}\right)\dfrac{\partial \Delta v}{\partial p}}{\Delta v}=\vartheta\left(\frac{\Delta z}{T}\right) \tag{2}$$

$$\mu \equiv \eta\left(\frac{\Delta z}{T}\right) \tag{2a}$$

$$\eta\left(\frac{\Delta z}{T}\right)\equiv \eta_0+\eta_1\frac{\Delta z}{T}+\eta_2\left(\frac{\Delta z}{T}\right)^2 \tag{2b}$$

$$\vartheta\left(\frac{\Delta z}{T}\right)\equiv \vartheta_0+\vartheta_1\frac{\Delta z}{T}+\vartheta_2\left(\frac{\Delta z}{T}\right)^2 \tag{2c}$$

$$\Delta z \equiv z^{II}-z^{I} \tag{2d}$$

$$\Delta v \equiv v^{II}-v^{I} \tag{2e}$$

Equation (2) is a partial differential equation having T and p as independent variables.

The order of the constants (η_0, η_1, η_2) and $(\theta_0, \theta_1, \theta_2)$ is that of decreasing magnitude.

(γ) Instead of taking AB, P_1, L, G in the case α) let us take respectively AD, P_3, S, G. On introducing v^I, z^I, v^{II}, z^{II} let I refer to S and II to G. in this case Equation (2) is still valid. Mathematically, the ordinary solution of the latter equation corresponds to the case β), whereas the case γ) delivers its singular solution.

Purely geometrically, the points P_1, P_2, P_3 represent the points on the opposite sides of the curves AB, AC and AD, respectively. They also represent the very points on these curves.

Physically, the term 'coexistent phases' immediately refers to any one of the following phenomena: The vaporization curve AB, the fusion curve AC and the sublimation curve AD. Mathematically the term in question generally refers to the equation of state by Engelbrektsson:

$$-T\frac{\dfrac{\partial v^I}{\partial p}+\mu\dfrac{\partial \Delta v}{\partial p}}{\Delta v}=\vartheta\left(\frac{\Delta z}{T}\right) \tag{3}$$

Mathematically, Equation (3) is a partial differential equation with the independent variables T and p, connecting the functions v^I, z^I, of the state I of the substance with the functions v^{II}, z^{II},

> *and z of the state II of the substance at the same values of the*
> *absolute temperature T and pressure p.*

In his supplementary treatise, the Introduction part of which is cited above, Karl considers in detail and deals with coexistent liquid and gaseous phases in the Part I, with coexistent solid and liquid phases in Part II, and with coexistent solid and gaseous phases in Part III.

To check Nils's theory, he uses the following standard experimental physical–chemical data of that time, which he denotes in what follows for brevity:

L-B. = Landolt-Börnstein-Roth: Physikalisch-chemische Tabellen (1912).

T. A. I = Tables Annuelles de Constantes et Données Numériques I (1910).

T. A. II = *** II (1911).

T. A. III = *** III (1912).

The final results by Karl truly speak for themselves, as Nils has mentioned in the Preface to their common publication. Nowadays, they would definitely need a careful thorough re-investigation.

The next question is: How Nils could manage to derive such a truly universal equation of state?

The answer could be found in his relevant publications. We shall publish the full list of them at the end of this chapter, but for the present we would just like to present a concise description of his train of thoughts.

In the Preface to their 1920 work, Nils mentions that he is introducing the empirical approach to thermodynamics.

What does this mean? Is he reconsidering the conventional thermodynamics? The answer—yes, he is. The next question—how he manages to do so?

The answer—first of all, he reconsiders the fundamentals of thermodynamics. The next question—what basis is he using to achieve this?

The answer—his second step is to reconsider the entire methodology—this is what he is writing in detail in Chapter I of their book.

Here we would greatly appreciate to present the concise 'lecture notes' of this chapter in the sincere hope that his part of the publication will be properly translated into English in detail and thoroughly commented by the specialists.

Chapter I.

Thermodynamics conditions in phenomenology, in mathematics and in the two basic laws of thermodynamics

1. *A definitive natural concept*
2. *A natural context*
3. *The relative phenomenal forms*
4. *Directions of thoughts at the expansion of the experience area*
5. *The general shape of a real pathway*
6. *Refinement of the scientific concepts*
7. *Limiting concepts in mathematics*
8. *Variation of a double integral*
9. *Variation of a double integral*
10. *The inverse concept at a generally taken variation*
11. *Graduation determined in contrast to approximation*

    The first and foremost, the above-cited paragraph headlines insist on the primary significance of the *pertinent classification* of the observables, irrespective of which particular phenomena are to be studied. The main point here ought to be the intrinsic dialectics of the observed phenomena. It is just this dialectics that ought to be one of the main methodological foundations of classifications. With this in mind, Nils clearly shows how to properly translate the relevant classification principles into the language of mathematics. According to Nils Engelbrektsson, the methodological basics for working-out the proper classification approaches are derivable in a straightforward manner from the work by the famous Swedish botanist, physician, and zoologist Carl von Linné (1707–1778).

12. *The basic laws in thermodynamics*

    Nils' approach definitely allows viewing the both laws in their intrinsic interconnection.

    Remarkably, this is in full accordance with the ideas of such a clear frontrunner in the thermodynamics field as August

Friedrich Horstmann, whom we have already met in Chapter I, although we find absolutely no references to Horstmann's work by Nils. This could be explained by the fact that the scope of Nils' interests was physics, not chemistry—and not even physical chemistry. Our conclusion ought to be—most probably, the true ideas were in the air, irrespective of the existence of the true international research community and Internet.

The only question remains but still unanswered: *Why do we then forget the colleagues who were the active, productive carriers of the true ideas*? Just a rhetoric question.

13. ***Application of the first basic law to the entropy concept***

Nils insists that the energy is the most general concept of thermodynamics. In fact, this is in a clear and dramatic contradiction with the standpoint of the Boltzmann–Planck adepts, who were praising the probability notion instead. *Was this just the actual reason why Nils himself and his efforts were never accepted?*

The following part of this paragraph is discussing in detail the notion of *some limiting concept, where the actual limit ought to be imposed from outside*. Finally, Nils concludes that **such a limiting concept might actually be viewed as entropy**. This is but in full accordance with ideas of August Friedrich Horstmann, which have been readily taken over by Gibbs, as we could learn from Chapter 1 here. On the other hand, this is in full accordance with the conclusions by George Downing Liveing (for more information about the details of Horstmann and Liveing's ideas see Chapter 5 of the present book).

The next chapters of Nils and Karl's 1920 work are dealing in detail with the mathematical inferences according to the above-mentioned principles. Nils is carefully analyzing the relevant contributions of two authors from Germany: Max Planck and Max Bernhard Weinstein. Everybody knows, who Max Planck was, but who was in fact the latter colleague we shall see in Chapter 3 of the present volume.

Remarkably, in Nils' detailed deliberations we find absolutely no references to the works by Josiah Willard Gibbs.

Nonetheless, to sum up, Nils' work was a brilliant development of Gibbs' efforts in the field of research concerning the phase rule. As

J. W. Gibbs had only a restricted time upon the Earth, like all of us, Nils and Karl were indeed the colleagues who could in fact bring the story about the aggregate states of the matter to the deservedly high position.

Peter Boas Freuchen, who was living just in the nearest vicinity to Nils Engelbrektsson and Karl Franzén, had absolutely no idea about the marvelous work by his Scandinavian colleagues—he was but extensively citing and analyzing J. W. Gibbs' works.

Moreover, the 'widely unknown' Nils Engelbrektsson was in his work actively using the results of the not less 'widely unknown' Max Bernhard Weinstein (see Chapter 3 of the book at hand).

Alas, it looks like that such are the natural vicissitudes of the 'true researchers' fate. The true researchers were, are and definitely will be working not for the publicity, but to get some interesting and important results.

There is an interesting point to be noticed at once, but we shall discuss this in more detail later on in this chapter.

Remarkably, during the above study Nils was just the theoretical physicist, who was planning and conducting his research, whereas Karl had added the definite chemical—or better to put it—physical–chemical/chemical–physical note, to verify Nils' theory.

Interestingly, in Chapter 1 of the present monograph we have seen that Karl Mohr was heavily criticizing Rudolf Clausius for having paid no attention to chemistry—meanwhile, Nils and Karl could have duly taken this point into account.

Remarkably, in Nils' publications we see no references either to Karl Mohr or to other adepts of the *different thermodynamics*. Howbeit, both Nils and Karl definitely belong to the forefront of the latter.

2.4 The Struggle of Nils Engelbrektsson for the Different Thermodynamics

As we have already mentioned above, from the beginning on the then Swedish academic circles seem to have completely rejected both Nils and then also Karl's results. Here is the English translation of Nils' critical lampoon on this theme, as published in the year of 1911.

This following publication was in effect the third one in Nils' publication list, to follow his lampoon criticizing the then Swedish school education system. But now we let Nils speak out.

The Royal Society for Science and Letters in Gothenburg and the Scientific Initiative

By Nils Engelbrektsson: Fil. Lic.

Lund, 1911

Håkan Ohlssons Editorial and Printing Company

Some time ago I have submitted a mathematical treatise to The Royal Science and Letters Society of Gothenburg. I have requested to consider this manuscript for publication in the Royal Society's media. A few months later I have received from the Royal Society secretary a notice in the case. The message has contained the following information:

'**Gothenburg, the 16/11 1910.** *The Royal Society, through its experts has undertaken a careful consideration of your submitted manuscript; but at the last meeting has unfortunately felt impelled to decide that it is inexpedient to publish it.*'

My immediate concern was to try taking advantage of the critical work carried out under the Royal Society's auspices.

Sure, my work had failed—possibly missing the deeper initiatives; well, I thought before, that it contains new interesting points—nota bene for mathematicians, but things that I had meant to be new and interesting could actually turn out to be well known. Were the Royal Society's experts able to determine, that my work lacks deeper initiatives to follow, they would have no other way than rejecting it, for the work in question had been reasonably unnecessary. Still, the rejection could in fact have a completely different reason as well; Royal Society's experts could have encountered some obscure points in my work, which appeared to require somewhat more detailed exposition. Hence, the rejection could definitely have good reasons for it.

It was of clear importance for me to know, what this rejection was in fact due to. It has become clear to me that there was something in my work, probably some points, which looked like being dim, so perhaps there even ought to be a task of reworking some parts of my thesis, which were affected by some detected errors. It has become clear to me that I should immediately take initiative, not wasting time and effort on recasting my work.

I needed therefore to ask the Royal Society for further information on the matter. I had asked what kind of objective reasons were there, which in fact underlie the Royal Society's rejection of my thesis. I took the liberty—for the sake of timesaving—to ask my question to the secretary personally.

The secretary's reply came promptly. His response consisted in that himself he was not the person with the necessary competence in the field involved, and that he therefore could not participate in solving the problem.

'Meanwhile, the case was submitted to the actual members of the Society's mathematical and physical class, that is, to a group of about 3 to 4 persons dealing with these matters. Who are these, you might easily find out in the state archive or address calendar. Their negative opinion concerning the publication might partly be owing to the fact that recently a (comparatively) large amount of mathematical work has been submitted to our indeed strongly limited yearbook. Surely you might even personally know some of them and this way easily deduce, the purely scientific reasons of which kind the experts were referring back to in fact, and whether there might be some possibility to revise the manuscript and then subsequently publish it. Unfortunately, I myself do not even know, what they think about this topic indeed. For the official protocol it had no meaning, and unfortunately it is occurring so often that we have to reject the publication of the manuscripts just because of the lack of space, and thus, in fact it should not be considered as an actual criticism of the manuscript's contents.

In the hope for having answered your questions as satisfactorily as I could, etc....'

The secretary mentions the lack of space, as one of the possible reasons because of which Royal Society has felt impelled to decline the publication of my thesis; and he maintains at the end of his letter that it might just be the reason already mentioned, which probably could have no serious meaning in this regard.

Secretary's highlighting of the space issue might in principle be understood as a couple of words of encouragement to consider the whole story just from that particular point of view. Hence I had to concentrate my attention on the exact meaning of the words about the lack of space, in trying to understand how I should interpret the rejection of my manuscript—to be due to the lack of space as the actual realistic reason, or just as one point in the list of a number of actual reasons.

Meanwhile, the space issue ought to be a topic for the scientific community at the preparatory phase, when separating submitted manuscripts of inappropriate prolixity from those of reasonable girth. The former kinds of manuscripts are normally sent back without further examination. The manuscripts of a reasonable comprehension are then studied further in terms of their quality. Is their quality excellent, if they represent realistic initiatives in their field—are the expressed thoughts not only verifiable, but also new and bold— if they deliver at least as clearly as possible formulated solutions of the tasks being central for or even major tasks in the field, then there is probably no difficulty as for their actual girth; they would surely be kept and then taken up as soon as the space for them is enough. Sure, the space ought to be available at some time, for the publication of the Royal Societies yearbook is not to be stopped soon. Instead, is the quality of this particular manuscript poorer than that of others available for comparison—would the former be less enterprising, less clearly formulated—then it would surely be sent back with the proper notice thereof as well. Finally, we might well assume that the scientific community, in sending the manuscripts back with the notice of their wretchedness as concerns the quality issues, would not like to inflate the happening to incredible limits of some 'big problem'—by giving itself a damn of revealing the very fact in the plaintext—but, instead, by reducing all the story down to the tiniest issue, by just mentioning that the manuscript in question does still have the same suitable girth as some writings of truly excellent value, which were submitted for refereeing at the same time.

The way the Royal Society's secretary had mentioned the space issue in my case was just to give an impression to the Royal Society's members that I have submitted a manuscript of inappropriate prolixity. In fact, my whole manuscript, if printed with the fonts of the standard size, would not even occupy some three sheets of paper. Hence, such a manuscript might well be considered to be of negligible girth, but not at all as something of truly unreasonably large scope. Indeed, then it ought to turn out that the Royal Society had not at all evaluated my story as something being of unreasonable girth, but, in essence, they were just considering the very content of my story. In case it could even be raised in their final announcement, the space issue should then just accompany their notice about the manuscript's acceptance. And the actual reasons for the very rejection could not be revealed in the final announcement—or even deduced based upon the latter, to duly separate them as something except for the essential space issue mentioned.

As to the instructions the secretary gives in regard to the discovery of those who handled the case in the capacity of expert, they seem at the first sight to be quite sufficient.

I do not know if there are any more official representatives of the science in question (mathematics) in Gothenburg than just teachers at public graduate secondary schools of the region, as well as lecturers in the field at the local School of Technology (Chalmers Technical University).

*Howbeit, it is in such a peculiar way that the high schools do summarize the concept of the scientific research possible within them. Consequently, it is truly not as simple and as practical to perceive it, as scientists might believe. The high schools tend to accept only such people in whom the scientific research zeal appears to be subdued to some certain extent. The high school teachers as themselves must have good ratings of their capabilities in many scientific fields but, along with this, their zeal for scientific research must anyway be definitely plumped down. In fact I have myself applied for a number of positions at the general secondary schools, referring to my skill as a scientist in my scientific field (and that is just only one field, nota bene!), but my application has been rejected as a basically non-competitive one. I would like to refer here to my booklet "The Royal Head Office for Secondary Schools and the Teacher's Personality" (**Our comment:** It has also been published in 1911). That means I have definitely to apply to the actual scientific research community in regard to the tough piece of my science, with the brand new research done only by myself, for, I guess, I should not think, that grammar school teacher might be entrusted with the examination of this work. And the 'modus operandi' ought to imply a peculiar circular process.*

Thus, with all this in mind, I have written in causa to a senior lecturer A. Söderblom working at the Chalmers Technical University. I had brought up the secretary's statement that my work has been rejected partly because of lack of space as mentioned and partly because of something else, most probably, of something in its content. I gave a lecturer Söderblom to understand that I would be pleased with learning about the good reasons for doing so, and asked him to let me know what were the actual criteria of examining the content and/or estimating the value of my work. Hence, are these the actual criteria for estimating my work? That ought to be the question I had no nerve to pose to the secretary, in view of the statement he had written and even underlined, that he did not know what exactly the experts were thinking about it. And that was just what I had to ask lecturer Söderblom about. I had asked him so distinctly that his answer—and that was just what I asked him about—ought to provide the barely factual information necessary for me. I asked whether it was something wrong with my whole initiative. Had the error been discovered, I requested him to let me know then at least about the most significant points.

My letter has been sent for about four months ago, but I have got no answer as yet. Answering to my questions would not seem to take longer time than that already elapsed. In fact, my questions should not sound so strange, for they are just what any scientist should always ask himself, if he decides to be involved into a critical work for scientific communities. Before he reports the results of his assessment to the scientific community, which had asked him to perform such a work, he should already have absolutely satisfactorily reported to himself, as concerns all these issues. Then, any different and/or better response should not be expected from him, as compared to the very report he has already delivered to himself.

I do not think I could somehow infringe on my critics' honor by just requesting him to reiterate his scrutiny. Reviews and scrutiny, over and over again, do surely and steadfastly belong to the practice of scientific progress. Definitely, I could have suddenly 'rocked up' in the long line of other investigative scientists at a critical time by expressing some initiative of mine, which represents a peculiar part of my professional activity and which allows to arrive at the correct judgment as to such a move of mine. Then, the latter means that in fact my approach should not be viewed as—which lecturer Söderblom might possibly take for granted—a pure impudence of mine, without taking into account my attachment to this theme, which I have chosen to be a part of my professional activity. Anyway, in our life it is not the usual practice to abandon our domains without any good reason. Should I then have reasons enough to act against the habits and practices in this case? Do I have no more competence to deal with the criticism, even when the problem concerns my initiative or strictly remains within the scope of my deepest expertise? How could lecturer Söderblom take these posers for already answered, even before his trying to do so?

Even the Royal Society's secretary, when notifying me that the Royal Society's experts have taken my submitted manuscript into the "careful consideration", has not presented the latter in the lecturer Söderblom's sense: that is, either as an irretrievable survey or as an ultimate practice, beyond which nothing more can be practiced. The secretary suggests instead that the Royal Society's critical work might well be useful, and especially for my case: he tells about the possibility of a revision starting with the Royal Society's critical comments. Even more: he assumes the possibility that the actual rejection of my manuscript should not be immediately taken as the criticism concerning the manuscript's contents. That might certainly be an actual task for a scientist to try finding out some details of this rejection, with respect thereto that the Royal Society community expresses by its decision the criticism just the manuscript's content, and therefore delivers the true evaluation of the job involved, purporting to be of general significance. Then I would probably tend to think: Oh, it is just me who would be almost ready to take up any scientific task. Well, I guess I would not immediately and completely refuse working on my theme, as lecturer Söderblom might probably wish, when he refuses to answer the questions in this case.

Otherwise, may perhaps, as lecturer Söderblom had not responded to my letter to him on the matter, he would simply like to stress this way that his silence would express his total absence of any interest in this particular case.

The question is then: Who exactly could have been connected with the matter in question? The only available source to refer to would be the information I have received from Royal Society's secretary: "...to the city residents being the members of the mathematical and physical class, that is, to some 3 to 4 persons who were involved in dealing with this matter." But who are these people? "Who are these, you will find easily in the municipal address book or address calendar," told me the secretary. What he did actually mean by such a disclosure is, however, not that easy to grasp.

The municipal address book for 1910 does not contain any more comprehensive classification in regard to the members of the Royal Society for Science and Letters at Gothenburg, than just the list of the main divisions' executive members (98 names) in addition to its honorary members. The Gothenburg's address calendar for the same year, although providing somewhat more detailed information, delivers the only division of the working members, namely: the scientific department (78 names) in addition to the literature department. But any subdivision of the scientific department into the members of "the mathematical and physical class" and/or other classes is missing. The calendar does not provide us with any more distinctive feature to enable me to recognize the 3 to 4 members of the mathematical and physical class, who would be more competent, as the secretary seems to think, when he sums up to himself all those who ought to be at home in the field involved.

That was really a peculiar countermove by Royal Society to send me to such a mystery tour for the purpose of looking for the expert names. The Royal Society should not need such a feature as a rule. Instead, the Royal Society ought to publish a particular short announcement in the local media (in H. T. and in G. A. B., at least) concerning the manuscripts submitted for refereeing, even before the very audit has taken place, where, in addition to the manuscript's topic (title) names of persons (usually two) are given, who are appointed to perform the relevant expert examination. The fact is that in my case no such announcement regarding my manuscript was published. Remarkably, the information that the manuscript submitted to the Royal Society has never appeared in local press. The names of the persons appointed to examine the scientific scope have never been published. I guess I would have the right to complain to the Royal Society about all these strange facts. The Royal Society seems to have caused me a totally unnecessary inconvenience in that I might in principle look myself for the experts to perform the refereeing.

"Surely you ought to know personally any of these gentlemen, and would then easily learn in such a way..." told me the secretary, alluding to some 3 to 4 expert members. May perhaps, it is in these words that the true hint ought to be contained. Indeed, I can get the best answers to my questions through a common conversation, in the form of a private communication. May perhaps, when considering anyone of the

3 to 4 expert members, I would have really no answer to reckon, if I do not accept this form of communication.

The only danger of any private communication ought to consist in that the latter might too easily turn into an utterance of private significance.

Any utterance of private significance might, and I admit this, target unspeakably high aims, sometimes even so high that it is overwhelming all the degrees of height. Any utterance of private significance might, and it is also very well known, might also target indescribably low aims. The aim of the work I have submitted to the Royal Society: 'The variational concept as the development form of the differential concept

(integrated form of the differential equation $N - \dfrac{dP}{dx} - \dfrac{dQ}{dy} = 0$ *)',*

can be interpreted in several relatively different ways: it might be seen as a real masterpiece of me myself—the very promoter, the very pioneer; the latter might perhaps be reduced in its importance by a severe censor to something much less than what I would first imagine. Anyway, it cannot be driven so high or so low as a private utterance. For it is a public utterance! Would it be itself coerced into disappearing from the public view, would it be entirely faulted down, then at least the criticism ought to be public. An author of a scientific work could not reach the necessary success so then his work can be rightly faulted down. However, when faulting the work down, the latter must not be criticized privately or subject to avoid having to pay for their own words; people would never let satisfy themselves with a criticism containing nothing more than just pure provisions; people would like to play a more open game, especially in cases where the fate of the scientific initiative is being decided.

What I have just said about the character of a manuscript presented to the audit contains no caveat against the Royal Society's rules. When the result of the review, there are then "purely scientific reasons" for or against; surely the secretary's letter to me assumes this as something self-evident. Meanwhile, purely scientific arguments for or against ought to be quite different from the private arguments for or against, because the former is expressed by the public words, so there are at least several words that might be presented to the public, in case anyone would be interested in such utterances and expects them. The Royal Society would most certainly hold out for the public character of the manuscript, but against the measures I have already told about earlier, i.e., the announcement via the newspapers concerning the manuscript's topic (title) to come out even before information about the manuscript's own development, its subject and formulation, how sustainable the development is.

With my manuscript the Royal Society would seemingly like to make an exception to the general rule. Sure, there are purely scientific reasons, which speak against it. And the Royal Society insists on this; the secretary's letter tells me black on white on it. Still, except for the very allegation I have not got anything in black and white; that is, the very learning about what these scientific reasons are does not seem to be up to me. Not only my considerations of the subject, but also my subject itself, have been assigned a character of a private utterance. They have not even bothered themselves to make it public in the usual way—using the local newspapers—in communicating that the manuscript treating such-and-such subject has just come to scrutiny.

My subject is a particular problem that is connected to the calculus of variations; the problem can be described by the following equation $N - \dfrac{dP}{dx} - \dfrac{dQ}{dy} = 0$. My own result in this regard might also be described with the same degree of brevity, as follows: I have just solved this equation in the common scientific sense.* The solution to this equation is unknown to the professionals, but the actual problem described by such an equation type is throughout well known. The acquaintance with the problem is included into the requirements for the A-grade in the test before getting the relevant Fil. Kand. (the PhD degree). When considering such a problem as a subject of some investigation it is throughout clear, that the latter would never bear marks of a purely private utterance; it will never suffer from the lack of objectivity. Rather, as a rule it should excite a quite general interest, likewise any modern mathematical problem striving for increasing the variety of essentially different tasks, on the basis of the general quest of mathematics and in parallel to any modern branch of the natural science.

As the Royal Society ought to treat my initiative as if it ought to separately exist by itself, and as if its investigation ought to lack any

---

*I denote this a solution in the common scientific sense, or a certain simplification of the present task to enable the substitution of the initial non-linear partial differential equation of the second order by a system of linear differential equations, of which the one ought to be of the first, and the remaining one—of the second order.

objective interest, the Royal Society does not seem to have tumbled to the actual sense of my initiative, when considering the whole topic's own nature.

If one takes into account my position as a comparatively isolated mathematician—I am working already several years in a lower educational institution—then it should be clear that it was of a particularly great interest to me to submit my professional report to a specialist medium, to present it to the expert members of the scientific community. This way I was hoping to get the experts' response in the form of a truly professional message, the message being detailed precisely to the extent of criticism, which is relevant to my manuscript, which is truly elaborate and rigorous. Furthermore, I would expect to get the experts' response in such a form that it really bears witness to their refereeing efforts and this way ought to be useful for my further working efforts, that is, I would expect to get such a response in the written form.

The manuscript I have submitted to the Royal Society for the purpose of the referee's audit has been accompanied by a cover letter in which I have expressly emphasized that the manuscript's handover should take place only under the preordained condition, namely under the premise that I would get the full clarity concerning the factual shortcomings discovered during the audit, in the event that my manuscript would not pass well the latter.

If the Royal Society considers the condition I put for the case of my manuscript's failure to be unreasonable, then the natural choice of the Royal Society would just be to send back my manuscript, enclosing the short notice that submitting the manuscript was connected to the conditions in principle precluding its consideration under the Royal Society's auspices. Instead, as I now get the notice that the manuscript has been accepted for refereeing and subjected to a thorough scrutiny, then I should rightfully conclude that the refereeing audit has been done by accepting the terms and conditions under which manuscripts are submitted for review. If it would not be possible for me to conclude this way, then I would not get any clear statement from the Royal Society in my case, at least not before the Royal Society would state loudly and clearly that for some reason they had found themselves obliged to break this trade agreement between us, together with their promise that I have received.

To break the trade agreement in question, to break a freely given promise, ought to belong to the modus operandi lying definitely under the Royal Society's dignity.

There is no doubt for me—for I have the distinct assurance by the secretary of the Royal Society that the critique of my writing has been thorough. The Royal Society could thus provide all the relevant information in detail. This way, the Royal Society would render their review very useful both for me and for others. Well, then the Royal Society should also do so. Otherwise, all the criticism—whatever sober, factual and precise it might have been—cannot be recognized as some word of crucial importance and extreme usability in my case. "Usability is the token of the truth", tells an outstanding Danish philosopher Harald Høffding (1843–1931).

May perhaps, the Royal Society guesses that I have no sufficient proof for viewing my interest as being of purely scientific nature, and hence my claim to benefit from the Royal Society's purely scientific arguments against my work might safely be rejected.

Fortunately, here I should not trigger any long-winded reasoning to defend the chastity of forces driving me. But, as soon as I would herewith like to inform the Royal Society, I choose to put in front of me another young Swedish mathematician striving for the same as me, a young mathematician heeded by his friends and teachers for his energy and eminent talent in this direction. Does the Royal Society believe the youth yet?

I have got a postal card from this mathematician. I would not like to disclose his name here; but I could only mention that the postal card from him is available for viewing by anybody who wishes to look at it. He writes me that he would greatly appreciate soundly diving into the ideas I have hinted in my written notice: 'A Report to the Mathematical Congress in Stockholm, 22–25 September. 1909', by assuming in this regard that now I can present a strictly mathematical proof for everything published only as preliminary assertions in my report mentioned above. Remarkably, in another letter to me he emphasizes that his interest in this matter is purely scientific. I have no reason to distrust him as for the nature of his interest. Moreover, I do not know him personally. And it is also well known to him that I am not on intimate footing with the officially leading Swedish mathematicians, (this is apparently clear enough from the introduction to my above-mentioned "written notice, etc."). He knows as well that so far he cannot count—at least immediately—on some tangible benefits through his cooperation with me, Therefore, I have no ground to distrust him when he states that his interest is purely scientific. With this in mind there is also no need for any further suspicions on his account.

In effect, my presently discussed manuscript concerning the partial differential equation $N - \dfrac{dP}{dx} - \dfrac{dQ}{dy} = 0$ deals with the development of one of the ideas I have alluded in my aforementioned "written notice, etc." Thus, my manuscript ought to deal with the problem mathematicians would definitely express their interest in. May perhaps, here I could not offer them much more than just my initiative as it is, just the originality of my thought. There is but not enough scientific truth, states one of the working members of Royal Society (namely, an outstanding Swedish philosopher Johan Vitalis Abraham Norström (1856–1916), the author of "Radicalism Once More"): "Individuality is the light source in the world of thoughts, but only thoughts resembling refracted rays we denote the scientific truth." And the Royal Society, for its part, ought to provide the interested mathematicians with something I cannot immediately provide them with. The Royal Society is expected to present them my initiative in a sharp critical illumination. The Royal Society is said to provide my work with some remarks of purely scientific content (just one remark, at least). Should the younger mathematician in connection with me, who considers my ideas to be his spiritual stove, be kept in complete ignorance concerning some important remarks made against these ideas? Is a clear-cut reticence against a person, who seems to ask quite unreservedly for learning some information about the scientific and/ or critical truth, so necessary indeed?

The Royal Society has told too much or too little, because the reaction of the Royal Society with respect to the submission of my above-mentioned manuscript to the Royal Society's yearbook sounds just like a standalone assertion that there are purely scientific reasons against the manuscript submitted. Had the Royal Society more to say (as something to defend their actual claim) in regard to their decision—then it seems a little strange that any foisting ought to be necessary to nonetheless get the necessary information. It seems as if this had been possible to arrive already at my first request I addressed to the secretary.

The above was just the beginning of Nils' life-long struggle. Below we would dare to republish the English part of his PhD thesis.

This is to be considered Nils' bequest, in that he could summarize everything what we would need in trying to promote the further development not only of thermodynamics, but also of any other branch of natural sciences.

In the citation to come below we have to 100% preserve the vocabulary, the grammar, and the style of the original writing by Nils.

Academical Thermodynamics or My Heat Engine Form:
A Dilemma

Uppsala, Sweden, 1946; Helsingfors, Finland, 1948

Setting down heat engine to be the generic name of generators and motors at large, we are entitled to call the present era the heat-engine epoch.

But from the fact of heat engineering turning out successfully we must not conclude traditive heat-engine science, i.e., academical thermodynamics to be very far advanced, nor yet conclude the said science to be capable of distinguishing declining from advancement. Of course, that science like others is voluminous to infinity; there is no end of it at all. However, if my analyzing these volumes is not lacking strength, academical thermodynamics has gone down so far as to stand below Mayer's point and below Carnot's. A practical operation getting on progressively, while the corresponding academical operation is falling off, is no uncommon thing. Just as heat engineering and traditive heat-engine science correspond to one another, so do the applicable elasticity form and the academical one.

Taking the academical elasticity form for the applicable one would be most dangerous; every schooled engineer knows that. The one he has for need's sake, the other for form's sake (to get his diploma). The common way of talking corroborates the testimony just given: our language—Swedish or English—much as it has its logic at its fingers' ends, doesn't object to us quite seriously if we venture to discriminate between practical problems and Academical.

The professors of thermodynamics are enraptured with the traditive heat-engine science, and the lecturers and doctors—in criticizing it—attack it very tenderly. The general public, however, begins to look rather alarmed. Not that people are running mad for some novelty of machine confronting them. People have common sense; they take any machine, old or new, ancient or modern, to be as innocent as the babe unborn. It is merely that people are feeling doubtful if the overwhelming machine deluge is the needful assistant of our herculean working, or a menacing giant to be repelled.

There is something uncertain about the machine evolution—people think so—from my touching upon where my heat-engine idea had to descend and stand in concreto, I proceed, now, to my main subject. I am to display in what follows how my heat-engine form has been dealt with in Academical quarters. I am to emphasize such points of Academical dealing with it that appear wrong and preeminently deserve notice.

It is for the Academical parties concerned to decide if there is to an only point emphasized such an objection as can well bear coming to light.

AD 1920, as we easily recollect, Einstein's cud was what everybody chewed. To uphold his and his like's hypothesizing Einstein prophesied physics never to be visualized.

A fact there was that told against the prophecy.

The fact that told was my book "Termodynamikens Grunddrag" with "Appendix" (by Karl Franzén), just published (at Lund). The analysation in the "Term. Gr.", albeit definitive as to its mathematical performance, wasn't meant definitively to set forth in words the analytical operation; the displaying of the results of operation, this and no more it had in view.

That the "Term. Gr.", in setting forth in words the analyzing, was lacking form acuteness didn't matter at all, in as much as it was about putting Einstein's prophecy to shame. All turned upon the very results of analytical operation, their possessing or lacking visuality. Franzén's Appendix included a collection of tables, which we should think the responsible Swedish criticism not to have been at a loss to catch sight of, for invisible it wasn't. By itself, in its entirety, it was placed at the end of the appendix.

To some extent it had, too; it took up eight sheets of paper. It was redacted in such a manner as to make each table capable of being understood by itself without acquaintance with anything else in the book; the heading concerned gave all needful information.

I take it for granted the responsible Swedish criticism contrived to glance over at least some two of the tables just mentioned—let us say, the first two; sense of duty can be a motive power in the world, moreover, human curiosity can impel man to go far.

The heading text was in English.

But one word, classification, rang like plain Swedish. A word like that mightily put the responsible Swedish criticism in mind of Linnaeus. The responsible Swedish criticism, of old familiar with Linné and Linné-celebration, felt sure

(1) *About the number of stamens in case of Campanula to be a constant figure, for the figure 5 fell short of expectation for no Campanula specimen examined,*

(2) *About the empirical figure-constant in mention to be non-individual, for, from examining the Primula individual it issues as certainly as from examining the Campanula individual,*

(3) *About the non-individual empirical figure-constant to be a tongue by which the Campanula makes known to Linné and his people that it can't be dark and mysterious throughout as it brings home the clear fact of its determination within the class of Pentandria. The responsible Swedish criticism, revolving in its mind those Linné-ideas, in the appendix, Table I and Table II a, came across the liquid-expansion formula*

$$v = i\left({}_0q_1 \; {}^{10}\log T + {}_{-1}q_0 \frac{1}{T} + {}_0q_2 \; {}^{10}\log^2 T + 1 \right)$$

To itself, I think, it was compelled to admit quantities i, ${}_0q_1$, ${}_{-1}q_0$, ${}_0q_2$ to deserve the name of constants, for, the mercury-expansion formula for interval $\left({}^{-20^0}\!\!\big/{}_{100^0} C \right)$, which in table I has the i, q-values ${}_0q_1 = -0.1279673$, ${}_{-1}q_0 = -1.606785$, ${}_0q_2 = 0.02520534$, i = 0.2415982, holds in full as interpolation formula, and even does as extrapolation ditto for interval $\left({}^{100^0}\!\!\big/{}_{300^0} C \right)$*

(2) *The figure constants q to be non-individual, for, the g-values in e.g., Table IIa ${}_0q_1 = -0.6919$, ${}_{-1}q_0 = -11.26$, ${}_0q_2 = 0.1240$ hold for different liquid individuals, carbon bisulphide, phosphorus trichloride, silicium tetrachloride etc.*

(3) *The non-individual constants ${}_0q_1$, ${}_{-1}q_0$, ${}_0q_2$, if Linné's botany has the least significance, to be tongues by which the heat-expansion process of e.g., carbon bisulphide makes known that it can't be dark and mysterious throughout, as it brings home*

---

*See Appendix "Term. Gr.", Table 17.

the clear fact of its determination within the Class A of Table II a. The clearness that the responsible Swedish criticism ran against in Table IIa and the following tables meant Linné's idea to proclaim its wish to be generalised and to revive in a higher form than that which Linné himself immediately gave. The responsible Swedish criticism didn't care for that aim to live which Linné's idea actually had. It joined in the Swedish and exotic exultation over the Einstein prophecies. Judging from appearances, the carelessness was voluntary, not unintentional.

(1) *Presentation copies of the "Term. Gr." were sent among the rest to the Physical Institutions of the universities at Stockholm, Uppsala, Lund, and to the Libraries of the Swedish universities and technical academies. The Swedish academical spirit of Investigation thus had the tabulation-work of the appendix under his very nose. The Spirit has seen it, I declare he has. He hasn't for all I know turned blind for gluttony, nor has he these last twenty years been continually sleeping, has he? If not, he has seen it, and no mistake.*

(2) *Einstein being quite busy prophesying his and his like's hypothesizing to be lasting for ever and ever, and the fact of the appendix never to be seen on earth, lo. there appeared this very fact like a bolt out of the blue. The Swedish academical investigation spirit couldn't help seeing it. What did the Spirit do, now? Did he honestly tell the truth, and call fact, and prophecy, swearing against fact, twaddle? Devil a bit he did.*

He called the prophecy a brand-new truth, quite the Nobel-Price Truth; that he did. The fact he hushed up, insinuating.* I cannot make of him but a cowardly liar. My tongue is coarse according to my being rather unfamiliar with academical language. Methinks, I can be taught, though.

I want a fine scholar, the responsible Swedish criticism, to tell me: This here cowardly lying, what is the academical name of it, now?

The academical greeting of the appendix tabulation-work—on the tongue-tip as disgusting as cowardice—grew still more loathsome, as the appendix-author came to taste it in full. The tabulation work

---

**See Nils Engelbrektsson, "Notiz von dem Entwicklungsgang meiner thermodynamischen Arbeiten" (A Notice about the Development of My Thermodynamic Works), p. 11: "Offene Sprache war etwas gefährliches..." (Open discussion was something dangerous...)*

established was Franzén's doing, not mine. In the first place, then, it was not I, but an irreproachable witness of mine, Franzén, which was bearing the responsibility for the tabulation work. The academical authorities, when they came to deal with the affair, weren't charged with giving the first-hand examination of a brand-new production, the "Term. Gr." They hadn't but to control an examination just submitted to them for inspection under the form of a determination of precision, the "appendix" tabulation-work, carried out with infinite endurance in the minutest details. They took the liberty of passing over the whole affair in silence. Their silences made Franzén happen to seem a false witness, and made me happen to seem using a false witness.

Artlessly, here springs up a comparison between two sorts differing toto coelo: meat to be thoroughly salted, production to be strictly criticized, on the one hand, a "spiritual great-power" to play the master overproduction, on the other. Labor, in point of principle, is collaboration; to be all alone is to be powerless. When from working alone I went into partnership with Franzén, and my partner appeared one upon whom our collaboration prevailed to make him withdraw from engagements of pecuniary interest, exert himself and do his utmost during some five years; if common sense holds good, my laboring must have gained by my having got a fellow-laborer. However, "spiritual great-power" playing the master over-production, common sense no longer holds. In fact, my laboring has lost, by once having been collaboration. I show it as follows. If I had declined Franzén's offer of service, and had met all alone with the academical criticism, the insinuating silence from academical quarters by common people would have been so interpreted as to qualify my production as a mere self-delusion; no worse interpretation would have been given to it. Now, on the contrary, Franzén in the appendix venturing to range figure columns by figure columns, and to make those figure columns tell in my behalf as plainly as figures can, of course, common people know what to think: "It was a rather brazen-faced imposture that one, practiced on our Swedish academical science some twenty years ago. Our Swedish academicians, if they hadn't expressed their contempt in so strong a manner as by preserving profound silence when they were required to speak, would have become a laughing-stock all the world over." In Karl Franzén's fate there is something rather queer: he comes out the man of matchless ill luck. If, from beginning to end, his figures had been incorrect, the academical scrutiny and judgment against Franzén would have been published, and at full length, too. And lo. It would have been extremely kind. (Is it that a Swedish scholar, maybe an amateur and one who happened to be misguided, ought to be quite left in the dark by the Swedish great-scholars? Does the general public think so?) Franzén's necrology, if there were a need of such, though, would have been most beautiful. Franzén would have been titled "a champion for Science," "a martyr," and all that sort of thing. His fair reputation Franzén would have preserved, depend upon it. This here Franzén would have been a lucky dog, indeed, but for his figure tables lacking faults. Why, I say. For Franzén's tabulating almost everything was pretty well arraigned; but for his having a run of ill luck, his tables couldn't but be erroneous. Among his drawbacks, as he was striving forward, there are two preeminently worthy of notice. One difficulty, which he always encountered, was about obtaining his material of investigation. He wanted experimental figures, and, if possible, from the very first hand. That might seem an affair easily to be settled. The Swedish academicians, as in the Swedish culturing papers they are instructing the general public, with indignation stand up against the pedantry and snobbery of those who insist upon, in every case, having the facts naked, as though facts weren't to be called stark naked when wrapped up in veils of imagination so thin as, let us say, Arrhenius's. One should think the Swedish libraries, in their giving suum cuique, might have had no obstacle to surmount, then: The spirited professional literature, containing experimental figures polished by thought to a nicety, might have been the prerogative of the academical men, as they preferred such to everything; the unwrought experimental figures, and the detailed statements just giving them out of the very first hand, might have been made easy of access to Franzén, as he was the only one wanting them indispensably. For the Swedish libraries, however, there were rules to be observed. Affording Franzén first-hand particulars wasn't according to rule, as those were kept for the academical men exclusively. Not even the second-hand particulars, the hand-books of Landolt-Bornstein-Roth type, could stir their stumps unless called for from academical quarters; to Hudiksvall, where Franzén filled the place of a teacher in a public school, and at leisure was assisting me, it was too long a way to go; those particulars couldn't go that way, really they couldn't. Third-hand particulars, after picking and choosing, could go that way, n. b. if it was about flying visits. Heart and hand, the libraries didn't hand over to Franzén but one publication, the "Vetenskapsakademiens i Stockholm Handlingar." With strong hypotheses and weighty reasons those Handlingar were brimming over, and I daren't say they were imponderables for him who had to pay the freight. Every now and then, in "Annalen der Physik" and in its French and English analogs, Franzén happened to find an experimental figure-value that wasn't improved by hypothesizing, nor even polished by thought. But never, in the Handlingar, from the first of Handlingar to the last, did he find one experimental figure-value that he could make use of.

You see your way, now.

When Franzén set to, he had third-rate experimental-figures to handle. He carried out his calculation. And we beheld lack of visuality in the calculation-results always to go hand in hand with inferiority as to figure-material employed. We now bought some handbooks of the Landolt-Bornstein-Roth type. We had got second-rate experimental-figures, accordingly. Using these figures, Franzén had to calculate as before.

The results of calculation were second-rate. The third-rate results, previously obtained, were to be slighted. First-rate material looked difficult to get at, for love or money. In the book-trade such was not to be had. To himself Franzén figured his prevailing upon Arrhenius to make him have such out of the libraries. And he wrote Arrhenius word. As a reply, Arrhenius had one of his myrmidons send Franzén a certain textbook (Jellinek), which, at the Swedish universities had out trumped another (Chwolson) according to the rule: omnia ruunt in pejus. Of course, if the hill will not come to Mahomet, Mahomet will go to the hill. Travelling Hudiksvall—Uppsala—Hudiksvall, sometime or other, enabled me to supply Franzén with some first-rate figure-material, gathered-up at the university library. Suffice to say: To some extent we contrived to manage Franzén being capable of slighting second-rate figure results in favor of such as were first-rate. As is shown, there was an academical contribution to Franzén's working. Franzén, on that account, was badly off, himself; and we should think his working results might have been pretty much impaired, too. Now, there was another difficulty Franzén had to encounter that seemed still more to threaten the validity of his working. Determination of precision like his, demands a first-rate calculating machine, you should think so. His wasn't. A banking-establishment at Hudiksvall chose to lend him a calculating machine that wasn't at all first rate, being long since dispensed with as unserviceable. Franzén speedily repaired it and set it going. However, Hudiksvall finding out Franzén going on most excellently by the aid of that machine rather ill spoken of, the bank didn't see but that the machine could be of use to the banking business still. And Franzén had to deliver it up, accordingly.

Fortunately, there were in Hudik more banking establishments than the one just mentioned. One of them had a calculating machine, considered quite useless and rejectable. Franzén got at it, took a survey of it, and had it go. A steadfast dependence on that machine one couldn't possibly have. Nor had Franzén. The machine repeatedly wanted close inspection, and that it had. Control of the very machine was antecedent to each control, performed by the machine, and also subsequent to. "I can hardly move my arm," said Franzén laughing. "Shoulder and arm are always aching, as I do not but turn the crank. When asleep, I am turning it, still. I'm as assiduous and over-particular with my asleep turning the crank, as though I was doing job-work.

Why, I say. There's no harm done, for all that. The figures delivered up from my turning the crank, upon the whole, are lacking faults."

As the Swedish academical criticism was driven to straits for Franzén's figures being certain in spite of all that, the old saying: Silence is golden: came into criticism's mind. And criticism made up its mind never to say a syllable of it, thus having Franzén "turn the crank" with a vengeance. The academical way of proceeding meant a violation of Franzén's rights, as manifest as shameless.

The Faculties, as they had to forbear scandalizing the public, couldn't openly own to being superior to criminal law.

It was criminal law that had openly to own to being subordinate to academical crime.

A frank criminal law was just exactly what Sweden wanted, and, for civilization's sake, it had to be hers in no time; there was some periculum in mora. Quick.

There stood the frank criminal law. Retroactive the frank law wasn't to the letter; its applying to Franzén's case retroactively was but a matter of course. The Faculties and all those sorts of things— like Caesar's wife—are above suspicion, the law ran like that. The law I've mentioned is still in force; in its peculiar way it has protected me for years and years; and it is protecting this very talking of mine most tenderly, as far—nota bene—as it can see in me the downright

Academician. The national spirit of Sweden, for all I know, isn't but the faculties' own. The feeling of Swedishry has grown just as lukewarm as are the faculties themselves.

I cannot well imagine, I'm enabled to make an impression on spirit and feeling like that. I have to try, notwithstanding.

In behalf of a Swede who made my affair his own, little minding who there was to stand security for it, I appeal to Swedishry. And I speak my mind like this:

> (3) Franzén was broken down for the mean trick the academical people had chosen to play him. He was snatched from my grasp previous to, but in part and superficially, making the acquaintance of my working. Just as Linnaeus visualized botany, so Franzén visualized thermometry and calorimetry. And there he was compelled to stop. As to the original action, Linnaeus and Franzén are equals.

Taking consequences and applications of action into account, the two actions look different, and the difference may be of very great moment. The difference in mention, however, decidedly argues in Franzén's favor.

Linnaeus was ennobled, and such meant the mob not to be entitled to, for that very action of his, spitting in his face or in the face of his children. How long shall the mob be entitled, to, but for that action of his, spitting in the face of Franzén and his children?

For Franzén and me it was requisite to publish the "Termodynamikens Grunddrag" under the form it had; we wanted the contemporaneous experimenting immediately and to the utmost limit to be directed on our undertaking. The subject, dealt with in "Term. Gr.," was the transformation of phases of any substance, i.e., the energy conversion that takes place at the physical limit. Extending the consideration to some new limit, e.g., the "electrical," the "chemical," could give definite results only on the proviso that a wide range of experimental investigation came into existence in the order prescribed by the very analysation. The Swedish academical criticism would have been able, in no time, to communicate to all the world the clear fact, made out by Franzén's tabulation work, and, at the same time, communicate to all the world such close examination of the said tabulation work as might have been criticism's own doing. Criticism chose to be silent about the whole of it. My analysis, in consequence, *got a new aim and direction. My results of analysation, already being visual forms (Franzén had brought about that), had to appear palpable forms, i.e., downright heat-engine forms, deduced from the heat-engine principles, motor principle and generator ditto. Whosoever chooses to try to make himself more and more easily understood, his path takes him from visuality to palpability, it cannot be helped; such criticizing a production, as does all that can be done, implies the extremely in composite to espouse the extremely real, ultima ratio to espouse cuts and blows—I refer to my printed papers:*

P. L. M.: Nils Engelbrektsson, The phase limit motor as heat engine in cases of boiling or melting (1926).

E. H.: Expansion by heat regarded as a function of the phase limit motor (1927).

I. F.: The integral form for the integrability condition of the entropy equation (1928).

D. F. T.: The deduction of the form of the thermometer (1928).

D.F.G.S.: Nils Engelbrektsson, The deduction of the forms for gas-expansion and solid-expansion at small pressures (1928).

N. E.: Notiz von dem Entwicklungsgang meiner thermodynamischen Arbeiten (1935).

These papers produce the palpable form of the analysation results, visualised by the tabulation work of Franzén. Just as the "Therm. Gr." was put under the nose of Swedish academical science research, so were these new papers of mine. The Swedish academical criticism had to take up a position towards the papers mentioned. The responsible Swedish criticism chose passing by each and all of them in silence. Down to the present day, criticism persists in passing by.

Production that tells is the offspring of unflinching will, and of need; the old sayings of people have it so. When the need was paradoxically sharpened, the production issuing could appear rather marvelous. See Timour, Tostenson, Cortez, Pizarro, Milton, Beethoven, etc.

"Our Swedish academical science research is so very much cramped for space just now," the faculties can't in chorus, "as not to be enabled to produce. Mind what we owe to science research. How tenderly should research duly be cared for? Up, once for all, in behalf of the liberty of science research." True it is (even though not generally known), academical science research, save for its rotten fruits, is barren. Such has been the fate of research since that day centuries and centuries ago, when, for the last time, Research was regularly collared

in order to be whipped at the whipping-post. It is not easy to name such support from without as cannot be dispensed with by an unflinching will, displaying itself in any creative form. We always can maintain, however, the proffered hand to expect from outer world a hand that accepts the offer. Being the surrounding company of production, criticism from without is what production never can do without. No production can create its outer surroundings; no production is all-powerful. A producer may practice self-criticism, and exercise his self-criticism even so far as to be tersest in style wherever richest in matter. If so he does, he may venture to maintain his production to be most easily graspable, and his demands upon those around him to be minimum. As the producer, in undertaking the task of criticizing his own production, with unswerving fidelity attaches maximum substance to minimum form, he shows a sense of form that has been duly exercised and does its very utmost.

Yet, the producer cannot in one operation be laconic and amplifying; he cannot at one and the same time try to suit a master, and try to be all things to all men. He can, if need be, exchange conciseness for amplification. For all that, as to each individual among those around him, he cannot himself decide the very point where this individual is at a loss to understand and wants an abundance of words for his instruction. The individual in question, if a sincere critic and a fellow-laborer, has to decide that point himself; none but he can. If, for any reason, he can't do that much, he is indocile. Academical science research crying out for help, the Swedish general public becomes all of a flutter with anxiety and crowds up Helter Skelter, unofficially as well as officially: Research is to live on the fat of the land. For Research must be paid its weight in gold, if less won't meet the case. Gold can be a blessing to him who gives, or, to him who takes; gold, too, can ruin one of them or both. If State and Patron, as they are distributing the material resources, were quite clear about where the corresponding will-resources are to be found, and in distributing could stick to these solely, State subvention and Patron age would always be fraught with blessing. Yet, quite clear about that, State and Patron can never be; for, at any rate, they are at a loss to find out him whose will is most determined. The will most determined has to be to the fore always, and such he cannot, unless he works with need for his main resource, the God-sent gift of his that is to be made the most of, and never to be set on show unnecessarily. Keeping geese no one likes better than the fox; when well-meaning sets to in good earnest, the fox is set to keeping the geese. I don't say, State's subvention or Patron's donation is quite of evil. I only hum away at the old Latin parturiunt rnontes et nascitur ridicuius mus, turning that Latin into an English as plain as plain can be, I say: The Kingdom of Heaven, when on sale, just costs the widow's mite. Dives didn't save himself, nor one of those around him; and he never will —Any authority among those of the State – such as any one of the Academical institutions—when all comes to all, isn't but authority pro forma. Pro-forma authority being substantial is a thing that may chance to happen; yet, pro-forma authority must never make pretense of being substantial. If it demands to enjoy your implicit confidence when, from some place or other, it is threatened with control or with criticism founded on facts; if it wants to get out of it scot-free, by simply referring you to its bearing a world-renowned name, and to its being universally trusted; if so it does, there's no good your controlling and criticizing it any more.

It has pronounced judgment upon itself: it is quite unworthy of your confidence.

(4a) *What the Swedish general public indispensably wants to know about scientific working and academical affairs isn't but this: Has a Swedish scientific producer, in spite of the privileged station of Swedish academical production, such support from without as is his alienable right having, i.e., the support of strict criticism?*

Furthermore: Has a Swedish scientific critic, in spite of the privileged station of Swedish academical criticism, such a platform for public speaking, as is his alienable right having, and without which the corruption of Swedish academical production is a danger that is imminent? By way of answer, Swedish academical criticism proclaims most solemnly: "A scientific production the academical parties concerned take up and examine strictly, regardless to from where it comes and without respect of persons" Swedish academical science research pro claims just as solemnly: "Academical production, if attacked, being rather capable of offering resistance, can bear with criticism tolerably well. Any criticism, founded on facts, without respect of person is most welcome." The question is now: Dare the holy ghost of Swedish academical science undertake reconciling the solemn declaration of academical 'practice in general to academical practice in my own affair? If he dares, I tell him {or any mouthpiece of his}: Jump along and try.

(4b) Swedish people are rather fond of academical science: their not having only analphabet among themselves is their immense delight and their pride.

But above all they indulge in the hope of being regarded as neither cruel themselves, nor likely to suffer any unnecessary cruelty to be committed before their very eyes. Keeping silence for years and years as to the proceedings of a scientific impostor is a way of academical criticism's hitherto unheard of, people are well aware of that. Let the impostor be ever so wretched and feeble, let his gain of his craft be ever so poor, unfailingly there is to be seen your academical criticism with the heavy artillery of science, being ready to lavish volley by volley upon him.

Warnings against poor devils trying to compete with science, with impressive force and warmth used to be given to such ignorant and self-deluded people as you are; indeed, they used to ring in your ears incessantly.

My heat-engine papers, consisting partly in my own discoveries, partly in my sweeping criticism of academic thermodynamic production; according to science, are all alike in one respect: they aren't to waste words upon. A reckless sort they are, though; they have shown they have a will of their own.

They fooled Franzén into pretty near killing himself with overwork. His character they picked to pieces into the bargain. It's a pity. Isn't it? Science cannot predict for a certainty none but Franzén to be deceived by that reckless set of mine. One more may be, many may be; people think so, people's heart misgives them. It is a thousand pities, really it is. Still no warning from academical quarters against that reckless set of mine, though. What's the matter, now? Why must science be so very cruel as not to warn against imposition, this bout, in its usual shrieking way? That dare-devil, the holy ghost of Swedish academical science research, may not have been frightened out of his wits by finding under his very nose my brand-new heat-engine discoveries and my icy examining of academical heat-engine truths. Or may he, though? Appearances are often deceitful. But this we know: A compassionate and undaunted fellow never had appearances so utterly against himself, as has the holy ghost of Swedish academical science research.

I proceed to display the main features of those heat-engine papers of mine which Swedish academical truth has chosen not to get at in *the rather shrieking way that is its own usually, but to get rid of, by sticking close to its rocks, insinuating.*

A heat engineer who contrives to catch what I have got to say in the following is likely to find out, my way is his very own. It's only that my heat-engine form is far in advance of modern heat engineering while academical heat-engine science is sauntering pretty far behind.

Where people of the present day meet with a limit confronting their imagination as an awful subject, there they can face a representative machine object of mine. Exempla docent. I make Physician equivalent to a repairer of machine. The heat-engineering expert has not taught me how to manage that. I'll but just sketch out a rough draught of the equivalence in question.

Academical medicine, in point of principle, labors under the difficulty of having its tongue forked. It has a desire of being saluted as an empirical discipline; yet, it is incapable of adhering in all vicissitudes to empiricism. The fatal stroke is vivisection and all belonging to it.

I call it fatal, not so much for it's being the absurdeste speculation that ever came into a physician's head, as for it's being the very sort of speculation and metaphysics. Surgical practice means essentially to save the subject whose body is the object to be experimented upon; it needn't judge the value of the subject, then. Vivisection starts with valuing. It finds out in no time, it has vile corpus on the operating table, and it goes to work ruthlessly. Torquemada, devout and cruel as he was, is said to have observed some moderation. Vivisection doesn't attach to academical medicine in the capacity of a survival, which modern medical science at heart abominates, and with all its might strives to set itself free from. To be sure, vivisection is an utterly modern stroke, and one that is ready to take its applause. Enough of that. I don't deny academical medicine to be capable of rendering tolerably good service in such and such a case of illness. I only deny it to be likely to be relied upon in point of principle, for it actually has two principles incompatible with one another.

Academical medicine, albeit inward-turned in the way mentioned, is an empiric that has made his experiments. Improvements within technics, analyses within physics, chemistry etc., frequently turned out medical discoveries, very worthy of respect. However, all medical discoveries made, have been isolated inventions. A method exhibiting promise, medicine is still in want of. Academical medicine never left a track, as did a bloodhound following his scent. Just as a chicken happened to get a grain or two in his gizzard, so academical medicine did. Zeal (and perchance some peckish feeling inside) was the thing existing and was the hard work that success was won by. Zeal with no judgment is Sancta Simplicitas, an old woman purveying firewood for the stake of Huss. Medical zeal, if an old squaw, is not the chief exactly. The very causes of diseases academical medicine knows very much better than I how to take seriously; I refer to the medical books. Empirical knowledge is definitive, not hypothetical.

I am going to give the form of empirical medicine as far as it serves my purpose present. Empirical medicine, as elaborate science, means instrumental medical method. The form of medicine is graspable with the hand, if taken as the boundary of the physician's instrument of precision.

The medical instrument in question is the physician's own body, taken in concreto—in case of the physician being a wonder of health— or else duly idealized. The living body—necessary, by abstraction, set free from—by way of deduction includes all different medical instruments and remedies.

And it is the unattainable ideal for any instrument of precision, medical or not. The ideal instrument I am speaking of must be such as to be capable of corresponding adequately to the defective instrument the sick one's body constitutes. The subject on sickbed is represented for the moment by its sense of pain; its outer surroundings the subject depicts by its body under the form of pains. The sick one doesn't find it easy to conceive his sense of pain to be the mightiest gift of his, more important to his integrity of life than his eye. But the physician must not be devoid of appreciation of the great moment attached to sense of pain, as the subjective limiting-point of the instrument the living body constitutes; the sense of pain of the ideal medical instrument must be acute. Physician's act of interference cannot possibly mean simply removing a patient's sense of pain. Far rather must a sufferer's sense of pain pass through the suffering as through a school, thus growing in a measure hardened; in such a way, only, can life appear as distinct as a grade of a grade series and attain a perfection of its own. What the physician interference has in view is taking the patient's place on the point of his being in danger of overstraining his sense of pain. That in life there is some field for medical assistance is obvious, in view of the fact that sense of pain always is strong up to a certain degree and never is immeasurable and all-powerful. As to the strength of the sense of pain, the patient represents the lowest degree, the idealized physician the highest, the real physician some degree between the two. The anodynes the physician is able to offer are alike inasmuch as being results of analysation. In their main species they are kept apart exactly so as are the analyzing senses. The variety of species and subspecies, subsumed under the main species, originates from the variety of combinations in which the analyzing senses are capable of co-operating. The dividing activity that arises, as the eye in front of the lensed diagnoses, as the physic melts on the tongue tip, as the hand with the lancet operates, this activity is analytic, and—being part of medical analysis—is definitively connected with medical discipline. The analytic operation, which, here it is about, is the differentiation process. At this process the patient is mostly a lifeless thing, mostly the object to be experimented upon, least of all the subject experimenting. In full possession of his senses he scarcely is. Communicate with the physician so, as subject normally communicates with subject, he cannot. That ear hears, when tongue is speaking, is normal correspondence between man and man; yet, at bedside of patient that sort of correspondence is rather doubtful.

Hand can squeeze hand, when speech is failing; this sort does better. Eye has its language, and it is rather easy of comprehension. Brute by this frequently got into communication with man; an eye at the breaking point—like that of a deer, mortally hit—was singularly eloquent. Yet, ultimately, it isn't tongue, ear, hand, eye that bring about the correspondence between subject and subject. When all comes to all, it is about a suffering subject, and about the subject, with pity inspired that stands by. In the last resort, then, it is the sense of pain that brings about the correspondence.

Sense of pain, as is well known, is a kind of feeling. What the differentiation process in medicine bears upon is clear, now. It bears upon the division by which sense of pain is deduced from sense of feeling as species from genus.

I have sketched the differentiation process in medical analysis. I proceed to the integration process, i.e., to medicine's rebuilding the body, set free from its defects by the operating knife or by proper representative of it. Sense of pain being momentous, one kind of feeling is momentous, and any kind of it may possibly be. Under sense of feeling we subsume—besides sense of pain—senses of temperature and pressure.

Eye we look upon as the pattern for an optical instrument; a fortiori, we are to regard temperature sense as the pattern for the special heat instrument that goes by the name of thermometer; and the combination of temperature sense and pressure sense as the pattern for a heat instrument or a heat machine in general. The integration process, the rebuilding activity, then, must mean the body to be organized as a whole of different machine types, steam engine, thermo-chemical machine, hydroelectric cell, etc.

Translating their speech into thermodynamic language, and applying the method of empirical thermodynamics to their integration problem, that is for physiology and medicine obviously the way to go, in order to master the task, I have called the integration process. In surveying existent results of physiological and medical analyzing, they are all of them to be found under the analytic activity I have called the differentiation process. The corresponding integration, up to this very moment, isn't but an operation in spe, and that's the fact. For enabling us to integrate, differentiation must have proceeded so far as to have the end organs of temperature- and pressure-senses analyzed and determined as regards their conformation and function.

The very discovery of the said end organs we owe to physiological analysis. As to the conformation and function of the end organs traditive physiology has brought about an analysis, slight as qualitative, and still more so as quantitative. We know the end organ of temperature sense to function systematically, we know hot-point to be opposite to cold-point. The hot- and cold-points we know to be glands containing some liquid, and that these glands are not ductless, we also know. We further know cold- and hot-point distribution in epidermis.

Cold- and hot-points, turned into thermodynamic language, are liquid-volume-thermometers. Analyzing cold- and hot-point functioning, then, means: 1) deducing the liquid-volume-thermometer from the general heat-engine principles (generator- and motor-principles), 2) displaying the deduced thermometer under its most general form, setting it out, consequently, in its mathematic form, 3) taking out of the general thermometer form the special one which presents a subjective temperature limiting-point (the physiological zero point) and thus immediately corresponds to cold- and hot-points. This thermodynamic analysis I have carried out. I refer to my papers N. E. and D. F. T.

(5a) *That human body is a heat engine and, properly speaking, the very ideal of heat engines, this fact modern medicine is not unfamiliar with. The illnesses that medicine strives to remove are, from some point of view, deficiencies of the body, or, plainly, machine defects. As far as this point of view extends, medicine falls inside the application range of thermodynamics. Every modern physician thinks his professionalism maintained by coming out a fine specialist in some point of medicine. Why, I say. The very species the specialist depends on is immediately called in question when viewed from the point of view I am exhibiting. Classifying illnesses, set forth in a strictly empirical way, i.e., as mere machine defects, cannot throughout be the arbitrary classification that is the traditive. A machine repairer, offering to mend a machine he isn't fully familiar with, and means not to make fully familiar with before repairing, that sort of repairer is confusingly like a humbug. The more the machine is an instrument of precision, the more the repairer is a humbug. A modern specialist in medicine is essentially in the predicament of a machine repairer like that. Isn't he?*

(5b) *Physiology and medicine, when all comes to all, are not wise virgins, but foolish. If they had preserved the idea of medical discipline, they should have remembered they had to wait for an instructor coming from the engineering line. But they have not had the forethought of discovery of the said end organs we owe to physiological analysis. As to the conformation and function of the end organs traditive physiology has brought about an analysis, slight as qualitative, and still more so as quantitative. We know the end organ of temperature sense to function systematically, we know hot-point to be opposite to cold-point.*

The hot- and cold-points we know to be glands containing some liquid, and that these glands are not ductless, we also know. We further know cold- and hot-point distribution in epidermis. Cold- and hot-points, turned into thermodynamic language, are liquid-volume-thermometers. Analyzing cold and hot-point functioning, then, means: 1) deducing the liquid-volume-thermometer from the general heat-engine principles (generator- and motor-principles), 2) displaying the deduced thermometer under its most general form, setting it out, consequently, in its mathematic form, 3) taking out of the general thermometer form the special one which presents a subjective temperature limiting-point (the physiological zero point) and thus immediately corresponds to cold- and hot-points. This thermodynamic analysis I have carried out. I refer to my papers N. E. and D. F. T.

(5c) *That human body is a heat engine and, properly speaking, the very ideal of heat engines, this fact modern medicine is not unfamiliar with. The illnesses that medicine strives to remove are, from some point of view, deficiencies of the body, or, plainly, machine defects. As far as this point of view extends, medicine falls inside the application range of thermodynamics. Every modern physician thinks his professionalism maintained by coming out a fine specialist in some point of medicine. Why, I say. The very species the specialist depends on is immediately called in question when viewed from the point of view I am exhibiting. Classifying illnesses, set forth in a strictly empirical way, i.e., as mere machine defects, cannot throughout be the arbitrary classification that is the traditive. A machine repairer, offering to mend a machine he isn't fully familiar with, and means not to make fully familiar with before repairing, that sort of repairer is confusingly like a humbug. The more the machine is an instrument of precision, the more the repairer is a humbug. A modern specialist in medicine is essentially in the predicament of a machine repairer like that. Isn't he?*

(5d) *Physiology and medicine, when all comes to all, are not wise virgins, but foolish. If they had preserved the idea of medical discipline, they should have remembered they had to wait for an instructor coming from the engineering line. But they have not had the forethought of preparing for receiving him. The modern physician or physiologist is not, as he ought to be, versed in mathematics and thermodynamic language. He isn't able, himself, to apply thermodynamics to physiology and medicine, nor is he likely to catch what I say, when, in D. F. T. and N. E. p. 48—55, I instruct as to the application most elementary. Or is he, though?*

Subjective limiting-point of instrument opposite and corresponding to objective, suchlike is double-Dutch to a blacksmith, making a dung-

fork. Yet, the blacksmith has made up his mind, a tool like his shall have a handle opposite and corresponding to its point. And he doesn't lose count, while at work. Not in such a way, as awkwardly to go and lose the difference between handle and point, and make the tool in its entirety of the same description; the point is meant for an object as lifeless as dung, the handle for a subject alive enough to try keeping the dung at arm's length. Nor in such a way, as by slovenly to bring into use a new principle in addition to the principles of point and handle. In acting so, the blacksmith is representative of any and everyone who knows his business thoroughly; the branches of knowledge being legion, yet, knowledge fundamentally is one.

A blacksmith in his forge, could organize, and even so as to throw his whole soul into his doing. Where he stood he represents the ideal of a workingman. No labourer, ultimately, can know any more than he, for, when all comes to all, no labourer can aim at a higher point than soulful working. I have to tell, and in detail too, of ignorance and soullessness. Previous to my telling about that, I have stated what is to be meant by real knowledge and soulful working. A heat engineer making a steam engine, in point of principle, sticks close to the blacksmith. Just as there is handle opposite and corresponding to point, so there is cylinder opposite and corresponding to boiler.

In both cases there are two aspects of principle, no more and no less: Empiricism implies analyzing; analysis implies instrumentation.

The heat engine is the true representative of instruments. In looking at the subjective side of instruments, this state of case is shown indirectly: Analysation of feeling, in discovering the representative sense (sense of pain), is discovering the two senses (temperature and pressure-senses) which the heat engine immediately refers to. In looking at the objective side of instruments, the said state of case is shown directly: On the objective side we have the energy species; and heat species and work species, which are met with in considering a heat engine, constitute the duality which is the principle of the variety of energy species. In heat engineering we may speak of an evolution, a proceeding from lower form to higher. But the higher form, as it came about, always happened to be a brand-new discovery, it never was a deduction of a special form of machine from the general form. And it always remained what it was from the very beginning; our definite knowledge about it being—today as it was yesterday—the minimum knowledge of its standing there an isolated fact, irrefrangible as solid iron. The flourishing of heat engineering did not mean the academical heat-engine science to take one only step in the right direction. The academical heat-engine (the traditive thermodynamics,) was, step by step, declining. Nernst taught me and all the world a heat theorem that I never had any practical way of returning thanks for, as I could never avail myself of the theorem by my applying thermodynamics, even including my applying it to the living body of animal or man. The world all about me, longing to return Nernst thanks for the lesson, did what it could, and a shade more. The textbooks took up extolling Nernst's theorem, and I can't see they have contrived to let it be standing beneath the principles: it was to be driven up, up among the very principles. I postulate (N. E., p. 22) the heat-machine principle to be the right limitation of the heat machine. Furthermore, I postulate the machine form always to oppose subjective limit to objective. Lastly, I postulate the machine form essentially compound to present subjective and objective limits by different parts of construction. I have, then, the heat-machine principle under the form of two independent aspects of principle, the subjective aspect (motor principle) and the objective (generator-principle). In point of principle, this is enough. The exaltation of Nernst's theorem is a step downward, rather noteworthy.

(6) *My putting Nernst's theorem (the third heat principle) on the shelf the Swedish academical heat-engine science has chosen to pass over in silence. Science ever saw a blacksmith at work; Science did, and no mistake. Only, the keen eye of science happened to overlook the working blacksmith's firmness of principle. That's it. Isn't it?*

As for heat machines, we should distinguish between left-hand side and right-hand side. On the left we have the non-ideal machine (which runs hot) and the ideal machine just starting or slackening speed. On the right we have the ideal machine keeping up its normal speed. The entropy figure increases, said Clausius about the left-side machine.

Speech like this was conciseness, and it was enough. Entropy is a maximum, Clausius would have said about the right-side machine, if he had consulted his sense of form (N. E., p. 26). But he said: the entropy figure is constant. The quantity that Clausius called entropy was confronting him as a function of a number of functions of state, unknown as to their form. This function was to be deformed, maximized, and in its very maximum form introduced as the entropy. See N. E., p. 26.

The operation that determines entropy is obviously a calculatory operation. See N. E., p. 27. Clausius recognized entropy as calculatory, but he chose to make a distinction between calculatory quantities and quantities experimentally attainable. He didn't conceive handling any calculatory instrument essentially compound to include two different limiting acts, opposed to each other as objective limit of instrument (differentiation) and subjective (integration); he didn't at all conceive calculatory operation to be instrumental. I have—in I. F. and E. H.— introduced the calculatory instrument essentially compound which works out the entropy for a certain motor. In I. F. there is to be found the limiting act (calculus of variations), which constitutes the objective limit of the calculatory instrument I am introducing. In E. H. we meet with the limiting act corresponding to the subjective limit of the said instrument.

On its subjective side the calculatory instrument has a common calculating machine (multiplications machine) and a coupling-device, connecting the calculating machine with the heat motor considered.

The heat motor considered is the steam-motor type, in a measure generalised, I have called it the phase-limit motor in cases of boiling or melting.

This is the very motor that has been chiefly dealt with by Clausius and his followers. I will sketch, now, what I have made out of this motor, and what Clausius and all the rest have made out of it. This being done, it is likely to be understood what means progressively getting on, and what means standing still or declining.

The limiting act, corresponding to the objective limit of the calculatory instrument I make use of, is carried out in I. F., p. 4–12, 26–27, 30–36.

The function of state that Clausius takes for entropy Clausius and his followers represent by a partial differential equation, the equivalent integral form of which they cannot produce. The function of state, that, loc. cit., I vary and maximize, and the maximum form of which I, in the following, introduce as entropy, this function I represent by a system of ordinary differential equations (p. 26, Form [32 a]; p. 35, System [44 a]), the equivalent integral of which I put close by (p. 27, Equation [40]; p. 36 Equation [46]). By transferring the calculatory problem from differential form to integral form, from objective limit of calculatory instrument to subjective limit, I have made one step forward in the calculation. For, differential (species) is to integral (grade) as something objective, indistinct, difficult of comprehension, to something subjective, distinct, and easy of comprehension.

The obtained integral form—being the analytic expression of the heat-machine form and being called, in what follows, the heat-machine law—is still undetermined, inasmuch as it includes two functions, which are arbitrary. Determining the arbitrary functions, met with in the heat-machine law, is the limiting act corresponding to the subjective limit of the calcular instrument I make use of. As to this matter, I refer to E. H., p. 5 and following pages. The heat-machine law, as far as it is undetermined, is to be determined so, as to express the amount of work, turned out by the heat motor, to be maximum, if the law were differently determined, the machine wouldn't be ideal. When the heat-machine law expresses the motor work, turned out, to be maximum, its expressing this fact must be such as to strike the subject as pungently as possible; if its expressing the fact mentioned were different, the heat machine would be a tool where—even if the edge were ideal—the handle was not.

For determining the arbitrary functions entering into the heat-machine law, I couple to the heat motor a common calculating machine, driven by hand or by some other sort of motor. My perception of the heat-machine law, as I am considering myself the subject, the heat-machine by its subjective limit (the motor) is referring to, is my perceiving the calculating-machine work, belonging hereto and going on at the cost of the subject. When the subject has chosen the forms for the arbitrary functions, in these functions (and in the machine law, accordingly) there remain no undetermined quantities save a number of constants, empirically determinable. Determining the constants is calculating-machine work, and it is carried out at the cost of the subject or its representative, the calculating-machine motor. One of couplings (2 a), (2 b), (2 c), I choose as connecting link between calculating machine and heat machine, and I give the arbitrary functions of the phase-limit-machine law the form that corresponds to the coupling chosen, See E. H., p. 7–8.

Being about coupling calculating machine to heat machine, the phase-limit machine, as is most easily understood, must be taken in concreto, thus a) at ideal gas point, b) at boiling point, c) at melting point; obviously, three different couplings here are met with.

Choosing couplings (2) means the calculating-machine work, spent in determining arbitrary functions (constants) entering into the heat-machine law, to be minimum. Choosing couplings (2), then, means the law, that expresses the heat motor to deliver maximum work, to express this by minimum work on the part of the calculating-machine motor. Choosing couplings (2), then, means the law, that expresses the objective motor to deliver maximum work, to express this in such a manner as to strike the perceiving subject as pungently as possible.

The phase-limit-engine law is, now, completely determined. Each of the special forms which are deduced from it, or, are—like the special forms in D. F. T.—capable of being deduced from it, is linear with regard to the empirical constants entering in, and, so far, in determining the constants, gives the calculating-machine motor minimum work.

See E. H., Equations (23), (41), (92), and (106), D. F. T., Equations (59), (75), (113a), (113b), (127), (134), and (137). And each form deduced, from the point of view of the heat property, which the form in question represents, puts an individual heat-machine object into the same class as some other individual heat-machine object.

The precision of classification corresponds to the precision of the experimental determinations used.

The classification principle is unitary, for it is the very phase-limit-engine law, viewed from an essential point. The point of view is as follows. The motor principle, expressing the heat machine to deliver maximum work, must express it by the machine law in such a manner as to strike the perceiving subject as pungently as possible. The tone by which the machine law strikes the perceiving subject as pungently as possible, with equal purity must resound in every form, deduced from the law; if it were otherwise, the machine-principle's voice would have a jarring secondary tone; it wouldn't ring sonorously in the subject's ear, then. Thus: in every form, deduced from the machine law, the calculation of empirical constants entering in must yield the calculating-machine motor minimum work. As to applying the forms, deduced from the machine law, to individual machine objects, we should remember: in every form, deduced from the machine law, as many as possible of the empirical constants entering in, must, as we pass from one individual machine object to another, remain quite unaltered, i.e., by our passing from individual to individual, the individual determination of those constants must be minimum, their class-determination eo ipso maximum. If the number of individual constants were above minimum and the number of class-forming (i.e., non-individual and in consequence, calculation-controlling) constants were below maximum, the control of calculation would require an extra contribution of calculating-machine work, and the work of the calculating-machine motor wouldn't be, as it ought to, minimum. When the perceiving subject has a concrete perception, it, as is well known, perceives by co-operating senses; the senses, in co-operating, form a grade-series. Species is to genus as concrete thing to abstract; then, perception of species may be concrete in a higher degree than perception of genus. The objective motor principle which the heat-machine law gives voice to—i.e., the principle of maximum work on the part of the objective motor—must be expressed so, as to strike the subject as pungently as possible, even in case of our suffering the stroke to hit co-operating senses.

Let us have a look at the forms deduced from the heat-machine law, and let us apply them to the individual heat-machine objects they refer to. We find the classification to be carried through exactly so, as required by the heat-machine law. Among the empirical constants entering in, one only is individual, all the others classifying. See E. H., Tables 2, 12, 14; D. F. T., Tables 14, 15, 19, 20. The phase-limit motor in case of boiling or melting, represented by couplings (2), is a generalization; the heat engine of steam-engine type, represented by coupling (2b), is the species which is generalized. In case of the phase-limit engine—the abstract thing (genus)—being the heat machine viewed, it is the subject's hand that grasps the calculating-machine work, lastly by its muscular sense, firstly by the motor of the calculation machine.

In case of the steam-engine type—the concrete thing (species)—being the heat machine viewed, then, too, it is the hand that grasps the calculating-machine work, but not only by its muscular sense or by its muscular-sense representative, the calculating-machine motor. Maybe, the calculating doesn't immediately touch upon the temperature sense of the hand—it neither feels heat nor cold—but a sterling representative of temperature sense appears in connection with the calculating-machine work. The representative in question is the water volume-thermometer. The water volume-thermometer is the thermometer essentially compound; to its objective temperature limit (the water boiling point) it unavoidably opposes a subjective temperature limit (the physiological zero point of temperature). See D. F. T., Table 4.

The tangible perceptibility of the forms deduced from the heat-machine law is united with bright clearness; just as temperature- and muscular-senses co-operate in perceiving them, so do eye and hand. The thermometer pattern (E. H., Equation [25]) by which the classification of liquid thermometers in Table 2 is performed, is a calculatory formula, and immediately refers to the calculating machine; but the classification mentioned can be performed with the same precision by means of an optical instrument, a burning glass. See D. F. T., Equation (64 b). The thermometer pattern (E. H., Equations [34] and [35]) by which the water thermometer is shown to correspond to its subjective and objective surroundings (D. F. T., Table 5) is a system of calculatory formulas; but the water-thermometer correspondence to subject and object can be made manifest by exchanging the calculatory formulas for an optical instrument with ocular opposite to objective. N. E., p. 47–48.

What is just said may be sufficient. If, as Clausius has it, entropy is a calculatory quantity, I think I have found out the calculatory instrument for calculating it. And I think I have managed to carry through the calculation as to the case considered, the phase-limit engine in case of boiling or melting.

Clausius and all the rest have given the entropy Integrability condition as a problem to be solved, a differential equation to be integrated and to be determined as to its integration limit. For solving the problem there are two steps to be taken, then 1) the integration of the differential equation, 2) the determination of the integration limit. If neither of these two steps is taken, the problem—treated in any way whatever—remains a thorough problem still. Clausius', Clapeyron's, Planck's hereto-belonging results of analysis mean the immediately given problem to be exchanged for a new problem, beautifully formed, yet, still incapable of solution. As for the said results of analysis, I must state that they have failed to bring the problem we are about one only step towards its solution, provided the way of solving it, is marked by the two steps just mentioned. The calculus of variations, as shown in the foregoing, is the instrument by means of which I take the first step of solving the problem.

Choosing to use this instrument in a rather ungainly way, we don't get but the result of Clapeyron. I. F., p. 12–15.

If, in using the instrument, we introduce an ungainliness of another kind, we don't get but the result of Clausius. I. F., p. 15–19. A third ungainliness corresponds to the result of Planck. I. F., p. 19–24.

Clausius' entropy idea is definitive; in as much as it makes a distinction between the heat machine on the left and that on the right I have given Clausius' calculatory idea its practical determination: I, completely, have solved the heat-machine problem. My limiting act corresponding to the objective limit of the heat engine is a calculus of variations; this act traditive Mathematics (Lindelöf) has given me gratis. My limiting act corresponding to the subjective limit of the heat engine means determining arbitrary functions by using the principle of co-operating motors (i.e., by coupling calculating-machine motor to heat motor); that is a stroke that traditive Mathematics did not give me the pattern for.

In D. F. T., I have modified my point of viewing the heat-machine problem, thus enabling myself to determine the arbitrary functions entering in, without the aid of such calculatory instrument as is unknown to traditive Mathematics. The entropy determination ("the second heat law") which is the act that necessarily limits the phase-limit engine, by Clausius and all the rest is suffered to refer to one actual machine limit (the phase limit), paying no heed to another actual machine limit (the engine vessel). Logically incontestable this procedure is not. For, as a matter of fact, a steam-boiler process isn't but partially independent of the very boiler; partially the process is dependent of the boiler. My entropy determination in D. F. T. is a limiting act that pays attention to the different limits a real heat engine requires. My entropy determination by this has a logical preference over that of my predecessors and, above all, a practical preference. The arbitrary functions, which appear in calculating entropy, I have been enabled to determine by specializing the machine limits. D. F. T., p. 13–17, p. 49–50, p. 61–62, p. 69–70.

My analysis in P. L. M., with regard to starting point and formation as little as possible differs from the traditive analysis hereto belonging.

In determining entropy, I neglect an actual machine limit (the engine vessel). In calculating entropy, I avoid using the calculus of variations as calculatory instrument. In P. L. M., then, I seem to stand on traditive thermodynamics' own ground. The result of my P. L. M. analysis is taking in full the first step of solving the heat-machine problem. My predecessors did not take that step. There must be, now—

all seeming consistency notwithstanding—some essential difference between the P. L. M. analysis and the traditive. I am to show the essential difference.

The thermodynamic potential, introduced by Massieu and Duhem, is a form that enables us to set right the heat-machine problem. Setting right a problem, as is well known, rather facilitates its solution. You should like to say, then: I ought to do justice to academical thermodynamics. If, as I have emphasized, there are two steps to be taken for solving the heat-machine problem, I should credit, not myself, but academical thermodynamics (Massieu and Duhem) with taking the first step. This sounds rather alluring. But there is a hitch somewhere.

Surely, academical thermodynamics has made a great number of the thermodynamic-potential form, but not at all from the point of its enabling us to set right the heat-machine problem. The property of the Massieu function, which makes it indispensable for setting right the heat-machine problem, is a property purely geometrical.

See N. E., System (E). To academical thermodynamics the geometrical property has seemed rather superficial and hardly worthy of notice; the dualistic transformation by which thermodynamic spaces see themselves really reflected in one another, academical thermodynamics did never exhibit under the form in which traditive geometry (Legendre) knows it.

Certainly, the forms in P. L. M. which display the heat-machine process as a problem set right—system (1), (2), (3), (9), and system (1), (2), (3), (35), accordingly—at first sight seem more difficult to get at than the traditive form (1), (2), (3), (5). Nevertheless, including the minimum number of unknown heat quantities, they must be laid stress upon as the very forms, which give the solution of the heat-machine problem.

Academical thermodynamics never laid stress upon this. I did.

The integration problem hereto belonging I, in P. L. M., solved by means of elementary calculatory methods, Lagrange and Monge's. See P. L. M., p. 7, N. E., p. 34 Anm. 1.

My pointing out the most superficial (geometrical-analytic and dualistic) property of the Massieu function to be the property of that function, preferably worth mentioning, is one essential difference between the P. L. M. analysis and academic thermodynamic analysis.

There is one more difference, which is essential. "The addition theorem of entropy," according to academical thermodynamics, "is *a necessary determination, of entropy." In determining entropy, my starting point—as distinct from that of Clausius'—is always so well defined as to get "the addition theorem" a priori removed.*

Very glaringly this strikes the eye, when calculus of variations is the instrument for determining entropy; insisting upon "entropy-addition-theorem" in an entropy calculation, based not upon phase entropies as known "addenda" but upon deformable and for the rest unknown quantities, would be contradictory. In I. E., where calculus of variations is the instrument made use of, I, beside and independent of my calculus of variations, have slipped in a calculatory instrument of a wholly elementary kind. See I. E., p. 28, Control, p. 36, Control.

Using the elementary instrument in question, I solve the problem of calculating entropy for the system of coexistent phases on which the heat machine stands, by introducing the least possible number of relations between the co-existent phases. In solving the problem, I, by one only relation, take account of the system phases and the phase partition. See loc. cit. p. 29, Definitions, p. 36, Definitions. This only and indispensable relation, obviously, cannot be "the entropy addition theorem."

My entropy calculation in P. L. M. and D. E. T. has in view exhibiting a certain analysis result, P. L. M. Equation (44) or D. F. T. Equation (34), by a calculation which is as easy to get at as possible, and where the deviation from traditive Thermodynamics as to linguistic usage and way of looking at things is minimal. My I. F. method—my I. F. control-method, too—is a direct denial of "the entropy-addition theorem;" my rejecting the theorem is unmistakable here. In P. L. M. and D. F. T. my criticizing the theorem is indirect. The theorem I don't find a use for, that is the criticism.

By introducing "the entropy addition-theorem" academical thermodynamics has given the motor principle ("the second heat law") under the form of a system of determinations incompatible with one another.

Determining entropy like that is a step backward from Carnot's point, not a step forward. The academical heat-engine form as elaborate science has proceeded so far, then, as to have utterly spoiled the heat-motor principle.

Max Planck cautions against introducing the machine heat under the form of a differential, as it cannot possibly be an exact one. To mathematical analysis non-exact ("Pfaffian") differentials don't seem a bit more strange than exact. On me—as may be found from I. F., P. L. M., etc.—Planck's caution has been wasted; I have adhered to good practice of mathematics.

The traditive Mathematical judgment—the equation—in point of principle expresses that the dilemma XY = 0, by choosing between alternatives X = 0 and Y = 0, is incapable of being unsound; i.e., the traditive judgment, in point of principle, denies the necessity of analysis, as far as mathematics is concerned. If the very depth of an analysis implied the analyzing subject's faculty of being worth speaking of as conscious, traditive mathematical analyses and mathematic analysing in general would be worthless altogether, the analytic tendency being incompatible with the traditive principle of mathematics. But the faculty in mention is not requisite. An animal ever on the alert, a playing child having all its eyes about it, may make an analysation outtrumping everything that was done in the way of analyzing. When academical mathematics displays its "theory of Pfaffian equations and systems," the theorising tendency, which is met with, expresses academical mathematics totally to fail to grasp the calculatory operation in question to be practical and instrumental.

The calculatory instrument—in Frobenius', Grassmann', Sophus Lie's hand—has its point, that notwithstanding; the results of analysis are standing there, and they do not vanish together with the theories. Chwolson, as must be admitted, has sided against Planck, and, in his well-known textbook, has excellently vindicated his view.

Yet, the intrinsic authority of Chwolson and his followers, in academical thermodynamics' eyes, in as much as I can see, has not been capable of counterbalancing the extrinsic authority represented by Planck and his followers. Planck's instruction about the machine heat is considered an important contribution to Thermodynamics. However, it means the mathematical—i.e., analytical—form of the generator principle ("the first heat law") to have been spoiled.

The heat machine I have produced has its strength in its being capable of symbolising the living body of animal or man. In its form present (the phase-limit motor in cases of boiling or melting) its strength is limited as follows: it idealizes the living body which is to be symbolized, and, in idealizing it, even goes so far as to make it correspond to a heat machine of the steam-engine type. My production in its elaborate state includes a criticism of the academical heat-engine science (the traditive thermodynamics); the criticism in question is essentially negative.

When a subject perceives a heat object by the instrument essentially compound which I have presented under the name of the phase-limit engine in cases of boiling or melting, the perception is brought about by an analysation the main features of which are the following.

The phase-limit engine in its limiting act is clept in vain: the motor principle is one act of limitation; the generator principle is the other act. The limit where the generator with its motor stands, immediately appears the limitation of water: water has subjective limit (physiological zero point of temperature) opposite to objective (temperature point of phase transformation). The subjects analyzing the machine object is not an immediate act of limitation; water, being representative of the subject, acts as intermediary. What the subject in its heat-machine principle immediately depicts is not the heat object itself, but such image of the heat object, which comes about by the water representation.

My way of perceiving things is not exactly that of Thales. Yet, my criticism on Thales is affirmative in the cardinal point. Evolution from Thales to the present day is characterized by its being an uninterrupted proceeding in the direction away from the naive view of Thales. By establishing the heat-machine form so, as is done here, this evolution is definitively stopped.

My production as to Thermodynamics looks like what I have here mentioned. The sharp-cut features under which, in P. L. M., E. H., I. F., and D. E. T., this production is confronting you, evince the affirmative criticism I have given it. My affirmative criticism on it is the protection I myself have been able to give it. Yet, it needs some protection from without, too. Labour, in point of principle, is collaboration. The strength of a strong one in the last resort depends upon a semblance of strength to be found in any one among those around him. Of greater importance than such, even a giant is not. No mortal man, in one and the same operation, can be he who gives and he who takes. I have sought for collaboration with people around me. And I can't say it has been received favourably. See N. E., Geschichte.

(7a) The faculties and, a fortiori, the professorships are subtle specialists; it is their ambition to be, and the public opinion pays them tributes of homage just for being. Viewed from the *specialist's point of view, some of the papers of mine that we are dealing with here are likely to be very especially inaccessible. A thermodynamics professor would rather take offence at I. F. He hardly could help but do.*

For I. E. is whispering in his ear, there's a talent—the talent of having an eye for calculatory forms—that he hasn't put to profitable use and made the most of. A mathematics professor was running the thermodynamics ditto close in their expressions of horror at E. H.

They couldn't well help doing.

For E. H. speaks clearly and to the point, and tells the two professors in so many words, there's a talent—termed sense of form—which they haven't put out to usury, but buried. Now, these worthies, being about to tell the general public what I. E. and E. H. have told them, couldn't twist their tongue to it. Silence, insinuation, remains the only course, open to them, if, henceforth as hitherto they will enjoy public confidence in large measure. It even is their way, of keeping clear of the eighth commandment. Is it?

"What a faculty is at a loss how to get into, and wants to get rid of by keeping dead silence, is likely to be rather preposterous, or, at any rate, to be pretty badly expressed," Swedish people feel it in their bones that such must be the case. This is the sore point. I dare differ from Swedish mind there.

Such operation of mine as the Swedish academical criticism was at a loss how to get into is likely to be such operation, as did its very utmost as to palpability of reason and tangibility of proof; such principle as the heat engineer, as far as his engine went, made use of without wasting words upon; such thought as for a babe was, as it were, nearest to hand; such invention as a brute carried out in a trice without sinning in the least against logic. Take my word for it: my instrument of co-operating motors and my deduction of heat forms by means of this instrument is what has set the academical parties concerned puzzling.

You have—if I'm not very much mistaken—your logic at your fingers' ends. Why, I say. You don't lay at my door this instrument of mine being spick-span new to academical thermodynamics; you don't suffer the unwritten law of heat-engineering, no heat machine can **manage by itself altogether, and tell the machinist to make himself scarce**, to be a cruel joke in heat-engine science (thermodynamics).

Nor do you lay at my door the just mentioned instrument being spick-span new to academical mathematics, as, owing to this instrument, I never failed in my determining of arbitrary functions appearing by integrating the partial differential equations confronting me.

You mind deduction—even as an immediate action of brain—to be but a mechanical operation. You mind such deduction as you call brilliant maximally to be the work of the very machinery of the brain, and minimally to be the work of imagination poking that machinery in order to start it.

(7b) *The instrument of co-operating motors—as stated in P. L. M., § 8—solves a problem that Clausius considered fundamental and that himself—much as he cudgeled his brains about it—was at a loss how to solve. Strange and offensive as- the said instrument may have seemed to a one-subject-man like the Swedish academical science research, that one-subject-man is likely to have known this: Keeping dark such statement as may turn out the positive solution of Clausius' fundamental problem, if the memory of Clausius is worthy of any deference, was unjustifiable. I never steadily denied, science research is a simpleton; I should think I never did. At any rate, I don't deny, simplicity played the first fiddle when my instrument of co-operating motors was the object experimented upon by S. research. But not even in the last-mentioned case can the conduct of S. research be put down entirely to the account of simplicity. Can it?*

The production of mine, contained in P. L. M., E. H., I. F., D. F. T., D. F. G. S., and N. E., includes a vast deal of thermodynamics applications that not even an academical specialist can have had any difficulty whatever of getting into. I am to set out some.

(8a) *The "rational" formulas of academical physics have their worth in the imagination of poor innocents. In practice—for the purpose of interpolating and extrapolating—they don't do at all.*

For interpolating, academical physics is possessed of formulas worth speaking of, viz. the "empirical," obtained by guesswork. See N. E., p. 71. In case of extrapolation, not even the "empirical" formulas hold; traditive Physics is possessed of no applicable extrapolation formula.

These are well-known facts, and indisputable.

Franzén has shown my analytical form for liquid volume-thermometers (D. F. T., Equation [59]) to hold good for any liquid, even including water, "the touch-stone for all liquid-expansion formulas" (Weinstein). Franzén imagined, a formula like the just-mentioned, as it never failed, when used as interpolation formula, is rather likely to hold as extrapolation ditto. He made the formula yield exactly the expansion of mercury from—20° C to 100° C, and then extrapolated up to 300° C. The event didn't fall short of expectation: there comes forth no systematic error owing to extrapolation. Furthermore: as to the difference, not unmentionable, between each extrapolated value and the corresponding experimental value given by Regnault-Brock, the difference in question is likely to be found within the bounds of experimental error, as, according to Callendar & Moss, the experimental results of Regnault-Brock aren't but indifferently to be relied upon.

The very subject of extrapolation academical physics never considered uninteresting; Weinstein (Thermodynamik und Kinetik der Körper) expressly complains of there being no applicable extrapolation formula in existence. Chappuis' mercury-expansion formula, corresponding to temperature interval (0°/100° C), Eumorfopoulos extrapolated up to 300° C. Eumorfopoulos, in starting extrapolation, quite overlooked that his extrapolation formula (the common cubic formula) in case of water and in many other cases, too, doesn't even admit of interpolation. "Extrapolation in such a case would be somewhat unreliable" (Callendar & Moss). And indeed it turned out an utter failure. See D. E. T., Table 1.

Why, I say. I've got a pretty thing to show you here. Eumorfopoulos's extrapolation attempt, hopeless as to its starting point and resulting a decided failure, directly and indisputably was admitted into the special literature of the subject, and in the textbooks is referred to, not without some fullness of detail.

Franzén's extrapolation attempt, incontestable as to its starting point and successful, indeed being the only successful attempt of extrapolating ever made, is insinuated to be beneath the special literature of the subject; for the swedish academical science research promptly refused to waste words upon it. How in all conscience has science research managed to refuse? Making the acquaintance of the academical truth concerning that would be rather amusing. Pleading ignorance or stupidity as an excuse in a case like this, may turn out ill.

It may happen that a schoolboy makes up his mind and tells science research how to carry out business in case of extrapolation. If such should happen, as well it might, people are likely to feel alarmed.

(8b) *Weinstein (loc. cit.) complains about the traditive physics-formulas not even admitting of differentiation.*

My analytic forms Franzén has subjected to a thorough examination as to their differentiability; so some of them a severe test has been applied. In this way, Franzén to a tittle determined the density maximum in case of water, and, in case of other liquids found their density maximum to be attained by supercooling; furthermore, strictly determined the temperature of liquid heat-capacity maximum; etc.

In D.F.T., p. 54 there is set out a test of Eranzen's that should seem rather significant. Table 13 (loc. cit.) furnishes evidence of my equation of state in case of vapours (equation [74]) admitting of differentiation and repeated differentiation with respect to temperature, and integration with respect to pressure. What we are dealing with here can not at all seem knotty to the specialist. If up to mischief here, he is likely to know perfectly well what he is about; pleading ignorance or stupidity will not do. By hinting Table 13 (as well as Table 1) to be quite beneath its own wasting words upon, Swedish academical heat-engine science has spit not only in my face and Franzén's but in the face of all of those great-scholars in whose eyes Weinstein is a name worth mentioning. A university-don spitting, for a moment, in the face of great-scholars of any description, having the moment before cringed to just that sort of people, and intending, next moment, to fawn upon that very sort again: suchlike, according to Hegel, is the Negro attitude. Is Swedish academical science research less black-a-vised than a black-a-moor, really? How runs the Academical Truth about the colour of science research?

(8c) *The heat capacity of water is a quantity that is often to be taken into account by engineering. The temperature function of water heat capacity has been determined experimentally by Lüdin; Callendar & Barnes; C. W. R. & W. E. Bousfield; Callendar. We have arranged the four determinations chronologically. Lüdin's determining of water heat capacity and the C. W. R. & W. E. Bousfield determining of it agree with one another, and form* *themselves into a distinct group G_1. The Callendar & Barnes determining of the quantity in question and the Callendar determining of the same, different as to the method made use of, agree with each other, and constitute another distinct group G_2. A function value, determined by G_1, and the corresponding function value, determined by G_2, as stated by Callendar, can differ with more than 1 %; and that is more than even technics can be satisfied with. The four determinations in question are all of them equally entitled to be determinations of precision. In consequence, such experimental determination, as is direct, i.e., wants water to be experimented upon, cannot be the way out of this difficulty. The following way offers itself, as it were, spontaneously. Experimenting upon water may be exchanged for experimenting upon the class-representative of water, Franzén having shown the heat capacity of a liquid to see its true image (burning-glass image) in the heat capacity of the class-representative of the liquid. The D. F. T. Table 16 tells us how Franzén brings about the examination of water-heat-capacity determinations given. To Franzén benzene (nitrobenzene) was an acceptable class-representative of water, for, objective limiting-point of the liquid phase, i.e., boiling point, in case of water and in case of benzene (nitrobenzene) is about the same. The precision of the benzene and nitrobenzene determinations Franzén had access to was satisfactory to him in view of the fact that the temperature function of nitrobenzene heat capacity was found comparatively clearly to depict the temperature function of benzene heat capacity.*

Franzén, D. F. T. Table 16 and Determinations 116, ascertains, whether any one of the water-heat-capacity determinations, represented by groups G_1 and G_2, can give such depiction of benzene-heat-capacity as in clearness surpasses the depiction of benzene-heat-capacity, given by nitrobenzene-heat-capacity. One of the determinations in question, viz. C. W. R & W. E. Bousfield's, can. Yet, it cannot for its entire temperature interval (0°/73° C), but only for the partial interval (0°/55° C). As regards the water-class-representative rejecting the Bousfield experimenting within the temperature interval (55°/73° C), it seems worthy of notice that this experimenting, according to a statement of the experimenters' own, could not be accurately established: "Between 70° C and 80° C there appears to be a considerable uncertainty, especially from 75° C to 80° C." In point of principle, the water-heat-capacity problem has got its solution by Franzén's just-mentioned discrimination of experimental determinations. The very result of discrimination, albeit expressively supported, is not indisputably correct, yet. In visualizing liquid-heat-capacity temperature-function at normal pressure, for want of a quite adequate representative of the class of water, Franzén adopted benzene and nitrobenzene as somewhat indifferent class-representatives. In calculating the empirical constants of the heat-capacity formula of the class representatives, Franzén didn't avail himself of the method of least squares.

Franzén not having admittance to the water-heat-capacity determination of Callendar, this determination has not got among the things discriminated. Putting right in a practical manner—or at least pointing out in express terms—just what was really deficient, would have meant probing to the bottom Franzén's result.

Insinuating silence is the slackest criticism existing. Pleading ignorance or stupidity is scarcely a feasible thing to do in the case we are dealing with. It no more admits of being done in the particular case we are about, than in the particular cases 8 a) and 8 b), afore mentioned.

The specialist, if up to mischief in his own specialty, cannot be considered so unconscious as a babe. Swedish academical science research hushed up Franzén's definitive discrimination of experimental water-heat-capacity determinations swearing against one another, and, by so doing, told the engineers, who happened very much to want the result of discrimination, to take care of their own concerns or go to Jericho. We should think, no such thing need have been done; but such thing has, and that's the fact. How runs the academical truth as to the reason why?

May be, you will read me my lesson rather gently and address me in some such way as this: "Science, in times of yore, was a system which a philosopher did his very utmost to compose and which the next philosopher made it a point to tear to pieces. Scientific criticizing, in those times, rather signified. Destruction closing in upon formation didn't mean philosophy of yore was mere words. Plato was not at a loss how' to bridle his tongue. Zeno contrived to spit out his. Words, issuing *from Fichte's form sense ever so logically, might as well have issued like streaming blood from a German soldier's heart. Hegel's Dialectics, if a carcass, hasn't a smell, for all that; tasting his philosophy of history still isn't exactly loathsome. Modern science is a different thing. Ancient religion was positive, modern science is. Criticizing ancient religion was heresy, criticizing modern science is. Each modern science being an accumulation of positive truths meant to be increased and increased, one turns out a scientist simply by placing implicit faith in the accumulation existent, and throwing one's own stone on the pile.*

Occasionally, a scientist's throwing his stone on the pile may seem to the mob the downright hocus-pocus, and set the mob shouting with delight; scientific deed like that is a positive truth, and nothing is more so. Criterium veri, now a day, is such decoration of a scientist's as makes the mob the entire world over shout with delight. As far as Swedish tongue goes, Nobel Prize is the criterion. Among all these piles that go by the name of sciences the pile named thermodynamics, in respect of fundamentals and method, holds a place of dignity. Such reception, as your thermodynamical papers have had from Swedish academical quarters, euphemistically speaking, is hard to be justified; of course, it is. However, it is most easily accounted for. These ones who throw their stone on the pile, all to a man, are following in the wake of one another.

Some time since, there wasn't in vogue but thermodynamics. Every one of the scientists, at that time, was a thermodynamicist, and none of them was anything else. In our days, optics is the pile that is all the mode. A score of years ago, you yourself saw and heard how it was in Germany. Optics was the only science cultivated, the only study prosecuted. Nothing but optics was capable of being listened to. Do you believe Sweden can be superior to Germany in taking vigorous measures? Now, there are, this very day, thermodynamics professorships. And a thermodynamics professor, of course, knows his thermodynamics lesson. In consequence, if you had checked your ambition so as to keep to the beaten track within the province of thermodynamics and throw your stone on the thermodynamics pile, you would have gained your end rather gloriously.

Unfortunately, you venture to criticize the very pile, and rather thoroughly as your criticizing refers to fundamentals, method and applications.

Not even as to any of its fundamentals, nay, not even as to the analytical (mathematical) form of any of its fundamentals, is that heat-engine form of yours dogmatically and faithfully determined.

Any academical man must shun dealing with you, if he can help it, then—I say.

The state of affairs is like this.

The faculties substantially are independent of the State; the State acknowledges them as independent.

Technically, Sweden is a realm, a limited monarchy; substantially, she is a genuine hierarchy: the State of faculties. The State of faculties, to assert itself, must be busy making people believe standing up for a sacerdotal caste-interest of its own, trifling as it may appear and mean as it may be, is standing up for the very soul of mankind. Such must be the policy of the State of faculties. High-minded that policy cannot be, anyhow; how could it? What a faculty fears is being compelled to stand up against you in public and yield you a criticism in black and white.

In interchanging ideas, intrinsic resources and extrinsic merits would be contrasted with one another. Their witnessing such would make people weaken in faith. When you are dead and gone, matters may take a different turn. When you are no more, the very substance of your papers is likely to be born again and dragged into the light of day, some academicians of established renown choosing to publish it in his own name and make your cumbersome and inconvenient name be buried in oblivion. A fruitless life you haven't led but so far as appearance goes, then. You don't forget, there are high-minded people existing: a dog, in risking his life, doesn't care a rap for his name being admitted into the chronicle. Yourself, do you mind show?"

Speech like this sounds rather alluring. Yet, there is a hitch somewhere. I am to show that, now.

To take a genuine lesson from a blacksmith's wielding his tool, such is out of the power of academical heat-engine science. The academical attitude, if it does not remain Thersitismus, immediately turns into apish imitation.

You are taken ill, let us say so. You burn with fever, or, perhaps, you are shivering with cold all over? All right. You turn to a physician of the faculty's, of course you do; he is the licensed practitioner. You are at a loss how to inform the doctor about your temperature, your words and gestures tell him little. No matter. The doctor is disposed to take such information straight from your temperature sense. Academical heat-

engine science proffers a clinical thermometer that is a fine invention if you like. And the faculty physician is just the man for using it. He has you set it in the anus. He triumphantly calls your attention to the anal wall agreeing with it, nay, fitting like a glove.

"Any lack of handle this instrument hasn't, I do declare. The anus musculature clasps the mercury bulb as pliably as ever hand closed round handle of tool."

Choosing to be rather polite, I don't say the faculty physician is the downright fool. He is; of course, he is; only, I don't say he is. I like better saying the faculty thermometer in the anus there, is a pretty regular humbug. A thermometer, if it knows anything, must know, temperature sense is, and muscular sense is not, the subject, which the thermometer itself is representative of. A thermometer has to discriminate between hot and cold accurately so as does temperature sense. I. o. w. a thermometer is of sane judgment only by being capable of dividing the expansion course of the thermometer liquid by a subjective temperature limiting-point identical with the physiological zero point. A thermometer is an instrument of precision and aids the temperature sense just as telescope aids the eye if it contrives more strictly than the temperature sense itself to point out the subjective temperature limiting-point. The water thermometer is the thermometer found to be competent to judge; the water thermometer discriminates between hot and cold quite in the way of temperature sense. The physiological zero temperature pointed out by the water thermometer, according to Franzén, is $35° C \pm 5° C$.

The water thermometer being the thermometer competent, the faculty's mercury thermometer makes an attempt at imitating it.

The faculty thermometer finds the objective limiting-points of water to be "fundamental points" of its own. As a matter of course, we must applaud the faculty thermometer for its clever way of exchanging mercury boiling- and freezing-points for water boiling- and freezing-points. If the faculty thermometer had the good sense to try to imitate the water thermometer in subjective limitation just as well as in objective, we never should take it into our head to call the fellow a monkey.

What ape has taught the faculty thermometer to yield a handle to people's anal musculature, just as a blacksmith's tool yields a handle to the blacksmith's hand? Physiological analysis is not the ape in question. Certainly, it is no merit of the physiological analysis that the correspondence between temperature sense and thermometer instrument has been definitely established. However, physiological analysis, as far as it has got up to the present, directly supports what I ascertained indirectly (i.e., by thermodynamical analysis).

The faculty thermometer dogmatically maintains, the anus is the representative place in case of thermal correspondence. The faculty gives us to understand, it can correspond adequately as well to its subjective surroundings as to its objective by wedging in there. The water thermometer prefers a strictly critical procedure. In determining the place where it can correspond adequately to subject and object, it carefully keeps aloof from arbitrary steps. Now, let us have a look at the steps it takes.

Just as temperature sense, according to physiological analysis, presents itself in the form of a system of which the components are hot-point and cold-point, so the water thermometer makes its appearance as the system of hot-water thermometer (35°/100° C) and cold-water ditto (0°/35° C). The senses being all of them real conformations the independence of any sense must be considered limited; a fortiori, the independence of any component of the system of a sense must be considered limited. The fact of cold-point happening to function for hot-point is not strictly paradoxical, then. However, academical physiology has chosen to name such cold-point functioning paradoxical.

The faculty thermometer in the anus there makes much ado about physiological analysis faltering in its speech as it expresses the relativity of the independence of temperature sense. The water thermometer, in making cold-water thermometer function for hot-water ditto, contrives to remove completely the paradoxical feature that attaches to the appearance of the process. This it effects by visualizing the process from beginning to end, so that optical correspondence can everywhere present itself representative of thermal. In short, the faculty thermometer turns a deaf ear to temperature sense inquiring about hot or cold, and makes it a point to give the anal muscular sense a helping hand. The water thermometer, in reply to the same inquiry, sets its hot-water component to the hot-point and its cold-water component to the cold-point. Moreover, in case of temperature sense apparently giving way and showing a heat fact of rather paradoxical appearance, the water thermometer allies itself with a new sense, nay, with the sense of all others that is of a nature to remove the paradoxical feature appearing. The new sense is the eye;

and the water thermometer, in putting eye to use, is no monkey face. Sense of temperature being the subject of perception, virtual image is a perception hitherto unheard of; a virtual heat image that makes its appearance may seem rather paradoxical, then. Accordingly, for removing the paradoxical feature of a virtual image appearing, it is requisite to show our being enabled everywhere to speak literary, i.e., to exchange a thermal point of view for an optical.

On its subjective boundary the water thermometer has found temperature sense to be its proximate subject, and, in the dilemma hot-point sense or cold-point sense, found sense of sight to be its closest subjective surroundings. In going to make it clear to the eye how a thermometer can correspond adequately to its own objective surroundings, the water thermometer points out the boiling point of water to be its objective limiting-point, and points out its closest objective surroundings to be all substances the boiling points of which coincide with the water boiling point. Formic acid and phosphorus oxychloride can tolerably well represent the class of water, as their boiling points are 101° C and 108° C respectively. The formic acid thermometer and the phosphorus oxychloride ditto are alike in form, but are different in size. The size of the one thermometer being given, the size of the other is determined in the following way. Each thermometer contains one unit of mass of substance. The temperature of both thermometer liquids is the same, e.g. 100° C, and is kept constant. Put the formic acid thermometer in front of a burning glass which magnifies it on the scale 85.53 : 81.06. The size of the oxychloride thermometer must be such as makes the oxychloride and the optical image of the acid cover one another. Now, we cool the liquids from 100° C to about 0° C, always keeping the temperature of the one equal to the temperature of the other. At any temperature whatever, the oxychloride and the optical image of the acid are found to cover each other. See D. F. T., Table 2, Class D, No 2 and No 17. We exchange the formic acid thermometer for a water thermometer and proceed as follows:

(a) *We set about it just so as in the case of formic acid, i.e., we, by magnifying on some scale, try getting an optical water image which at any temperature whatever within the temperature range (0°/100° C) precisely agrees with the phosphorus oxychloride.*

We acknowledge that the undertaking comes to nothing.

(b) *We magnify on the scale 85.53 : 48.10. The oxychloride and the optical image of the water are found approximately to cover each other within the range (35°/100° C).*

   See loc. cit. Table 2; Table 4; Table 5, lines 3 and 7.

(c) *We introduce a second water thermometer. This thermometer we magnify on the scale 85.53 : 51.77 (on about the same scale as used for the first water thermometer, accordingly).*

   In magnifying, we manage to get a virtual image. The oxychloride and the virtual optical image of the water approximately cover each other within the range (0°/35° C); however, for range (0°/35° C) the depiction of oxychloride by water is still vaguer than for range (35°/100° C). See loc. cit. Table 5, lines 2, 3, 6, 7, Columns 5, 6, 7d). Our liquid-thermometer pattern, i.e., Equation (64) is deduced from the heat machine of steam-engine type.

   This heat machine is a compound instrument, the boiler-cylinder system. Consequently, our thermometer pattern must comprise a compound thermometer-instrument, a thermometer system. In depicting optically, we must then manage to introduce the principle of a compound optical instrument. The hot-water thermometer, by functioning within interval (35°/100° C), gives us an objective image. The cold-water thermometer, the functioning of which immediately refers to interval (0°/35° C), we make function within interval (35°/100° C), i.e., we suffer the virtual image the cold-water thermometer affords, to be the ocular image which makes the rather vague objective image clearer to the eye. We oppose one objective thermometer to one ocular thermometer i. o. w. we assign to the two thermometers importance as 1 : 1. This compound thermometer is our mean-water thermometer and is denoted nq-water thermometer. The nq-water thermometer depicts the phosphorus-oxychloride thermometer better than the hot-water thermometer does depict, but not as precisely as does the formic acid thermometer. See loc. cit. line 4, d" oppose 13 objective thermometers to 7 ocular, i.e., we assign to hot- and cold-water thermometers importance as the temperature ranges of hot-water and cold-water 65: 35. The ensuing mean-water thermometer is denoted m_1-water thermometer. It depicts the phosphorus-oxychloride better than the nq-water does depict, but still not as precisely as does the formic acid. See loc. cit. line 5. d3). We oppose *x objective thermometers to y ocular thermometers, and we determine x and y so that the ensuing mean-water thermometer, the m_1-water thermometer, depicts the phosphorus-oxychloride thermometer just as precisely as does the formic-acid thermometer. We find x : y as about 7 : 2. The water thermometer, in order to correspond adequately to its outer surroundings, the class representative of water, has come out a compound thermometer instrument, hot-water components: cold-water components being as 7 : 2 or 3,5 : 1. It proceeds to finding on the body surface a spot where temperature-sense functions like a compound thermometer instrument and has hot-points: cold-points as 3.5 : 1. There and nowhere else it corresponds adequately to the proximate subject it has.*

The body, as an answer to the question where thermometer is to be placed most properly, turns its face against the thermometer. Face answers as to the spot representative. Tongue from first to last is answering; physiological analysis has put down the answer.

(A) Certainly, the number of hot- and cold-points is large (about 280.000). But arbitrarily the thermometer cannot be placed, if it wants to correspond to any one of those points. The body cares for that, as there are such parts of the body surface scattered about as lack hot- and cold-points altogether. Tongue speaks up as to that: considerable parts of its upper side are thermally indifferent, its lower side entirely is. On the body surface, hot-point has cold-point within range generally. Yet, such-like doesn't mean, a compound thermometer instrument is generally met with on the body surface.

(B) The single hot-point is the type of end organ of hot-sense, generally met with.

(C) The single cold-point is the type of end organ of cold-sense, generally met with.

(D3) The hot-points making up but 12 percent, of the cold-points doesn't preclude hot-points from being in the majority on the very face, the representative spot, the spot where the senses congregate for corresponding to the outer surroundings: tongue with vigor supports, it doesn't.

Let a : b be as number of hot-points: number of cold-points. For tongue dorsum a : b is as 1 : 1. (See d1). (D2) For tongue border a : b is as 2 : 1. (See d2). (D3) For tongue tip a : b is as 3 : 1. For the lips a : b must be greater than 3 : 1, the hot-points lying so close together that physiological analysis is at a loss to isolate them and sum them up. Between tongue tip and lips the end organ of temperature sense appears a compound instrument of precision.

(See d3). Between tongue tip and lips the thermometer competent must be placed to correspond adequately to sense of temperature, then.

The water thermometer, by asking its way as regardless as is just mentioned, contrives to see its way and place itself.

Does the faculty thermometer follow in its footsteps and stick to its heels?

Tongue calls the fellow loudly. Eye is eagerly bent upon guiding his steps from beginning to end. But he has chosen to be located in the anus there, and he is steady after all. He can't be got to move from the spot. Not he.

The water thermometer, always taking up a critical attitude regarding things, cannot consider it's functioning exhausted by what is mentioned in the foregoing. In the foregoing we have made out, that boiling point is the objective limiting-point. We, now, set that special case aside; we want a general view of water-thermometer proceeding. Let freezing point be the objective limiting-point.

In the case we are considering, water-class or water-thermometer outer-surroundings is the comprehension of such substances as have water freezing-point for their own freezing-point, or very nearly so.

The liquid-thermometer pattern, in the case we are about, is the E. H. Equation (93). Any class-representative of water turns out a depiction of a water-class representative chosen; the depiction scale turns out the only individual feature of the depiction. Water itself cannot depict nor be depicted. See E. H., Table 12 b) Lines 2, 3, Columns $_0q_1$, $_{-1}q_0$, $_0q_2$. From this we are by no means to infer thermometer pattern E. H. (93) to fail of success in case of water. It is only that water thermometer E. H. (93) as well as water thermometer D. F. T. (64) wants a systematic form for its corresponding adequately to its objective surroundings. In case of D. F. T. (64) our starting point was water of boiling-temperature.

In cooling this water, we contrived to determine a subjective limiting-point of water viz. the physiological zero point of temperature. In the said case the systematic form of water thermometer was the system of hot-water thermometer and cold-water ditto, corresponding *to the system of independent senses hot-point sense and cold-point sense. In case of E. H. (93) our starting point is water of freezing-temperature. In this case we take no interest in the system of hot-point sense and cold-point sense. The system interesting to us is the system of cold-point sense and pressure sense, for, at great pressure, the thermometer pattern (93) is found, on a certain scale, to depict water by its class representative.*

(See loc. cit. a). Columns $_0q_1$, $_{-1}q_0$, $_0q_2$. Instead of connecting cold-point and pressure senses we may connect cold-point sense and taste or smell. Let the water thermometer contain water which is sweet, salt, sour, or bitter; in all cases, the thermometer liquid is found adequately to correspond to the class representative of water. See loc. cit. b). Aqueous solutions. Columns $_0q_1$, $_{-1}q_0$, $_0q_2$.

The cold-water thermometer finds, it corresponds adequately to its objective surroundings by suffering great pressure to be put on the thermometer-water, or, by suffering the thermometer-water to taste or smell like an aqueous solution. It finds, it in this manner corresponds adequately to its subjective surroundings, too. For, as physiological analysis tells us, just the acting together of cold-point sense and pressure sense is what makes us feel wet (i.e., feel subject of water-thermometer).

In this connexion it is worthy of notice that the temperature points of the nipple are cold-points almost exclusively, and that glans penis hasn't but cold-points; furthermore, as to nipple and glans penis, that functioning under rather great pressure is a characteristic feature of both; lastly, that such nipple- and glans-functioning is a generic determination of sense of temperature and, in consequence, essentially corresponding to the subjective surroundings of the sense in question.

Temperature sense and pressure ditto are not the only senses, which bring about their functioning by the instrumentality of water-thermo-meter. Tongue, by all it tells you it has to deal with, makes you understand it is dealing with some aqueous solution. And nose, wheresoever it points, points in the direction of an aqueous solution, if the testimony of physiological analysis is to be depended upon. Obviously, the tongue-and-nose-functioning just spoken of, in contradistinction to the above-mentioned nipple- and glans-functioning, essentially corresponds to the objective surroundings of temperature sense (i.e., to the water-thermometer instrument).

The water thermometer, wanting directions as to its locating most properly, asks the question: "Where on the body surface is a sense to function, when the sense in question wants to correspond adequately to its objective surroundings?" The senses take up the chorus: "Face is the place representative of a sense whatever, if subject corresponds to object adequately by the instrumentality of the sense in question."

The faculty thermometer adheres to its dissentient opinion. It maintains, the anus is of all places the place representative of a subject that wants to correspond adequately as well to the other subjects it is dealing with, as to its object. According to the faculty thermometer, subject, if a relevant matter, is muscular sense. Elephant trumpets into its ear "Muscular sense, if it turns up in good Ernest, speaks exactly so, as do the other senses," Rhinoceros, who, as is well known, hasn't any manners, never had any, and never will have, is tactful and debonair just this once, and in his very plainest English tells the faculty thermometer "Organ of muscular sense, if muscular strength of a pachyderm does signify, is very much the same as nose." I have called the faculty thermometer a monkey. I might as well call it a donkey.

For that little ass, upon the five senses unanimously expostulating with him against his sitting in the anus there can't be prevailed upon to move.

The Faculty of Medicine's clinical thermometer, to tell the truth, is as innocent as a paschal lamb. I have had it come out as sin-burdened as a scapegoat. It's, I can tell you, a matter of form. The arms I wield must be steadily turned against the Faculty of Arts. It may be that the Faculty of Medicine is hell incarnated. Yet, I am not at it just now; there are formal obstacles. It may seem awfully ruthless to follow the Faculty of Medicine through thick and thin and to second that faculty per fas et nefas, leaving for the same faculty's clinical thermometer but such moral support as the anal musculature can afford.

But let us be courteous, or die.

(9a) *You are in your senses, and you can tell where the senses are to be found. You see and hear, and you are indebted to eye and ear for being enabled to. Nose is sensitive to odours, and you have taste just as far as tongue goes. A pet, if its pain is your own, you protect as the apple of the eye. If on the alert for someone, you turn the face towards him, and not the abdomen.*
You know that much.

Indeed, you know more; the study of optical instrument was actually the way to be well acquainted with eye, so you are told. Just as the invention of optical instrument could reveal the conformation and function of eye, so my discovery of water-thermometer competence may reveal the conformation and function of temperature sense; you can but think, it may. The water-thermometer aiming at face and not at anus, when it intends aiding temperature sense in corresponding adequately to its object, you don't find preposterous; on the contrary, judicious and easy to comprehend, as face is the place where the other senses, all of them, correspond adequately to their object. Your imagination holds out a prospect as follows. In years to come, some renowned academicians will put forth as an academical truth: the faculty's clinical thermometer must be disengaged from the grasp of iron wherewith the anal musculature clutches it, and be forwarded to the place due to it as a representative of the thermometer competent—Why, I say, you are rather sanguine? Physiological analysis has set forth already a good deal in my behalf; so, it has, and no mistake. The faculty, in its own thorough way, has looked into the inside of these results of physiological analysis, and has contrived to translate them into the King's English. The triumphant issue of the violent contest for translating is, that the faculty thermometer sticks where it sticks and where it has stuck since thermometry began, some centuries ago. Maybe, my own translation of the hitherto belonging analyses—those of physiology and those produced by myself—is as in composite as possible and as striking as possible: it is only a matter of copying what I have written. But it is rather difficult to understand an academician being enabled to display a true copy of my interpretation of results of mine by not taking the copy but after my being dead and gone, and by publishing it for his own original. Isn't it, indeed?

(9b) *Observation of nature, such n. b. as tells, is an occurrence of recent date, and that is academical truth. There are two ways, differing utterly, of viewing nature, both ways equally primeval, equally modern, and that's the truth. We'll just sketch out the ways.*

A beaver, as enigmatical nature is abruptly confronting him, is not at a loss how to behave. Shy he isn't exactly; imagination does not serve him the trick of vivifying Nature and having that new subject play the master over the subject he is himself. He takes it coolly, he does. His keen eye reduces the "natural phenomenon" to a real limiting point, and eo ipso to a real problem. He tries to make out the species of the problem. A certain species is the very thing to him, as he has got some talents of his own. The object his special talents afford, he elaborates with consummate skill. He makes up a complete form, the famous beaver dam— The grossindustrieller and the beaver are engineering-line men; apparently widely different, yet, essentially, very much alike. The grossindustrieller is particular about his raw material. He doesn't mind plunging into a vast abyss to obtain what answers his purpose. The masterpiece issuing is finished goods or semi-manufactures, made in Germany—Having a celestial thing lie in the mud beneath him, to Atlas was υπερμορον, you know.

Your giants, the Großindustrieller and the beaver, may dig their treasure from the mud beneath. The very mud they do not fetch, for all that. The mud is nothing to them. Nature is to keep its secrets, insofar as those two are concerned.

I'm going to show the power of the heat-machine idea descending, the heat machine realized being the steam-engine type at large. The piece of machine product, turning out here, is the palpable form of empirical things, their species and degrees, their grade correspondence and species ditto, expressed as a classification unified.

Obviously, here is still the same point of view as in the cases of grossindustrieller or beaver.

The "natural phenomenon" and limiting point, here to be dealt with, is the coexistent phases in case of melting or boiling (sublimating). There the phase-limit-generator has to be placed. And there it has to be determined. Genus proximum is the universal generator principle, i.e., the equivalence of machine heat and machine work. Differentia specifica are the machine-heat expressions (8) and (31) in P. L. M. Now, the phase-limit-generator is established.

The phase-limit-motor has its determination of motor species from (44) in P. L. M. The phase-limit-motor is the upper point of the phase-limit-machine, i.e., the point close to the subject directing. The phase-limit-generator is the lower point of the phase-limit-machine, i.e., the point close to the object. From the upper point the different pieces of machine product are surveyed.

The number of original heat instruments, surveyed from where the subject directing is standing, represents the main pieces of machine product. The volume thermometers are to be mentioned here, and so are the calorimetric instruments.

Consider the case of a certain volume thermometer being worked out so, as to attain a stamped form. The phase-limit-motor shares in this business according to the universal motor principle: maximum output of work; and, in consequence, makes the subject directing share in the same business according to the principle: minimum output of work.

In making this statement of principle, the phase-limit-motor defines itself an objective motor, and fixes the ties, which attaches it inseparably to the subject it has above. These ties are the arbitrary functions $\eta\left(\dfrac{\Delta\Phi}{T}\right)$ and $\vartheta\left(\dfrac{\Delta\Phi}{T}\right)$, entering into motor form (44). The subject works hand, or (more generally) by a certain motor M that may be called the subjective motor—the calculating machine which from experimental determinations affords the numerical values of the arbitrary constants entering into η and ϑ. The numerical values in mention made enter into η and ϑ, the ensuing process refers to the objective motor.

As to the proper form of functions η and ϑ, and the proper number of empirical constants, being to enter into those functions, it is a question which the subject at the calculating machine is capable of settling without any doubt. For the subject cannot possibly mistake the form of function and the number of constants corresponding to its own minimum output of work. See the foregoing, p. 20–24. Here, we have set forth the objective point of view. The objective motor is the one that has the control everywhere. The object it controls directly. The subject it controls indirectly. It makes the subject take one step upwards and be standing, not at the point of the phase-limit-motor, but at the point of the calculating machine. The taking of the step in question, translated into objective language, runs like this: The phase-limit-motor sees itself compelled to take the objective motor principle under its universal form for the only guide to go by; the phase-limit-motor, accordingly, being guided directly, the subject at the calculating machine indirectly: The principle of calculating-machine motor:

minimum output of work; owing to this, appears an idea, conceived, as it were, by the very object. This sounds paradoxical, about so, as does the saying: necessity is the mother of invention. Such as it is, it is palpably graspable.

The incomposite train of machine-working ideas means maintaining from beginning to end the subject as the only guide to go by. At the calculating machine M the subject guides directly, minimum output of work being the motto. At the phase-limit-motor the subject guides indirectly, maximum output of work being the motto. The subject descending lower down, and realizing a certain instrument of those surveyed, the management remains unchanged; in case of motor M being made use of, the management is direct; in case of the phase-limit-motor being made use of, the management is indirect.

Such piece of machine product as lies lower, may turn out more faintly stamped than such as lies higher. But each piece of machine product to a certain degree bears the stamp of the subject directing. In virtue of this stamp and in no other way, a "natural object" is strongly marked as to its form. The heat-machine idea descending into the shape of a real subject, the real subject realizes it 1) immediately as the category of objective motors and generators, and 2) ultimately as the definitive limit of such as is in existence. The real subject moving downwards is a show of strength; for the subject, low as it may be standing, inflexibly links in with the idea it has above.

The professorship of thermodynamics makes a point of inspecting any industrial works all the world over. Such visit at each time is a festive occasion. The whole of the industrial establishment in question parades before the professorship. The very machines are, as it were, a going. The professorship is by no means intending to learn engineering skill, in order to apply the lesson to academical thermodynamics.

A real subject being capable of, in sober earnest, directing any machine work, and affording a stamped form to any machine object, that is what he cannot well imagine, in spite of all he saw and sees. The power of the heat-machine idea he never did believe in. Upwards he cannot look; by far, he cannot. He looks downwards; most awkwardly, in this direction lies powerless imitation. The professorship sees generator and motor at work. The very generator, the very motor, he does not see, for all that. He sees depths, abysmal depths, the professorship does.

He sees a brand-new subject. He sees Nature. Shouldn't he be submissive, once in a way? Prove the submissive naturalist? Of course, he should. The generator principle proves nothing to him. He chooses adopting a natural law: preservation of the energy of Nature: it seems to him that this statement is judicious and not at all preposterous. He terms it the first heat law. The motor principle, too, he exchanges for a natural law—dispersion of the energy of nature—this statement he terms the second heat law. The two heat laws pocketed, our hero exults at having revealed two momentous secrets of Nature. If I tell you: the beaver and the grossindustrieller are representative of the princely Oedipus, the academical natural philosopher is representative of the vulgar Theban devoured by the Sphinx, am I wrong there, anyhow, as to the significance of actual academical science?

(9c) *Analysis is a cross to be borne. Analyzing is searching with one's senses fully on the qui vive. A dog, possessed of nose, knows that much. A vulture, possessed of eye, knows. Either duly puts his talent out to usury. Neither chooses to complain. The academical scientist has nose, eye, and some talents more. In counting his talents, he has his praeterea censeo: the academical worker's heart-rending penury. Academical mathematics rejects the principle of analyzing, and, in rejecting, acts straightforwardly; the principle of judgment, divulged by the theory of equations, runs: a calculatory dilemma is sound: academical physics, chemistry etc. also reject the principle of analyzing, yet, in rejecting, don't act straight-forwardly; the scientist in mention taking a liking to the name of empiric. A dilemma confronting him is unsound, he lays stress upon his granting that: his bill of costs for investigation purposes has Sweden all but make up her mind to go a begging. The exertion of analyzing the dilemma he always manages to spare himself from, rather having resource to speculation. Imagination whispers in his ear the dilemma, just confronting him, to be a Nature catastrophe; the catastrophe to be regulated by a natural law never-to-be-forgotten; the hand that brings about the catastrophic action and makes law for it, to be belonging to Nature, the mysterious subject he contrived to reveal, after all. The dilemma does not shake, on the contrary, strengthens, the naturalist's faith. The very name of empiric he cannot be prevailed upon to give up, though The Swedish expression—**den yttersta domen**—denotes the highest* *judgment form (logic disjunction), or else, the lowest judgment form (logic division). In a real dilemma, you are tempted into resorting to the highest judgment, this being as close at hand as liberty of choice, arbitrariness. Resisting temptation like that, you have to extricate yourself from the dilemma by taking your refuge to the lowest judgment; you have to make your way by making inquiries as meek and lowly and unreserved as a "soulless beast," a "dumb brute;" you have to analyse, sparing no pains at analyzing. Academical mathematics is arbitrary procedure frankly set up as the norm. Academical physics, chemistry, & co. love the name of science research very dearly; their **yttersta dom**, if their own word does signify, is the form of questioning. In action, they come short, promptly. Without reserve, the academical scientist cannot stand out questioning. Areal difficulty, as he encounters it, turns out the Nature-secrets-investigators dilemma, not that investigator's. Face to face with sober reality, the academical crusader's crusade is straightway at an end. Am I the one who pronounces sentence of annihilation on physical analysis, handed down? Or else, are the academical physicists themselves those who pronounce this sentence on their own attempts at analyzing? Let us see.*

Expansion by heat is a chapter of physical analysis, dealt with by unnumbered academical physicists.

Their analyzing, according to academical criticism (e.g., Einstein), has come to nothing—there is no species visualized, no grades. Neither are substances shown to correspond to one another, nor are phases. Hypothesizing is the only hope remaining. In saying so about the traditive literature hitherto belonging, academical criticism happens to overlook a pearl lying in the mud, that pearl, although a little one, being bright enough and most easily seen.

The pearl in question is "Gay-Lussac's volume law"; it refers to the ideal-gas phase, and there clearly and distinctly depicts the expansion-by-heat of any substance by the expansion-by-heat of another substance. The academical critics venturing to declare, as Gay-Lussac's bright volume-law is confronting them, "no species visualized, no grade," their criticism is a 'practical joke. How they manage, in so joking, to keep clear of the eighth commandment, I cannot tell. Who can?

(9d) *Mendelejeff's formula* $v/v_0 = \dfrac{1}{1-kT}$ *by aid of one only empirical function cannot be adopted accordingly but under the constant k approximately determines the expansion-by-heat of liquids in general. The academical critic judges: All or nothing: He expects form of a polynomial of a strict determination, not an approximate. The academical truth is then: Mendelejeff's analyzing comes to nothing.*

Avenarius' formula (modified by Mallet & Friedrich), $v = a + b\lg(A - t)$, determines rather strictly the expansion-by-heat of 25 substances, and contrives to bring about correspondence between them, as far as one, b/a, of the empirical constants, entering into the formula, is concerned. The academical critic judges: All or nothing. 25 substances, that's not all substances, that's nothing. Correspondence as to one empirical constant out of three given, that's not complete correspondence, that's nothing: The academical truth is then: Avenarius's (Mallet & Friedrich's) analyzing comes to nothing. The academical critic's yttersta Dom is obviously a disjunction; his truth is—here as always—arbitrarily chosen. The empiric's yttersta Dom is a division. He doesn't judge: Mendelejeff or Avenarius. He judges: Mendelejeff and Avenarius. His criticism is a piece of analyzing work which he carries out himself. In Avenarius's statement he comes across a dogmatical feature; he removes it. In Mendelejeff's statement he comes across a dogmatical feature; he removes it. He forms a grade series where Avenarius's pure statement is present as superior grade in relation to Mendelejeff's pure statement as subordinate. He forms another grade series where Mendelejeff's pure statement is present as superior grade in relation to Avenarius's pure statement as subordinate. These two-grade series together bring about the solution of the problem of liquid expanding.

I cannot better criticise Mendelejeff's and Avenarius' analyzing than by adding for my criticism the very little piece of analyzing work that has those two pearls of analysis glisten. I denote phases: ideal-gas, vapour, liquid, solid, respectively by roman I, II, III, IV, my criticism, and runs like this. Gay-Lussac's "volume law" settles the question of temperature zero-point. Temperature, as far as phase correspondence is sought for, must be "absolute" temperature T then, and cannot be Celsius temperature t. Phase correspondence must be sought for at liquid freezing-point as well as at liquid boiling-point, solid superheating and vapour supercooling being undisputable facts, both of them. Mendelejeff's rational function $v/v_0 = \dfrac{1}{1-kT}$ *cannot be adopted accordingly but under the form of a polynomial of 1/T.*

$$v_M = A_1 + \frac{B_1}{T} + \frac{C_1}{T^2} + ... \quad \text{(Mendelejeff's pure statement)}$$

Likewise, Avenarius' (Mallet & Friedrich's) logarithmic function v = a + blg(A − t) cannot be adopted but under the form of a polynomial of lgT:

$$v_A = A_2 + B_2 \lg T + C_2 \lg^2 T + ... \quad \text{(Avenarius' pure statement)}$$

In either statement, a constant term, A_1 resp. A_2, is present. The two different grade series to be formed cannot differ as to a term, which is common to the pure statements we are dealing with. Consequently, the first term of either grade series is a constant

$$A' + + + \qquad\qquad \text{First grade series}$$

$$A'' + + + \qquad\qquad \text{Second grade series}$$

Let Avenarius' statement be superior grade, and Mendelejeff's subordinate. Then, two terms (corresponding to B_2 and C_2) must be taken from v_A, and one term (corresponding to B_1) from v_M. Liquid expansion-form is accordingly

$$v^{III}(T) = A' + B_2' \lg T + C_2' \lg^2 T + \frac{B_1'}{T} \quad \text{Liquid form } (\alpha)$$

Again, now.

Let Mendelejeff's statement be superior grade, and Avenarius's subordinate. Then, two terms must be taken from v_M and one term from v_A. Liquid expansion-form is accordingly

$$v^{III}(T) = A'' + \frac{B_1''}{T} + \frac{C_1''}{T^2} + B_2'' \lg T \quad \text{Liquid form } (\beta)$$

In choosing to declare Mendelejeff and Avenarius' analyzing results to be forms incapable of visualizing liquid expansion-by-heat, academical criticism happens to overlook liquid form (β).
The overlooking is rather fatal:

(1) *Form (β) with precision represents the course of expansion of any liquid whatever,*

(2) *Form (β) depicts the liquid expansion of any substance by the liquid expansion of another.*

(3) Form (β) holds for infinite pressure as well as for normal,

(4) Form (β), under the form (β duo), represents solid-crystalloid expansion,

$$v^{IV}(T) = a^{//} + \frac{b_1^{//}}{T} + \frac{c_1^{//}}{T^2} + b_2^{//} \lg T \quad \text{Solid-crystalloid form (β duo)}$$

(5) Form (β) depicting liquid expansion optically, i.e., keeping three divisions $B_1^{//}/A^{//}$, $C_1^{//}/A^{//}$, $B_2^{//}/A^{//}$ independent of individual substance, form (β duo) makes solid crystalloids correspond by keeping two divisions $b_1^{//}/b_2^{//}$, $c_1^{//}/b_2^{//}$ and one multiplication $a^{//}b_2^{//}$ independent of individual substance,

(6) Forms (β) and (β duo) make phases III and IV correspond, by keeping independent of individual phase the two divisions $b_1^{//}/b_2^{//}$, $c_1^{//}/b_2^{//}$, kept independent of individual substance.

In choosing to declare Mendelejeff's and Avenarius' analyzing results to be forms incapable of visualizing liquid expansion-by-heat, academical criticism happens to overlook liquid form (α).

The overlooking is rather fatal:

(1) Form (α) with precision represents the course of expansion of any liquid at finite pressure,

(2) Form (α), interpolated for some temperature interval, admits of extrapolation to a rather great extent; and no systematic error,

(3) Form (α) admits of differentiating, repeated over and over,

(4) Form (α) depicts the liquid expansion-course of any substance by the liquid expansion-course of another,

(5) Form (α) exhibits any liquid expansion as objective image of hot-water expansion (35°/100°C), and as ocular image of cold-water expansion (0°/35°C); i.e., form (α), generally making object correspond to object, in case of water "the touchstone for the expansion formulas respecting liquids" makes object correspond, not only to another object, but to subject.

Mendelejeff's being one academical attempt at analyzing liquid expansion, Avenarius's being another; academical criticism has chosen to slight either. We've slighted neither academical attempt. We just have given either a touch and set both glistening like genuine pearls. Other touches, and they'll sparkle brighter still. You just wait and see.

We intend to confront (α) with an academical attempt at analyzing real-gas expansion. Callendar's is the attempt in question. Callendar's *pointing out the liquid volume to be present in the vapour expansion-formula p(v − b) = RΘ is a strong point of his. However, in his statement there is a dogmatical feature, which should be removed.*

(1) *The validity of p(v − b) = RΘ for a perfect, or pluperfect, gas at high temperature being "practically certain,"*

(2) *The correspondence of phases (superheating and supercooling) being an indisputable fact,*

(3) *The said correspondence being known as one that goes extremely far; things being so, we do not judge cautiously, if we take the equality of quantity b and liquid volume v^{III}, observed by Callendar, to be in any way accidental. We judge most cautiously if we try formula p(v − b) = RΘ for one that makes phases liquid and vapour really correspond.*

This admitted of, we are to try the formula for any temperature and pressure, and not only for "temperatures where vapour pressure is small," Furthermore, b being "often negligible," we should put $b = v^{III}$ (T) + η, where v^{III} (T) is the liquid temperature-function, and η is a quantity independent of temperature.

Callendar's pure statement is accordingly

$$v^{II} \ (T) = KT + v^{III} \ (T) + η \qquad \text{Callendar's pure statement (γ)}$$

According to (γ) and (α), the vapour expansion form is

$$v^{II} \left(T\right) = KT + η + A' + B'_2 \lg T + C'_2 \lg^2 T + \frac{B'_1}{T} \qquad (δ)$$

Academical Criticism, slighting Callendar's analytical attempt p(v − b) = RΘ, eo ipso has overlooked the analytical forms (γ) and (δ).

The overlooking is rather fatal:

(1) *Vapour expansion-form (δ), interpolated for a short temperature interval, admits of extrapolation to a rather great extent, furthermore admits of differentiation, repeated, and repeated over again.*

(2) *Form (δ) clearly and distinctly depicts the vapour expansion of any substance by the vapour expansion of another substance,*

(3) *The depiction still holds well, in the case of one of the substances, or both of them, being below triple point.*

(4) *Take S for the substance to be experimented upon. Let S be under pressure p, above that of triple point p_s. Let T_b be S-liquid boiling point, corresponding to pressure p. Denote by (a_s) the*

S-liquid thermometer-form (α), determined for temperature interval T_b–T_l. Denote by (δ_s) the S-vapour thermometer-form (6), determined for temperature-interval T_L–T_b. Forms (α) and (δ_s), according to (γ), being representative of one another within temperature limits l and L; the expansion of supercooled S-vapour is determined with precision, if maximum number of (δ_s)-constants is determined from (δ_s), and minimum number of (δ)-constants determined from (α_s); likewise, the expansion of superheated S-liquid is determined with precision, if maximum number of (α)-constants is determined from (δ_s), and minimum number from (α_s).

(5) Let the substance to be experimented upon be carbon dioxide under triple point pressure π_{CO2}.

The liquid temperature-interval T_b–T_l is a single temperature point, the triple point temperature T_{co2}-Form (δ_{CO2}), i.e., the CO_2-vapour thermometer-form (δ), determined for interval T_L–T_b, affords the numerical values of coefficients K, $\eta + A'$, B'_2, C'_2, B'_1. Relation $v^{ll} = KT_{CO2} + v^{lll}$ (T_{CO2}) + η affords the numerical value of η. Eliminating between form (δ_{CO2}) and the corresponding (γ)-form displays the determination of precision of carbon-dioxide liquid v^{lll} (T) when superheated under triple point pressure to temperature point T_L.

Only one question more, and I have said. The chapter just examined, being the one upon expansion by heat, is examined and reexamined, over and over again, by unnumbered academical eyes. If I tell you: the academical critic's criticism about the academical attempts at analyzing ideal gas-, vapour-, liquid-, solid crystalloid-expansion is a falsification from beginning to end: am I wrong there, now?"

2.5 Conclusion

That was the story of Nils and Karl, of their desperate struggle for the advent of a different thermodynamics against the academic science of their time.

Independent of Peter Boas, who had clearly called for the serious revision of the conventional foundations of thermodynamics, Nils and Karl could do their very best to carry out this highly ambitious project. They did succeed in accomplishing this, irrespectively of what was the official reception of their results. The story just told was developing in the Scandinavian countries.

In effect, Nils was successfully continuing the traditions of his famous relatives—Engelbrekt Engelbrektsson (1390–1406), the actual founder of the today's Sweden and Olav Engelbrektsson (1480–1538) in Norway, the powerful proponent of the Catholic Church in his country.

As we know, Engelbrekt was brutally killed, whereas Olav had been not less brutally expelled from his native country. Meanwhile, Nils and Karl were neither killed, nor expelled—but are, instead, wearing their oblivion veils. To our mind, the latter is the worst what could happen to a scientific research worker, philosopher, journalist, and – the last but not the least—to a patriot—to a creative soul, in a couple of words.

Nonetheless, we hope that the monograph at hand could finally pay on all the accounts in question.

We noted here that Nils was hardly criticizing Albert Einstein and his allies all over the world. In Chapter 3 of the volume at hand, we shall continue discussing the actual Einstein's contribution.

Importantly, Albert Einstein should not be "demonized," likewise whoever. Not correct is also making use of the 'national' criteria to estimate professional achievements of whomever. Anyway, the end of the XIX-th century and the beginning of the XX-th were devoted to the so-called "revolution in the natural science." Many researchers of different nationalities from countries all over the world were actively and enthusiastically taking part in this revolution, with Albert Einstein being just among them. One should not "demonize" the revolution itself as well. It had brought many positive results.

What was and is but a clear injustice ought to be handing out the veils of oblivion galore according to the erroneous (but exceedingly revolutionary!) principle: "Кто не с нами, тот против нас" (Who is not with us is acting against us). The monograph at hand is trying to reconcile this gap.

In Chapter 5 of this monograph we shall also try to look behind the curtains of that revolution in aiming at more or less clear picture of what happened to thermodynamics at that time. Without doing any harm to those who is not more without us for the long time, we would greatly appreciate a clear picture of the then events in thermodynamics and the tightly related fields to get a clear picture of the actual happenings to enable an instructive analysis of the latter.

2.6 Complete List of Publications by Nils Engelbrektsson

1. Nils Engelbrektsson (**1909**): *Meddelande till matematiska kongressen i Stockholm (22–25 Sep. 1909).* Håkan Ohlssons Boktryckeri, Lund, Sweden.
2. Nils Engelbrektsson (**1911**): *K. Läroverksöfverstyrelsen och lärarepersonligheten.* A. Lindgren & söner: Göteborg, Sweden.
3. Nils Engelbrektsson (**1911**): *K. Vetenskaps- och Vitterhetssamhället i Göteborg och det Vetenskapliga Initiativet.* Håkan Ohlssons boktryckeri: Lund, Sweden.
4. Nils Engelbrektsson (**1915**): *Krigskulten.* Uddevala, Sweden.
5. Nils Engelbrektsson (**1916**): *En skandinavs tankeliv.* N. P. Pehrssons förlag: Göteborg, Sweden.
6. Nils Engelbrektsson (**1916**): *Entropieekvationens $dS = \dfrac{dQ}{T}$ integrationsteorem och kontaktsprincipen.* Håkan Ohlssons Boktryckeri, Lund, Sweden.
7. Nils Engelbrektsson (**1917**): *I. Über die Bestimmtheit des zweiten Hauptsatzes der Thermodynamik. Lösung mittels des zweiten Hauptsatzes von Problemen bei denen der Entropiesatz keine bestimmte Lösung gibt. (Anwendung: Herleitung der Temperaturfunktion des Dampfdruckes. Analytische Definition des kritischen Zustandes bei Substanzen). II. Zusammenhang zwischen Wärmefunktionen bei einer Substanz, die einem thermischen Prozess unterworfen ist und dem unveränderlichen Gewicht der Substanz. (Anwendung: Herleitung des Gesetzes von Dulong-Petit und der Regel von Trouton).* Håkan Ohlssons Boktryckeri, Lund, Sweden.
8. *Termodynamikens grunddrag,* af Nils Engelbrektsson; Appendix: *Verification of Engelbrektsson's equation*

$$-T\,\frac{\dfrac{\partial v^{l}}{\partial p} + \eta\left(\dfrac{\Delta z}{T}\right)\dfrac{\partial \Delta v}{\partial p}}{\Delta v} = \vartheta\left(\frac{\Delta z}{T}\right) \text{ (the rational equation of state}$$

of bodies) for saturated and unsaturated vapors, liquids at finite and infinite pressures, and isotropic solids—by Karl Franzén. Håkan Ohlssons Boktryckeri: Lund, Sweden, **1920**.

9. Nils Engelbrektsson (**1926**): *The Phase Limit Motor as Heat Engine in Cases of Boiling or Melting*. A. Lindgren & söner: Göteborg, Sweden.

10. Nils Engelbrektsson (**1927**): *Expansion by Heat Regarded as a Function of the Phase Limit Motor*. A. Lindgren & söner: Göteborg, Sweden.

11. Nils Engelbrektsson (**1928**): *The Integral Form for the Integrability condition of the Entropy Equation*. A. Lindgren & söner: Göteborg, Sweden.

12. Nils Engelbrektsson (**1928**): *A Thermodynamic Study: The Deduction of the Form of the Thermometry*. A. Lindgren & söner: Göteborg, Sweden.

13. Nils Engelbrektsson (**1928**): A Thermodynamic Study: The Deduction of the Forms for gas-expansion and solid-expansion at small pressures. A. Lindgren & söner: Göteborg, Sweden.

14. Nils Engelbrektsson (**1935**): *Notiz von dem Entwicklungsgang meiner thermodynamischen Arbeiten*. A. Lindgren & söner: Göteborg, Sweden.

15. Nils Engelbrektsson (**1948**): *Akademisk matematik eller mitt räkneinstrument—ett dilemma; Academical Thermodynamics or my Heat Engine Form—a Dilemma*. Centraltryckeriet: Helsingfors, Finland.

2.7 Nils and Karl's Struggle with 'Swedish Academics'. Was It Their Pure Phantasy?

In this chapter we have learned about Nils Engelbrektsson's irreconcilable position with respect to what he denotes as the 'Swedish academics'. Was it Nils' pure phantasy in combination with his personal bad temper?

Our investigations show that the answer to the above poser is by far not so immediate.

Let us open the well-known and renowned Swedish academic journal '*Arkiv för Astronomi, Mathematik och Physik, utgiven av*

Kungliga Svenska Vetenskapsakademien', Volume 16, Number 3 of the year 1922. We find many interesting papers over there, with some of them of definite interest and relevance for our present discussion.

The first and foremost contribution of interest for us is the paper by Prof. Dr. Carl Wilhelm Oseen (1879–1944) published in German and entitled '*Eine Methode, die Zustandsgleichung der beliebigen Flüssigkeiten oder Gasen exakt zu berechnen*' (A method to calculate exactly the equation of state for diverse liquids and gases) [1]. As this is just the topic of our present chapter, it is of immense interest to present here our full translation of this chapter into English.

A Method to Calculate Exactly the Equation of State for Diverse Liquids and Gases

By C. W. Oseen

Communicated on March 9, 1921 by Ivar Bendixson and Helge von Koch.

Anyone familiar with modern Swedish mathematics knows the place in which it takes the problem of representing an analytic function by some unified mathematical expression in the widest possible part of its existence realm. If I had to describe the history of this problem here, I should have to remember how Mittag-Leffler devoted almost all his energy to the solution of this Weierstrassian problem. Of his successors in this field I would primarily have to recall Helge von Koch, whose extension of the Mittag-Leffler problem is of fundamental importance for what follows. However, I do not want to talk about the mathematical details here. The purpose of these lines is to show that a straightforward application of the results obtained by the colleagues mentioned should be sufficient to obtain a mathematically exact and, I believe, physically useful method of solving the problem that has for several decades been subject to the most zealous effort of a great number of scholars. Their actual task has been to infer the equation of state for a gas or a liquid, whose atoms interact with each other through known, but largely arbitrary forces.

For the sake of simplicity, I will limit myself here to a monatomic gas or liquid. I denote by x_i, y_i, z_i (i = 1, 2 ... N) the coordinates for the center of the i-th atom. For the sake of simplicity, I also denote all these position coordinates q_i (i = 1, 2 ... 3N).

Now I cast $\xi_i = m_i \dfrac{dx_i}{dt}, \eta_i = m_i \dfrac{dy_i}{dt}, \zeta_i = m_i \dfrac{dz_i}{dt}$, or in a shorter form $p_i = m_i \dfrac{dq_i}{dt}$ (*i* = 1, 2 ... 3N), where m_i stands for the mass of the *i*-th atom. In what follows I assume all the atoms to be identical, so that $m_i \equiv m$ throughout here. With this in mind for the kinetic energy we get the expression

$$\sum_{i=1}^{3N} \frac{1}{2m} p_i^2 .$$

Next, we introduce the potential energy, P, which we assume to be dependent on the variables q_i, but not on the variables p_i, and/or on the time. If we denote the total energy with E, then we get for the latter

$$E = \sum_{i=1}^{3N} \frac{1}{2m} p_i^2 + P.$$

We might therefore represent the state of our system as a point in the 6N-dimensional Euclidean space, by assuming q and p to be orthogonal coordinates for some point of the space involved. Of course, the arbitrary point does not stand still in the 6N-dimensional space. It moves, but during the movement it must always remain on the so-called energy surface, that is, on the (6N – 1)-dimensional manifold:

$$E = const = E_0.$$

Statistical mechanics proves that the probability for the arbitrary point to remain within some surface element dS of the energy surface has the value:

$$\frac{1}{\dfrac{\partial V}{\partial E_0}} = \frac{dS}{\dfrac{dE}{dn}} ,$$

if by the V we understand the 6N-dimensional volume enclosed by the energy surface $E = E_0$, and if the $\dfrac{dE}{dn}$ would stand for the gradient of the total energy function E, taken along the outward normal to the surface element dS of the energy surface $E = E_0$.

Now I would like to define what we ought to mean by the 'physical state' of our gas or liquid. Let v be the volume that the system occupies.

We decompose *v* into a large number of physically smaller, but equally large volume elements dxdydz = do. Besides, we consider the $\{\xi, \eta, \zeta\}$–space. We also break the latter into equal volume elements $d\xi d\eta d\zeta = d\tau$. We assume that the elements do and $d\tau$ are chosen in such a way that, at least in general, there are a large number of atoms 'belonging' to the element do*$d\tau$. By the latter statement we understand that whereas spatially some atom itself lies within the element do, the corresponding 'dynamical' point in the $\{\xi, \eta, \zeta\}$–space, whose coordinates are given by the momentum coordinates of the atom involved, that is, by $\xi = m\dfrac{dx}{dt}, \eta = m\dfrac{dy}{dt}, \zeta = m\dfrac{dz}{dt}$, lies within the element $d\tau$. Bearing this in mind, to define the 'state' of our system I take out one of the atoms, for example, the N-th one. Then the remaining (N – 1) atoms must be distributed in some way among the elements do*$d\tau$. Accordingly, we denote by $n_{do d\tau}$ the number of the (N – 1) atoms, which do belong to the element do*$d\tau$. Hence, knowing the state of our system should now mean that we ought to know the numbers $n_{do d\tau}$ for all the volume elements. Further, it should also mean that we do definitely know, in which spatial element do the N-th atom is located, as we do know the region of the $\{\xi, \eta\}$—plane, on which the dynamical point with the momentum coordinates $m\dfrac{dx_N}{dt} = p_{3N-2}, m\dfrac{dy_N}{dt} = p_{3N-1}$ must lie—for example, within a quadratic quadrilateral $d\xi d\eta$. This means that the actual value of the last, N-th atom's dynamical coordinate, that is, $m\dfrac{dz_N}{dt} = p_{3N}$ cannot be chosen just at will. The information concerning the rest of the (N – 1) atoms ought to unequivocally determine it, since the total energy of the system under study should have the constant value E_0.

Bearing this entirety in mind, it is now straightforward to find the probability of a "state of the matter." In such a case, we would only have to look at the integral:

$$\dfrac{1}{\dfrac{\partial V}{\partial E_0}} \int \dfrac{dS}{\dfrac{dE}{dn}},$$

which extends over the available area of the energy surface E = E_0 and whose values lie anyway within the actual variability ranges of

the coordinates describing the desired state of the matter, that is:
$q_1, ... q_{3N}, p_1, ... p_{3N-1}.$

First, it is convenient to ask for the probability that a certain number of atoms $n_{do d\tau}$ does belong to each element do*dτ. From the probability thus obtained, we might find the one we are looking for, if we multiply the former probability by the number indicating how many mutually congruent surface elements might be found on the energy surface $E = E_0$, which does in fact correspond to the number of different ways to distribute the rest of the (N − 1) atoms among the elements do · dτ. The desired number is, how one can easily find and as is well known,

$$\frac{(N-1)!}{\prod (n_{do d\tau}!)}.$$

As a result, the expression for the desired probability might be cast as follows

$$\frac{(N-1)!}{\prod (n_{do d\tau}!)} \frac{1}{\frac{\partial V}{\partial E_0}} \int \frac{dS}{dE}.$$

Now I substitute the surface element dS with the following expression:

$$\frac{dq_1 dq_2 ... dq_{3N} dp_1 dp_2 ... dp_{3N-1}}{|\cos(n, p_{3N})|}.$$

As

$$\cos(n, p_{3N}) = \frac{\partial E}{\partial p_{3N}} : \frac{\partial E}{\partial n},$$

If cos (n, p_{3N}) > 0, but

$$\cos(n, p_{3N}) = -\frac{\partial E}{\partial p_{3N}} : \frac{\partial E}{\partial n},$$

If cos (n, p_{3N}) < 0, we arrive at the following expression

$$\frac{dS}{\frac{\partial E}{\partial n}} = \frac{\prod_{i=1}^{3N} dq_i \prod_{i=1}^{3N-1} dp_i}{\frac{\partial E}{\partial p_{3N}}} = \frac{m \prod_{i=1}^{3N} dq_i \prod_{i=1}^{3N-1} dp_i}{p_{3N}}$$

$$= \frac{m \prod\limits_{i=1}^{3N} dq_i \prod\limits_{i=1}^{3N-1} dp_i}{\sqrt{2m\left(E_0 - \sum\limits_{i=1}^{3N-1} \frac{1}{2m}p_i^2 - P\right)}}.$$

With this in mind, we ought to arrive at the following expression for the "probability of the states of matter," W:

$$W = \frac{(N-1)!}{\prod(n_{dod\tau}!)} \frac{\sqrt{\frac{m}{2}}}{\frac{\partial V}{\partial E_0}} \int_{(6N-1)} \cdots \int \frac{m \prod\limits_{i=1}^{3N} dq_i \prod\limits_{i=1}^{3N-1} dp_i}{\sqrt{2m\left(E_0 - \sum\limits_{i=1}^{3N-1} \frac{1}{2m}p_i^2 - P\right)}},$$

wherein the integration is to be extended over a certain surface element of the energy surface, which is selected to be such that it corresponds to a certain way of realizing the above-defined 'state' of our system. I believe to be consistent with Boltzmann's way of thinking [and also not conflicting with the Gibbs' train of thoughts] when I define the entropy, S, of the system by the equation:

$$S = k\log W + k\log\left(\frac{\partial V}{\partial E_0}\right),$$

Meanwhile, we ought to take into account that our definition of the 'state' allows for indeterminacy that we might apply to our advantage. For the state to be determined, among others, there must be a definite area on the $\{\xi, \eta\}$–plane, within which the point $\xi = p_{3N-2}$, $\eta = p_{3N-1}$ must lie. With this in mind, without appreciably altering the physical meaning of the term 'state', we might shape this region as a stripe parallel to the ξ–axis, whose width in the direction of the η–axis is equal to dp_{3N-2}. This means that in our integral we perform the integration over the variable p_{3N-1} in the whole range, in which the denominator takes a real value. In such a case we get a somewhat simpler expression for W:

$$W = \frac{(N-1)!}{\prod(n_{dod\tau}!)} \frac{\pi\sqrt{\frac{m}{2}}}{\frac{\partial V}{\partial E_0}} \int_{(6N-2)} \cdots \int \prod\limits_{i=1}^{3N} dq_i \prod\limits_{i=1}^{3N-2} dp_i.$$

If we now let the value of p_{3N-2} be undetermined, then we get:

$$W = \frac{(N-1)!}{\prod (n_{dod\tau}!)} \frac{\pi m \sqrt{m}}{\frac{\partial V}{\partial E_0}}$$

$$\int_{(6N-3)} \cdots \int \left(\sqrt{E_0 - \sum_{i=1}^{3N-3} \frac{1}{2m} p_i^2 - P} \right) \prod_{i=1}^{3N} dq_i \prod_{i=1}^{3N-3} dp_i.$$

One can obviously continue transformation in this way. Any number of variables p_i can be left completely indefinite, if only this number is very small compared to 3N.

Whichever of these expressions one chooses, one constantly struggles against the same difficulty, namely that complicating the form of the energy surface. There is one case, when this difficulty disappears. If the expansion of the atoms in the space might be neglected and no or only attractive forces between them are effective, all the values of q_i that lie within the given bounds correspond to the points on the energy surface. If, on the other hand, the atoms do have spatial extent, this would no longer be the case. When performing the integration, note that two atoms might never penetrate each other. Even more difficult is the task when repulsive forces between the atoms are effective, and the latter are considered punctiform/dot-shaped. The gain of the enormous efforts, which van der Waals, Boltzmann, Planck and their numerous successors have applied to this problem, is an approximate treatment of the case, when the atoms are treated as impenetrable spheres. There is but nothing known, as far as I can see, about the models regarding the atoms as point-like centers of force, among which repulsive forces act.

I have come to the point where, I believe, modern mathematical methods must be used. Let us now look at the function:

$$f(z) = \frac{1}{\sqrt{1-z}}, f(0) = +1.$$

We might find a polynomial series in the form

$$\sum_{n=1}^{\infty} P_n^{(1)}(z),$$

which on the whole plane, except for the part of the real numbers axis lying between the point +1 (including this point) and the infinity, adopts the same value that f (z) obtains when z is starting to move from the coordinate center and finally approaching the mentioned point without crossing the just mentioned part of the real numbers axis. To put it in more detailed way, the polynomial series assumes the values obtained by f (z) when z starts from the coordinate center, follows the real numbers axis up to the neighborhood of the point z = +1, renders a semicircle around that point in the positive direction, and finally follows the real numbers axis up to the point in question.

We can also find another polynomial series

$$\sum_{n=1}^{\infty} P_n^{(2)}(z),$$

with just the similar properties, except that on the line between the point +1 and the infinity it adopts the same values as the f (z) when z is surrounding the point, z = +1, but now in the negative direction. Thus, let us take a look at the following series

$$\sum_{n=1}^{\infty} \frac{1}{2}\left[P_n^{(1)}(z) + P_n^{(2)}(z)\right] = \sum_{n=1}^{\infty} P_n(z).$$

For the real values of z that are less than +1, this adopts the same value as f (z). For z > +1 it takes the value 0.

The convergence of the series is uniform within each part of the real number axis that does not contain the point +1. By the way, it can easily be demonstrated that the integration by line of the series is allowed even if the integration range includes the point +1.

Now let us consider our first expression for W.

With this entirety in mind we get, if $\dfrac{1}{2m}\sum\limits_{i=1}^{3N-1} p_i^2 + P < E_0$:

$$\frac{1}{\sqrt{E_0 - \dfrac{1}{2m}\sum\limits_{i=1}^{3N-1} p_i^2 - P}} = \frac{1}{\sqrt{E_0}}\sum_{n=1}^{\infty} P_n\left(\frac{\dfrac{1}{2m}\sum\limits_{i=1}^{3N-1} p_i^2 + P}{E_0}\right),$$

whereas the series should go to zero, if $\dfrac{1}{2m}\sum\limits_{i=1}^{3N-1} p_i^2 + P > E_0$.

As a result, nothing stands in our way to extend the integration range over all the values of q and p, which lie within the limits

Nils and Karl's Struggle with 'Swedish Academics'. Was It Their Pure Phantasy? **283**

determined by our definition of the 'state of matter'. Nor is there any difficulty in performing the integration.

W is then obtained in the form of some series, which is certainly convergent. In the mathematical sense, the problem is therefore solved.

If we want to employ our second expression for W instead of the first one, then we have to introduce into the integral a function that can be represented as the following polynomial series of our variables

$$\frac{\frac{1}{2m}\sum_{i=1}^{n}p_i^2 + P}{E_0}$$

which takes the value +1 if this variable is less than +1, but goes to 0 if the variable is greater than +1. However, no sensible simplification could be gained by applying this second term.

Of course, when working out the theory of practical physical use, one does not employ a series of polynomials, but rather calculates an approximation using a finite polynomial. From the point of view of saving our labor, it is of the utmost importance to reduce the arithmetical work as much as possible by a suitable choice of the expression for W, as by the application of the best possible approximation method. The mathematical investigations of the recent years on the approximate calculation of a real function by a polynomial [due to the Russian-Soviet mathematician Sergei Natanovich Bernstein (Сергей Натанович Бернштейн, 1880–1968)] ought to be of the greatest relevance in this regard.

As a particular advantage of the method outlined here, I want to emphasize that it is quite independent of the molecular disorder hypothesis*. I believe that the following brief consideration is apt to shed some light on the reliability of this hypothesis.

Let us consider a gas whose atoms, either of finite spatial extent or point-like, do attract each other, if they are sufficiently close to each other, but on the other hand exert no force on each other, if the distance between two arbitrary atoms exceeds a certain limit. This

---

*See M. Planck, 'Über die kanonische Zustandsgleichung einatomiger Gase' ('On the Canonical State Equation of Monatomic Gases'); Sitzungsberichte der königlichen preußischen Akademie der Wissenschaften, 1908 [2]. See there especially page 641 at the top.

case can be treated with van der Waals, Boltzmann's and/or Planck's methods. The obtained result is all the more accurate, the larger is the volume of the gas. I want to assume this so great that if the atoms were equally distributed in it, they would be outside their mutual spheres of action. If we now consider a volume containing a large number of atoms, we find that the number of atoms in it is just the one, which we would obtain if the atoms were distributed uniformly throughout the gas volume, and that their velocities distribution obeys Maxwell's distribution law. Let us now ask what is the probability that the atom no. 1 is situated in a volume element do_1, the atom no. 2—in the do_2 and so on, so forth. On the other hand, we let the corresponding velocities be indefinite. We get the desired probability, if in the expression

$$\frac{\sqrt{\dfrac{m}{2}}}{\dfrac{\partial V}{\partial E_0}} \int \cdots \int_{(6N-1)} \frac{\prod\limits_{i=1}^{3N} dq_i \prod\limits_{i=1}^{3N-1} dp_i}{\sqrt{E_0 - \sum\limits_{i=1}^{3N-1} \dfrac{1}{2m} p_i^2 - P}},$$

we extend the integration range over the q-variables to certain areas do_i', which in case the atoms are punctiform, coincide with do_i, but otherwise are smaller than do_i. Instead, the integration range for the p-variables should be extended over the whole permissible range, that is, over the whole realm, for which the integrand has real values. Under such an assumption the integrations over the p-variables might be accomplished immediately.

We get

$$\frac{K}{\dfrac{\partial V}{\partial E_0}} \int \cdots \int_{(3N)} (E_0 - P)^{\frac{3N}{2}-1} \prod\limits_{i=1}^{3N} dq_i.$$

K stands here for the constant dependent on m and N. We now assume that the volume elements do_i are of the same order of magnitude as the volumes of the atoms, that is, larger than the latter, but still smaller than the spheres of action of the atoms. We first arrange the volume elements so that the spheres of action of the atoms fall completely outside each other. Moreover, we might set it up in such *a way that in every physically small part of the gas volume there would be just as many volume elements do_i as in fact required by their sizes. Under such circumstances, we ought to get $P = 0$, and we might then cast the probability W_1 we are looking for as follows:*

$$\frac{KE_0^{\frac{3N}{2}-1}}{\frac{\partial V}{\partial E_0}}\prod_{i=1}^{3N}do_i'.$$

Here we arrange the volume elements in such a way that in every physically small area there is still the number of them that corresponds to their actual sizes, but that the do_i belonging to the same area lie as close as possible to each other, without their intervening into each other. We might for example, consider the case that the atoms aggregate 10 to 10 each, but that these aggregates are evenly distributed throughout the gas volume. Under such assumptions, the potential energy P should adopt a negative value, the absolute magnitude of which depends on the type of forces acting as well as on the distribution of the atoms.

Denoting the probability of such an arrangement W_2, we obtain:

$$\frac{W_2}{W_1}=\frac{1}{\prod\limits_{i=1}^{N}do_i'}\int\limits^{(3N)}\dots\int\left(1-\frac{P}{E_0}\right)^{\frac{3N}{2}-1}\prod_{i=1}^{3N}dq_i.$$

We obviously have $W_2 > W_1$ and this difference might be very large due to the high value of the exponent $\left(\dfrac{3N}{2}-1\right)$. From this it seems clear to me that the proposition that the most probable distribution of the gas atoms is (under certain conditions) the most uniform, should not be understood in such a way that the volume elements of the same order of magnitude as the sphere of action must always contain just the same number of atoms, irrespective of the method of how we have chosen them.

Accordingly, the number of atoms in such a volume element must be greater if it is chosen in the neighborhood of an atom than if it is *assumed to be far away from all atoms. In other words, that means that the hypothesis of molecular disorder is inaccurate*.*

Once the function W and thus the entropy S are determined as a function of the quantities, which characterize a 'state' of the system under consideration, one has to determine these quantities in such a way that S obtains its greatest possible value. With the help of thermodynamic relations, one can set up the state equation.

I would not like to go in for these questions here. I wish to emphasize solely, that an assumption that is not uncommonly encountered in the literature, namely that the most probable distribution of the atoms is always the most uniform, is not correct. It can be shown directly from entropy, without applying the equation of state, that under some circumstances a van der Waalsian gas has to be distributed between two different phases, in order for S to obtain its greatest value.

First of all, it is important to learn, who was Prof. Dr. Carl Wilhelm Oseen, the author of the above communication. The details about him might be found on the Internet site entitled 'The Dictionary of the Swedish National Biography' within the frame of the Swedish state archive [4]. The text over there is in Swedish, so we just place below our English translation of it.

Carl Wilhelm Oseen: A Biographical sketch

Born: 1879-04-17, Lund Cathedral Assembly, Skåne County

Death: 1944-11-07, Engelbrekts Assembly, Stockholm County

Carl Wilhelm Oseen was early intellectually developed. During his school time in Halmstad, he was eagerly self-studying, especially in the fields of history and mathematics, and was still continuing to hang back, even after his academic qualifications, as to which of the both subjects he should choose for his professional activity. Deciding for his final research focus might be his student trip to Germany, Göttingen in the winter of 1900-01, where he could mainly attend lectures on partial differential equations by the renowned

---

* In his well-known treatise '*The Virial of a Mixture of Ions*' Phil. Mag. XXIII, 1912, Mr. S. R. Milner emphasizes that the hypothesis of molecular disorder is inappropriate for a system in which attractive forces are acting [3]. Mr. Milner's expression for the probability differs from the above in that the factor $\left(1 - \dfrac{P}{E_0}\right)^{\frac{3N}{2} - 1}$ is replaced by the expression $e^{-\frac{P}{kt}}$. The two expressions are mathematically equivalent if $|P|$ in relation to $|E_0|$ is small. However, for a liquid, this assumption is not valid, not even for a mixture of atoms or ions in a free space.

mathematician David Hilbert. Most probably he was also influenced by another outstanding mathematician in Göttingen, Felix Klein, and, during his later visit, by the hydrodynamicist Ludwig Prandtl. Oseen's teacher in Lund, Albert Victor Bäcklund (1845–1922) had also exercised a great influence. Oseen's doctoral thesis was entirely in line with the research focus developed by Klein and his friend and co-worker, Norwegian mathematician Sophus Lie, as well as that by Bäcklund.

Oseen was working in the field of pure mathematics until 1906, up to his two publications on fluid movements. This year he could have acquired his 'scientific manhood', as he recalled about this himself. Being inspired by the results of Dutch theoretical physicist H. A. Lorentz, he posed an important problem regarding fluid movements. The actual novelty in Lorentz's research was consistently taking into account the viscosity of the liquid. This had usually been neglected in the previous work, mainly owing to very complicated mathematical treatments. Oseen's ingenious suggestion in the framework of Lorentz's approach consisted in considering the time dependence of fluid movements. This great work was an important part of Oseen's qualification proof, when he sought a professorship in Uppsala in the autumn of 1907.

In the late 1910's, Oseen had returned to the issue of viscous fluid dynamics, and was further developing the work carried out in the 19th-century by the renowned British physicist George Stokes on the resistance to the movement of a ball in a fluid. This resistance is a consequence of the viscosity of the liquid. Bearing this in mind, Oseen could amend a significant error that Stokes did and set the equation that in the standard work in the field is nowadays named after Oseen. From this equation, a correction could be found for the Stokes formula, which is valid if the velocity of the ball is low enough or the viscosity of the liquid is sufficiently high. Using this equation by him, Oseen could also pioneer explanation of a familiar phenomenon, namely that behind a body moving in a fluid a wake should occur. Mainly his disciple Hilding Faxén was continuing the work in this direction, besides Oseen himself.

A not completely unrelated area within theoretical physics, where Oseen had arrived at an internationally recognized result was the theory of anisotropic fluids, later commonly called liquid crystals. These have become nowadays an increasingly important part of the standard technology as the components of digital clocks, pocket calculators and flat screen TVs. Liquid crystals represent an intermediate form between ordinary liquids and ordinary solid crystals in the sense that their properties are not the same in all directions, as opposed to the usual liquid. The latter phenomenon has been discovered in the 1880s, whereas in the 1920s Oseen could pioneer developing the consistent theory of such systems. He published a detailed summary of his research in this field in 1929 in the Volume 20 of the renowned German research journal '*Fortschritte der Chemie, Physik und Physikalischen Chemie*'.

These are primarily the aforementioned efforts that rendered Oseen's name known in the theoretical physics all over the world, but he could also pay important contributions to many other areas. He was an unusually versatile mathematical physicist with a solid knowledge within and understanding of the classical mathematical analysis, combined with a sure physical sense for possible simplifications of mathematical formalism necessary to achieve a useful final result. He contributed to the classics of physics, elasticity theory and geometric optics, but also to the theory of relativity and quantum mechanics, among the areas of modern physics. Thus, Oseen could become a pioneer of modern theoretical physics in Sweden in the 1910s and 1920s. Of particular importance in this context was a series of lectures he held at 1919 summer courses in physics in Uppsala, which had then been published under the title of '*Atomic Exhibitions in today's physics: Time, Space and Matter*'. The lectures deal with both the theory of the special, as well as general relativity and the so-called old quantum physics, based mainly on Niels Bohr's atomic model from 1913. Although much of their contents soon became obsolete, these lectures do represent a rewarding and interesting reading: They provide us with a historical background to the development of physics during the first decades of the 20th century, written with a personal and fresh style, typical of Oseen.

As a member of the Swedish Royal Academy's Nobel Committee for Physics, Oseen could pay another significant contribution to the modernization of physics in Sweden, this time with the important international affiliation. He was the first modern theoretical physicist in the committee since its accession in 1900. In his first membership year, the 1922, Oseen could take over a delicate task of motivating a Nobel Committee to lend the Prize to Albert Einstein, who had long been nominated by a large number of known physicists. Due to difficulties both inside and outside the committee regarding the scientific evaluation of Einstein's theory of relativity, the case had been 'drawn in the long run'. Through a careful, well-motivated and well-written investigation, Oseen could persuade other members of the committee that Einstein was a worthy Nobel Prize candidate, and they had finally approved the proposal: Einstein was designated as the recipient of the 1921 reserved price. Another Oseen's effort of international importance during the same period was to initiate the International Congresses for Applied Mechanics, which were then actively contributing to the restoration of the international scientific research relations after the First World War.

Oseen's versatility was not limited to the field of physics. He successfully combined his early interest in history with his professional activities in physics and adjoining areas in a number of biographical works. In addition to a large number of shorthand sketches describing lives and works of outstanding scientists, including his teacher Bäcklund, he wrote a larger biography of Johan Carl Wilcke. Another Oseen's important mission in the Swedish Royal Academy of Sciences consisted in that he edited Scheele's left-hand paper, which had finally been published in 1942 to celebrate the 200th anniversary of the chemist's birth. Like many other outstanding naturalists in his generation, Oseen had a great interest in philosophy, especially in relation to science. His installation lecture in Uppsala in 1909 was entitled '*The question of the will of freedom, viewed from a scientific point of view*'. Plato was his favorite philosopher. Oseen's lecture at the closure of the presidency in Swedish Royal Academy 1935 had the title of Plato's idiom and mathematics.

Literature and art were also within the area of Oseen's interest. He was holding lectures in both classical and modern literature, being also a performing artist with a number of oil paintings, watercolors and drawings as a result.

Apart from his intellectual capacity and sharpness, pupils and friends of Oseen have emphasized his willpower and ability to work as his characteristic features, which were manifesting, for example, in that Oseen was never confused by the professional challenges perceived and could quickly manage any painstaking and complicated mathematical calculation. He vehemently claimed the requirement for freedom of thought and expression. Truth must be sought unconditionally. As an academic teacher, Oseen was inspiring by his compassionate and interesting lectures.

In 1982, the Swedish Royal Academy has released a medal devoted to the memory of Oseen and designed by the sculptor Léo Holmgren. The medal's inscription reads: "*Arcana umorum revelavit anisotropicorum*" (He revealed the secrets of the anisotropic liquids).'

The above is a portrayal of an outstanding Swedish theoretical physicist and pedagogue. Meanwhile, of particular interest for our present discussion is Prof. Oseen's involvement into the research on the universal thermodynamic equation of state.

His paper on the theme, the English translation of which have presented earlier here, has also received an attention of the international research community. Prof. Dr. Max Trautz, whom we have already met in Chapter 1 of this monograph, has published in 1933 a mini-review in the field, where he i. a. clearly expresses his thoughts and estimates as for the relationship of the modern physics to the actual physical chemistry [5].

We present here the English translation of the relevant short excerpts from Prof. Trautz' mini-review.

Molecular Theory of Gases

Max Trautz,

*Heidelberg, Physical-Chemical Institute of the University,
April 30, 1933*

Fundamentals: Most recently new interesting problems have appeared for the molecular theory, which stem from the novel facts found and verified experimentally: Temporal aspects in connection with tuning heat capacities at constant volume, C_v, the external field effects on the transport variables, the interconnections between the latter ones for the ideal gases and the critical temperature, which appears to act as a 'midpoint' of the temperature scales.

In the recent time, the theory was developing the formal analogy between the gas theory and radiation theory; see, for example, the 'impact cross section (Wirkungsquerschnitt, WQ)' approach by Philipp Lenard (1862–1947), the works by Arnold Sommerfeld (1868–1951) and George Jaffé (1880–1965). The theoreticians have managed to prove a theorem, which is at least approaching the desired quasi-ergodic rule [T1]. Meanwhile, this result loses its actual significance due to the introduction of the Uncertainty Principle by Werner Heisenberg (1901–1976) [T2].

Still, any feasible, viable definition of the physical 'body' instead of the dispatched notion of the 'mass point' is unavailable for the present; furthermore, other axiomatically introduced and employed fundamentals of the statistical mechanics, despite their truly subtle refurbishing are still bearing their largely hypothetical character [T3]. Any noticeable changes due to future improvement of the theory at the present point are probably to be expected solely for proximity effects at the smallest scale.

Remarkably, in the recent years several statistic-theoretical approaches have appeared, specifically, the one due to Boltzmann as well as those owing to Bose, Einstein, Fermi, and Dirac [T4], of which the Bose–Einstein and Fermi–Dirac approaches are valid for actual atoms and molecules in extreme temperature ranges only. All these approaches have been studied with the special reference to their compatibility, their consequences. As a result, the past formulation of the partition function has been retained, and there is no principal change in the most basic fundamentals of the statistical approach, as soon as the relevant studies involve atoms and molecules themselves (instead, for the protons one has to use the Fermi–Dirac statistics from now on).

Modeling atoms and molecules is presently of the utmost—de facto, but not de jure—purely mechanical [T5]. Nowadays, there are four basic model types, which have already been carefully and thoroughly checked for their gas-theoretical consequences:

1. Rigid elastic balls;
2. Rigid elastic balls with mutual attraction (van der Waals molecules [T6]);
3. Centers of force, capable of both mutual attraction and repulsion (the forces are functionally dependent on $\propto r^{-s}$, where r stands for the distance between the centers, the s = *const* is the force exponent. If $s = 5$, then we speak about Maxwell's molecules [T7]);
4. Centers of force, capable of both mutual attraction and repulsion (forces are functionally dependent on $\propto r^{-s}$, where the constants s are in general possessed of different values for the both types of forces; for the present such a model has been successfully checked for crystals by Max Born (1882–1970) and for gases by John Edward Lennard-Jones (1894–1954) [T8].

The newest theory has started considering the models 3 and 4 from the quantum–mechanical standpoint, in studying the collisions among the molecules and their consequences in terms of wave mechanics. The corresponding electro-dynamical models are also mostly centrosymmetric, that is, they might be considered as a kind of 'representative balls', whereas the models based upon the wave mechanics are in part axially symmetric [T9].

Further, the 'chemical' models, which are not mechanically formulated for all the possible proximity effects, involve only the classical number of impacts and contain all the rest in the attitude and transition probabilities, which are not given by a simple Boltzmann factor, but by the exponential of the following form: $e^{-\int \frac{qdT}{RT^2}}$ [T10].

The applications of the modern molecular theory are involving gas analysis (friction, thermal conductivity), gas discharges (impact cross section research), isotope separation (diffusion), and much else.

Thermal quantities, equation of state: Theoretical physics started to consider the search for a general equation of state as somewhat hopeless and more or less descriptive, owing to its tightest connection with the pertinent models. In the conventional inferences the principle of molecular disorder used to enter, that is, the average spatial uniform distribution of the molecules. However, it is only valid for the models without attractive forces [3], for where the latter are acting, the groups of the particles should form, and possibly even two distinct phases of the matter. Instead, the general approach by C. W. Oseen [1], published in 1921 does not depend on the molecular disorder hypothesis but leaves in exchange the molecular model choice open, and by means of the most recent mathematical methods, mainly owing to the Swedish school, leads to a conclusion that if the entropy is known (as a function of the state variables), this ought to yield the general equation of state; such an approach appears to be somewhat reversed, as compared to the ways one usually proceeds with statistical–mechanical interpretation of thermodynamics.

In practice, the invaluable efforts by Johannes Jacobus van Laar (1860–1938) remain to be the most important contribution [T11]. In the most cases the approach is significantly phenomenological, so that an enormous amount of experimentally observed facts is being condensed into a few relatively simple functions, and the physically important information is forced into van der Waals' parameters a and b, which are numerically fitted to the facts. Since b_0 and b_∞ remain finite, the generalized Model 2 or the unified Model 1 and 4 ought to constitute the theoretical basis. In such a way, or even fully empirically, much of the recent work in the field follows van der Waals' methodology.

It is possible to achieve even a quantum–mechanical interpretation of the van der Waals' forces [T12, T13] by introducing dipolar effects of the atoms, with the polarization occurring in the cases of proximal interactions. To describe the situation classically, this would require permanent di- or multi-poles; see for example, the works by Peter Debye, Willem Hendrik Keesom (1876–1956), and several other colleagues [T14].

Meanwhile, physically, the short-period zero-point movements of the charges in the atoms or molecules should also create dipoles, so that attraction ought to take place as a result, which might be approximated as $\propto r^{-6}$ on average. The present theory also leads to the essential conclusion that, although the atomic field in itself does reach to infinity, under certain conditions, in contrast to the classical conception of the force effects, it is fundamentally no longer possible beyond definite distances. A quantum–dynamic correction of the equation of state only becomes noticeable in systems like He and H_2, that is, physically, where the de Broglie's wavelength falls within van der Waals' range. For the not yet accessible area of degenerate gases the poser about the relevant meaningful statistics remains to be practically unanswered.

References to Prof. Trautz' mini-review

1. G. D. Birkhoff, *Proc. Natl. Acad. Sci.*, USA, **17** (1931) 656; J. von Neumann, ibid., **18** (1932) 70; E. Hopf, ibid., **18** (1932) 93.
2. W. Heisenberg, *Die physikalischen Prinzipien der Quantentheorie*; Hirzel, 1930.
3. Ralph Howard Fowler (1889–1944), *Statistische Mechanik*, übersetzt von O. Halpern und H. Smereker, Akademische Verlagsgesellschaft, 1931, which does discuss the fundamentals in the field but in a too concise way. The book by

Richard von Mises (1883–1953) presents much more detailed account in the field; one might also recommend reading the book by Richard Chace Tolman (1881–1948), *Statistical Mechanics, with Applications to Physics and Chemistry*, Chem. Catal. Co., New York, 1927, as well as the book Edward Uhler Condon (1902–1974) & Philip McCord Morse (1903–1985), *Quantum Mechanics*, McGraw-Hill-Book-Co. 1929; for reading about the corresponding mathematics, one might recommend the work by David Hilbert (1862–1943), *Mathematische Annalen*, **72** (1912) 262–577.

4. Albert Einstein (1879–1955), *Berliner Berichte*, **1924**, 261, **1925**, 1 and 18; Satyendra Nath Bose (1894–1974), *Zeitschrift für Physik* **26** (1924) 178; **27** (1925) 384; Enrico Fermi (1901–1954), *Zeitschrift für Physik* **36** (1926) 902; Paul Adrien Maurice Dirac (1902–1984), *Proc. Roy. Soc.* London (A) **112** (1926) 661 and '*Probleme der modernen Physik (Sommerfeld-Festschrift)*'. Hirzel 1928, p. 31 ff.; Alfred Landé (1888–1976), '*Vorlesungen über Wellenmechanik*'. Akad. Verlagsgesellschaft. 1930, p. 29 ff.; Arthur Erich Haas (1884–1941): '*Materiewellen und Quantenmechanik*', 2. Auflage. Akad. Verlagsgesellschaft. 1929.

5. Max von Laue (1879–1960), *Naturwissenschaft*, **20** (1932) p. 915.

6. Johannes Diderik van der Waals (1837–1923), *Over de Continuïteit van den Gas- en Vloeistoftoestand*, Seminal Thesis, Universität Leiden, 1873.

7. James Clerk Maxwell (1831–1879), *Phil. Mag.*, **35** (1868) 212, *Sci. Papers*, **2** (1890) 72.

8. J. E. Jones, *Proc. Roy. Soc.*, London (A) **106** (1924) 441 and 709.

9. Peter Debye (1884–1966), *Leipziger Vorträge*, Hirzel, 1928/1930; *Polare Molekeln*, Hirzel, 1929.

10. A. F. Horstmann's molecules or Max Trautz' works on the kinetics of chemical reactions in the *Zeitschrift für physikalische Chemie*, 1909–1911, as well as 1917, 1918.

11. J. J. van Laar, *Sechs Vorträge über das Thermodynamische Potential und seine Anwendungen auf chemische und physikalische Gleichgewichtsprobleme.* Vieweg 1906; *Die Zustandsgleichung der Gase und Flüssigkeiten.* L. Voss 1924.

12. M. Born, *Moderne Physik*. Berlin, Springer 1933; Fritz London (1900–1954), *Zeitschrift für Physik*, **63** (1930) 245; *Zeitschrift für physikalische Chemie*, **11** (1930) 222; *K. Wohl* (*who was serving as a Privatdozent for physical chemistry at the University of Kiel – at least, in the years 1928–1930, whose destiny but remains largely unknown*).

13. *Zeitschrift für physikalische Chemie*, 1928, 1929; *Bodenstein-Festschrift* (a special issue of the *Zeitschrift für physikalische Chemie*)—J. E. Lennard-Jones and J. J. van Laar are not contributing in this special volume, but it does nonetheless contain a lot of valuable information).

14. P. Debye, *Physikalische Zeitschrift*, **21** (1920) 178; W. H. Keesom, *Physikalische Zeitschrift*, **22** (1921) 129; Hans Falkenhagen (1895–1971), *Physikalische Zeitschrift*, **23** (1922) 87; Fritz Zwicky (1898–1974), *Physikalische Zeitschrift*, **22** (1921) 449; Henry Margenau (1901–1997), *Physical Review*, **36** (1930) 1782.

Prof. Trautz' mini-review contains much more interesting and important information, but we have chosen to present here its part, which seems to be most relevant to our present discussion.

We are grateful to Prof. Trautz for his thorough and thoughtful analysis of the then field of the physical chemistry. He has clearly described the actual effects of the physical revolution taken place in the first quarter of the XX-th century, by carefully noting its actual impact, successes and apparent drawbacks with regard to the physical chemistry/chemical physics.

Of immense importance for our present discussion ought to be Prof. Trautz' fetching and keen estimate of Prof. C. W. Oseen's contribution to the field. We recognize the undoubted importance of Prof. Oseen's effort to provide us with the valid and basic statistical–mechanical fundamentals of the universal thermodynamic equation of state. We sincerely regret to conclude that this particular effort could not lead to any noticeable success, due to the largely hypothetical nature of the then statistical mechanics together with quantum physics, as Prof. Trautz could show us.

To 100% we agree with Prof. Trautz' conclusion that any success in trying to formulate the valid theory of the universal thermodynamic equation of state requires first of all a careful re-thinking of the basic fundamentals in thermodynamics. We might only add here that a reasonable re-formulation of the statistical mechanics ought to be in order as well. The work by Dr. Georg(e) Augustus Linhart, whom we shall meet here later on, namely in Chapter 4 of this monograph, should give interesting and important stimuli for such projects. This is why we would like to express our sincere gratitude to Prof. C. W. Oseen for skillfully posing the very problem of building-up statistical-mechanical foundations for the universal thermodynamic equation of state.

Howbeit, the actual authors and promoters of the true and valid universal thermodynamic equation of state are undoubtedly Dr. Nils Engelbrektsson, who had combined the classification approach introduced by his and Prof. Oseen's famous compatriot Carl von Linné (1707–1778) with the mathematical approach by the famous French mathematician, physicist, chemist, and statesman Gaspard Monge, comte de Péluse (1746–1818) to formally mathematically infer the universal thermodynamic equation of state, as well as Karl Alexius Franzén, who could successfully check experimentally and verify the validity of the theoretical inferences by Nils Engelbrektsson.

2.8   Supplementary Notes

[Note 1]: Nils Engelbrektsson has written his 'Outline of Thermodynamics' in Swedish. Instead, Karl Franzén has prepared his Appendix in English. Whereas there are no more publications authored or at least co-authored by Karl Franzén, we present here those by Nils.

[Note 2]: Fig. 2.1 is nothing more than just a standard general representation of the conventional phase diagram. But the story told us by Nils and Karl is truly unique. Nobody upon the World has carried out such a study, although the problem of physical–chemical peculiarities among different aggregate states was and is by far not novel and massively investigated. It is this point that ought to be the very zest!

2.9 References

1. Carl Wilhelm Oseen (1922): *Eine Methode, die Zustandsgleichung der beliebigen Flüssigkeiten oder Gasen exakt zu berechnen.* Arkiv för Mathematik, Astronomi och Physik, utgivet av K. Svenska Vetenskapsakademien. Band 16, Häftet 3, sidorna 1–12.

2. Max Planck (1908): Über die kanonische Zustandsgleichung einatomiger Gase. Sitzungsberichte der königlichen preußischen Akademie der Wissenschaften zu Berlin, pp. 633–647.

3. S. R. Milner (1912): *The Virial of a Mixture of Ions.* The London, Edinburgh and Dublin Philosophical Magazine and Journal of Science, **XXIII**, pp. 551–578.

4. Bengt Nagel (1992–1994): *C. Wilhelm Oseen* – Svenskt Biografiskt Lexikon, Band 28, sida 395:
   https://sok.riksarkivet.se/sbl/Presentation.aspx?id=7816

5. Max Trautz (1933): *Molekulartheorie der Gase.* Die Physik in regelmäßigen Berichten. Erster Jahrgang. Seiten 131–140.

Chapter 3

Max Bernhard Weinstein (1852–1918)

The Lamp must be replenish'd, but even then
It will not burn so long as I must watch:
My slumbers—if I slumber—are not sleep,
But a continuance of enduring thought,
Which then I can resist not in my heart
There is a vigil, and these eyes but close
To look within; and yet I live, and bear
The aspect and the form of breathing men.
But grief should be the instructor of the wise:
Sorrow is knowledge, they who know the most
Must mourn the deepest o'er the fatal truth,
The tree of knowledge is not that of life.
Philosophy and science, and the springs
Of wonder, and the wisdom of the world,
I have essayed, and in my mind there is
A power to make these subject to itself ...

Manfred, Act I., by Hon. George Gordon Byron (1788–1824)

A Different Thermodynamics and Its True Heroes
Evgeni B. Starikov
Copyright © 2019 Pan Stanford Publishing Pte. Ltd.
ISBN 978-981-4774-91-8 (Hardcover), 978-0-429-50650-5 (eBook)
www.panstanford.com

This photo of Max Bernhard Weinstein stems from the collection of the Humboldt University's library in Berlin, but its exact date is not known. Since Dr. Weinstein should have started his activity at the Friedrich-Wilhelms-Universität zu Berlin as a *Privatdozent* in physics and geography in 1886, then the above photo might be dated in the time period of the year of 1886.

3.1   The Biography of Max Bernhard Weinstein

To begin with, there is the following Internet reference to his biography:

http://www.sammlungen.hu-berlin.de/dokumente/8466/

Meanwhile, from this source, we learn nothing more than that Dr. Max Bernhard Weinstein started working as a *Privatdozent* at the Friedrich Wilhelm University (nowadays: Humboldt University) of Berlin in the fields of physics and geography.

There is definitely much more to the story! Fortunately, there is also a German book [1], which does communicate somewhat more information about him. The archive data show:

- He was a senior official, namely, the privy governmental councilor/advisor (*geheimer Regierungsrat*), as well as a deputy director of the German Imperial Commission for Calibration (*Kaiserliche Normal-Eichungskommision*).

- He was also active as a patent expert in the German Imperial Patent Office.
- He had become a titular/honorary professor of the Berliner University to hold lectures in physics, geography and natural philosophy.

But this is by far not enough as well. Finally, we could be capable of fetching the dossier of the official record concerning Prof. Dr. Max Bernhard Weinstein and his wife/widow Luise, maiden name Krüger, in the German State Archive, Berlin (Record token: R 1501/211857). So, let us now have a look at this.

At the then time dossier was a part of records by '*Das preußische Staatsministerium*' (the Prussian analogue of the British Home Office or the Department of the Interior in the USA) and represents '*Ruhegehaltsakten*' (the record of retirement pay/old-age pension) for the family of the official privy governmental councilor Dr. M. B. Weinstein. The fact that such a dossier was maintained in the German 'Home Office' shows that Dr. Weinstein was a Prussian/then German State Servant (in German: *Beamter*).

The dossier delivers a truly detailed personal and professional history of Max Bernhard, including his autographic CV, as well as official autographs/typewritten copies of his testimonials and credentials.

At this very point we would like to express our sincere gratitude to Mr. David Hamann, the company '*Recherchedienste*', Berlin, Germany, for his invaluable professional help with decoding the documents handwritten using the ancient Gothic Kurrent script.

Noteworthy, endowing M. B. Weinstein the honor to be a Prussian/German State Servant was triggered by the Chief Imperial Commission for Calibration in Berlin (*Die Kaiserliche Normal-Eichungskommision zu Berlin*). The latter commission turned out to be his primary working place during his entire professional life. First of all, he had to present his CV in response to this endowment. Second, he should anyway hand out the copies of his most important testimonials.

This is how we might consider his actual biography in full detail using the information from the first hands. Now, let us see, what exactly Dr. Weinstein would like to tell us about his life?

He was born on September 1, 1852, in Kowno, the then guberniya city in the Russian Empire (now Kaunas, Lithuania), into a devout mosaic Jewish family in just two months after the death of his father, a merchant in Kowno. Consequently, he had spent the first thirteen years of his life in Kowno. Then, in 1865, he was adopted his uncle B. Weinstein, who was an owner of a private carpentry in the then eastern Prussian town of Insterburg (now Chernyakhovsk, a district administrative center of the Kaliningrad Oblast in Russia).

M. B. Weinstein in the year 1910 (from the Humboldt University collection)

The uncle had then enabled him to take elementary preparatory courses in German, Latin and French during one year and thereafter, in the year 1866, sent him to '*Kgl. Gymnasium zu Insterburg*' (The Royal Gymnasium of Insterburg) for continuation of his general education. Max Bernhard might therefore enter the regular educational stages.

Meanwhile, soon it had become clear that due to his educational level it was no more necessary for him to spend the official two preparatory gymnasium years, so he could just be able to start studying in the following gymnasium classes—and was at last capable of graduating in the year of 1874, by obtaining the eligibility certificate for entering universities. Max Bernhard had presented the hand-written copy of this document. Now we would just like to have a look at the English translation of this document.

3.1.1 M. B. Weinstein's School Education

"*Eligibility Certificate for the Alumnus of the Insterburg Gymnasium*

Bernhard Weinstein from Kowno, 21 years old, of Jewish confession, the son of a deceased Kowno merchant H. B. Weinstein, spent 8 years in our gymnasium, 2 years of the First Class.

I. *Moral Performance and Diligence: Although it was apparently rather difficult for him to follow the school ordinance, his demeanor could never give rise to any rancorous objections; his diligence for certain disciplines, for example—in mathematics, in German—though very exerted and laudable, was from time to time missing to some extent, but—generally speaking—nonetheless satisfactory. To be accolade separately are his voluntarily undertaken and duly performed works in mathematics and German.*

II. *Knowledge:*

1. *in the religion: He was contended about his mosaic belief during the religious lessons.*

2. *in the German language and literature: Due to his lively interest and exerted diligence, he could gain good knowledge in the history of literature and philosophic propaedeutic. In his essays he could demonstrate a definite felicity in ordering of the material, clarity in developing his thoughts, correctness in his locution. His overall knowledge—according to the result of his assessment test—could be referred to as "Good."*

3. ***in the Latin language***: *He is capable of quickly and properly translating the works of the authors included into the school curriculum, moreover, he is sufficiently versant with the metrics and grammar, so that he might apply the both without fail. His Latin exercitation was good, the essays satisfactory, his interpretation and scansion of a fragment from Horatian work suggested to him during the exam was good, and, considering his good initial Latin conversation capabilities, it is possible to estimate his overall knowledge with the rating "Good."*

4. ***in the Greek language***: *He was mostly capable of properly translating the works by the authors included into the school curriculum and could gain rather satisfactory knowledge in the grammar. As soon as the result of his oral assessment was satisfactory, then it is possible to estimate his overall knowledge as "Satisfactory."*

5. ***in the Hebrew language***: *Although he is not enough proficient in the phonetic and morphology rules of this language, his vocabulary is definitely exceeding that indicated by the school curriculum, and hence his overall knowledge in this field might deserve the rating "Satisfactory."*

6. ***in the French language***: *Although he might volubly and nimbly translate the works of some authors included into the school curriculum, in the application of the grammar he is unsure to the extent that his overall knowledge— especially corresponding to his unsatisfactory assessment test result—might only be estimated as "Unsatisfactory."*

7. ***in the mathematics***: *In view of his truly earnest and exerted diligence, he could gain the knowledge by far exceeding that envisaged in the school curriculum, and as soon as solving simpler tests is by far not difficult for him, in this field he clearly deserves the rating "Good."*

8. ***in the history and geography***: *Although he wasn't either regularly or intently participating in the conventional reports and repetitions, he could nonetheless demonstrate both adequate diligence and clear apprehension, so that his resulting knowledge would definitely deserve the rating*

"Good," which has also been proven in the full sense by the result of his oral assessment test.

9. **in the physics**: *Through his truly brisk diligence he was always tending to exceed his knowledge area beyond the school curriculum, and as soon as he is clearly capable of the pertinent mathematical foundation, it is possible to estimate his knowledge in the field as "Good."*

10. **in the gymnastics**: *He was regularly taking part in the exercises and could acquire the "Satisfactory" education in the field.*

The signatories of the assessment commission are now granting this Eligibility Certificate to Bernhard Weinstein, because he is about to graduate our Gymnasium and going to study medicine.

We are disbanding him herewith and sincerely hope that in continuing his studies he would face the necessary fervor and sobriety, even in the fields, which might initially seem to be of no interest for him, and would live up to the name of our gymnasium.

Insterburg, July, 10, 1874.

The Royal Assessment Committee composed of the following signatories:

Dr. W. Schrader, Provincial Schools Inspector,

Dr. Krah, Director of the Gymnasium,

Dr. Lange, Senior Gymnasium Teacher,

Mr. Preuss, Senior Gymnasium Teacher,

Dr. Schmidt, Gymnasium Teacher,

Mr. Ehrhardt, Gymnasium Teacher,

Mr. Wilke, Gymnasium Teacher,

Dr. Rumpel, Gymnasium Teacher."

3.1.2 M. B. Weinstein's Higher Education and Prussian Naturalization

To continue his education, Max Bernhard had chosen the University of Breslau (now the Wroclaw University, Poland—a truly prestigious university in the then Eastern Prussia, with a renowned school in physics and mathematics—capable of producing such outstanding scholars as Prof. Dr. Max Born (1882–1970), for example. Meanwhile, Max Bernhard started studying medicine in spending one semester at that faculty, but had finally decided to follow his penchants and switched to the mathematical–physical faculty.

After one year of studying in Breslau he had moved to the University of Berlin and could during six semesters attend the lectures of such outstanding teachers in the fields of physics and mathematics, as Prof. Dr. Helmholtz, Prof. Dr. Kirchhoff, Prof. Dr. Weierstraß, and Prof. Dr. Kummer.

Thereafter, in the year 1878, Max Bernhard could acquire the Prussian citizenship after duly passing muster of the Russian Imperial State, the citizen of which he was remaining all the time from his birthday on. His naturalization certificate was signed on March 23, 1878, at the town of Gumbinnen in the eastern Prussia (now Gussew, the administrative center of Gussewsky District in Kaliningrad Oblast, Russia) by the then president of the Prussian Imperial Government, Karl Friedrich Viktor Graf von Westarp (1864–1945).

This document communicates the following information:

"By signing this document the President of the Prussian Imperial Government certifies that the Russian tributary, student of Physics Baruch Mendel, bearing the name of "Max Bernhard," Weinstein, born on September 1, 1852, in Kowno, has acquired the Prussian citizenship based upon his solicitation and for the purpose of his settlement in Insterburg.

The document at hand ought to start justifying from the moment of its delivery on that only the above-mentioned person is eligible to enjoy the fill set of rights and duties of the Prussian citizen."

Just half a year later M. B. Einstein could acquire the position of an assistant at the Chief Imperial Commission for Calibration (*Kaiserliche Normal-Eichungskommision*), where he was continuing to work at least up to the time of writing his CV (about the start of December, 1883). Here he was dealing in part with the technical and in part with the scientific work, whereas his main theme was 'alcoholometrics' (that is, skillful determination of the alcohol content in liquids).

Meanwhile, he could have published the results of his mathematical and physical work in the *Metronomic Proceedings of the Chief Imperial Commission for Calibration*, in *Wiedemanns Annalen der Physik* (nowadays this is just the Wiley-VCH journal *Annalen der Physik*), as well as in *Hoppes Archiv für Mathematik*.

[Please see Supplementary Note 1, Section 3.5, Page 375]

And finally, in the year 1881, M. B. Weinstein could manage to reap the degree of the philosophy doctor at the University of Berlin—and is referring in his CV to the pertinent copy, which is also detectable in the dossier of his official records.

After the official application to obtain the status of the Prussian State Servant (including all the credentials and testimonials) by Dr. M. B. Weinstein on December 3, 1883, the positive decision by the Prussian Home Office had come on December 22, 1883, with the detailed statement of grounds signed by the then home/interior secretary of Prussia, the vice chancellor in Otto von Bismarck's cabinet, Karl Heinrich von Boetticher (1833–1907).

It has been sent out to Dr. Weinstein and then duly registered by the Prussian Home Office/Department of Interior in January, 8, 1884, together with all the relevant supporting decisions during the years of 1884 and 1885.

3.1.3   M. B. Weinstein's Primary Working Activity and His Social Portrayal

This is why, M. B. Weinstein's performance in the Chief Imperial Commission for Calibration ought to be of extreme interest, as soon as the Prussian government had estimated the five years of Weinstein's professional activity in such a truly honorable way. To this theme there are even more records in his dossier.

1. On May, 17, 1894 the plenipotentiary ambassador of the Kingdom of Italy, Graf Lanza, has communicated a decision of the Italian government to the then Prussian Foreign Minister, Adolf Marschall von Bieberstein (1842–1912), that they would like to lend the *Knight's Cross of the Italian Crown*, a distinguished Italian order of the then time, to the following employee of the Prussian Royal Calibration Commission—in recognition of the excellent alkoholometer and thermometer specimens and the pertinent directions for use of them he had made available to the relevant specialists in the Italian Kingdom:

Dr. Max Bernhard Weinstein, the councilor of the Prussian government being also the professor of the Berliner University.

Knight's Cross of the Italian Crown.

Moreover, his colleagues at the commission, who were also participating in this work, namely, Dr. Georg Schwirkus as well as Mister H. F. Wiebe had also gotten the due Italian accolades along with him.

The list of Weinstein's distinctions goes on, while the next one of the orders he obtained is demonstrating his truly outstanding personality. Undoubtedly, the Italian order was anyway serious distinction without fail; meanwhile, the following accolade came from a special country for M. B. Weinstein—from his actual original country.

2. On February 13, 1900, the plenipotentiary ambassador of the Russian Empire, Baron von der Osten-Sacken, has communicated a decision of the Russian Imperial government to the then Prussian Foreign Minister, Bernhard Heinrich Karl Martin von Bülow (1849–1929), that they would like to lend the order of St. Sviatoslav, the 2nd degree, a distinguished Russian Imperial order to the following employee of the Prussian Royal Calibration Commission—in recognition of the services he showed up to the employees of the Russian Imperial Ministry of Finance during their business trip to Germany:

Dr. Max Bernhard Weinstein, the councilor of the Prussian government being also the professor of the Berliner University.

To the Russian visitors Prof. Dr. Weinstein had explained important details of the procedure to verify the areometric devices, before performing measurements on various petroleum products.

Order of St. Sviatoslav, the 2nd degree—a distinguished Russian Imperial order.

[Please see Supplementary Note 2, Section 3.5, Page 375]

To continue this theme, in the M. B. Weinstein's official records there is also a very instructive letter of communication between the then chief of the Prussian Imperial Commission for Calibration, Wilhelm Karl Arthur von Jonquières (1857–1931), and the then Secretary of Prussian State, Arthur Adolf, Count of Posadowsky-Wehner, Baron of Postelwitz (1845–1932)—concerning the dossier records about M. B. Weinstein and the situation at his working place. The letter is dated 15.04.1904 and W. K. A. von Jonquières is expressing his standpoint as follows (we present here the English translation):

"At his main working place—in the Prussian Royal Calibration Commission, as long as I am a chief of this organization—Weinstein is frequently displaying an utter elusiveness, but never getting into debt with a lack of professional thoroughness. Although Weinstein ought to be truly encroached owing to such side-way jobs of him as his services in the Royal Patent Office, his professorship at the University of Berlin, his publishing activities, he is nonetheless capable of working very quickly, while his personal utter forms and correctness appear sometimes to be hurting, irrespective of his definite eloquence. And this is just the point where he clearly ought to be controlled. But his impersonal working style is throughout true to type.

His competence and experience in the fields of interest for the Calibration Commission (physics, chemistry) are definitely thorough and versatile; but our organization has nothing to do with the electrical engineering he is dealing with at the Royal Patent Office.

I haven't noticed any vagueness of Weinstein's scientific standpoints or in his technical thinking—either in his written works or in his oral reports to the conferences of the Calibration Commission. He is definitely a pliable specialist always ready to hear to the opposite opinions—the feature not omnipresent without fail, when considering conventional engineers and technicians.

Despite his apparent scarcities, Weinstein ought to occupy his working place at the Calibration Commission to the fully satisfactory extent, although our requirements are clearly different from those he would have to comply with at his other positions, like his expert-referee's activity at the Royal Patent Office. It is true that my predecessor at the post of the Commission's President, Dr. of Law, His Excellency Privy Governmental Councilor Carl Hauß (1855–1942), must have estimated his activity in our Commission in a different way, namely only unfavorably. Meanwhile, I can share his opinion not in the least.

[Please see Supplementary Note 3, Section 3.5, Page 375]

Obviously, President Hauß was discussing with me the above outlined theme several times. The problem we were discussing consisted in whether Weinstein should himself solicit a consequent release from his extraneous part-time activities. However, he had declined this —and that was fully justified, to my mind."

What could be immediately deduced from the above material is that the professional life of Prof. Dr. Weinstein was truly extremely hard.

Obviously, he was stubbornly striving for becoming a researcher in the field of theoretical physical chemistry/chemical physics. In fact, he was definitely not a standard research worker, as we can clearly see from his publications list—especially, his definite philosophic fervor has to be stressed here. He was subsequently digging into the very depths of the natural sciences—by reaching the glorious areas of the natural philosophy.

Meanwhile, his actual working place had nothing to do with the latter entirety, for his actual working duties in the commission ought to be of purely engineering nature. Furthermore, to successfully get more opportunities for following the topics of his own choice, he should have to simultaneously exert enormous pressures on himself along with duly fulfilling his official working duties.

Sure, such a life was a heavy burden in every imaginable sense—just in duly summarizing the whole story:

1. Incapability to get a regular working place in the higher academic sectors (i.e., at the university faculties). In Europe such places were, are and definitely will be reserved for the native squads, irrespective of the foreign applicants' actual qualifications. Moreover, Germany ought to be traditionally very difficult country for foreigners in the ethno-political sense, and the roots of this situation lie deep in the history of the country. The first immediate idea that Weinstein's difficulty consisted in his Jewish origin doesn't apply here to 100%.

   In effect, M. B. Weinstein was a ***foreigner*** in Germany (irrespective of his formal naturalization plus perfect mastering the German language), for he was born and spent his early childhood outside Germany, and so did his parents, and so on, so forth. He could reach Germany solely by immigration. Historically, Germany as it is embodying the dreams of Otto Eduard Leopold von Bismarck-Schönhausen (1815–1898) is one of the most difficult countries in the discussed sense. Remarkably, even today a person born, say, in Hamburg and having spent his/her early childhood over there will silently not be considered belonging to the Native Squad in Bavaria, Baden or Swabia (Württemberg)—and *vice versa*.

   Finally, to get a lucrative working place requires good local connections/strong local contacts wherever possible. As a bright example of the situation: Max Planck, an ordinary professor at the University of Berlin, in his last memoir, recalls that he could finally get a desired professorship at Kiel University only (he was born and had spent his early childhood in Kiel) only due to strong local contacts of his father [2].

   Thus, the ultimately topmost position M. B. Weinstein could anyway reach at the university was *Privatdozent*—and he had indeed managed to do so, owing to his enormous 'swotting'.

   Well, this entire deliberation ought to fully explain, why Weinstein was and is remaining detached from any kind of

"The Berliner Geist" [1], notwithstanding his actual professional profile. Meanwhile, in attentively reading the book [1], we recognize that there was but much more to the story, and we shall revert to this point later on in this chapter.

2. At M. B. Weinstein's active time there were no official foundations supporting scientific research activities financially, like we have them nowadays all around the world. This is why, due to the complete absence of the 'stinking rich relatives'—he should be loaded—and even over-loaded—by always considering how to financially support entire working activities of his actual interest. So, he could indeed get money in addition to his rather high salary at the calibration commission—by his sideways activity as a patent referee. In addition to supporting his scientific works, he had to provide financial support to his widow mother and to the widow mother of his wife.

3. M. B. Weinstein's tightest involvement into the multifaceted sideway job activity was definitely not in compliance with the first line managers' attitude at his actual working place. He could manage avoiding direct scandals at the expense of his truly professional and selfless services at the working place, but was earning a stubborn malicious gossip among the highest authorities. Hence, it was only a definitely humane position by his second Chief, W. K. A. von Jonquières that could more or less stabilize M. B. Weinstein's working—and, consequently, his over-all living situation.

4. Meanwhile, our actual entire human resource is never endless to our sincere regret! Consequently, M. B. Weinstein had finally got seriously ill—being only 52 years old. In his dossier we read the following official note by W. K. A. von Jonquières in response to M. B. Weinstein's enquiry (this official note, which is in fact an application to the home/interior secretary of Prussia, is dated by February 27, 1905—we present below its English translation):

"Berlin-Charlottenburg,   *February 27, 1905.*

– Post Office Charlottenburg 2 – Werner Siemensstraße, 27/28.

Official Registration number: 493/05.

From the Director of the Imperial Normal Calibration Commission

To the Home/Interior Secretary

With my obedient request of your highly inclined consideration I would greatly appreciate to hand over the following application by the Privy Governmental Councilor Weinstein for the financial support to defray his expenses in connection to his very serious illness.

Since October, last year, Weinstein is suffering from a serious lymphadenitis, which caused him to periodically stay away from his working place with only short visiting the latter. According to the medical report Weinstein has attached to his application, the current actual state of affairs is characterized by several chirurgical operations and hospital stays. The huge costs of such a medical treatment ought to represent an additional heavy burden for him, in combination with the steady financial support he is making available to his destitute relatives.

After he had quit his side-way activities at the patent office by the April 1, last year, his salary is currently amounting to 6900 Mark, plus 900 Mark to cover his renting flat.

It holds true that Weinstein is a knowledgeable and skillful colleague. Awarding him the title of Privy Governmental Councilor on December 7, 1903 has recognized his activity at our Commission.

The autographic signature by W. K. A. von Jonquières"

The positive decision by the home/interior secretary concerning the above theme was also obtained, and M. B. Weinstein was continuing to work in somewhat less restrained manner, due to his quitting the sideway activity as a patent referee. He could then more properly concentrate his efforts on the scientific research, as we can recognize from his publication list. But to the time point in question his health was already significantly undermined and deteriorated—and he had to start applying for prolonged vacations time—due to his mounting health problems. And as a consequence, by the beginning of the year 1918 he was basically spending his time away from his regular working place at the Commission, namely at Bad Elster—the well-known spa resort in the southern part of Saxony—where he could write his farewell scientific papers—and his last departure took place on March 26, 1918.

The dossier does contain the copy of M. B. Weinstein's death certificate, but there is absolutely no information about the main reason of his death. It seems to be a throughout logical deduction, if we assume that the most probable main reason of his death could be the notorious 'Spanish flu', the pandemic just started its terrible global travel at the beginning of the year 1918 and lasted till 1920.

The copy of the above-mentioned death certificate lists the names and the destinies of M. B. Weinstein's closest relatives, namely those of his parents and of his wife. There is also a statement about the absence of the underage children of the defunct person.

3.1.4   M. B. Weinstein's Familial Situation and His Clique

Therefore, we learn that Weinstein's real name was Boruch Mandel. His father's name was Mandel. Hence the source of his German-equivalent name "Max Bernhard" was the name of his father: "Mandel ≡ Max," and of his uncle on the spear side: "Boruch ≡ Bernhard." Sure, it was just his uncle, who could manage saving the Jewish half-orphan from the then Russian Empire with its notorious 'Love–Hate' for its citizens of Jewish origin.

The name of M. B. Weinstein's mom was Hanne (her maiden name was not more known at the time).

Finally, from the above-mentioned death certificate we learn that M. B. Weinstein was married, his wife's first given names were Auguste Pauline Luise, while her maiden name was Krüger.

In dossier the record concerning the marriage certificate issued on March 17, 1891, in Berlin, follows the death certificate. The former document gives more information about M. B. Weinstein's relatives and his personal connections (the deponents during his marriage ceremony):

The address of the newborn family ought to be Berlin, Urbanstraße, 1 (Berlin Kreuzberg, now—district Friedrichshain-Kreuzberg).

The information about the relatives of the newborn couple:

1. Hanne, M. B. Weinstein's mom, who married a merchant by the name of Zion after the death of her first spouse, the father of the bridegroom, a merchant by the name of Mandel Weinstein, who died before the moment of his son's birth. She died in Königsberg before the moment of her son's marriage.

2. Auguste Pauline Louise Krüger, born on March 9, 1860, in Guben, district of Guben (in southern Prussia) without profession, protestant confession, a daughter of a workman by the name of Ernst Krüger, who died before the moment of marriage of his daughter, while living in Crayne, district of Guben, and of his wife Auguste, maiden name Döring, living in Guben by the moment of marriage of her daughter.

[Please see Supplementary Note 4, Section 3.5, Page 375]

> *The newborn couple has definitely no other relatives.*
> *Both deponents*:

3. A doctor of philosophy, Heinrich Gerstmann, 37 years old, living in Berlin, Schöneberger Ufer, 17.
4. A teacher and doctor of philosophy, Leopold Levy, 39 years old, living in Berlin, Großbeerenstraße, 8.

[Please see Supplementary Note 5, Section 3.5, Page 375]

The rest of M. B. Weinstein's records dossier is devoted to the payment of 'surviving descendants' pension to the address of Mrs. Weinstein after the departure of her husband—up to her own departure on March 20, 1936.

Most probably the married couple had no common children, which for the most part seems to be a fortunate development, if we take into account what started to happen in Germany to the fellow citizens of Jewish descent just after the year of 1933.

To sum up, we have learned about a very interesting colleague, whose life was full of endeavors, successes, failures, and miscarriages. He could achieve an unambiguous recognition for his professional qualities, and now the natural question that might be posed follows.

3.2   Why Should M. B. Weinstein and His Work Be of Interest for Us?

The answer is throughout clear: Because of his research and pedagogic work in the field of natural sciences and natural philosophy, with the special reference to thermodynamics and kinetics. The attentive reader might but immediately react:

> *But wait a bit! What are his works, for goodness sake?*

Aren't the monograph's author trying to excite an artificial interest for some strange person just mentioned by another not even less strange person (Nils Engelbrektsson, see Chapter 2)?

To correctly answer the both above posers, let us now come to considering the actual research works by M. B. Weinstein.

First of all, after the due research, we immediately recognize that Prof. Dr. Weinstein's publication list consists of 34 original publications by him. This looks like truly impressive and ought to duly reflect all the 35 years of his truly multifaceted professional experience (see *this list at the end of the chapter*).

To our mind, reading his publications ought to bring the same effect as breathing-in a pure fresh air or drinking a pure fresh water at a gulp—so clear, professional and unbiased is his writing, whatever its theme. The volume of the present book at hand doesn't allow carrying out any detailed analysis of Weinstein's papers and books. Like in all the cases considered here, a good project ought to be translating all the materials into English and republishing them—for the sake of promoting thermodynamics, kinetics, as well as the natural philosophy.

Thus, here we would solely like to do the minimum preliminary work: First, we shall briefly discuss the full listing of M. B. Weinstein's publications, and then consider the referee's reports concerning several of his seminal works.

3.2.1 M. B. Weinstein's Research and Pedagogic Work

Surely, of separate interest ought to be M. B. Weinstein's translating and editing activity. Specifically, he performed an authorized translation into German and managed publishing the latter, as concerns the cornerstone work on the theory of electricity and magnetism by an outstanding British physicist and mathematician, James Clerk Maxwell (1831–1879)—in two volumes (Ref. [1] in Weinstein's publication list), as well as the fundamental work by a renowned French physicist and mathematician, Jean-Baptiste Joseph Fourier (1768–1830) on the analytical theory of heat—in a separate volume thereafter (Ref. [2] in Weinstein's publication list).

If to assume that the above translations could help forming the basis of his university lectures, then he was teaching something like the course in general physics, namely—theoretical physics of the mechanical motion, heat, electricity, and magnetism. The publications on his own works (and lectures) are presented in his publication list to the full extent, and in the chronological order (see the remarks in bold indicating the field/topic of every work in his publication list, see Section 3.4, Page 371).

Now, the answers to the posers at the beginning of the Section 3.2 is hopefully clear: In fact Prof. Dr. Weinstein turns out to be an outstanding specialist in the field of physics, physical chemistry, physical geography, and natural philosophy.

Howbeit, at the first glance, we clearly deal here with a talented and proactive physicist and natural philosopher, whereas of special interest for the readership of the book at hand are his works on thermodynamics and kinetics. So let us have a look at what is the actual resonance at his work.

3.2.2 The Over-All Resonance in Regard to M. B. Weinstein's Research Work

First of all, at the beginning of this chapter we have already noticed a strange notorious silence all around M. B. Weinstein's name. Meanwhile, in the English literature we encounter the following remarkable notice, we cite [3]:

> "… *As word of Einstein's accomplishment spread in Germany, more sensational articles appeared in newspapers. They kindled darker reactions to Einstein's theory, fuelled by political unrest and the duress of war. One popularizer, Max Weinstein, claimed that general relativity from its earlier isolated position and made it into a 'World Power' controlling all laws of Nature. He warned that physics and mathematics would have to be revised. Popular reactions to Weinstein's writings prompted Wilhelm Foerster, emeritus professor of astronomy at the University of Berlin, to urge Einstein 'to find a way of addressing the German public' to allay anxiety and skepticism stirred up by 'doubts about previously held basic tenets of our knowledge of the world.' He attributed this agitation to an almost psychopathic state among the populace. To illustrate, he noted: 'Some are happy that you had now put an end to the global confusion caused by the Englishman Newton, etc. Surely you will find words free of scholarly jargon to introduce the German public to a sound and sober-minded interpretation of your so extremely important ideas and problems, but there really is a need for this now' [4] …*"

In fact, the contents of the book by J. Crelinsten has caused a number of serious questions among the truly briefed colleagues in the field [5], but here we would like to note another rather interesting and remarkable point.

After reading the above cited deliberations the questions immediately arise: Was the "popularizer" Weinstein a true psychopath? Or, may perhaps, he was a wild bandito trying to organize numerous psychopaths in trying to attack the "brightest genius of all the times and all the peoples?"

Meanwhile, a detailed consideration of the story told us by J. Crelinsten, shows that Prof. Dr. Wilhelm Julius Foerster (1832–1921) has indeed written the letter in question. He was a professor of astronomy at the University of Berlin and an outstanding German astronomer. Remarkably, at the time of writing the above-cited letter (25.03.2016) Prof. Foerster was already emeritus and had been principally involved in and fully absorbed by the purely political activity. On the one hand he was well known as a fierce censor of any kind of nationalism/jingoism, but on the other hand, he was without fail sharing the over-all 'patriotic excitement' in his homeland during the First World War.

On the other hand, Albert Einstein, though being born and having spent his childhood in Ulm, Germany, was at that time just a Swiss citizen living in Germany and trying to get the permanent residence permit. With this in mind, he was immensely active in many different fields including research work, with the purely political activities being clearly chiseled among others. Consequently, it was first of all the political activity that had enabled Albert Einstein's approach to Prof. Dr. Foerster. For the actual details of that story please see the pertinent historical works [6, 7].

Howbeit, of interest and importance for us but ought to be the sheer absence of any strict professional evaluation of A. Einstein's research activity by Prof. W. Foerster. Moreover, after reading attentively the J. Crelinsten's reference [4], at its page 276 we note the comment stating that

> "[2]*Max Weinstein. A recurring theme in Weinstein's publications is the view that general relativity has removed gravity from its earlier isolated position and made it into a 'World Power' (Weltmacht) that controls all laws of nature, necessitating a revision of physics, as well as of mathematics (see, e. g., Weinstein 1914)."*

Therefore, in Weinstein's case we are dealing not with *one popularizer*, but with the serious professional systematically working and actively publishing in the field. The 1914 paper by Weinstein in question was indeed published in a scientific-popular journal. Weinstein's idea behind such a publication gets to 100% clear, if we take into account his tremendous pedagogical efforts. Please see M. B. Weinstein's publication list at the end of this chapter.

To sum up, in no way the deliberations by J. Crelinsten can be applicable to analyzing and estimating Weinstein's activities and contributions. This is why, later on in this chapter we shall revert in much more detail to the paper by M. B. Weinstein concerning A. Einstein's work mentioned by J. Crelinsten. Anyway, our sincere gratitude should go to the account of Jeffrey Crelinsten for his having fetched this clearly noticeable popular paper by Weinstein!

Further on, our readership would also definitely ask about the German literature. Is there more detailed information about M. B. Weinstein? At the beginning of this chapter we have already cited the German book [1] devoted to the destiny of philosophic faculty in the University of Berlin, at the end of the XIX-th and the beginning of the XX-th century.

Let us open this book and look for the name of Max Bernhard Weinstein. What we get?

Only two fits pointing to the exceedingly short notes ultimately buried into the wealth of ulterior information (in the sense of telling the minimum about M. B. Weinstein—and even without presenting any portrayal of him), so that as a result of reading them we remain with a very confused impression:

1. **Page 108.** The text quoted below is its English translation. We shall note beforehand that this particular page corresponds to **Part II,** entitled *Formierung* (The Formation); **Chapter 7,** entitled *Fortsetzung der Philosophie aus dem Geiste einzelner Disziplinen* (*Continuation of the Philosophy Out of the Spirit of Other Disciplines*); **Paragraph (e),** entitled *Die 'Wiedergeburt der Philosophie' durch die Naturwissenschaften* (*The 'Revival of the Philosophy' Through the Natural Sciences*):

"... In those years a physicist Max Bernhard Weinstein had started his teaching activity.[103] His postdoctoral qualification under the leadership of Hermann von Helmholtz has been successfully terminated in the year 1886.

[Please see Supplementary Note 6, Section 3.5, Page 376].

From the year 1895 on he becomes a titular/honorary professor.[104] His lectures are connected with the natural philosophy as a whole, and in particular with "materialism and idealism," "monism and dualism," and with the possibility of a general natural-philosophic paradigm.

Later on, Max Planck (1858–1947) attracts Albert Einstein (1878–1955) to the academy (*The Prussian Academy of Sciences, Berlin*) as a professor without any official permission to teach. From 1913 till his ouster in 1933 Einstein was being active at the Friedrich Wilhelm's University (*now Humboldt University of Berlin*); he is regularly delivering public lectures without any formal connection with the philosophic faculty. According to his own avowal, 'he is feeling a tight relationship to the philosophic thought'. That is, all of his great discoveries 'stem from the natural-philosophical ideas'. The natural philosophy ought to be a 'metaphysical heuristics' for him.[105] And only one philosopher in Berlin was somehow communicating with Einstein—namely Prof. Dr. Ernst Cassirer (1874–1945).[106] Moreover, Prof. Dr. Hans Reichenbach (1891–1953), who had in 1926 successfully completed his postdoctoral training in Berlin, started a systematic reappraisal of the modern physics. This way the emancipation- and differentiation-trends between the natural sciences and humanities could be consistently and repeatedly counteracted with the attempts of philosophic communication." [Please see Supplementary Note 7, Section 3.5, Page 376]

2. **Page 134**. The text quoted below is its English translation. We shall note beforehand that this particular page corresponds to the same book's part as the **page 108** and belonging to its **paragraph (f.),** entitled *Ein Blick ins Vorlesungsverzeichnis* (A Glance into the Catalogue of Lectures):

"As we have already mentioned above, Max Bernhard Weinstein (1852–1918), who had successfully completed his postdoctoral studies in Berlin in 1886, was between the years 1906 and 1915 regularly delivering lectures about the natural philosophy and related topics..."

The only feeling to arise and, consequently, the conclusion to be drawn, after reading the books [1, 3] and thinking over all the above citations, ought to be as follows:

Not only was Prof. Dr. M. B. Weinstein never embodying the "Berliner Spirit," instead—he was playing a definitely conservative, or, one might even denote it—a definitely negative role in the functioning of the Philosophic Faculty of Friedrich Wilhelm University of Berlin. He was forcefully hindering the fruitful contacts between the philosophies of natural sciences and humanities. To sum up, down with Prof. Dr. M. B. Weinstein.

*Meanwhile, the actual productive worker in the field of natural philosophy ensuring the positive outcome was Albert Einstein. The then Philosophic Faculty of Berliner University had apparently no mood and/or no competence to recognize such a **true star** in the field, by giving A. Einstein no official permission to teach.*

Fortunately, only the wise intervention by such prominent specialists in the field of natural science as Max Planck, an active member of the then Prussian Academy of Sciences, could at least partially improve the otherwise quite wobbled situation.

Remarkably, the later developments could duly restore the desired balance after the intervention of Prof. Dr. Ernst Cassirer, an outstanding Berliner philosopher, a serious specialist in the fields of epistemology, philosophy of culture, and philosophy of symbolic forms. But the final enlightenment of the whole situation could be established after the advent of Prof. Dr. Hans Reichenbach, a renowned physicist and, along with this, a nameable specialist in the field of Natural Philosophy, the grandparent of the Berlin School in the latter field.

The readership might get a funny feeling at this point. Meanwhile, the briefed readers would immediately recognize clear dissonances in the above-cited representation.

First of all, Albert Einstein was never a professional philosopher—the fact clearly excluding him from the list of those capable of getting the official license to teach at a philosophical faculty anywhere all over the world. Max Planck was a renowned theoretical physicist, but also not a professional philosopher (although, sure, he was definitely interested in the natural philosophy, but this is not enough for claiming the pertinent professional capabilities). Thus, the belated attempts to crassly demonstrate Einstein's philosophic skills (see the both works by Prof. Dr. Bernulf Kanitscheider [8, 9]) look like a forceful nisus to prove the improvable and hence seem to be hardly credible.

Well, one might in principle try to personally persuade professional philosophers to properly recognize some amateurish strivings—and this is most probably just what could be successfully manageable in the case of Prof. Dr. Cassirer. Meanwhile, the year of Cassirer's publication (1921) is somehow coincident with the year of winning the Nobel Prize by Albert Einstein, whereas Prof. Dr. Reichenbach had successfully finished his postdoctoral studies in Berlin in the year of 1926, that is, only five years later—and only thereafter he was in fact free to choose both the field and the character of his professional activity. Were the both named colleagues the *outstanding clairvoyants*—or just adequately following the actual trends?

Howbeit, all the above considerations reinforce our wish to consider below the above story in much more detail. Meanwhile, the main point immediately recognizable here ought to be a definite professional (and/or personal???) **conflict** between Albert Einstein and Max Bernhard Weinstein, as a result of which the former guy is a **definite star, genius**, etc.—whereas the latter remains only marginally known, irrespective of his tremendous research and publishing activity.

Apart from all the stories to follow, a significant purely human dissonance still remains—both Einstein and Weinstein were active and proactive research workers. The achievements of the both are undoubted. So what are then the roots of such a final result?

Furthermore, the both names are tightly connected to the field of thermodynamics and the related research areas, so that it is throughout clear that the present book wishes to disclose the relevant details and to attempt a thorough analysis of the Weinstein's standpoints—in comparison to those of Einstein and his allies. Bearing all this in mind, what should we undertake right now?

Our next move ought to be looking for the actual opinions of M. B. Weinstein's contemporaries as concerns his research work. Specifically, our second move would be divided into three following steps:

- The work of M. B. Weinstein in the field of natural sciences;
- The philosophical work of M. B. Weinstein;
- The details of the conflict between M. B. Weinstein and Albert Einstein and its implications in regard to thermodynamics and statistical physics.

3.2.3 The Work of M. B. Weinstein in the Field of Natural Sciences

A significant number of such reports could be fetched in the publication *Archiv für Mathematik und Physik* (AMP), which has been initiated and edited by outstanding German mathematicians and pedagogues, Prof. Dr. Johann August Grunert (1797–1872) and, after his departure, by Prof. Dr. Reinhold Hoppe (1816–1900), to trigger useful and productive discussions as for all the necessary modalities of the successful educational classes in mathematics and physics.

First of all, the authorized translations carried out by M. B. Weinstein (Refs. [1] and [2] in his publications list) have been refereed by Prof. Dr. Hoppe himself (in the years 1885–1887):

1. As to the translations of the books by J. C. Maxwell, the referee has especially stressed a careful re-consideration of all the mathematical calculations in the original English and French publications—so that, as a result, the corresponding German publications are containing a significantly more detailed account of the relevant inferences and calculations—and, moreover, they are free from typos in their original issues.

2. Further, Prof. Dr. Hoppe has refereed M. B. Weinstein's book about theoretical metrology (Ref. [6] in his publications list). The book by Weinstein ought to significantly widen the conventional scope discussed in the books by his predecessors. In his book Weinstein presents not only clear practical recipes of how to properly treat ubiquitous experimental observation errors, but also the detailed theoretical basements of the recipes involved.

3. The Weinstein's book *Introduction to the Higher Mathematical Physics* (Ref. [11] in his publications list) had been reviewed by Emil Aschkinass in Berlin, who presented the following criticism in the AMP, vol. 9, pp. 281–282, 1905 (we translate his report into English):

"The book represents a compendious introduction into the entire field of mathematical physics. As its foreword states, 'the book is written not only for pupils, but also for teachers', although actually reading the book reveals its adequacy rather for the latter than for the former ones. The

point is that the actual treatment of the bulky topic chosen appears to be succinct to the extent that only readers truly versed in the field discussed might get enough information. For such a readership the book might indeed be considered a review of at least some help.

Further, the book's preamble states that 'the grass roots of each theory are considered with the special rigor', and the initial part of the book is dealing with these grass roots by considering their natural-philosophic and practical significance in detail. Indeed, the book's author apparently hopes that it is just this part of his book that might be truly welcome, because, as strange as it might sound, 'there is still a trend to be not mindful enough of duly penetrating the very grass roots'. In this connection the referee must confess that, to his mind, it is just this agitating the grass roots and the epistemic analysis of the physical notions that are unsatisfactory almost everywhere. Specifically, almost every page of the chapter entitled 'The Subject of Mathematical Physics and its Basics' contains fragments to be definitely disputable and sometimes the book author's parlance even lacks the necessary precision. To support such a conclusion, I would like to allege several examples here:

*When discussing the notion of Space, the following statement crops out: 'The three dimensions of space are even by the Nature or, in other words, the location of each particular area in space ought to be completely determined by the three straight lines having a common point in this area ... **But physics dictates (emphasized by the referee!)** that any movement beginning in our space along a straight line, should not be forced to bend by the space itself, so that any straight movement in our space happens just informally, that is without any constraint.'*

Then, the following statement ought to sound even in a much stranger way: 'Die Zeit ist immer die nämliche' (that is, 'the time is always the particular one'). The word 'immer' (always) but does mean to wit 'zu Allen Zeiten' ('at all the times')—that is, to sum up, we arrive at the following statement: 'the time is at all the times the particular one'. And then comes another sentence stating: 'Der Raum ist überall derselbe' ('the space is everywhere the same'). It seems that the book's author erroneously tends to connect something substantial with the notions of Space and Time, which the conventional physics solely employs as a kind of abbreviations to describe certain interrelations among the phenomena.

Further let us just site the following statement: 'If we wouldn't like to assume that the Time affects different substances in different way, so there is no more choice than just to imply that this is not the Time through the influence of which the changes do happen, but there are some pertinent causes for the changes in question. We would call the latter Other Causes and will discuss these later on here.' Hence, to sum up, it is never the Time itself, but it is some particular Cause owing to which the change comes!

Just one more statement should also attract our attention: namely, that 'any uniform movement happens informally, without any constraint. **This is why (emphasized by the referee!)** *the time stretches are normally represented by straight lines'.*

The energy notion is introduced as follows: 'the energy is nothing more than just what we normally denote as a mechanic work. But since physicists have to differ among versatile sorts of work, in connection with different kinds of phenomena, it is convenient to use a loanword as a unifying term'.

Remarkably, to define further physical notions the book's author has mostly chosen a similar largely imprecise form.

When speaking of thermodynamics, the author starts using the notion of Temperature without any explanation. But later on we come across the following explicit statement to the theme: 'for this variable θ we have used the term temperature, without specifying how we should measure this variable. Here we might give no additional information to this topic.' Nonetheless, or fortunately, the author does provide the readership with 'some further information' in the following sentences.

To sum up, we would only like to note that the book's parlance contains a noticeable flippancy in the linguistic sense."

[Please see Supplementary Note 8, Section 3.5, Pages 376–377]

One more referee of M. B. Weinstein's work was Prof. Dr. Ernst Pringsheim (1859–1917), a renowned German physicist at the Universities of Berlin and Breslau.

Concerning the 1st volume of M. B. Weinstein, *Thermodynamics and Kinetics of Bodies. General Thermodynamics and Kinetics* (Ref. [13] in his publications list), Prof. Dr. Pringsheim has written the following (AMP, series III, volume III, pp. 66–67, 1902, we present the pertinent English translation):

"The book is written with an outstanding assiduity and acuteness, it contains a lot of interesting and original material. But it is definitely not

to be considered a teaching book for thermodynamics. Extraordinary are the ordering and the presentation of the material, as well as the delight in choosing the topics to consider.

As to the material ordering, the book demonstrates a promiscuous, balled-up introduction of thermodynamics and kinetics. In accordance with this, the both Basic Laws of the mechanical heat theory are first introduced thermodynamically, and straight after the latter move – the both are derived based upon the kinetic hypotheses. Meanwhile, the author could not to 100% succeed in attempting a parallel application of thermodynamic and kinetic approaches. The basic difference between the former and the latter forces us to their subdivision, which is anyway desirable, for the respective degrees of the insight delivered by the both of them are quite different from each other.

As for the material presentation, the book reflects a clear striving not to depict the relevant stuff in some purely plausible manner, but to try a detailed critical verification of the methods to use in regard to their actual value and strictness, to reveal their applicability limits and flaws—and, finally, to reveal, where they actually stand in need of their modification, as well as the aspects of their principal modifiability. To my mind, it is the latter criticism that clearly proves the book's main merit and zest, for these are just the aspects that ought to stimulate the readership to criticism. Without following such a stimulus here in detail, I would just like to point out the inference of Maxwell's distribution function. The hypothesis introduced by the author that the deviations of the three velocity components from the average value are independent of each other looks like throughout plausible, like the Maxwell's original hypothesis that the components in question are fully independent of each other themselves.

Among the problems considered in the book, a special place is occupied by the equation of state derivation, which the author is already studying in detail for a longer time. In the third chapter of his book under the review the most general equations of state have been derived from the virial theorem under two conditions, namely: (a) a hypothesis that the substance is continuous and (b) a hypothesis of molecular consistency, whereas the calculation is carried out assuming two cases, namely: (a) that only external forces are influencing the system and (b) that mutual internal intermolecular collisions also cause a thrust, which is different from the external forces. These inferences contain an enormous work behind them, the actual fruits of which are in fact very modest for the present. The latter point could be especially recognizable in the last chapter of the book, where a careful consideration of all the available experimental results has been undertaken by the author in an attempt to compare the theoretical results with the data on realistic gases. One might hope that the physical sense of the author's work, being a generalization of the earlier considerations by van der Waals, Clausius et al., could make its mark in a more definite way in some future. The book at hand might possibly stimulate some broader experimental investigation, which the author envisages in the preamble to his book, and for which he hopes to win the assistance of the state, or, more specifically, of the Imperial physical technical institute.

To finalize the present report, I would like to present a short list of the book's chapters."

In the AMP, vol. 11, pp. 93–94, 1907 Prof. Dr. Pringsheim has reviewed the further actually published volumes of the thermodynamics and kinetics book by M. B. Weinstein by the time in question (Refs. [14] and [16] in his publication list). Prof. Dr. Pringsheim's later report sounds as follows (we present here the relevant English translation):

"The second volume of the book under consideration begins with a chapter about the absolute temperature. This chapter contains detailed descriptions of the thermodynamic and thermo-kinetic temperature scales, then a comparison among the absolute and the conventional scales. The novel temperature scale based upon the radiation theory— the accrual of which was solely connoted at the moment of the book's publication—is only lightly touched in a commentary. Then the chapter follows, which is devoted to liquids, and then— the one about gases. Like in further devolution of the story at hand, we clearly see in the latter chapters that the equations of state studied in full detail in the first volume of this book are still playing a very modest role in the actual applications of thermodynamics, and that the kinetic ideas are still occupying a rather tiny flat in the whole building of the general heat theory.

The next chapter entitled 'Thermodynamic Mechanics and Irreversible Processes' presents from the beginning on a theory of thermodynamic equilibria, and then the Gibbs' phase rule, the notions thermodynamic potentials, entropy and energy are considered in detail. Then, subsequent to the latter story, comes the theory of conversion rates, as well as that of the processes, where considering the time variable ought to be essential.

The last chapter finally considers the theory of mixtures and solutions, assuming that the latter both are rather dilute.

The latter are also the subject of the third volume's first part. Here, the van der Waals theory is considered in full detail, to enable a throughout check of its theoretical and experimental basics, which, as they are, definitely require a further non-trivial verification, owing to the significance and the role of the theory under study in promoting the scientific progress. The thermodynamics of electricity and magnetism composes the final part of the volume considered. The study on these topics ought to be continued in the second part of the third volume, where the electrolysis theory should be presented.

Like the very first volume, the entire set of the recently published further parts of the book under review is written with extreme accuracy, plus detailed usage and critical analysis of the carefully collected theoretical and experimental literature sources. On many pages there are analyses of the author's own studies, and practically every string of the book at hand is letting on about the extreme diligence of the author, that he has full mastery of and, using his diligence and spiritual energy to the full, managed gaining the deepest insight into the topic he is describing. Meanwhile, the useful features expected from the book's potential readership are by far not scant. And the referee sincerely hopes that the readership would find the mood and time enough in comparison to the earlier periods, to study this book in even a more precise manner than before. That the book would then provide the readership with a rich source of the proper guidance and stimuli ought to be of no doubt for the present referee."

The last, but not the least, of special interest ought to be the communication Prof. Dr. Weinstein has published in the commemorative volume dedicated to Ludwig Boltzmann's sixtieth birthday on February 20, 1904 (Ref. [15] in Prof. Weinstein's publication list). Below we publish our English translation of this communication and the reaction to it by the German colleagues.

"*Entropy and Internal Friction*

(Published in: Commemorative volume devoted to the 60th birthday of Ludwig Boltzmann, 1904, pp. 510–517)

By B. Weinstein, in Charlottenburg

Szily, Clausius, and Boltzmann have proven that, in applying the known principles of mechanics to the dynamics of molecules in physical bodies, it is possible to infer a rule correspondent to the Carnot–Clausius law for reversible processes under some certain conditions. Specifically, let dε be the portion of energy fed into the system during some change in the dynamic state of the latter and defined by the average livening force T, and let i be the time length of the change involved. Bearing this in mind, we have

$$\frac{\delta\varepsilon}{T} = 2\delta\left[\log(\bar{T}i)\right]$$

From the proof of the above principle I have published in my Thermodynamics book (B. Weinstein, Thermodynamik, I, p. 31 f.) the role of the magnitude i as an anterior time length. In effect, the latter parameter ought to be solely incidental to the particles dynamics. It should be sourced in such a way that any elementary dynamic act should ensure that the participating particles would as a result level up their velocities with a vengeance. In other words, this should ensure that every particle involved might attain any possible velocity. It is this way that the lower velocity limit is imposed. Now let us denote τ the time between two subsequent collisions of some particle. On the other hand, if there will be no collisions during some time, we might speak of an average oscillation period and set i = vτ.

As t should have a tiny magnitude, the n will be huge, so that we may recast the principle in question as follows:

$$\frac{\delta\varepsilon}{T} = 2\delta\left[\log(\bar{T}v\tau)\right]$$

According to the conventional hypotheses, we interpret the energy input as a heat uptake dQ, whereas ϑ stands for the absolute temperate and J stands for the Joule's mechanical equivalent of heat. We thus arrive at the useful relationships.

$$\delta\varepsilon = \delta Q, \quad \bar{T} = R'\vartheta$$

With this entirety in mind we get the following expression for the entropy change δS

$$\delta S = \frac{2R'}{J}\delta\left[\log(R'\vartheta v\tau)\right]. \tag{1}$$

In my book already mentioned above I have used the latter equation, though recast in somewhat different form, to infer the formula for the internal friction in gases. Remarkably, the latter expression delivers an extremely good approximation to the experimental data (l. c. I, paragraphs 32 and 41), in that it becomes possible to clarify the dependence of the internal friction on the number of atoms and the atomic mass. In my inference I am calculating the magnitudes of v and τ.

Considering the former parameter, I would guess that there ought to be no other approach to its theoretical inference as compared to that already published in my book just cited. Indeed, if we denote V to be specific molecular volume, then v should be proportional to nothing else than $V^{-2/3}m^{-2/3}$, just where m stands for the atomic mass. Then, the magnitude of τ might be derived from the well-known formulae for the average velocity of gas molecules.

In the second volume of my book, I have presented a theory of the solid state, which delivers a very good approximation for the experimental data as well (l. c. II, paragraph 63, 64). The formulae of the mentioned theory are more general than those used in the theory of gases, in that the latter are the particular cases of the former. With this in mind I would greatly appreciate using here the more general toolbox. My actual aim ought to be (a) pointing out the grounds for the friction formulae being of importance to my mind; (b) looking for some plausible way of improving the formulae of interest.

Let us assume that we might properly describe the dynamics of a molecule without any collision by an A', the average path length, and t', the time being underway. Then the path length A and the time being underway τ would properly describe the situation between two successive collisions of the molecule in question. Furthermore, let us denote N as the number of molecules in a mass unit and assume that the molecules are involved into the oscillations of the simplest possible forms. As a result, we get the following relationship

$$\vartheta = \frac{1}{R'}\frac{Nm}{2}\frac{A^2}{\sin^2\left(\frac{2\pi\,\tau}{\tau'\,4}\right)}\left(\frac{2\pi}{\tau'}\right)^2\left(1+\frac{\sin\left(\frac{2\pi\,\tau}{\tau'\,4}\right)}{\frac{2\pi\,\tau}{\tau'\,4}}\right).$$

As for the parameter $\left(\dfrac{2\pi}{\tau'}\right)^2$, it is possible to show (see my book mentioned above) that this expression is representable as a function of $v^{1/3}$ —and as a consequence we might write down

$$\left(\frac{2\pi}{\tau'}\right)^2 = \psi\left(v^{1/3}\right).$$

The both above equations might be used to calculate τ. Indeed, let us introduce the following denotations

$$\frac{1}{R'}NmA^2 = \mu, \quad \frac{\pi\tau}{\tau'} = \varphi.$$

With this entirety in mind we immediately arrive at the following relationship

$$\vartheta = \mu\psi\frac{1}{1-\cos\varphi}\left(1+\frac{\sin\varphi}{\varphi}\right)$$

The solution of the latter transcendent equation might be cast below in the series form

$$\tau^2 = \frac{(\tau')^2}{\pi^2}\sum_{n=1}^{n=\infty}\frac{A_n\mu^n\psi^n}{\vartheta^n}.$$

Here $A_1 = 4$, $A_2 = 0$, $A_3 = 0.09$, $A_4 = 0.016$. Thus, in substituting the value for $\frac{(\tau')^2}{\pi^2}$, we get

$$\tau^2 = 4\sum_{n=1}^{n=\infty}\frac{A_n\mu^n\psi^{n-1}}{\vartheta^n}. \tag{2}$$

The latter formula can be substituted into the entropy principle to recast it

$$\delta S = \frac{R'}{J}\delta\left[\log\left((R')^2\vartheta^2a^2v^{-4/3}m^{-4/3}4\sum_{n=1}^{n=\infty}\frac{A_n\mu^n\psi^{n-1}}{\vartheta^n}\right)\right]. \tag{3}$$

Herein a stands for the constant of proportionality for v.

On the other hand it is known that for gases

$$\delta S = \frac{1}{J}\delta\left[\log(\vartheta^{Jc_v}V^R)\right]. \tag{4}$$

Here the denotations c_v and R have the conventional meanings of the heat capacity at the constant volume and the universal gas constant, respectively. Therefore, for gases we should immediately obtain

$$\delta\left[\log(\vartheta^{Jc_v}V^R)\right]=R'\delta\left[\log\left((R')^2\vartheta^2a^2V^{-4/3}m^{-4/3}4\sum_{n=1}^{n=\infty}\frac{A_n\mu^n\psi^{n-1}}{\vartheta^n}\right)\right]. \quad (5)$$

To sum up, all the story turns out to matter to the magnitude of the parameter R'. We might find the latter parameter in the equation for the internal potential energy. According to the theory presented in the third chapter of my thermodynamics book mentioned above, we might express the internal potential energy U as follows

$$U = U_0 - pV + \frac{5}{3}R'\vartheta - \frac{2}{3}F.$$

Here U_0 is the initial value of U, whereas the term F denotes a function describing the effects of molecular collisions. If the equation by Boyle–Gay–Lussac is valid for gases, then we immediately get pV = Rϑ. The resulting expression for the potential energy ought to be then

$$U = U_0 + \left(\frac{5}{3}R' - R\right)\vartheta - \frac{2}{3}F.$$

To sum up, we get

$$c_p - c_v = \frac{R}{J}$$

$$R' = \frac{3}{5}Jc_p. \quad (6)$$

This demonstrates that R' should normally never be a constant. However, if we force it to be a constant, then we get from Eq. (5) above

$$(C\vartheta^{Jc_v}V^R)^{\frac{1}{2R'}} = 2R'\vartheta aV^{-2/3}m^{-2/3}\sqrt{\sum_{n=1}^{n=\infty}\frac{A_n\mu^n\psi^{n-1}}{\vartheta^n}}.$$

Here C is just a proportionality constant. Now, further transformations would deliver the following result

$$\sqrt{\sum_{n=1}^{n=\infty}\frac{A_n\mu^n\psi^{n-1}}{\vartheta^n}} = \frac{C^{\frac{1}{2R'}}}{R'am^{-2/3}}\vartheta^{\frac{Jc_v}{2R'}-1}V^{\frac{R}{2R'}+2/3}.$$

Here we would just like to substitute the natural simplification

$$\frac{C^{\frac{1}{2R'}}}{R'am^{-2/3}} = B$$

If now we take into account Eq. (6) and that $\dfrac{c_p}{c_v} = k$, *we immediately arrive at the following result*

$$\sqrt{\sum_{n=1}^{n=\infty} \frac{A_n \mu^n \psi^{n-1}}{\vartheta^n}} = B\vartheta^{\frac{5-6k}{6k}} V^{\frac{9k-5}{6k}}. \tag{7}$$

In the first approximation we retain only the first term of the above sum and get the result as follows:

$$2\sqrt{\mu} = B\vartheta^{\frac{5-3k}{6k}} V^{\frac{9k-5}{6k}}.$$

If we now use the latter definition of μ, then we get

$$2A\sqrt{\frac{Nm}{R'}} = B\vartheta^{\frac{5-3k}{6k}} V^{\frac{9k-5}{6k}}.$$

In the above expression the parameter 2A stands for the average path length, which we denote l and arrive at the following result

$$\bar{l} = B\sqrt{\frac{R'}{Nm}}\,\vartheta^{\frac{5-3k}{6k}} V^{\frac{9k-5}{6k}}. \tag{8}$$

Now we get the answer as for the magnitude of the parameter R': $R' = \dfrac{3}{5}c_p = \dfrac{3}{5}(mc_p)m^{-1}$. *Taking into account that* $mc_p = c'_{p'}$ *we immediately arrive at*

$$R' = \frac{3}{5}c'_p m^{-1}.$$

Finally, in taking into account that Nm = 1, we recast the expression for the average pathway

$$B\sqrt{\frac{R'}{Nm}} = \frac{mC^{\frac{5m}{6c'_p}}}{\frac{3}{5}ac'_p m^{-\frac{2}{3}}}\sqrt{\frac{3}{5}c'_p m^{-1}} = m^{\frac{7}{6}}\frac{C^{\frac{5m}{6c'_p}}}{\frac{3}{5}ac'_p}\sqrt{\frac{3}{5}c'_p}.$$

Moreover, we have to introduce the molecular volume Y = mV and carry out the following transformations to get the following preliminary conditions

$$\sqrt{\frac{3}{5}c'_p}\,a = \left(\frac{\pi}{8}\right)^{-\frac{1}{2}}(R)^{+\frac{1}{2}}e^{-\alpha}\quad,\quad C^{\frac{5m}{6c'_p}} = e^{\beta m}. \tag{9}$$

Here R is the gas constant, so that for the average pathway length we finally get

$$\bar{l} = \left(\frac{\pi}{8}\right)^{\frac{1}{2}} (R)^{-\frac{1}{2}} \vartheta^{\frac{5-3k}{6k}} Y^{\frac{9k-5}{6k}} e^{\alpha+\beta m} m^{-\frac{2k-5}{6k}} . \tag{10}$$

To this end, we just arrive at the same equation for the average pathway length, which has been inferred other way in my thermodynamics book already cited here (l. c. I. p. 207, Eq. (16) and Eq. (17)). This opens our way to all the expressions for the proven and verified formulae for the internal friction given in my book.

In the second approximation we take into account that $A_2 = 0$, so that Eq. (7) could be recast in this manner

$$\sqrt{A_1 \frac{\mu}{\vartheta} + A_3 \frac{\mu^3 \psi^2}{\vartheta^3}} = B\vartheta^{\frac{5-6k}{6k}} V^{\frac{9k-5}{6k}} .$$

Then for the average pathway holds the formula

$$\bar{l} \sqrt{1 + \frac{A_3}{A_1} \frac{\mu^2 \psi^2}{\vartheta^2}} = (B)\vartheta^{\frac{5-6k}{6k}} V^{\frac{9k-5}{6k}} .$$

We get thus the following expression for the magnitude (B)

$$(B) = \left(\frac{\pi}{8}\right)^{\frac{1}{2}} (R)^{-\frac{1}{2}} e^{\alpha+\beta m} m^{-\frac{2k-5}{6k}} . \tag{11}$$

Let us note that the value of the ratio A_3/A_1 is actually about 0.02. Then, we might even get some relationship in relation to the function ψ depending only on V, although it is in general not defined,

$$\frac{\mu \psi}{\vartheta} = \frac{1}{R'} NMA^2 \left(\frac{2\pi}{\tau'}\right)^2 \frac{1}{\vartheta} .$$

Now, let us note that in comparing the average livening force for the two physical situations: (a) when the molecular oscillations are not dying away and (b) contrariwise, the average livening force ought to be greater in the former case. This poses the clear mathematical requirement that the ratio $\mu \psi / \vartheta$ be less than 1 in general. If we would still retain the second approximation and denote l' the first approximation for the average pathway length, the we get for the relevant second approximation l'' the following relationship

$$l'' = l' \left(1 - \frac{1}{2} \frac{A_3}{A_1} \frac{\mu^2 \psi^2}{\vartheta^2}\right). \tag{12}$$

Here μ^2 is proportional to A_4; now we would like to investigate everything solely in explicit connection with the temperature dependence of the first approximation l', so that we get

$$l'' = l'\left(1 - D\vartheta^{\frac{10-12k}{3k}}\right).\tag{13}$$

As the ratio k should always be greater than 1, the temperature-dependent term in the above expression should be decreasing with the temperature increase. Moreover, as the parameter D introduced above should always be positive, hence the approximation l'' should more briskly increase with the temperature than the approximation l'.

Now let us consider the friction coefficient. To complete solving the latter problem we have to calculate the magnitude of $\bar{u}$, that is, the average molecular velocity. In general, the magnitude we are now looking for is expressed as the ratio 2A/τ, so that we use Eq. (2) to get

$$\bar{u} = \frac{A}{\sqrt{\sum\limits_{n=1}^{n=\infty} \frac{A_n \mu^n \psi^{n-1}}{\vartheta^n}}}.$$

If we take into account the explicit magnitude μ, we arrive at the relationship

$$\bar{u} = \frac{1}{\sqrt{\dfrac{Nm}{R'\vartheta}}}\frac{1}{\sqrt{1 + \sum\limits_{n=3}^{n=\infty} \dfrac{A_n}{A_1}\dfrac{\mu^{n-1}\psi^{n-1}}{\vartheta^{n-1}}}}.\tag{14}$$

Further, if we define some number z having value around 0.3 and d to stand for the density, then it is well known that the general expression for the friction coefficient would be cast as follows:

$$\rho = zd(\bar{l})\bar{u}.$$

In our case we make use of Eq. (7) defining the average pathway length to get

$$\bar{l} = 2A = \frac{B\vartheta^{\frac{5-3k}{6k}} V^{\frac{9k-5}{6k}}}{\sqrt{\dfrac{Nm}{R'}}\sqrt{1 + \sum\limits_{n=3}^{n=\infty} \dfrac{A_n}{A_1}\dfrac{\mu^{n-1}\psi^{n-1}}{\vartheta^{n-1}}}}.\tag{15}$$

Then, we express the desired friction coefficient as shown below

$$\rho = \frac{zB}{\dfrac{Nm}{R'}}\frac{B\vartheta^{\frac{5}{6k}} V^{\frac{3k-5}{6k}}}{1 + \sum\limits_{n=3}^{n=\infty} \dfrac{A_n}{A_1}\dfrac{\mu^{n-1}\psi^{n-1}}{\vartheta^{n-1}}}.\tag{16}$$

With this in mind, we consider the first and the second approximation for the above magnitude

$$\rho' = zBR'\vartheta^{\frac{5}{6k}}V^{\frac{3k-5}{6k}}, \tag{17}$$

$$\rho'' = \rho'\left(1 - \frac{A_3}{A_1}\frac{\mu^2\psi^2}{\vartheta^2}\right). \tag{18}$$

This immediately boils down to

$$\rho'' = \rho'\left(1 - 2D\vartheta^{\frac{10-12k}{3k}}\right).$$

Here D is dependent on V only and should always be positive. Howbeit, with the temperature increase the approximation ρ'' is growing anyway faster than the approximation ρ'.

Now, in considering the ratio k, we recognize that its value ought to lie between 5/3 for the monatomic gases and 1 for the molecular gases (the gas particles consist of any number of atoms). Then 5k/6 amounts to the values between 1/2 and 5/6, so that the dependence of the friction coefficient on the temperature lies between the functions $\vartheta^{\frac{1}{2}}$ and $\vartheta^{\frac{5}{6}}$. Remarkably, according to the relevant theory by Maxwell, which is in apparent accordance with the experiments, the latter functional dependence ought to be proportional to ϑ itself, so that Maxwell's ρ itself would be stronger dependent on temperature than our ρ'. This ought to be in consistent with the fact that our ρ'' is stronger temperature-dependent than our ρ'. Therefore, it is logical to realize that in our case the third approximation would be stronger temperature-dependent than the second one, and so on, so forth.

To sum up, although in my book based upon seemingly satisfactory deliberations I have had to conclude that it is difficult to expect the complete accordance with the experiment (l. c. I, p. 330 ff.), the above theoretical considerations could be viewed as a definite upgrade of the theory presented earlier. Specifically, the main point of the amendment deals with the temperature dependence. As concerns the relevant functions of pressure, density, molecular weight and atoms number, they could be viewed as much more consistent with experiments, just as I could state in my book (l. c. I, pp. 321–336).

(MS received for consideration on September 25, 1903)"

We would also like to present here the English translation of the referee's report to the above communication by Prof. Dr. Siegfried Valentiner, published in the *Beiblätter zu den Annalen der Physik*, volume 29, p. 640, very soon after the publication of the above-cited fragment, that is, in the year 1905.

"The author uses a truly general relationship between the collision-free average time period τ and the average pathway length A for a molecule experiencing two successive collisions with its surrounding, while this relationship has already been inferred and published in the second volume of the author's thermodynamics book. Here the author starts with establishing the relationship between the well-known expression for the entropy change known for the ideal gas and that including, inter alia, the explicit τ dependence based upon the general mechanical considerations. Using the latter relationship the author infers the new expressions for the average molecular path length and velocity. Finally, the author substitutes the latter both into the known formula for the internal friction coefficient for gases to get the new form of the temperature dependence of the friction coefficient. The author believes that his present result might be expected to have a better accordance with experiments as compared to his previous relevant results inferred under not so general conditions and published in his thermodynamics book. Moreover, the resulting temperature dependence ought to have the same good accordance with the experiments as the pressure, density, atom number and molecular weight dependencies inferred and published earlier by the author."

My immediate comment: The above referee's report has been delivered by Prof. Dr. Siegfried Valentiner (1876–1971). At that time he was a research assistant in physics at the Physikalisch-Technischen Reichsanstalt (nowadays, Bundesanstalt, the German metrological institute, PTB). In the year 1910 on he took over the position of the ordinary professor in physics, as well as the rector of the Bergakademie Clausthal in Lower Saxony (nowadays: Technical University of Clausthal).

We might note here that the report is fully unbiased. We are fully sharing the referee's opinion and note in addition what could not be immediately recognizable at the time of publication of both the work and its review.

M. B. Weinstein's work demonstrates skillfully and clearly that any kind of reasonable further development of the seminal ideas by Ludwig Boltzmann and Josiah Willard Gibbs was, is and remains to 100% possible—without sacrificing the basic physics to the *red herrings* of the *global indeterminism*, the *global domination of the pure chance*. We have started discussing this important methodological theme in Chapter 1 here and shall continue this discussion in Chapter 5 and the notes thereto.

To sum up, any attentive inspection of Prof. Dr. Weinstein's publication list demonstrates that he was a highly qualified theoretical physicist, with truly intensive involvement into the fundamental research on thermodynamics and kinetics. Further, it is immediately observable that such topics were definitely 'out of vogue' at his time, but in the present book we are dealing with the unique cases, where **not the researchers are choosing their favorite topics—but, instead, the topics are choosing their proper researchers.**

There were also some further critical reactions concerning Prof. Dr. Weinstein's work, which had caused a lively discussion, see the Refs. [34] and [35] in his publication list at the end of this chapter. The criticism had come from Paul S. Epstein.

Remarkably, Prof. Dr. Paul S. Epstein (1883–1966)—at the then time working at the University of Munich under the guidance of Prof. Dr. Arnold Sommerfeld (1868–1951)—was pleading the correctness of Max Planck's interpretation of the third basic law of thermodynamics, by rebutting the attempt by M. B. Weinstein to look for the plausible physical interpretation of this very important physical rule. We shall see later on (see Chapter 4 of the book at hand) that Dr. George Augustus Linhart could prove (somewhat later, namely, in 1922) that unattainability of entropy's zero isn't a novel separate basic law, but just a consequence of a much more basic dialectic principle—the dialectic law of unity and struggle of opposites. But, to our regret, Prof. Dr. Weinstein was no more among us at the time point of Linhart's seminal publications.

[Please see Supplementary Note 9, Section 3.5, Pages 377–378]

To sum up, the professional opinions as to the scientific research activity of M. B. Weinstein were by far not uniform. Still, his main victory ought to be the fact that Nils Engelbrektsson and Karl Franzén could manage to verify his inferences, as well as the compliance between the latter and the experimental results (as we already know from Chapter 2 of the book at hand).

Remarkably, from the above-presented referees' reports we could just recognize M. B. Weinstein's striving for the deepest possible consideration of the physical–chemical problems, without restricting his studies to purely mathematical reasoning. Noteworthy, in Chapter 2 here we could see that Dr. Engelbrektsson was in fact following the same direction as Max Bernhard—by submitting the physical–chemical facts to a thorough logical analysis before performing the pertinent mathematical exercises.

This is just where Weinstein's activity definitely abuts the field of the natural philosophy. It is the fact that ought to be of a separate interest for us. Moreover, he was holding regular lectures on natural philosophy at the Philosophy Faculty in the then Friedrich Wilhelm University (now known as the Humboldt University) of Berlin. This way, he was striving to provide us with the fullest possible analysis of his diverse results. Hence, it is this way that we ought to consider his natural philosophy course book (Ref. Number 17 in his publications list).

Before trying to analyze M. B. Weinstein's philosophic standpoint, we would first like to present here the English translation of the referee's report concerning his philosophy lecture book just mentioned above. The referee's report below has been published in the AMP, Series III, volume XIII, pp. 252–253, 1908.

> *"The tasks of the present author ought to be a depiction of the basic notions and laws of scientific research, helping the readership to reveal the actual interrelationships among the latter both, as well as trying to reveal the cognition sources of our mental facilities. Whereas the natural sciences make up the main topic of the book at hand, the last two chapters of the latter are devoted to the poetry and to the everyday life as well. The work suggests mostly author's own contriving—plus, where it comes to reviewing the works by other colleagues, the conclusions the present author could himself internally batten onto; along with this, the book's contents under study are truly comprehensive and versatile; the universe, as it is being reflected in our experiences, is represented in the book tellingly and interpretively, without roping the over-all mimesis to some preconceived doctrine. Such a way of representing the material is*

definitely preserving from biases, but on the other hand, it might also bedevil the readership's insight, especially since the reiterations and references to some later considerations are strikingly numerous in the book.

The first six lectures, that is, about one sixth of the whole book, deal with the conception and classification of the general basics, with the psyche and processes of its functioning, to which the external and internal worlds are traced. Then, after discussing the notions of the cognition and perception, the actual Space, Time and Causality are set against the basic notions of Timeliness, Spatiality and Substantiality. Lectures about the processes, phenomena, about the unity and maintenance of the world as a whole are setting up the inference of the natural-philosophic considerations. The three-dimensionality of the space, the theory of the matter, the energy principle, the entropy notion, as well as the principle of the least action are considered in a rather detailed manner by the author.

Due to the clear independence and plurality in choosing the topics to consider and discuss, it is difficult to assign this book to any particular philosophic system. Meanwhile, the book's basic philosophic idea ought to be the transcendental idealism by Immanuel Kant: The Space, Time, Substance and Causality are then just the formal products of our psychic activities to encompass our experiences; further, in trying to trace intuition and thinking forms to some common roots, the book's author ought to follow Schopenhauer. Still, the actual interrelationship between the Empirical and the Beyond Worlds remains unclear in the book at hand, which is anyway the case for the Kant's results as they are. The author is nonetheless clearly repudiating the materialistic–mechanistic perception of the world.

Definitely, the aim of the book to be pellucid ought to be achieved without any doubt; it is clear everywhere, what is meant, and where the deliberations are not enough to convey the exact representations to the inexpert readership, for example, as it comes to the entropy notion and the law of the least action, the author is noticeably tending to switch to discuss the theme in a rather superficial manner. Contrariwise, it is much less clear, whether the mission of the book's author to present something useful to the pundits as well could also be considered to be duly accomplished. Who is choosing the historical–critical way of the material's presentation, like, e.g., Ernst Mach in his studies on mechanics, ought to be capable of both instructing the laymen and engrossing the existing insights. But in comparison to the latter studies, Weinstein isn't acting truly inquiringly in his book; instead, by such discords as the essence of the psyche, space, matter he is solely listing the most generally accepted opinions and finally owning up more or less clearly to some particular one of them, without explaining all the pros and contras of his final choice. With this in mind, the credit of the sophrosyne should definitely belong to the book's author; but such an approach would not be fruitful from the standpoint of a principles researcher.

The referee would like to join the author of the book in reproaching those who is trying to write popular stories about scientific themes without any pertinent expertise and thus cheating their readership. As compared to the latter approach, the referee would like to point out the clarity and sterling quality of the book at hand. To put it figuratively, the author offers flowers, but **'to remain in the picture, the weeds are rampant in the orchards surrounding the Garden of the Scientific Research, like the attics are surrounding the Mainz Cathedral, and these are just the weeds that are being hewed by the unbidden persons'** *(Page 6). Wouldn't such a praise of the book be more graceful, when it's expressed by the readership?*

Berlin,

P. Johannesson"

The first impression of the above writing–a big serious philosopher has written a truly graceful referee's report about M. B. Weinstein's philosophy book. But, indeed, who was the guy named P. Johannesson from Berlin, for goodness sake?

[Please see Supplementary Note 10, Section 3.5, Pages 378–381]

To sum up, the professional activities by M. B. Weinstein were not only just visible to the collegium, but also actively estimated in a versatile and definitely constructive manner. In this way, he was an active, proactive—and thus—a definitely noticeable member of the scientific research society in a very interesting epoch—namely, at the time of the well-known revolution in physics. Hence, before we continue our present report, it might be of considerable interest to have a closer look at a couple of further collegial reactions, but now especially in regard to M. B. Weinstein's philosophical works.

3.2.4   Philosophical Work of M. B. Weinstein

Now we would like to try drawing Max Bernhard Weinstein's portrayal as a philosopher to analyze the conclusion we have drawn after reading the books [1, 3]. In fact, the authors [1] try insisting on Dr. Weinstein's intensive involvement into theology, by referring to just one of his philosophy books, namely *The View of World and Life Emerging from Religion, Philosophy and Natural Sciences* (see Ref. [19] in the publication list of Max Bernhard).

Meanwhile, the topic of primary interest and attraction for Max Bernhard ought to be just the natural philosophy. Indeed, this was just the topic of his lectures at Berliner University (see Ref. [16] in his publication list). Interestingly, this is just the book refereed by Paul Johannesson, as we have seen above.

An outstanding German physical chemist, natural philosopher, Prof. Dr. Friedrich Wilhelm Ostwald (1853–1932) had also reviewed the same book by Weinstein.

Below we present the English translation of F. W. Ostwald's standpoint:

Annalen der Naturphilosophie, Volume 7, pp. 190–191, **1908**: Max B. Weinstein (**1906**): *Die Philosophischen Grundlagen der Wissenschaften. Vorlesungen gehalten an der Universität Berlin.* Verlag von B. G. Teubner, Leipzig und Berlin. The Referee: Friedrich Wilhelm Ostwald:

> *"Again, we are dealing here with the unique case, when a scholar working in the field of the natural sciences, namely a representative of the theoretical or mathematical physics, rises his voice to present his general views on the 'interrelationship between the science and nature' to a broader readership circle. It has been repeatedly stressed how enjoyable ought to be such a publication trend, with the latter clearly identifying the widespread feeling that the philosophers employed at the universities are not using to offer their audience what the latter is expecting from them. Meanwhile, there is a danger that the recreational philosophers representing the field of natural sciences might insufficiently or even wrongly delineate many things, which are lucubrated enough in the very professional philosophy. But on the other hand, their impartiality, which is not marred by any of the countless traditional biases, ought to represent a definite boot, when trying to avoid the artificially arisen pseudo-problems that sometimes render certain investigations so fruitless in the professional philosophy.*

Howbeit, the present work may perhaps not be considered one of the excellent specimens in its genre. Indeed, the author expresses his hope to have communicated not an insignificant amount of particularities and moreover gained new standpoints; nonetheless, the rapporteur could not encounter such points during his probably non-exhaustive reading attempts, so that he is not capable of discussing them here. But since in this respect the author also refers to the in-depth study of his work, then his recommendation might definitely be borne out at this point."

To sum up, Prof. Dr. Ostwald has clearly stated here the professional standing of M. B. Weinstein, by pointing out that it ought to be of crucial importance, when workers in the natural-scientific fields dare to take their time to duly share their philosophic standpoints with the interested audience. Moreover, Prof. Ostwald stresses the positive effects of a clear-cut philosophic impartiality of the 'recreational philosophers' as compared to the professional ones. This is why, to our mind, the report of the above referee ought to be viewed as a generally positive one.

Along with this, the criticism by Prof. Dr. Ostwald is truly peculiar: On the one hand he was softly querying the 'absence of important particularities and novel standpoints' in the book, and on the other hand, he had definitely supported the book author's suggestion 'to go in for in-depth reading of his book'. To logically put both suggestions together, we might translate Prof. Dr. Ostwald's conclusion as follows: 'The author isn't communicating anything new as compared to my own publications, but an in-depth reading of his work might be recommended'. Hence, in summing up, the philosophical standpoint of M. B. Weinstein was in full accordance with the 'energetics', one of the powerful adepts of which was just Prof. Dr. Ostwald. He was actively struggling against what he was denoting as a 'scientific materialism'—in fact being nothing more and nothing less than just the revolutionary attempts to revise the whole physics:

Wilhelm Ostwald, professor of chemistry at Leipzig University (1895): *Die Überwindung des wissenschaftlichen Materialismus. Vortrag gehalten in der Dritten allgemeinen Sitzung der Versammlung der Gesellschaft Deutscher Naturforscher und Ärzte zu Lübeck am 20. September 1895*. Verlag von Veit & Comp., Leipzig, Germany.

We would like to present here an English translation of the introductory part of this lecture:

"*Overcoming the Scientific Materialism*

A lecture held during the Third General Meeting of the German Natural Scientists' and Physicians' Society in Lübeck on September 20, 1895

By Wilhelm Ostwald, a professor of chemistry at the University of Leipzig

At all times the complaint is being lodged that we are in acute shortage of mutual agreement as to the most important and fundamental problems of humanity. It is only nowadays that the complaint concerning one of the most important problems is almost hushed; though some contradictions still ought to remain, it is nonetheless possible to conclude that a relatively wide consensus could be achieved in relation to conveying the external appearances, just as it seems to be the case in our scientific century in comparison to any other time period. Indeed, being either a mathematician or a practical physician—virtually any scientifically thinking person would answer the question the actual world's "internal design" in such a way that all the things ought to be composed of moving atoms, and that these atoms as well as the forces among them, ought to be the ultimate realities behind all the individual phenomena. In hundredfold repetitions you can hear and read the statement that there is no other way to understand the driving forces in the physical world than solely to ascribe them to the "mechanics of the atoms;" and then the notions of the matter and motion appear as the ultimate terms on which the entire variety of the natural phenomena must be based. It is such a stance that might be viewed as the 'scientific materialism'.

*I would greatly appreciate expressing herewith my sincere persuasion that generally speaking the standpoint outlined above is in fact untenable; that such a mechanistic world view doesn't fulfill the purpose for which it has been formed; that it enters into contradiction with the undoubted, well-known and well-recognized truths. So, the conclusion to be drawn therefrom, can be undoubtedly formulated as follows: the scientifically untenable standpoints must be abandoned and replaced by other ones, namely, by the better ones, as soon as possible. And the natural question to be asked at this point, whether such a better standpoint ought to exist at all, I guess I might answer positively. What I would greatly appreciate to tell you here, my **highly esteemed assembly**, might be taken into two logical parts, a destructive and a constructive one. Anyway, destruction is always easier than building, and, with this in mind, the inadequacy of the persisting mechanistic viewpoint should be easier to prove, as compared* *to verifying the adequacy of the newer one, which I could describe as the energetic standpoint. But if I would immediately emphasize that this new conception had already gotten its opportunity to be successfully verified on the basis of unbiased contemplations and riding roughshod over difficulties of its practical evaluation in some particularly favorable areas of the experimental science, then this might bring, if not the immediate persuasion that what I'm speaking about here ought to be absolutely true, then at least some vague feeling that the material I'd like to present you now may perhaps deserve its careful consideration.*

Possibly, it wouldn't be superfluous if I could start with emphasizing from the outset that my sole concern for today is a purely scientific discussion. Basically, I would like to expressly refrain from drawing any possible conclusions based upon the result of our meeting in regard to other— e.g., ethical and religious—affairs. I'm not doing this because I disregard the importance of such conclusions, but because my results have been obtained regardless of such considerations, purely on the ground of the exact sciences. Moreover, for the 'processing of the soil' in question, the view ought to be true that anyone, who does put his hand on the plow, but is still looking back, is clearly not cut out to be working in this realm. It is in hurting anyone or for one's sake that the scientific researcher shouldn't be obliged to report what he has found, and we must always rely on the Force we are looking for—to the extent that we recognize: sometimes, it is throughout possible that temporarily—but never permanently—we might still be declined from the rightest path.

I cannot deny that my present endeavor brings me at odds with the viewpoints of those who have accomplished truly great achievements in science, and to whom we are all looking up in awe. Please don't interpret this as my arrogance, when in such an important matter I am still in a direct in contradiction with them. Possibly, you wouldn't consider it arrogance as well, when some sailor having the service in the pole basket, in shouting 'Surf's Ahead!' might change by the trajectory of the large ship, where he is only a tiny serving member. He has the duty to report what he sees, and he would oppose his obligation to act if he would skip reporting. In this sense, it is just my very duty, which I would like to fulfill today. No one of you is under obligation to change your academic trajectories right upon my sole exclamation 'Surf's Ahead!'; each of you might instead wish to scrutinize, whether I'm indeed on the verge of something realistic, or I'm just fooled into believing by a simulacrum. But since I'm sure that the particular nature of the scientific work I am busy with would allow me to instantly recognize certain phenomena even more clearly than they might appear to show up from other standpoints. Then, with this in mind, I have to consider it an injustice, if owing to some external reasons I would leave unsaid what I have in fact seen."

For a more detailed discussion about Prof. Ostwald's philosophical belief system, see the English translation of his monograph [12].

Moreover, a more recent detailed general study of the then academic situation in Germany could be found in the monograph [13].

Howbeit, the above Ostwald's citation makes it 100% clear, what exactly was the actual philosophical standpoint of M. B. Weinstein, to which natural-philosophic school he was actually belonging.

Finally, to clarify the above-mentioned [1] story around the philosophic monograph by M. B. Weinstein we would like to point out that he was indeed criticized for his later book (Ref. [19] in his publication list). The referee was rebutting M. B. Weinstein's standpoint as to the role of Pantheism. The criticism was published by an outstanding German theology specialist, Prof. Dr. Otto Kirn (1857–1911), a lecturer in theology at the University of Leipzig by the time of the publication discussed below.

Howbeit, the very fact of criticism isn't of any peculiar importance—this does happen in the academic world every day— and, to wit—many times a day. Anyway, the main point should be that the criticism is always remaining within the frame of scrupulous fairness.

The exact reference to the rebuttal in question is as follows: Theologische Literaturzeitung, Nr. 26, pp. 826–828, **1910:** Max B. Weinstein (1910): *Welt- und Lebensanschauungen. hervorgegangen aus Religion, Philosophie und Naturerkenntnis.* The Referee: Otto Kirn.

Indeed, at the end of his rebuttal Prof. Dr. Kirn clearly states that (we translate his statement into English as follows)

"...Weinstein appears to come out of the closest on being the admirer of Spinoza and Kant. But the latter fact does not stop him at pondering Ernst Haeckel with high encomia (page 449). Weinstein estimates his "Welträtsel" (page 446) as an "excellent book written with the highest degree of noblesse." If we might agree with such a conclusion by Weinstein,

then his own book would deserve the same estimate; meanwhile, the same degree of the philosophic dilettantism ought to be the main speaker in the both books. In fact, we do believe that our worldview is not determined by the natural-scientific research alone, but anyway by moral foundations and religious beliefs as well. Thus, who would indeed like to learn about the achievements of religion and philosophy in the mentioned field ought to look for a more skillful guidance than that represented by the writings discussed."

First of all, the above citation does clearly expel the work of M. B. Weinstein from the field of the theology—as well as even from the field of general philosophy—for he was in fact a definite specialist in the field of theoretical physics, being sincerely interested in the natural philosophy. This is anyway clearly reflected in his publication list.

The same might also be concluded in regard to the work by Prof. Dr. Haeckel mentioned in Prof. Kirn's rebuttal above.

[Please see Supplementary Note 11, Section 3.5, Pages 381–384]

3.3.5   Conclusion: The Einstein–Weinstein Controversy

What we might borrow from this entire consideration ought to be the tremendous intensity of struggle among the 'revolutionists' and 'counter-revolutionists' in the natural sciences at the end of XIX-th and beginning of XX-th centuries. All of the protagonists of the book at hand, including M. B. Weinstein, were clearly representing the 'counter-revolutionary' train of thoughts. Thermodynamics was one of the central fields involved.

As we well know, the 'revolutionists' could defeat the 'counter-revolutionists', or, in other words, the 'materialism' could vote out the 'idealism', if you wish. As a result, the standard handbooks in natural sciences are mostly describing the then 'revolutionary' standpoints, whereas the 'counter-revolutionary' ideas are practically never mentioned. The carriers of the latter ideas are mostly wearing their veils of oblivion. So that, solely the brightest, the most proactive 'counter-revolutionists' are remaining on the scene—i.e., the colleagues like Wilhelm Ostwald, Ernst Mach. To our sincere regret, conventionally there are not many more items in the latter list. Still, it is throughout important to retain a complete overview about the picture. Otherwise, there is a destructive trend to absolutize the achievements of certain colleagues. This but nips the progress of scientific thought in the bud.

With this in mind we come now to considering the Einstein–Weinstein controversy.

Now, reverting to Prof. Dr. Weinstein's activities, he was also serving as 'a sailor on the pole basket' to timely announce, 'Surf's Ahead', like Prof. Dr. Ostwald. In this connection we would like to discuss here Weinstein's very illustrative popular paper, of which we present the pertinent English translation below (*highlighting of the text below is ours*):

The Relativity Theory and the View of the World

M. B. Weinstein (1914): *Die Relativitätslehre und die Anschauung von der Welt.*

Himmel und Erde, v. 26, pp. 1–14.

"*Of all the most recent discoveries only the detection of the X-rays might be comparable in its importance with the arrival at the relativity principle. And these both discoveries do ideally complement each other, because, while the X-rays are now known to be capable of revealing to us the world of the tiniest particles, the relativity principle does lead us instead to the big picture of the universe, it creates views, it opens prospects that ought to be immediately related to the Universe as a whole. Moreover, these two discoveries do have another great resemblance to each other: Indeed, the careless usage of X-rays has definitely led to the most serious diseases, whereas thoughtless working with the relativity principle—to the strangest errors and distortions ever in science. Well, we ought to gradually learn how to apply the X-rays in the correct way, and the same should be true with respect to making use of the relativity principle.*

Meanwhile, in using the X-Rays we could still have rendered a much better and more obvious service to the humanity and the science as a whole—than by introducing the relativity principle. Yes, many serious researchers do recognize a virtual disaster for the healthy development of science through the introduction of this principle and demand a return to the earlier conventional views. And they are right to the extent that the importance of this principle has often been so immeasurably extended, without any prior consideration, that along with truly fatuous assertions a starkly obnoxious intolerance comes frequently up with respect to the dissidents, which to some extent even resembles the medieval religious coercion. It is only nowadays that a much more reasonable and cooler view of the relativity principle ought to be in the offing, and the author might even dare to impute some credit to him himself with respect to the more correct valuation of this principle eventuating in what follows.

Only eight years have passed since the discovery of the relativity principle; still, during this relatively short time the theories thereof have truly mushroomed from the ground. Fortunately, we have in effect only two focal points for such theories, namely the work of the discoverer of the principle, Einstein, and that of the great mathematician Minkowski. And everything else we might leave aside by all means, without any tremor.

Meanwhile, the theories of Einstein and Minkowski ought to fundamentally differ from each other. Specifically, although the very senses of Minkowski's and Einstein's relativity theories are quite dissimilar in comparison to each other, the former—you cannot even say essentially contains, but it is better to put it this way—can essentially contain the latter as its special case. Remarkably, the Einstein's theory tends nowadays more and more to address itself to the Minkowski's one. Most probably, only the Minkowski's theory would lastly remain. And the readership ought to learn from what should be discussed below, the state of affairs in the field involved looks like so tense—in fact.

To clarify everything to the extent required by the importance of the subject, I have to go in for some aspects in more detail. Moreover, such an introduction ought to be necessary in so far that, except for the professional physicists and mathematicians, probably only few readers do actually know, what the whole story is actually about and what the relativity principle ought to purport in effect.

With this in mind let us assume that something physically sensible (we denote the latter 'the guise') is propagating through some physical substance (we denote the latter 'the carrier')—like, for example, sounds propagate through the air, or light propagates through the Aether. Then, the propagation rate of the guise through the carrier should turn out to be different, depending on the fact whether the latter is at rest or it is motile. First of all, it's throughout clear that the movement of the guise ought to be influenced by the movement of the carrier. Therefore, the propagation rate of the former will be greater if the movement of the latter occurs in the direction of the propagation, or less, when the guise's and the carrier's movement directions oppose each other. Meanwhile, in any case, the absolute value of this propagation rate difference ought to be equal to the absolute value of the carrier's speed. This type of rate difference is called the kinematic or phoronomic one.

Further, the carrier's movement might also intrinsically change the properties of the carrier's substance itself, and this way the carrier's movement would also indirectly influence the guise's propagation rate, for the latter is definitely dependent on the properties of the carrier's substance, in which the propagation takes place. Hence, this ought to be the second type of the propagation rate change, which we would call the material or physical change. The direction and the magnitude of such a change cannot be so easily predicted, as in the purely kinematic case; in fact, the determination of the both ought to require non-trivially difficult theoretical and experimental studies, carrying out which doesn't basically guarantee any secure result. And it is exactly here that the one of the remarkable achievements owing to the relativity principle is much in evidence—because the mentioned additional laborious research efforts might be made dispensable by simply applying the principle involved.

Further, let us assume that an external observer monitors the propagation of the guise through the carrier. Then, it will be important, whether the observer is at rest or moving himself. If both the carrier and the observer are at rest, we might view the rate of the guise's propagation with respect to the observer as the absolute rate. Instead, if the observer is moving, while the carrier rests, then the former might either approach the guise or recede from it. Then, the relationship between the observer and the guise ought to be the same as if the observer is at rest, while the guise would approach him or recede from him. To sum up, in such a case there ought to be solely a kinematic change in the propagation rate.

If now both the carrier and the observer are moving quite the same way, then any kinematic change in the propagation rate is not available, because the observer and the guise are always keeping their initial disposition with respect to each other, but owing to the carrier's movement a material/physical change in the guise's propagation rate might probably arise. But if the observer and the carrier are moving in various ways, then the material/physical change mentioned would also be accompanied by the kinematic/phoronomic one, which ought to arise owing to the difference in the observer's and carrier's movements. Finally, if the observer were at rest, while the carrier is moving, then the both kinds of rate change would be possible, like in the latter case. To sum up in general, there are two types of the rate changes; the one kind is the kinematic/phoronomic, to be caused by the difference in the movements of the carrier with respect to the observer, and the second kind, the material/physical one, owing to the movement of the carrier itself. The former is taking place with respect to the observer, while the second one—in relation to the carrier in its given state, e.g., the hibernation of the carrier. And the two relativities in question are quite different from each other; they do have not a scintilla of anything in common to each other.

Let us now consider how the stars are spreading their light as an example. The Earth is experiencing a threefold movement, namely, the daily one around its axis, the annual one around the sun—and the one together with the sun through space through the whole universe, which in effect isn't clearly known to us. All these movements might readily be described as a movement using the known laws of mechanics. The same is more or less true in regard to the movements of all the stars surrounding us in the Universe. If now a celestial body would be giving some light, then the propagation speed of that light ought to be changed in itself materially/physically. And it is one of the basic assumptions of—we must here underline: the older/previous— Einstein's relativity theory, namely that once the light has come into the open space, into the 'Free Aether', so to speak (physically, into the vacuum), then this material/physical change ought to disappear.

To sum up, it should be truly no matter, which particular bodies emanate light, whatever movements are experienced by those bodies as a whole and/or within their own substance, the propagation speed of the resulting light should still be the same everywhere and in all the possible directions in the free space—hence, it is with this in mind that one might consider the latter speed a kind of universal constant.

Now, let the light in question reach the air layer surface of the Earth and be captured there by a diopter or a photographic plate, then we need to be concerned only with the kinematic change of the light propagation speed, stemming from the respective movements of the emanating celestial body, and of the Earth itself. Then the one part of the propagation speed change, resulting from the increase or decrease of the distance between Earth and the emanating star, would just lead to the Doppler's effect, while the other part of this change, namely the transverse movement between the Earth and the emanating star—to the aberration. These both effects are well known to the astronomers and physicists.

In its following course the light beam penetrates into the Earth's atmosphere, then into a telescope, and finally into the observer's eye. In addition to the kinematic change the Doppler effect and the aberration immediately arise–plus the material/physical changes owing to the air, telescope and eye, which are totally different from the free ether (that is, physical vacuum), and hence they ought to slow down the light propagation. Therefore, the light's "refraction quotient" would as a result deviate from its basic value of 1, also due to the additional contribution from the Earth's motion. The resulting new phenomenon is the Fresnel–Fizeau's dragging effect, whereas that was Fresnel, who has discovered its law. Remarkably, this ought to be true not only for the light of the stars, but also for each kind of light in general, that is, even for that coming from the conventional terrestrial sources.

And now let us take two light tractions originating from two luminous bodies upon Earth, and force them by mirrors, lenses, etc. to propagate through the same room and in the same direction. Then the two light tractions would interfere in this area and produce regular strips of varying brightness. Since the Earth is moving, the light tractions would have to experience the both types of the light's propagation rate changes. Therefore, as the tractions involved come from different sources, that is, from different directions, which are, moreover, different from the direction of the Earth's motion, then the expected changes should anyway be different for each of the both light tractions in question.

Bearing all this in mind, it is clear that the resulting interference phenomenon must anyway be different from the form it would exhibit, if the Earth might be at rest. And again, it has to offer quite another picture, if one turns the two light sources in such a way that their rays propagate against the direction of the earth movement. And the latter story is just the description of **the Michelson's experiment**, which has become so important—**and practically calamitous—for the scientific research as a whole**.

This experiment, as it had been performed by Michelson and copied almost slavishly by others, has had in fact the startling result that what was foreseen, did not occur; that is, the

interference phenomenon remained in fact the same, whatever the directions you might wish to turn the two beam tractions in question to.

Hence, the Earth's motion has not been exhibited this way. Gradually it has been realized that the very experiment, as Michelson has performed it, ought to be difficult at all to interpret. Most recently, I have demonstrated in my book ("The Physics of Moving Matter and the Relativity Theory"—M. B. Weinstein (1913): "Die Physik der bewegten Materie und die Relativitätstheorie; Verlag von Johann Ambrosius Barth, Leipzig, Germany") that, as a result, Michelson had not seen what he probably wanted to see, owing to the experimental set-up he chose. Instead, he was dealing with a completely different interference phenomenon, which is even theoretically incapable of helping detect the movement of the Earth motion—or the latter was of such a little significance that the experimental set-up chosen was not sufficient to recognize it. Already the experiment itself has led to the most remarkable views, and the one of them is the relativity principle. In fact, the most obvious assumption would be that the substance where the light propagates, the **aether**, not only moves together with the Earth but also does this just the same way as the latter. Then the kinematic change of the propagation rate would no longer be applicable and what would remain thereafter would solely be the material rate change, which according to the Fresnel's law is indeed very low in air.

The great physicists Maxwell and Heinrich Hertz have assumed that the Aether contained in the bodies is firmly connected with the latter and ought to move together with the bodies—just the same way the latter move. And this is just the cornerstone of their theories of the electromagnetic phenomena, which, as we expect, the light phenomena belong to as well. Fresnel has also fully accepted the idea that the **free aether** rests under all circumstances, and the aether contained in the physical bodies does move indeed together with them, but only with a definitely lower velocity, as compared to the velocity of the body itself, namely the $\left(\dfrac{n^2-1}{n^2}\right)$th part of this speed, where **n** stands for the refractive index of the relevant body (so that's just the essence of the above-mentioned Fresnel's law). And this ratio for the air ought to be so much less than 1 that it amounts only to about

(1/l0000)-th of the Earth motion's velocity, so that the Aether could safely be regarded as dormant.

   With all this in mind, it is throughout clear that any movements of the aether could not explain the result of Michelson's experiment. The recent electron theory of H. A. Lorentz even deprives the aether of any kind of movement, and attributes the absolute dormancy to it. At Fresnel's time this immobility of aether has been thought to be required for explaining the aberration. Then Stokes has worked out a theory that ought to deduce the aberration out of the movements of the aether. And in this theory H. A. Lorentz has again revealed contradictions. But I myself was able to show that a different, but nonetheless complete treatment might eliminate these contradictions. Hence the very phenomenon of aberration is in no way helpful in trying to decide, whether the aether moves or not. But the electron theory has nowadays gained such a power that just for its sake the aether is currently considered immobile. So, this is why the result of Michelson's experiment cannot be explained in terms of aether's mobility.

   Bearing all this in mind, H. A. Lorentz has arrived at a different assumption, namely, that all the bodies, when they move, could theoretically be flattened along the line of their movement, so that they might be reduced along this line from the both sides as a result. This shortening for the Earth due to its movement through the space (around the Sun and together with the Sun) would amount to approx. 6 cm, and, for all other bodies, up to approx. (1/200 000 000)-th part of their size in the direction of their movement together with the Earth. Now, if the carrier of the light propagation, i.e., the aether, rests, then the influence of the Earth's motion would first of all be exhibited in the fact that the way of the corresponding light traits from the one point of the Earth to the other would now be extended, now shortened, depending on the disposition of this way in regard to the direction of the propagation. And finally, H. A. Lorentz, in line with his assumption just mentioned, was able to show that, since the light propagation way through the Michelson's equipment must be shortened, *a compensation* should always take place between the changes in the light propagation path of the beam due to the over-all Earth's motion on the one hand, and those owing to the virtual 'shortening of the apparatus trajectory' on the other hand. To sum up, the latter ought to be just a purely physical explanation, based upon the realistic conditions.

Then came Einstein and posed us the following questions: Whether the aether would move or not move—should this anyway be clarified in our experiments? Could the guises, which propagate through the aether, be moving in such a way that the pertinent laws wouldn't at all be dependent on the movement or non-movement of the carrier's body? Just to conclusively answer these posers, he had invented the principle of relativity, which in its original form would in general amount to as follows:

The form of the laws, according to which the natural phenomena pass off, ought to be exactly the same for the observer moving together with the carrier body, as well as for the observer resting with respect to the carrier body, the only difference between the both observers would be that the static observer would monitor some particular aspects of the guise in much different way than the moving one, while they both might also well be observing other aspects of the guise under study in one and the same way.

The former group of observables include the space and time measurements performed for the guise, and therefore such variables as volume, temperature, motion moments (impulses) forces, etc., whereas to the latter group belong such quantities as pressure, entropy, etc. Einstein has also immediately determined, how would all those variables change, when we transfer our standpoint from that of the moving observer to that of the one at rest, and in particular the formulas he has derived for the space and time measurements, have become famous, and play the leading role in the relativity theory.

The consequences of this doctrine are now very strange indeed. All the accessible mass media expatiate on Einstein's doctrine by delivering copious reports usually without any correct insight into the nature of the latter. For example, there were statements like as follows: The observer at rest ought to perceive a moving body (e.g., a human) in such a way that the latter would be more and more expanding thwartwise in regard to the direction of the actual motion, irrespectively of whether the moving body approaches the observer or diverges from him—further, the clock moving together with the body would be more and more slowing down, the temperature of the moving body ought to rise more and more—and so on, so forth.

Originally, the Einstein's doctrine was a purely mathematical production, since the laws of the natural phenomena might and *should be expressed mathematically. And initially the attention was only concentrated on these laws, on their mathematical expression, but never on what the phenomena themselves offer to our senses. Sure, the entire first version of the Einstein's theory consists solely in fixing how the radiation lengths should be measured between two points: if the one of these points at rest or moves differently from the other one. Still, this theory is in fact tailored solely to the light propagation, albeit it was immediately extended to all the phenomena in general. Hence, what is basically possible to extract from it ought to be nothing more than just the mathematical theorems about the light propagation, just like from the body's geometry follows nothing about the body itself, except for the information about the dimensions and location of the latter.*

But right after the enunciation and acceptance of this principle, those colleagues were quite right, who thought that it represents a purely mathematical principle with only a shadowy meaning, and it might only intercede into the reality area, as soon as it were interpreted physically. Einstein himself had triggered the expected physical interpretation by considering the size of the moving body and the swing of the moving clock; the both should indeed be exhibited other way to the observers at rest, as compared to the situations with the moving ones. And all other guises ought to exhibit quite the similar behavior. For example, this could in fact be a body that is as cold as the moon, but when it does move, to the stationary observer it ought to appear warmer and warmer all along; moreover, the greater its speed, the higher should be its temperature increase for the eyes of the observers at rest, at some point it would become red-hot, blazing like the Sun, it might then appear to be hot beyond any measure. Hence, the stars ought to send us their shine not because they are actually glowing, but because they are moving very quickly. But, regrettably, all the spectacle wouldn't be available for the observers moving together with the objects under study. This way it would be throughout possible for us to put a radiant sky into practice—solely by bringing together all the movements of actually very cold empyrean bodies. Sure, the speeds we know to be possible under the heaven's tent should definitely not be enough to actuate the scenery—for, according to Einstein's formulas, they ought to approach the speed of light propagation, i.e., be around 300000 km per second, while the actual maximum speed of a star known to us amounts to only ca. 600–700 km per second. If we had much higher speeds, then tiny specks of dust would appear to us at rest upon Earth as huge glowing sunny behemoths, finally covering the whole sky with huge shining disks, whereas the attraction forces exerted by the latter would grow into infinity—and so on, so forth. And the whole story ought to be valid only for the observers at rest, but not a little bit, truly none of the masquerade would emerge for those at rest on the moving carriers, i.e., for those co-moving. Sure, everything is due to the principle of relativity.

*But even these conclusions could not yet satisfy the **Relativiker** (from here on—this term ought to serve as an identifier for the true adepts of the Einstein's relativity theory). Moreover, the view of the world was moved to the observer in case the latter is capable of conceiving everything as a true veritableness. This way a breakneck jump was made from the observers to the things themselves. These are the things themselves that are exhibited one way to one observer and show up quite differently to another one, if the things are moving differently to each of the both observers.*

Here the reader might readily recall the Kant's transcendental idealism. According to the latter, we tend to conceive the observables only in accordance with our basic conceptions. Hence, the things themselves ought to have no spatiality and no temporality, and then we ought to observe the world around us in space and time, because spatiality and temporality are unique parts of our innate basic conceptions. For us, there is no other way to look at things surrounding us. Hence, for us the actual world remains hidden in itself, because we are not capable of grasping our surroundings as they actually are, owing to the uttering specificity of our basic conceptions. Now, the reader should also agree that an actual turbulent world surrounding us is in fact hidden from us due to its intrinsic movement, for it exhibits itself to us only in accordance with its motions, its dynamics, and there ought to be absolutely no valid way to somehow unravel it. Hence, the world is possessed of the countless numbers of faces—and at this point I would even have to say—the world makes here the one of these faces, and there—the other one. So that, just a very few of the world's properties/features could be uniquely recognized by all the observers to be one and the same observable, whereas the countless number of other observables ought to be always changing the appearance from observer to observer, whatever observed object it could be.

Oh, indeed, what a kind of strange world, which we are living in! And what properties, what observables might we then actually put our reliance upon—even if we take such seemingly sturdy properties as pressure and entropy, of which the latter is meanwhile the most dangerous discovery of science as well, for it ensures so nicely the **ultimate end of the world** *(of all the processes available in the world)!*

Recently, Einstein has abandoned his first relativity theory of the kind outlined above in favor of an even more general theory of the same kind. The latter, so far as it has become known at present, just falls into line with Minkowski's theory, so that it does not need to be dealt with separately. But the reason for why Einstein has been forced to seek a new theory is of a separate extreme interest, as well as because of erroneous assertions against Minkowski's theory itself.

The formulas describing the conversion of various observable magnitudes, when coming from one observer to another one, contain the propagation speed of light in the free space. Meanwhile, Einstein has always considered the latter to be unchangeable, and those formulas would indeed have hardly any value, if it were changeable. Now, on the other hand, it is a consequence of Einstein's doctrine, and also of the modern electron theory, that the inertia of the body—or, we might also say: its mass—is determined by its energy content, if the latter magnitude is divided by the squared magnitude of the light's propagation speed in free space. Hence, any change of the energy content, for example by radiation, or heat gain, or by chemical processes, or through movement, or ... so on, so forth ... ought to accordingly promote respective changes in the inertia. And this should really represent something very strange, for normally we are used to view the inertia mass of the body as an absolutely unchangeable parameter. But one must already get used to such assertions of the modern science, which tends to be qualifying virtually everything. Apart from the inertia (mass) we ought to characterize a body with its weight as well, which describes the effect of the gravity on the body. The latter is now well known to be exactly proportional to the former, as highly accurate experiments of the Hungarian physicist Eötvös could demonstrate. Einstein has called this proportionality between the inertia of a body and its weight an equivalence law. The latter implies that the weight of a body, i.e., the effect of the general Newtonian attraction on the latter, is throughout dependent on its energy content that varies accordingly, e.g., decreases, when the energy content decreases.

Further, it is clear that the attraction force between one body, such as the Earth, and another one ought to decrease with increasing the distance between the both bodies. With this in mind, let us remove the second body from the first one, so that the weight of the former ought to decrease in relation to this process. But, if nothing else happens to the body in question, then its energy content has apparently undergone no change, and hence the body should have kept its weight. If nonetheless the latter result is still not the case, then we have to look for the change in the energy content of the denominator, by which the body's energy content is to be divided, in order to finally preserve the resulting value of the body's inertia, i.e., to get the same body's weight value at the end of calculation. Fine! But, as noted above, this denominator is nothing more than the squared propagation speed of light in the free space. Therefore, the equivalence law requires, unlike the first theory of relativity, this speed to be variable, so that the gravity and the speed of light should somehow interact with each other.

Time and again we deal with an amazing achievement of modern physics, the sense of which I have tried to explain to the readership, but apparently in somewhat unscientific way, as we have seen above, since the precise considerations are of rather difficult kind. Surely, Einstein's earlier relativity theory couldn't stand the pace with this achievement. Nonetheless, the former hadn't been completely abandoned, although its actual value was noticeably diminished.

And now let us turn to the other relativity theory—namely, to that by Minkowski. In the science it is already long known that the mathematical expressions of its laws might often contain certain vagueness. One always uses the example of the Galileian laws of motion in this respect, and refers to them as nothing less than just the Galileian relativity principle, for these laws allow to determine the actual speed of a movement only up to some arbitrary constant speed; hence, it is always possible to change any velocity by some positive or negative speed supplement having a constant magnitude and a constant direction, without introducing herewith any alteration to the mathematical form of the Galilei dynamics laws. Should we have calculated a velocity for some motile body using the Galileian equations, we are still not sure whether this body has another uniform speed except for the already calculated variable. For example, this is absolutely true for the speed of the Earth. Meanwhile, this entire story is surely applicable for the Galilean theory of motion only. There is but still another theory of motion, namely that by Lagrange; and the speed is not of such importance for the latter, compared to the momentum of motion (impulse), as the current author could have demonstrated. And it is just this circumstance that should be of crucial importance for the proper interpretation of the relativity principle. So that, bearing all this in mind, let us now directly approach the very Minkowski's theory.

As our space has three dimensions, through which all other directions can be determined, hence, any law of motion might be decomposed into three directions, or, as we use to say, into the three coordinate axes. There are now variables that can also be decomposed into the three directions, and therefore fully defined for any other direction, as soon as their values are given in these three directions. Such variables, namely: lengths, velocities, accelerations, forces, we call vectors. The calculation for all these variables is throughout the same; and if we are changing the above-mentioned coordinate axes, e.g., by somehow rotating them around their starting point, then everything ought to change in exactly the same way. It is then said that all the corresponding quantities (vectors) are covariant.

In Galilei's theory of motion, the laws would now express that in every direction in space the acceleration is equal to the accelerating force. And since accelerations and accelerating forces are thus covariant, it follows that the Galilei's equations of motion would not change when you rotate your coordinate system arbitrarily. Or, to put this in other words: from the Galilei's equations it is never apparent, to which exactly coordinate system they refer, they do have exactly the same shape for all the possible coordinate systems. And it is just here that one of the main points of the Minkowski's theory resides.

Remarkably, besides the three dimensions in space, we do have another one, namely the dimension in time. Since truly long ago we have become accustomed to treat the dimension in time just like each of the three space dimensions; we consider the time as it is to be without any influence on the world, like the space as it is. And Minkowski goes one step further to consider the space and time together as a world's area of four dimensions. Hence the world's coordinate system ought to be possessed of four axes, the three space axes plus the time axis, all of which are perpendicular (or, mathematically speaking, orthogonal) to each other. Therefore, every realistic movement ought to be decomposable in the three parts regarding the three spatial axes and in an additional part with respect to the time axis. And it is just this latter part of the velocity that ought to be the most peculiar theoretical feature for us. Indeed, Minkowski has in fact never explicitly specified what ought to be the actual physical meaning of this part, but nonetheless it clearly follows from his theory that he augments the three spatial speeds to the extent that the resulting over-all velocity in the whole space-time area ought to be smoothly ensuing and under all the possible circumstances ought to be equal to the speed of light in the free space. Moreover, such an over-all velocity would never get, or we can even note—might never get—any other value different from the latter one, whereas the conventional velocity in the three-dimensional space itself may change at a whim. To sum up, in the over-all space-time area only one type of movements ought to be possible: namely, the uniform movement with the speed of light in the free space, whereas all other dynamics forms do never exist and even might never exist—still, here I would just have to add: the latter conclusion might only be possible under the certain conditions of the time computation. And therefore the entire story told has the following mathematical background:

The mathematics might operate with real and imaginary numbers.

The former group of numbers is such that it is always possible, either exactly or approximately, to take the square root of any one of them. E.g., if we take 1 then its square root is equal to 1 or if we take 2 then the square root would be approximately equal to 1.41 and so on, so forth.

Meanwhile, for the latter group of numbers, for example, for such of them as -1 or -2 taking the square root ought to be sheer impossible; indeed, there is no number that can represent $\sqrt{-1}$ or $\sqrt{-2}$. For such numbers any conventional representation is impossible at all, the amounts like $\sqrt{-1}$ or $\sqrt{-2}$, etc. ought to be inconceivable. In fact we perform our conventional calculations with the real numbers, but if we would do the same using the imaginary numbers, then the real numbers would be inconceivable for us.

With all this in mind, Minkowski describes all the space dimensions using the conventional real numbers, while the time dimension is expressed in his theory by the imaginary numbers. He has to go this way in order to derive the Einstein's theory as a special case of his theory.

There is also one more important aspect here, which is related to the following:

It is conventional that any motile body would have some peculiar trajectory in the three-dimensional space, but now the Minkowski's theory suggests ascribing some trajectory in the artificial space–time area to it and denotes this as the "world line." Now, if we would always like to set the time axis in the four-dimensional space in such a way that it touches this world line, then the both would be running in parallel to each other, then the dimension of the time axis could be referred to as the proper time. It is just this proper time axis and just this specific imaginary dimension that the very constancy of the body's over-all space–time velocity refers to. Hence, that the latter ought to be a kind of universal world constant; that it also equals the speed of light in the free space, should just be the particular results, which are actually not related to the Minkowski's theory. The whole world of the bodies, whether they are actually at rest or somehow moving, would proceed in the four-dimensional space-time are with the constant velocity (the speed of light in the free space) and in the constant space-time direction. To sum up, in this artificial space–time area all the differences between rest and the motion, as well as among all the possible kinds of motion ought to disappear. And the latter statement ought to be indeed a deucedly peculiar conception of the universe, which, according to Einstein, should follow from Minkowski's theory.

Meanwhile, the Minkowski's relativity principle just lutes as follows and conveys the following important idea: All the natural laws ought to keep their mathematical expression unchanged, if we would set up a different space–time coordinate system in place of some initial space–time coordinate system. The mathematical expressions of the natural laws represent sequences of different pertinent members, which are added to or subtracted from each other. And hence the members of these mathematical expressions ought to vary in a covariant manner, if we rotate any coordinate axis system in the space–time area. The Minkowski's relativity principle is hereinafter nothing more than just the mathematical covariance principle, which ought to be in entire accordance with the long-known and aforementioned covariance principle of the Galilei's mechanics in the three-dimensional space, but now extended to the four-dimensional space–time region.

Bearing in mind the entire story we have just told above, the following reasoning would appear to be straightforwardly in line with deriving the actual relativity theory: If a body is at rest in the space, then the time is drifting past around the latter, just like a river's stream by a stationary boat standing near the berth at a river's coast. We might equally interpret the same story in such a way that the time is at rest, but the body is moving uniformly in the time. Resting in the space is then synonymous to the uniform motion in parallel to the time. And vice versa: the uniform motion in parallel to the time is equivalent to resting in the space.

If now the body in question is moving in the space, then we might simultaneously rotate the four-dimensional space–time coordinate system to follow the body's movement in such a way that the time axis would be always in parallel to the body's world line. As a result the body might be considered being at rest at any point of its world line with respect to the coordinate system chosen, so that the body's time is nothing more than just the proper time (see the definition of this notion above). With this in mind, we should be eligible to apply the entire set of the pertinent natural laws to study the body at rest. And, as soon as the mathematical form of all the natural laws ought to be independent of the coordinate system chosen, according to the Minkowski's covariance principle, then we are eligible to track all the physics of the motile matter to that of the matter at rest. To sum up, it is just opening the latter possibility that ought to be the actual main aim of the whole relativity theory.

Minkowski definitely belongs to the row of prominent mathematicians and outstanding thinkers. His life was lasting only 47 years, but even during this relatively short time he was nonetheless able to achieve a lot. And his untimely departure is a definite bereavement for the science as a whole, and everybody interested in the intellectual activity is mourning him, even without having any personal contact to him. In his work comprising a bit more than 50 pages he has summarized his principle in regard to mechanics and electrodynamics. This paper of him contains an overwhelming set of important investigations, which are in fact very difficult to follow and realize, so that we ought to spend several months to study all of them in detail, even if all of us would be skillful experts in mathematics.

The Minkowski's theory might rather be compared to a sturdy, self-consistent building, where it is never possible to get rid of any brick. The contradictions seemingly present in the Minkowski's theory, which are in fact connected with the field of the Einstein's theory, could solely be explained by relentless willing to introduce some seemingly justifiable squalors. But any contradiction would immediately disappear, if we consequently apply the Minkowski's theory just in its original form worked out by an ingenious mathematician. Indeed, to his original theory belong: The world line, the velocity of the universe in the space–time area, which ought to be the universal constant indeed, as could be concluded from the experimental results other than those with the velocity of light propagation in the space. The velocity of the universe in the space–time area is truly independent of any kind of 'external circumstances, e.g., gravitation, unlike the Einstein's "world constant"—yet again, after all.

We have already mentioned that the Einstein's theory is just a particular case of that by Minkowski. But now it would be instructive to compare the both in more detail. If fact, Einstein has employed the formulas by Minkowski, but Einstein uses his own interpretation of the resulting formulas, which has nothing to do with the original theory by Minkowski.

For example, Minkowski speaks everywhere of coordinate system rotations, whereas Einstein mentions spatial coordinate systems locomotion. Specifically, what should never be exhibited in any observable event, according to the true relativity principle, does appear in the Einstein's considerations as a trivial locomotion in the space, whereas Minkowski speaks in this case of uniform rotations in the imaginary space–time area. To ensure theoretical equality in this case, one would apparently have to imply that the latter rotations would be exhibited as locomotion in the conventional space. In my book "The Basic Natural Laws" I have analyzed this situation in more detail to show how this might be possible.

To sum up, this "rotation-locomotion inconsistency" is just the second point, apart from the inconsistency in the "world constant" interpretation already mentioned above, where Minkowski's theory is essentially different from that by Einstein. And the third very important point ought to be the throughout simplicity and clarity exhibited by the conclusions from the Minkowski's Theory. Indeed, what we are normally accustomed to use as the conceptions of space and time— the same basic notions are duly transferable into the space–time area. This is why for all the observers the length in the conventional 3D space is in the 4D space (space–time) the same 3D length, but just in combination with the relevant "proper time" interval, etc.

Hence, we would just need to get accustomed to using the combined 4D space–time area instead of the conventional 3D space plus a separate 1D time—and to transfer all the mathematical expressions for the natural observables and laws, which are valid in the latter conventional area, to the former one. This is how we might approach a unique 4D world with the time being just a separate orthogonal axis in the coordinate system, so that all the observables would turn into the 4D objects as well, where the time is just the fourth dimension of the latter. Still, despite its apparent four-dimensionality, the Minkowski's Universe ought to be quite different from the 4D worlds of our earlier spiritualists, for the fourth dimension is just the time all of us well know, instead of some mysterious novel dimensions.

Withal, there are some aspects of Minkowski's theory, which are truly difficult to realize. Specifically, the clear external complication ought to be that it is quite impossible for us to visualize the four axes perpendicular/orthogonal among each other, to visualize the angles between a "spaceline" and a "timeline." It is also not obvious to calculate the time in some imaginary units in combination with the conventional calculation of distances in the real units. Indeed, as in the Minkowski's formulae the time is always multiplied with the World Constant, which has the velocity dimensions, then the resulting product would have the length dimensions. Hence, the time calculation would then, at the first glance, perfectly correspond to the length calculation—but, meanwhile, the "length of the time interval" should thus be imaginary, and it is then not quite clear, how to combine the real length calculations in the conventional space with this imaginary length calculation in the time direction.

Well, I have already mentioned that, in effect, all the above problems arise, when one is stubbornly trying to incorporate the Einstein's theory into the original Minkowski's framework. And if we stop trying to couple these both theories, then, according to the original prescriptions by Minkowski all the calculations are the same in all the pertinent dimensions. There is but still a true intrinsic difficulty in connection with the Minkowski's theory, which I wouldn't like to discuss here in detail—but solely a concise problems listing still ought to be in order.

Interestingly, it is noteworthy that uniting the conventional time and space ought to produce serious epistemic problems. The both are in fact the reification of the basic notions for the spaciousness and temporality, of which the former stems from feeling our externally visible body, whereas the latter arises from realizing our internal spiritual activity in its permanent 'runway' (for more detailed discussion about this interesting and important theme, please see my book "Die philosophischen Grundlagen der Wissenschaften" ("The Philosophic Basics of the Sciences"). And if we intend to interpret such truly separate reifications as a unity, as a communion de facto—and not just formally, but truly in substance—then one would have to prove the existence of something in us, which might ultimately unite the both of our spiritual activities—the permanent physical feeling, producing the spaciousness—the permanent 'runway' awareness, producing the temporality—into something truly indecomposable. This produces a compulsory requirement that the both spiritual activities just mentioned ought to somehow turn out to be a unique activity, like our physical feeling taken alone produces a reification of the indecomposable three-dimensional space.

*Are we possessed of such a spiritual activity to unite the both mentioned ones? If so, then why should we be notified about it at all? And—why should we **be notified** about this **right now** at all? Indeed, if something like space–time–area would already be substantially existent, and therefore intrinsically present in us as a reification of some relevant basic notion, then there would be no need to issue some additional statements about the theme, for there would be no other way than to perceive the space–time as a unique unity. Likewise, we actually perceive our visible surrounding in no other way than as a unique three-dimensional space, but not as a set of three one-dimensional ones, or a combination of two one- and one two-dimensional spaces, for our physical feeling ought to be coherent.*

The above exposition would actually go round for a short inkling, for actual treatise ought to occupy a thick book volume—indeed, we would then actually have to dive into the depth of our spiritual activity. Still, such difficulties shouldn't remain absolutely unmentioned, for so many colleagues just flout such topics in a truly sniffy manner.

There remains but a single last, but truly not the least notice. Indeed, the theories appear to be mathematical expositions, where the main point ought to be the proper expression of some rule/law.

This point is frequently remaining disregarded. We have to clearly distinguish the very process/phenomenon from the actual lawfulness of these processes/phenomena. Indeed, the relativity principle is solely conveying the idea that if the realistic process/phenomenon shouldn't be changed by some additional dynamics, the lawfulness of the process under study ought to remain unchanged as well. In this respect, we know processes in the systems, which might somehow be afterwards brought back into their initial states. The lawfulness of such events dictates in fact that after the process perfectly completes its full working cycle the entropy of the system remains unchanged. This ought to be true even in the event of some additional dynamics, but the true poser is: Would the resulting system still be remaining reversible under such conditions?

This way we would like to underline here that the relativity principle is possessed of some physical sense, except for its purely mathematical expression. And this is connected not only with the conservation of the mathematical form of the processes under study, but also with the conservation of underlying physical phenomena/ processes themselves. Is the latter result indeed the case?

To what extent should the relativity principle be valid? Even if we wish to ascribe the widest possible applicability to it, we have to admit that it is not totally valid in general. Well, the most credible form of the latter principle is surely due to Minkowski, but even this perfect formulation might not be extended to the Universe as a whole, as far as we are capable of estimating the current factual situation. Indeed, this principle is valid for the Universe of nothing but a separate individual. If the individuals are somehow interacting with the observable, then the principle fails to be valid. The field of mechanics in this regard delivers the best example.

Indeed, the principle is fully compatible with the Galilean mechanics, for the latter describes separate massive particles. If we transfer our consideration to large extensive bodies, then the Galilean mechanics becomes invalid, and we have to apply the Lagrange's theory. And the relativity principle is valid for the latter, as far as I could have proven, but not for the truly general case. Well, one could even still crave a gimmick to productively apply the Galilean mechanics to such truly general cases; one had to introduce the internal pressure to allow for theoretical disassembling of a given large extensive body into tiny pieces to be treated separately from each other to enable the proper usage of the Galilean mechanics. But it is noteworthy that a gimmick as it is would still always remain solely a gimmick—but how it is correspondent to the real nature we don't know. Moreover, the mentioned gimmick is not always valid—for example, it is definitely not applicable to stiff, rigid bodies, and if it is nonetheless used in such cases, then the corresponding results are of truly apocryphal value.

In my book about thermodynamics I have discussed a number of truly peculiar examples, as for the theme. The internal pressures involved might be viewed as tensors, whose mathematical behavior is similar to that of vector products, or squared vectors (vectors power two). With this in mind, one might believe that the relativity principle ought to be valid in the Universe, where the observable might be properly described either using pure numbers without any definite direction (using scalar mathematics, if we apply to the proper terminology), or using vectors, or even using tensors. Minkowski was agreeing with such a standpoint. But, as far as I could see, our Universe goes far beyond the mentioned mathematical structures.

And, to finalize our discussion, I shall pose a question, where even just the very fact of sole posing it would cause the Relativiker to burn the person just posing it to death—without any hesitation.

Namely, is the relativity principle necessary at all?

In other words, is it necessary to introduce such a principle into the scientific research practice, for otherwise we would not be capable of realizing the actual course of some particular processes/phenomena?

My personal answer would be negative. Among those phenomena/processes, which we are currently aware of, there are none of those that might be realized solely on the basis of the relativity principle. There is no possibility for me to present here the full listing.

There is definitely no doubt, that the relativity theories are definitely very significant and pregnant with the deepest meaning, and especially the Minkowski's theory ought to undoubtedly rank among the marveling results of the human spirit. But its inevitability is still awaiting the pertinent proof. And when the theories like that by Einstein are getting more and more complicated, to the extent that it is possible to clarify their results solely using the formulae being as long as your arm, then a rightful distrust is fomented, with all the above in mind. To my personal mind, nature as it is can never be forced into some artificial scheme."

This fully professional, clear and immensely forceful report is in fact just what Prof. Ostwald has called the exclamation 'Surf's Ahead!'. To M. B. Weinstein's mind, it was just Albert Einstein's research activity that was playing the role of the 'surf' to be avoided.

We learn from the 'apologetic' literature (by 'apologetic' we denote the sources duly glorifying the great theorist of all times and all peoples, Albert Einstein, like the book [3], for example) that M. B. Weinstein was nothing more than one of the numerous detractors (may perhaps—just envious colleagues, may perhaps—inveterate anti-Semites), who stubbornly tried to undermine the struggle of Albert and its allies for the purity and liberty of the scientific research.

Meanwhile, M. B. Weinstein does definitely belong to a totally separate group of 'dissidents' in regard to the 'struggle by Albert & Co.'. After learning more about Max Bernhard's life and work quite different feelings arise after hearing to his exclamation 'Surf's Ahead'. Indeed, the latter signal by Max Bernhard is also followed by his intelligible and comprehensive description of the form, location, nature and the trend of the 'surf' he had just detected. The time since the year 1914 has gone, the scientific research was, is and will be in progress—many secrets have meanwhile become clear, lots of posers waiting for their exhaustive answers. And, as usual, a lot of 'dirty linen' has accumulated.

Well, the 'total spring cleaning' on the theme isn't the main goal of the present volume. But it still ought to be of lively interest for us—at least to try monitoring, how the **true thermodynamics** duly created and clearly described by our protagonists could nonetheless get lost. And for the scientific research as a whole it is of considerable importance to steadily detect errors and learn from them.

This is why we shall duly react to Max Bernhard's exclamation—right now and just here, in trying to clarify our problem. We shall skip numerous scathing reports about Albert Einstein and his works, but hold to the serious professional contributions, like the most recent publications by Allan Lightman, Jim Baggot, and Bernard H. Lavenda [14–17].

Meanwhile, it is just Prof. Dr. Lavenda, who dares to boil the whole story down to an essence [17]:

"...The single individual who created general relativity is Einstein, with a little more than a bit of help from Poincare. And not only did Einstein create general relativity; he also created his own cult that now pervades almost all of physics. The Einstein myth is that new physics does not need new experiments, and can be replaced by thought, or Gedanken, experiments. It has been the rallying cry of string theorists who see it as the way of completing Einstein's dream of a unified theory without getting bogged down in details like experimental verification, even though there is a lot more to unify now than in Einstein's day.

*... So it appears that not only Einstein's theory of general relativity is at the root of the present dilemma physics is faced with, but also his **deus ex machina** for arriving at his preconceived conclusions."*

[Please see Supplementary Note 12, Section 3.5, Pages 384–385]

... But, wait a bit, isn't the above conclusion just what Prof. M. B. Weinstein would like to tell us about, isn't this just the most of what he wanted to warn us against ... A closer look is then due concerning the details of Albert Einstein's role in dealing with thermodynamics and statistical physics. It is interesting to hear to what Prof. Dr. Lavenda is telling us in his most recent very interesting and instructive book on the perspectives of thermodynamics (we would like to present a bright fragment of this book, to be found at the page IX over there, with a little bit modifying its general appearance, to point out some important ideas) [18] (*highlighting of the text below is ours*):

"... Notwithstanding the enormous success of Clerk-Maxwell's 'heat', he fell prey to Tait's confusion about entropy. This is revealed in his 1873 letter to Tait where we read:

... Only lately under the conduct of Professor Willard Gibbs that I have been led to recant an error from your [Sketch], namely that the entropy of Clausius is unavailable energy while that of Thomson's is available energy. The entropy of Clausius is neither one nor the other.

*Textbooks are passed down from one generation to the next, while public lectures and writings die with their audiences. **Tait, in effect, institutionalized the scientific mafia, like no one before him.** He replaced Clausius entropy, and its unending tendency to increase through thermal interactions, by Kelvin's principle of the minimum dissipation of energy, whereby all the available energy of the universe* *must diminish without end. The universe rather than suffering a heat death would be frozen out in an 'energeticalless' state.*

Once scientists become prisoners of a single doctrine, there is no longer any room for a Carnot. ..."

It is truly a harsh claim against Peter Guthrie Tait! Is Prof. Lavenda right here indeed? This question ought to be of serious importance and crucial significance for pertinent estimations of the actual trends of thermodynamics' development, of the thermodynamics' 'newest history', so to speak. And what about the work of Albert Einstein and his allies?

We started this important discussion in Chapter 1 here and shall continue it in Chapter 5 here. Remarkably, this discussion will not be based upon the full picture, before we consider the achievement of Dr. George Augustus Linhart in the next Chapter.

3.4 Complete List of Publications by Max Bernhard Weinstein: The General Topic of Each of the Items is Indicated

1. James Clerk Maxwell and Dr. Bernhard Weinstein (1883): *Lehrbuch der Electricität und des Magnetismus: in zwei Bänden. Autorisierte deutsche Übersetzung von B. Weinstein.* Springer, Berlin, Germany.

2. Jean-Baptiste Joseph Fourier und Dr. Bernhard Weinstein (1884): *Analytische Theorie der Wärme. Dt. Ausgabe von Dr. B. Weinstein.* Springer, Berlin, Germany.

3. Bernhard Weinstein (1882): Über die Bewegungsgleichungen von Lagrange. *Annalen der Physik*, v. 251, pp. 675–680. (**General Mathematical Physics**)

4. Bernhard Weinstein (1884): Zur Berechnung des Potentials von Rollen. *Annalen der Physik*, v. 257, pp. 329–360. (**General Mathematical Physics**)

5. Bernhard Weinstein (1886): Untersuchungen über Capillarität. *Annalen der Physik*, v. 263, pp. 544–584. (**General Physics**)

6. Dr. Bernhard Weinstein, Privat-Dozent an der Universität zu Berlin und Hilfsarbeiter bei der Kaiserlichen Normal-Eichung-Kommission (1886): *Handbuch der Physikalischen Maaß Bestimmungen. Erster Band. Die Beobachtungsfehler, ihre*

Rechnerische Ausgleichung und Untersuchung. Axel Springers Verlag, Berlin, Germany. (**Theoretical Metrology**)

7. Dr. Bernhard Weinstein (1889): Kapillaritäts-Untersuchungen und ihre Verwerthung bei der Bestimmung der alkoholometrischen Normale. *Metronomische Beiträge, Nm. 6, herausgegeben von der Kaiserlichen Normal-Eichung-Kommission.* Axel Springers Verlag, Berlin, Germany. (**Geophysics, Practical Metrology**)

8. Dr. B. Weinstein (1890): Über die Bestimmung von Aräometern mit besonderer Anwendung auf die Feststellung der deutschen Urnormale für Alkoholometer. *Metronomische Beiträge, Nm. 7, herausgegeben von der Kaiserlichen Normal-Eichung-Kommission.* Axel Springers Verlag, Berlin, Germany. (**Geophysics, Practical Metrology**)

9. Dr. Bernhard Weinstein (1898): *Physik und Chemie. Gemeinsätzliche Darstellung ihrer Erscheinungen und Lehren.* Axel Springers Verlag, Berlin, Germany. (**General Physics, Chemistry, Physical Chemistry, Chemical Physics**)

10. Dr. Bernhard Weinstein (1900): *Die Erdströme im Deutschen Reichstelegraphengebiet und ihr Zusammenhang mit den Erdmagnetischen Erscheinungen.* Friedrich Vieweg und Sohn, Braunschweig, Germany. (**Geophysics**)

11. Dr. Bernhard Weinstein, Universitätsprofessor (1901): *Denken und Träumen. Dichtungen.* Ferdinand Dümmlers Verlagsbuchhandlung, Berlin, Germany. (**Natural Philosophy**)

12. Dr. Bernhard Weinstein, Universitätsprofessor (1901): *Einleitung in die höhere mathematische Physik.* Ferdinand Dümmlers Verlagsbuchhandlung, Berlin, Germany. (*An Introduction Into the Higher Mathematical Physics*)

13. Prof. Dr. Bernhard Weinstein (1901): *Thermodynamik und Kinetik der Körper. I. Band: Allgemeine Thermodynamik und Kinetik, und die Theorie der idealen und wirklichen Gase und Dämpfe.* Friedrich Vieweg und Sohn, Braunschweig, Germany. (*Thermodynamics and Kinetics*, Volume I)

14. Prof. Dr. Bernhard Weinstein (1903): *Thermodynamik und Kinetik der Körper. II. Band: Absolute Temperatur. Die Flüssigkeiten—die festen Körper. Thermodynamische Statik und Kinetik. Die (nichtverdünnten) Lösungen.* Friedrich Vieweg

und Sohn, Braunschweig, Germany. (*Thermodynamics and Kinetics,* Volume II)

15. Prof. Dr. Bernhard Weinstein (1904): Entropie und innere Reibung, in: *Festschrift Ludwig Boltzmann gewidmet zum sechzigsten Geburtstage 20. Februar 1904*, pp. 510-517; Stefan Meyer (1872-1949), Editor, Verlag von Johann Ambrosius Barth, Leipzig, Germany. (**Entropy and Internal Friction. How to establish the Interrelationship between the Thermodynamics and Statistical Mechanics?**)

16. Prof. Dr. Bernhard Weinstein (1905): *Thermodynamik und Kinetik der Körper. III. Band, I. Halbband: Die verdünnten Lösungen. Die Dissoziation—Thermodynamik der Elektrizität (Erster Theil).* Friedrich Vieweg und Sohn, Braunschweig, Germany. (*Thermodynamics and Kinetics*, Volume III., Part I)

17. Prof. Dr. Bernhard Weinstein (1906): *Die Philosophischen Grundlagen der Wissenschaften. Vorlesungen gehalten an der Universität Berlin*. Druck und Verlag von B. G. Teubner, Leipzig und Berlin, Germany. (**Natural Philosophy**)

18. Prof. Dr. Bernhard Weinstein (1908): *Thermodynamik und Kinetik der Körper. III. Band, II. Halbband: Thermodynamik der Elektrizität (Zweiter Theil)— Elektrochemie.* Friedrich Vieweg und Sohn, Braunschweig, Germany. (*Thermodynamics and Kinetics*, Volume III, Part II)

19. Bernhard Weinstein (1908): *Entstehung der Welt und der Erde nach Sage und Wissenschaft*. Verlag von B. G. Teubner, Leipzig und Berlin, Germany. (**Natural Philosophy**)

20. Prof. Dr. Max Bernhard Weinstein (1910): *Welt- und Lebensanschauungen hervorgegangen aus Religion, Philosophie und Naturerkenntnis*. Verlag von Johann Ambrosius Barth, Leipzig, Germany. (**Natural Philosophy**)

21. Dr. Max Bernhard Weinstein (1913): *Die Physik der bewegten Materie und die Relativitätstheorie. Dem Andenken Hermann Minkowskis*. Verlag von Johann Ambrosius Barth, Leipzig, Germany. (**Theoretical Physics: Minkowski's Relativity Theory)**

22. Bernhard Weinstein (1914): Zu Minkowskis Mechanik. Die Weltkonstante, die Systemmechanik. *Annalen der Physik,*

v. 348, pp. 929–954. (**Theoretical Physics: Minkowski's Relativity Theory**)

23. Prof. Dr. Max Bernhard Weinstein (1914): *Kräfte und Spannungen- Das Gravitations- und Strahlenfeld*. Friedrich Vieweg und Sohn, Braunschweig, Germany. (**Geophysics, Theory of Radiation**)

24. M. B. Weinstein (1914): Die Relativitätslehre und die Anschauung von der Welt. *Himmel und Erde*, v. 26, pp. 1–14. (The popular paper translated here.)

25. Max B. Weinstein (1916): Zur Strahlungstheorie. *Annalen der Physik*, v. 354, pp. 363–372. (**Physics: Theory of Radiation**)

26. Max B. Weinstein (1916): Über die innere Reibung der Gase. I. Der erste Reibungskoeffizient. *Annalen der Physik*, v. 355, pp. 601–654. (**Physics: Internal Friction of Gases**)

27. Max B. Weinstein (1916): Über die innere Reibung der Gase. II. Der zweite Reibungskoeffizient, die Gustav Kirchhoff'schen thermodynamisch-hydrodynamischen Gleichungen, die Maxwell'sche Gastheorie. *Annalen der Physik*, v. 355, pp. 796–814. (**Physics: Internal Friction of Gases**)

28. Max B. Weinstein (1916): Über die Zustandsgleichung der festen Körper. *Annalen der Physik*, v. 356, pp. 465–494. (**Physics: Thermodynamic Equations of State**)

29. Max B. Weinstein (1917): Über die Zustandsgleichung der festen Körper. *Annalen der Physik*, v. 357, pp. 203–217. (**Physics: Thermodynamic Equations of State**)

30. Max B. Weinstein (1917): Das Nernst'sche Theorem und die Wärmeausdehnung fester Stoffe. *Annalen der Physik*, v. 357, pp. 218–220. (**The Third Basic Law of thermodynamics, the Theorem of Nernst**)

31. Max B. Weinstein (1917) Über die Zustandsgleichung der festen Körper. *Annalen der Physik*, v. 357, pp. 506–526. (**Thermodynamic Equations of State**)

32. Max B. Weinstein (1917): Das Nernst'sche Theorem und die Wärmeausdehnung fester Stoffe. *Annalen der Physik*, v. 358, pp. 47–48. (**The Third Basic Law of thermodynamics, the Theorem of Nernst**)

33. Max B. Weinstein (1918): Zur Prüfung der Annahmen über die thermodynamischen Potentiale. *Annalen der Physik*, v. 360,

pp. 497–526. (**Theoretical Analysis of the Thermodynamic Potentials**)

34. Paul S. Epstein (1917): Bemerkung über das Nernst'sche Wärmetheorem. *Annalen der Physik*, v. 358, p. 76–78. (A rebuttal of M. B. Weinstein's paper number 32 in this list)

35. Max B. Weinstein (1917): Zu Hrn. Epsteins Bemerkungen über das Nernst'sche Theorem. *Annalen der Physik*, v. 359, pp. 79–80. (M. B. Weinstein's response to the above criticism)

3.5   Supplementary Notes

[**Note 1**]: *We could indeed fetch all of M. B. Weinstein's publications to compile his publication list, except for those in the 'Hoppes Archiv für Mathematik', which is nonetheless containing a number of interesting referees' reports as to his books, and we shall consider them in more detail later on below, and most possibly this is just what he would like to mention in his CV.*

[**Note 2**]: *Prof. Dr. Weinstein would hardly get any Russian Imperial order, if he would stay in the Russian Empire and retain its citizenship.*

[**Note 3**]: *Dr. Hauß was an administrative lawyer active in the supervisory authority for patent matters in the Prussian Ministry of the Interior since 1896 and since 1902—the president of the Prussian Patent Office.*

[**Note 4**]: *Now it becomes clear which relatives of the Weinstein's couple were in need for the financial support: They were just the mothers of the both.*

[**Note 5**]: *The both deponents were physicists, whereas Dr. Gerstmann was a specialist in astronomy/geophysics. Most probably, he was working together with an outstanding German astronomer, Prof. Dr. Friedrich Simon Archenhold (1861–1939), for he was publishing the works by Dr. Gerstmann in the journal "Das Weltall" he edited.*

Meanwhile, Dr. Levy was teaching (most probably, mainly physics) at the school for craftsmen, and his lecture about thermometry could be fetched in the periodical edited by the German clock/watchmakers association.

[**Note 6**]: *From the time point of successful termination of the postdoctoral qualification it is in principle possible to perform the official teaching activity (**in German**: Auf Grund der erfolgreichen Habilitation erteilt die Universität die Lehrbefugnis auf Antrag der habilitierten Person).*

[**Note 7**]: *The above **Ref. [103]** informs us about the 'Habilitationsakte UAHU, Phil. Fak. 1212, Blätter: 250-264' (Berlin's University Records of Postdoctoral Qualifications, Philosophic Faculty, Volume: 1212, Pages: 250–264), and then gives the (**by far incomplete!**) list of Weinstein's publications.*

*The above **Ref. [104]** informs us about the official registration of M. B. Weinstein's professorship, Berlin's University Records, Volume: 1436, Page: 229.*

*The above **Ref. [105]** directs us to a German compilation book [8] and see the same in the monograph [9].*

*The above **Ref. [106]** directs us to the book by Ernst Cassirer in German [10]; meanwhile, there is also an English translation of this work [11].*

[**Note 8**]: *To answer the poser, who was in fact the guy by the name of "E. Aschkinass" in Berlin, is not quite easy. There is practically no consistent Internet information about him. But still, there is an interesting German book [8.1]. Here we find the following personal information about Emil Aschkinass (we present the English translation here):*

"Aschkinaß, Emil, physicist in Berlin and at the Technische Hochschule (Technical University) Charlottenburg, born in 1873 in Berlin, among other things, could prove the compliance between the cathode rays' deflection and the ionic theory, could determine the absorption spectrum of the water, was studying the action of the Becquerel rays on bacteria."

*Remarkably, there is but **absolutely no information** about Prof. Dr. Max Bernhard Weinstein in this book, although it is duly presenting not only biographies of outstanding Jewish specialists having roots in the German-speaking room. Indeed, we might find an information about Mister Dunec (-Lurie), Elia, a pilot from Russia, who died in Berlin, in the April of 1913, during his plane crash, and who could be distinguished by his outstanding bravery in defense against the banditos, during the furious Jewish pogrom in the year of 1905 in*

Slonim **(now a town in Byelorussia, in the Grodno district, just nearby Lithuania, where M. B. Weinstein was born**).

By the way, **we shall not find in this book the name of Albert Einstein**, *who was born in Ulm, Germany, and—by the time of the book's publication—very active in Berlin as a physicist, chemist, philosopher. But he was a Swiss citizen that time. Was this the reason to skip his name?*

Moreover, in the German scientific periodicals of that time we might find a number of publications by E. Aschkinass. There is finally a German book [8.2]. This is the authorized German translation of the original work about the radioactivity by an outstanding, world-class experimental physicist from New Zealand, Nobel Prize winner in chemistry (1908). Emil Aschkinass is introduced there as 'Prof. Dr. E. Aschkinass, Privatdozent an der Universität Berlin'.

The above information clearly shows that Prof. Dr. Aschkinass was most probably a hot rival of M. B. Weinstein. Being born in Prussia, namely in Berlin, he had much more purely technical possibilities to make his professional career—and was definitely striving to work on the topics in vogue.

Meanwhile, there is truly no more information about 'Prof. Dr. Emil Aschkinass', who ought to be just 60 years old, as Adolf Hitler came to power in Germany.

References to Note 8

8.1 Nathan Birnbaum (1864–1937), Ernst Heppner (1913): *Juden als Erfinder und Entdecker. Die Biographien jüdischer Erfinder und Entdecker.* Veröffentlichung der Henriette-Becker-Stiftung, Welt-Verlag, Berlin-Wilmersdorf, Germany.

8.2 Ernest Rutherford (1871–1937) and Emil Aschkinass (1907): *Die Radioaktivität,* Springer-Verlag, Berlin, Germany.

[Note 9]: Howbeit, the further publications by Paul S. Epstein had rather long time apparently nothing to do with thermodynamics, till his sudden return to the theme, most probably, in parallel to the work of his teacher, Arnold Sommerfeld [9.1–9.3].

Meanwhile, the thermodynamic treatise by the apprentice and that by the teacher are nothing more than duly summarizing the then conventional standpoints concentrated on the treatises by Planck, Boltzmann and Gibbs. We know that the latter both colleagues had no more opportunity to finish their works, due to their untimely departures. However, to our sincere regret, neither Arnold Sommerfeld, nor Paul S. Epstein provide the readership with any kind of detailed search for the plausible directions of further development of Boltzmann's and Gibbs' ideas.

Jesse W. M. Dumond has described P. S. Epstein's biography in full detail [9.4].

Remarkably, the name of Paul Sophus Epstein is also familiar to us from the diaries recorded by Dr. Georg Augustus Linhart. Indeed, in the second part entitled "Amid the Stars" he only vaguely mentions P. S. Epstein as just the one "amid the stars," without giving any further convincing details (see Chapter 4 for more details).

References to Note 9

9.1 Paul S. Epstein (1922): The evaluation of quantum integrals. *PNAS*, v. 8, pp. 166–167.

9.2 Paul S. Epstein, Professor of Theoretical Physics, Californian Institute of Technology (1937): *Textbook of Thermodynamics*. John Wiley & Sons, Inc., New York, USA; Chapman & Hall, Ltd., London, U. K.

9.3 Arnold Sommerfeld (1956): *Thermodynamics and Statistical Mechanics (Lectures on Theoretical Physics, Vol. 5)*. Academic Press Publishers, Inc., New York, USA.

9.4 Jesse W. M. Dumond (1974): *Paul Sophus Epstein: 1883–1966. A Biographical Memoire*. National Academy of Sciences, Washington, D. C., USA.

[**Note 10**]: *My dear readers, in trying to answer this poser any googling the Internet would regretfully be in vain. The maximum you'll find would be some professional connection between a guy named Paul Johannesson and Rev. Lars Olof Jonathan Söderblom, an archbishop of Uppsala, Sweden and a renowned theology specialist (1866–1931), but the detailed analysis of the result clearly shows that the latter has nothing to do with our current topic.*

Most probably, Mr. (Dr.???) Paul Johannesson was a teacher at the Sophien-Gymnasium in Berlin, and most probably even its director– here is the reference to the published collection of his speeches in connection with many different gymnasium events during several years (the following publication was a supplement to the Annual Report of the Sophia Real Gymnasium, Berlin, Easter Time, 1906) [10.1]. Paul Johannesson was most probably teaching philosophy, whereas his main scope of interest was concentrated on the Natural Philosophy— here is the reference to another supplement to the Annual Report of the Sophia Real Gymnasium, Berlin, Easter Time 1896 [10.2].

The latter publication is a concise but throughout detailed critical analysis of the work by Hon. Isaac Newton on the general formulation of the law of inertia and its use in the fields of physics and natural philosophy. From both the physical and philosophical site, the following works by different outstanding authors are carefully analyzed [10.3– 10.11].

Remarkably, in the above-mentioned volume of the 'Philosophical Magazine' there is also a remarkable communication by Ludwig Boltzmann of interest to our main topic [10.12].

Although the latter work is also directly related to the problem of dynamics, Paul Johannesson has not even mentioned it, which suggests that basically he was sharing Ernst Mach's standpoints: In the above-cited short communication by him, Prof. Boltzmann is continuing his dispute with Prof. Mach, so that Boltzmann's main idea is just to call the colleagues' attention to the emerging novel approaches in theoretical physics, without fully skipping the old and good conventional ones. A good poser: Why P. Johannesson had skipped that Boltzmann's work?

Meanwhile, in his personal 1893 letter to Ernst Mach, Ludwig Boltzmann is presenting the zest of his novel theoretical idea as follows:

"I do believe that the very impossibility of Perpetuum Mobile is nothing but a purely empirical judgment, which might at any moment be debunked through some empirical evidence in some yet unknown cases. That I deem the latter to be hugely impossible for the so-called First Basic Law, but by far not that impossible for the so-called Second Basic Law is just my purely personal, subjective, indemonstrable opinion."

The above is just our English translation of the German citation according to the monograph [10.13].

Now we know that Ludwig Boltzmann's above suggestion boils down to acknowledging the basic leading role of the probability notion, which ought to underlie all the actual physics—indeed, as the entire matter in the universe is consisting of a huge number of small particles (atoms/molecules) furiously moving with respect to each other—apparently, this is just the very case for the mathematical statistics to be of help to theoretical physicists). We shall consider this important point in Chapter 4 here.

Reverting to the analysis of the reports in connection with M. B. Weinstein's works, the publication by Peter Johannesson we have mentioned above is dealing with the following works by truly renowned authors (see the references to their works: [10.14, 10.15]).

Moreover, Paul Johannesson is discussing in his publication several reports on the theme by Ludwig Lange appeared in 1885 and 1886 in the *Philosophische Studien* bulletin by Wilhelm Maximilian Wundt, (1832–1920).

Johannesson's work in question was not remaining unnoticed. Indeed, we find several referees' reports: a mostly positive and constructive estimate by Prof. Dr. Constantin Gutberlet (1837–1828) in his journal *Philosophisches Jahrbuch*, vol. 9, pp. 453–457 in 1896, as well as another throughout positive opinion about the same Johannesson's communication in the pedagogic journal under the auspices of Prof. Dr. Ernst Mach, *Zeitschrift für den physikalischen und chemischen Unterricht*, vol. 10, pp. 255–257 in 1897.

Therefore, to sum up, Mr. (Dr.) Paul Johannesson was himself undoubtedly competent enough to provide us with truly valuable estimates of M. B. Weinstein's works.

References to Note 10

10.1 Paul Johannesson (1906): *Schulreden*. Weidmannsche Buchhandlung, Berlin, Germany.

10.2 Paul Johannesson (1896): *Das Beharrungsgesetz*. R. Gaertners Verlagsbuchhandlung Hermann Heyfelder, Berlin, Germany.

10.3 Prof. Dr. Carl Gottfried Neumann (1870): Über die Principien der Galilei-Newton'schen Theorie. B. G. Teubner, Leipzig, Germany.

10.4 Prof. Dr. Ernst Mach (1872): *Die Geschichte und die Wurzel des Satzes von der Erhaltung der Arbeit*. J. G. Calvésche

K. u. K. Univ.-Buchhandlung Ottomar Beyer, Prag, K. u. K. Österreichisch-Ungarisches Reich.

10.5   Prof. Dr. James Clerk Maxwell (1878): *Matter and Motion*. D. Van Nostrand Publisher, New York, USA.

10.6   Prof. Dr. Heinrich Streintz (1883): *Die physikalischen Grundlagen der Mechanik*. B. G. Teubner, Leipzig, Germany.

10.7   Prof. Dr. Ernst Mach (1889): *Die Mechanik in ihrer Entwickelung historisch und kritisch dargestellt*. F. A. Brockhaus, Leipzig, Germany.

10.8   Prof. Dr. Leonhard Weber (1891): Über das Galiläische Princip. Verlag der Haeseler'schen Buchhandlung, Kiel, Germany.

10.9   Prof. J. G. MacGregor (1892): *On the Fundamental Hypotheses of Abstract Dynamics*. Science, Volume 20, pp. 71–74, 1893; Proc. Roy. Soc. Canada, Volume 10, pp. 3–22, 1893.

10.10   Prof. Oliver Lodge (1893): *The Foundations of Dynamics*. Philosophical Magazine and the Journal of Science, the Fifth Series, Volume XXXVI, pp. 1-36.

10.11   Prof. J. G. MacGregor (1893): *Hypotheses of Dynamics*. Philosophical Magazine and the Journal of Science, the Fifth Series, Volume XXXVI, pp. 233–265.

10.12   Prof. Dr. Ludwig Boltzmann (1893): *Methods of Theoretical Physics*. Philosophical Magazine and the Journal of Science, the Fifth Series, Volume XXXVI, pp. 37-44.

10.13   K. D. Heller (1964): *Ernst Mach: Wegbereiter der modernen Physik*. Springer-Verlag, Berlin, Heidelberg, Wien, New York, p. 27.

10.14   Eugen Karl Dühring (1833–1921) (1873): *Kritische Geschichte der allgemeinen Prinzipien der Mechanik*. Verlag von Theobald Grieben, Berlin, Germany.

10.15   Rudolf Herrmann Lotze (1817–1881) (1882): *Grundzüge der Naturphilosophie*. Verlag G. Hirzel, Leipzig, Germany.

[Note 11]: *By the way, mentioning here an outstanding German biologist, naturalist, physician, and philosopher, Ernst Heinrich Philip August Haeckel (1834–1919), and his book 'Die Welträtsel' (in English: 'The Riddle of the World') [11.1] ought to be of crucial significance for our present theme.*

Indeed, in the mentioned book by him Prof. Dr. Haeckel was, inter alia, discussing in detail the very significance of the second basic law of thermodynamics:

Most probably, in going such a remarkable step, while being not a physicist by diploma, but still a natural philosopher by nature, Prof. Dr. Haeckel would greatly appreciate to express the physically true idea about the dialectic relationship between the first and the second basic laws of thermodynamics, according to which the Ockham's razor is simply 'cutting the second basic law away'—or, to express the actual state of affairs in a better way, the Ockham's razor in fact dictates merging the both First and Second Laws to produce the unique basic law of energy conservation and transformation, with the latter both aspects to be dialectically related to each other).

At this very point we should mention the work [11.2], for it delivers a clear-cut philosophic analysis of the true dialectical reasoning while not treating thermodynamics at all.

Remarkably, Prof. Dr. Mohr has at his time just expressed similar ideas (Chapter 1 of this book, Pages 90–179, for more details).

Remarkably, an immediate colleague of Prof. Dr. Haeckel in Jena, one of the eminent and respected physicists of his time, Prof. Dr. Felix Auerbach (1856–1933), as well as other outstanding authors clearly visible in the field have joined his train of thoughts as well [11.3–11.7].

In Chapter 5 of the volume at hand we shall consider in more detail a truly worldwide heated discussion on the theme 'What is Entropy?' triggered in part by the book of Swinburne [11.7], but—to our sincere regret—incapable of bringing the debaters to any satisfactory consensus at last.

On the other hand, the renowned Russian theoretical physicist, Prof. Dr. Orest Daniilovich Chwolson (Орест Даниилович Хвольсон) (1852–1934) subjected Prof. Dr. Haeckel's writing to a withering (and, to a great extent, even arrogant!) criticism from the 'the conventional physical' point of view—while several times humbly stipulating in patches something like [11.8, 11.9]: 'Well, howbeit, but I'm not a philosopher'—thus causing an immediate natural question to his address.

*According to Prof. Chwolson, 'it is throughout impossible to get rid of the second law', because the latter as it is ought to be one of the very basic natural laws. Chwolson maintains that thermodynamics is anyway dealing with the **two** basic natural laws—the first one* *deals with the conservation of energy, whereas the second one—with the entropy increase. What is the actual physical sense of the latter process—Prof. Chwolson was never discussing either in the above-cited libel and its humble sequel, or in his serious physics handbooks [11.10], which have even been bestsellers in Russia/USSR and also translated into German, French, and Spanish.*

Remarkably, Ludwig Boltzmann demonstrated a similar attitude, as we have already learned in the Supplementary Note 10. Indeed, he was actively looking for the ways to circumvent the basic laws (Was his chief idea to theoretically guard 'the very possibility of the holy existence' of the perpetuum mobile?). Then, Max Planck had also joined him, as soon as the both of them had encountered some 'different thermodynamics' in the form of energetics.

Howbeit, it is the latter marvelous duet of messmates that was rather demonstrating quite personal intolerance with respect to any kind of dissidents, than a purely idealistic common drive to finding the 'absolute truth'. It is just the latter dilemma that seems to be the actual theme of the fragment from Prof. Ostwald's lecture we have cited in this chapter. Furthermore, this is just what M. B. Weinstein was attentively monitoring as well.

Meanwhile, Prof. Chwolson had clearly taken to the 'novel physics', that is, the physics by Boltzmann and Planck. Indeed, he tried to do his best in looking for the rational nucleus of the 'revolutionary physics' and underlining its positive impact [11.11].

Remarkably, Nils Engelbrektsson was also appealing to Prof. Chwolson's work in pointing out its definitely positive impact.

Now we might clearly recognize that Prof. Chwolson was sincerely and stubbornly trying to fulfill a serious and important task: Namely, to somehow reconcile the drastic gaps between the 'Revolutionary' and the 'Counterrevolutionary' physics. On that difficult way he did but such a noticeable mistake, as his apparently arrogant attack against Haeckel's book clearly demonstrates. But, howbeit, 'Errare humanum est, sed in errare perseverare diabolicum'.

References to Note 11

11.1   Ernst Haeckel (1899): *Die Welträtsel. Gemeinverständliche Studien über monistische Philosophie.* Verlag von Emil Strauß, Bonn, Germany—surely to mention here should also be its

English translation: Ernst Haeckel (1929): *The Riddle of the Universe*. Watts & Co., London, Great Britain.

11.2 Armand Maurer (1996): *Ockham Razor and Dialectical Reasoning*, in: Mediaeval Studies (Pontifical Institute of Mediaeval Studies), v. 58, pp. 49–65.

11.3 Felix Auerbach (1899): *Kanon der Physik: Die Bergriffe, Principien, Sätze, Formeln, Dimensionsformeln und Konstanten der Physik nach dem neuesten Stande der Wissenschaft systematisch dargestellt*. Verlag Veit, Leipzig.

11.4 Felix Auerbach (1902): *Die Weltherrin und ihr Schatten. Ein Vortrag über Energie und Entropie*. Verlag G. Fischer, Jena, Germany.

11.5 Felix Auerbach (1910): *Ektropismus und die physikalische Theorie des Lebens*. Wilhelm Engelmann, Leipzig.

11.6 František (Franz) Wald (1889): *Die Energie und ihre Entwertung*, Wilhelm Engelmann, Leipzig.

11.7 James Swinburne (1904): *Entropy; or Thermodynamics from an Engineer's Standpoint and the Reversibility of Thermodynamics*. E. P. Dutton & Co., N. Y., USA.

11.8 O. D. Chwolson (1906): *Hegel, Haeckel, Kossuth und das zwölfte Gebot: eine kritische Studie*. Verlag Friedrich Vieweg u. Sohn, Braunschweig.

11.9 O. D. Chwolson (1908): *Zwei Fragen an die Mitglieder des deutschen Monistenbundes*. Verlag Friedrich Vieweg u. Sohn, Braunschweig, Germany.

11.10 O. D. Chwolson, ouvrage traduit sur les éditions russe et allemande par E. Davaux (1906): *Traité de physique*. Tomes 1–5. Librairie scientifique A. Hermann. La librairie de S. M. le roi de Suède et de Norvège, Paris, France. *(The third volume of this treatise is devoted to the detailed representation of thermodynamics and its foundations)*.

11.11 O. D. Chwolson (1925): *Die Evolution des Geistes der Physik, 1873–1923*. Verlag von Friedrich Vieweg & Sohn AG, Braunschweig, Germany.

[Note 12]: [*sic*] *Meanwhile, Hermann Minkowski simply disappears here! This statement alone stimulates us to continue this discussion! It is truly difficult to try finding some particular*

colleague or even a group of them, who could be pleaded clearly guilty for the catastrophe of thermodynamics.

3.6   References

1. Volker Gerhardt, Reinhard Mehring, Jana Rinder (1999): *Berliner Geist. Eine Geschichte der Berliner Universitätsphilosophie*. Akademie Verlag GmbH, Berlin, Germany.

2. Max Planck (1968): *Scientific Autobiography and Other Papers* (English Translation). Citadel Press (Kensington Publishing Corp.), New York, USA.

3. Jeffrey Crelinsten (2006): *Einstein's Jury: The Race to Test Relativity*. Princeton University Press, Princeton, NJ, USA.

4. Wilhelm Foerster (1832–1921), A Letter to Albert Einstein. In: *Collected Papers by Albert Einstein*, Translated into English and edited by Klaus Hentschel & Ann M. Hentschel. Volume 8. Part A. *The Berlin Years: Correspondence 1914–1917*, see Document 204, pp. 275–276, Princeton University Press, Princeton NJ, USA, 1998.

5. Jürgen Ehlers (2007): *Book Review. Einstein's Jury: The Race to Test Relativity*. Classical and Quantum Gravitation, v. 24, pp. 5313–5314.

6. Jürgen von Ungern-Sternberg and Wolfgang von Ungern-Sternberg (1996): *Der Aufruf "An die Kulturwelt!": das Manifest der 93 und die Anfänge der Kriegspropaganda im Ersten Weltkrieg*. Franz Steiner Verlag, Stuttgart, Germany.

7. Siegfried Grundmann, Ann M. Hentschel (2005): *The Einstein Dossiers*. Springer-Verlag, Berlin, Germany.

8. Bernulf Kanitscheider (1989): Albert Einstein, in: *Klassiker der Naturphilosophie*, Gernot Böhme, Ed. Verlag C. H. Beck, München, Germany.

9. Bernulf Kanitscheider (1988): *Das Weltbild Einsteins*. Verlag C. H. Beck, München, Germany.

10. Ernst Cassirer (1921): *Zur Einstein'schen Relativitätstheorie. Erkenntnistheoretische Betrachtungen*. Bruno Cassirer Verlag, Berlin, Germany.

11. Ernst Cassirer (1923): *Substance and Function, and Einstein's Theory of Relativity*, Open Court Publishing Company, Chicago, USA.

12. Wilhelm Ostwald (1902): *Vorlesungen über Naturphilosophie*. Verlag von Veit & Comp., Leipzig, Germany; The English translation of this

work: Wilhelm Ostwald (1910): *Natural Philosophy*. Henry Holt & Co. New York, USA.

13. Frederick Gregory (1977): *Scientific Materialism in Nineteenth Century Germany*. D. Reidel Publishing Company, Dordrecht-Holland/Boston-U.S.A.

14. Alan Lightman (1993): *Einstein's Dreams*. Pantheon Books, New York, USA. (The German translation by Friedrich Griese should also be mentioned here: '*Und immer wieder die Zeit: Einsteins Dreams*. Hoffmann und Campe Verlag, Hamburg, Germany')

15. Harald Fritsch (2008): *Sie irren, Einstein!* Piper Verlag GmbH, München, Germany (The English translation to be mentioned here: Harald Fritsch (2011): *You are Wrong, Mr. Einstein! Newton. Einstein, Heisenberg and Feynman Discussing Quantum Mechanics*. World Scientific, Singapore)

16. Jim Baggott (2013): *Farewell to Reality: How Fairytale Physics Betrays the Search for Scientific Truth*. Constable & Robinson Ltd, London, U. K.

17. Bernard H. Lavenda (2015): *Where Physics Went Wrong*. World Scientific Publishing Co. Pte. Ltd., Singapore, Hackensack, London.

18. Bernard H. Lavenda (2010): *A New Perspective on Thermodynamics*. Springer, New York, Dordrecht, Heidelberg, London.

Chapter 4

The Work and Life of Dr. George Augustus Linhart (1885–1951)

> Hence! Avaunt! He's mine.
> Prince of the Powers invisible! This man
> Is of no common order, as his port
> And presence here denote: his sufferings
> Have been of an immortal nature—like
> Our own; his knowledge, and his powers and will,
> As far as is compatible with clay,
> Which clogs the ethereal essence, have been such
> As clay hath seldom borne; his aspirations
> Have been beyond the dwellers of the earth,
> And they have only taught him what we know—That knowledge is not happiness, and science
> But an exchange of ignorance for that
> Which is another kind of ignorance.
>
> *Manfred*, Act II. By Hon. George Gordon Byron (1788–1824)

A Different Thermodynamics and Its True Heroes
Evgeni B. Starikov
Copyright © 2019 Pan Stanford Publishing Pte. Ltd.
ISBN 978-981-4774-91-8 (Hardcover), 978-0-429-50650-5 (eBook)
www.panstanford.com

Georg(e) Augustus Linhart at around 1923

4.1 Biography

George Augustus Linhart was born on May 3, 1885 near Vienna, Austrian-Hungarian Empire, and died on August 14, 1951 in Los Angeles, CA, USA. To our sincere regret, this is but the only statement we might begin describing his biography. While it was possible to collect a representative amount of official information about the guy from the US archives, our own attempts to get at least a bit of the archive information about the guy in the Austrian State Archives were in vain. Taking into account that the monograph's author has no Austrian citizenship might reveal the reason of the failure, but the professional Austrian connections of the author were somehow reluctant to help as well. So, the story seemed to head for the clear blind alley, if we could not get the copies of G. A. Linhart's diaries, for which he could get the proper copyright, but could never publish them. There are in effect two volumes of them:

1. George Augustus Linhart: *Out of the Melting Pot*, 1923.
2. George Augustus Linhart: *Amid the Stars*, 1950.

These volumes contain the full personal description of Linhart's CV from his earliest days on, so that we have decided to republish some important parts of them in a chronological succession.

GEORGE A. LINHART

Georg(e) Augustus Linhart at around 1950

Out of the Melting Pot

By

George A. Linhart

PUBLISHED IN THE UNITED STATES OF AMERICA

1923

A few copies of the rough notes of the contents of this book were printed in 1921-22, in sections, under the copyright title: "Lessons in Americanization. A Crystal from the Melting Pot."

DEDICATION

To the Philadelphia Central High School boys, my first American pals;

To my classmates of the University of Pennsylvania;

To the Dennisville High School boys and girls, who stood firmly by me in the hours of trouble and taught me the meaning of true friendship;

To the hundreds of Yale freshmen and sophomores, who taught me more in two hours than I taught them in two years;

To the New Haven High School boys and girls, and especially to the "Golikells," who were so hungry and thirsty for knowledge that they made me go like— them;

To my sweet young friends of Simmons College, and especially to the Q. Chem. class with Miss C. as the guiding spirit;

To my pupils and friends at the University of Washington, where I first learned the meaning of co-ed;

To my colleagues and pupils of the University of California, where I spent five of the most fruitful years of my life;

To my "buddies" of the Specialists Company at Fort Winfield Scott, where we fought our battles with Algebra, Trigonometry and with Chinese frozen beef;

To the boys and girls of the Liberty Union High School, Brentwood, California, where the girls taught me the delicate art of grafting paint onto powder to produce peaches;

And finally, to the boys and girls of Eureka High School and Junior College, where I have just spent the happiest two years of my life;—

To all these boys and girls I lay bare my heart and soul, for they have made me what I am. They have given me my new life and are keeping me young and happy; they are my Fountain of Perpetual Youth. If Ponce de Leon came back to earth he would find the "Fountain of Youth" not in some hidden fairy woods, but in the public schools of the United States of America.

C O N T E N T S

Chapter I

CHILDHOOD

Since the purpose of this story is to show to what heights a homeless, foreign, ignorant tailor-boy may rise in free America, I shall skip over the days of my childhood abroad rather hastily.

Of my origin I know nothing authentic, save what I heard, at the age of ten, from the lips of a dying old man who had been a faithful watch dog of my unfortunate mother. It is a story of the usual unfortunate clash of religious creeds. Briefly: a Bohemian student, with no definite creed of his own, and a Polish girl, just out of a convent school, met and fell in love. And when nine months later their secret tie came out in substantial form, the girl's father shot and killed the youthful lover, and before the latter had quite expired the girl bride killed herself in the presence of father, husband, and her faithful servant. The baby boy, the cause of it all, suddenly disappeared from

1

the scene of death, only to suffer the
agonies of life. Be that as it may. All I
know is that at the age of three I found
myself a member of a poor, wretched
family of eight.

At the age of seven I was sent to school,
but was withdrawn as soon as the cold
weather had set in, ostensibly because I
had no decent things to wear, but in
reality, to help support the family. This
had been foreshadowed several weeks before by the gift, from "father", of a steel
thimble without a top. By the way,
thimbles without either rims or tops are
cruel things, anyway; at least, such was
the impression one of them made upon
me on the first day of my apprenticeship.
If I pushed the eyelet with the inside of
the thimble the needle slipped off, tearing
along my finger with the speed of light-
ning, it seemed, and imbedded itself about
a quarter of an inch into the flesh near
the second joint. A similar difficulty I
experienced when I tried to push the needle
with the outside of the thimble, save that
it would tear into my finger up to only
the first joint. In either case, pushing a
needle through two thicknesses of heavy,

2

stiff cloth was surely a painful operation; but the master tailor, standing at the table about three feet away, pretended not to notice my bleeding finger and chuckled inwardly.

One day my bleeding fingers got on the master tailor's nerves, and, after he had kissed the vodka bottle a considerable number of times, he stepped forward and exclaimed, "What are you trying to do? Fool the king?" alluding to the horrible injuries which many a man inflicted upon himself to evade military service. "Here, let me show you a trick," and he tied my finger with a piece of thin cord so that the fingertip projecting from the thimble nearly touched my palm. This certainly was a clever trick, but while it helped me out of one difficulty I soon encountered another, for after I had become accustomed to the string I often attempted unconsciously to straighten my finger. This sudden impulse generally drove the point of the needle into the forefinger of my left hand and would instantly remind me that my finger was tied, and so it naturally made me slow down, but the master tailor would not tolerate that.

3

"Speed!" he cried, "more speed!" I speeded up and let my fingers go on bleeding; there was no danger of my soiling anything, for I was learning my lessons first on rags.

By the end of a month I had acquired considerable skill. I could take about thirty stitches per minute and, as the thread grew shorter, nearly fifty; but that was not all I had learned in this short period of time. I could fell, backstitch, run stitch, blind stitch, baste, and could also stitch two pieces of broadcloth together so that they looked like one piece. This was no idle achievement, for in repair work we often had need to sew up old button holes in such a way as to leave no sign of a seam. But while I had gained in one respect, I was losing something more valuable, my strength and eyesight. "Father" aroused us at five in the morning, and without a bite to eat I had to crumple up between the high chair and the sewing table so that my knees and chin were only about six inches apart. In this position I worked at top speed until about nine when breakfast was served on the sewing table.

4

In prosperous times our breakfast consisted of a plate of barley soup and bread,
while during the dull season—and most
of the year was dull with us as to income
—a piece of dry bread and a clove of gar-
lic had to suffice. After this sumptuous
breakfast, in bites between stitches, I
worked away until noon. Dinner, if we
had any at this hour, consisted usually
of boiled potatoes sprinkled with a suggestion of cheese or, when potatoes were
high, of stiff corn meal mush served in the
same style. I then worked until dark,
when everybody stopped for half an hour.
It was during these intervals of rest that
I realized what effect the cramped position at my work and the continuous strain
upon my eyes were having upon me. Objects no longer looked clear to me, and
I found walking erect very painful. The
end of my left forefinger had become so
calloused that I could stick a needle into
it an eighth of an inch and feel no pain.
The point of a needle would frequently
remain in the finger without my knowing
it until I attempted to take another
stitch. Thus I struggled and starved for
three long years; then, at the age of ten,

when I suddenly discovered who I was,
I ran away to a large city where I secured a position as journeyman's assistant
in a dressmaking establishment.

After a week's trial the madam decided
that I was worth ten dollars a year and
full keep; that is, board and clothes, both
money and clothes to be awarded at the
end of the year. In the meantime if I
should happen to need any clothes I was
told I could wear the wornout clothes of
her only son. In spite of these luminous
prospects I worked hard and tried to
learn all I could, and before the winter
had slipped by I was fairly able to hold
my own with the two journeymen, young
men in the twenties who were earning
five dollars a week apiece. (Five Austrian
dollars were then equivalent to two American dollars and were the wages of a
journeyman.)

In constant fear of falling into the
hands of the police, for I had no passport,
I never left the house. After my day's
work, from seven in the morning to seven
in the evening, I would sit on the front
steps awhile, when the weather permitted,
and then turn in for the night. One even-

ing in May, while sitting in my favorite spot, with my face in my hands, trying to forget the past and thinking hazily of the future, I was startled by a familiar voice calling my name. I looked up and saw my "father."

"I have just arrived," he said calmly, as if I had been expecting him. "I have something to talk over with you. Let us go down the street where we can talk without interference."

"How did you know where to find me in this big city?" I ventured to ask.

"Oh," he replied with a smile, "I will tell you that later. At present we have something more important to talk about. You see," he went on, "'brother' is nearing his twentieth year, and he is not going to enter Franz Joseph's army if I can help it.

"After your disappearance I fell to thinking about our future and soon came to the conclusion that it could not be as rosy as I had pictured it. With you away from home and the other children still too young, who will help me support the family? After much deliberation, I decided to follow your example and leave

7

home to seek work as an itinerant tailor among the well-to-do country folk. To carry out my plan, which I shall presently disclose to you, I left home as soon as the severe winter was over and soon found plenty of work among the rich land owners. The pay is good and the food is excellent, and that, I thought, would especially appeal to you. Also, I haven't slept inside a house since I left home. Stacks of fresh hay on high lofts have been my beds ever since. I am sure that the outdoor sleeping, the good food, and the fresh mountain air will quickly fill out your pale, hollow cheeks and paint them pink. So you'd better come along and help me save 'brother' from the hands of the militarists. We can save enough in six months to send him to America."

The last sentence he spoke with finality and without the shadow of a doubt. He waited a few minutes to see what effect it had upon me, and finding no opposition to his plan he took my silence for yes.

"By the way," he resumed in flattering tones, "I dare say you are a first-class tailor by now and up on all the latest fashions. I am sure you will make quite

8

a hit with those country ladies when they learn that you have just come from Przemysl. Now, coming back to my scheme. After we get 'brother' over to America we can put in another summer in the country and save enough money to ship you off next, then the two of you will soon fetch over the rest of the family, and after that—to hell with Austria and the army."

"Sh! You might be overheard, and then it will be to hell with us. This town is full of spotters."

"Well?" he resumed in a quieter voice, "what do you say? Is it a go? Of course, you are your own boss, you know. If you would rather stay here and work away the best part of your life for ten or twelve dollars a year, why—"

"No, no," I interrupted, "I rather like your scheme, but on one point I cannot agree with you. You see, I don't have as much confidence in your oldest son as you have. You send him to America and you will never hear from him. He will find some woman and marry and forget all about us. It seems to be the rule, judging from the several cases at home

9

and from the few I have heard about here.
Now, my suggestion is this: you go first,
then take your oldest son across, after
awhile send for me, and lastly the three
of us can save up enough in six months
to fetch the rest of the family If this
suggestion appeals to you I am with you,
heart and soul. If it does not, I shall try
to be contented with my lot here."

After a few moments' silence he said,
"I like your plan, but, you see, we haven't
much time left, and we can't wait until
the last minute."

"We don't have to wait until the last
minute," I interrupted. "Supposing it
takes us six months to save up enough
money for your voyage and one month
for you to get across. Don't you think
that in three months' time you will easily
be able to put away at least forty dollars?
This means one hundred Austrian dollars,
and that amount would be more than
sufficient to buy a ticket for your son,
but even if you don't save up so much,
you can do what others have done before; I mean, have some of your old friends
there vouch for you and so secure a ship
ticket on terms."

10

"That sounds more plausible," he said meditatively. Then: "When can you leave?"

"Right now.' I have come here empty handed and so I shall go. The sooner we get away from here the better for me. As to my wages, my she-boss never expected to pay them anyway."

"Very well, then," he concluded, "we have just thirty minutes to wait for the next train."

11

CHAPTER II

THE JOURNEYMAN

It was a good three hours run on this slow line, and when we finally reached our destination it was past midnight. Incidentally, it was my first ride in a train, and during the entire trip I was a very sick boy; but the fresh, cool mountain air soon revived me. A fifteen minutes' walk on a soft trail through a refreshing pine woods brought us to a stack of sweet scented hay under a large straw canopy supported by four high poles. This hay mound we climbed by means of a rope ladder, and we soon forgot the world and all our troubles.

I was awakened rather early the next morning by a storm of confusing noises. The clear mountain air was just ringing with sounds of birds, hens, pigs, calves and cows. It seemed to me that every living thing was praying—or crying—for food, and it made me hungry too, for the bit of supper of the evening before I had

12

reluctantly ejected on my maiden trip. My stirring about in the lofty couch soon awakened my new old partner, and shaking the hay out of our hair we descended from our nest.

"Now," I challenged, "I am ready for that breakfast you so teasingly outlined last night."

"Very well," he replied with an assuring smile, "you shall have it, but not with the ladies; not until I trim you up a bit. We shall have breakfast in the little town this morning. It is about four miles from here, a fifteen minutes' run by train."

It was a very sumptuous breakfast, indeed. Hot biscuits, scrambled eggs, coffee, fresh cheese with cream and freshly churned buttermilk, and a generous portion of each. After this breakfast he took me into a store and bought me a new outfit from head to foot, and then we started for our new home. By nine o'clock we were back in our village; but instead of going home to our hay stack "father" led me along a mere suggestion of a brooklet in the woods, and presently we came upon what looked to me like a water hole about ten feet across and about three

13

feet deep. This was the source of the brooklet. Here he stopped, put on his spectacles and produced a pair of shears.

"Now, the first number on the program," he said jestingly, "will be a haircut and then we shall begin with the delousing process. You see," he went on explaining seriously now, "the people out here are a bit fussy. It makes them shudder when they find a louse creeping on a brand new garment—and I dare say you have brought plenty with you from Przemysl. Why, there is one creeping on your cap right now, and a handsome one, too. I bet I shall mow some of them down as I plow through that bushy, blond crop of yours."

When this operation was over he pulled out of his back pocket an article which looked like a cake of soap—a thing that hadn't soiled my body since the day I ran away from home. This was the month of May, but I hadn't had a bath since the previous August when I had literally kissed the town-creek good-bye. Of course there were indoor public baths in Przemysl for winter bathing but I never had the price, and, besides, I had always kept shy of public places for fear of the police, and there was, of course, no bath tub in the home of my employer.

Turning my glance now from the tarry cake of soap in his hand to the chilly water hole, I began to shiver.

"Don't hesitate, boy," he urged, "the sun is coming up hot and strong. Come on! Strip! It will soon be dinner time."

Talk about old time hazing! I never had gone through such bodily torture before or since. It was a kind of tar sand soap used for scrubbing horses, and the more he rubbed me with it the more I itched. Then, without a second's notice, he hurled me into the keen, ice-cold water hole. I floundered like a duck when struck by a bullet, and as soon as I caught my breath I rushed out and began jumping and shrieking like a wild man, and then we both laughed. Just the same, I could not help feeling that I had been stripped of an essential part of me, for I felt very naked, indeed. Having neglected to buy a towel, I jumped around until the sun dried me, and then for the first time in my life I went through the strange thrills of getting into an outfit of brand new

15

things. In the meantime "father" gingerly collected my cast-off apparel and hurled the bundle among the thick of the trees. Finally we started for our haystack and arrived just in time for dinner. We ate, of course, with the servants, but before entering their dining room "father" introduced me to the ladies as his 16-year-old little son. "Little, indeed," remarked one of the ladies smiling. "I should have guessed him to be twelve at most," and I had just passed my eleventh birthday.

Before a month had slipped by I became convinced that "father" had not exaggerated in the least when he assured me in Przemysl that we should have plenty of work, good eats, and money to save. Another thing that greatly pleased me was "father's" indifference to the vodka bottle.

"Oh, I don't care for it now," he would say. "In fact I find that I can do without it altogether. It's the good eats and plenty of it, and especially the delicious, refreshing buttermilk that makes me forget vodka. I suppose when I return home I shall go back to kissing the bottle. But of course," he caught himself, "I shall

16

never go home to stay; only to say good-bye. I wonder what kind of stuff they drink in America. They say there is no such drink as vodka in America; but I hear that they have plenty of good beer and that it is much cheaper than here."

As "father" could neither read nor write, the folks at home hadn't the least inkling where he was and, much less, that "father" and I were together. Great and joyful, therefore, was the surprise when, early in September, we came home to-gether remarkably prosperous in money and in looks. In the three months of itin-erant tailoring we had laid aside one hundred and fifty Austrian "gulden", an amount they had never seen before.

For the moment I was the hero. In recounting his three months' adventure, "father" had repeatedly acknowledged that the laying aside of this enormous sum of money (equivalent then to sixty American dollars) was due chiefly to my efforts and, especially, to my proficiency at the trade. But like many a hero before, and since, I soon fell into disfavor, at least with "father", for no sooner had we recovered from the effects of the re-

17

union than I began planning his immediate departure for America. Perhaps I
was a little too urgent; if so, it was because
"father" was kissing the vodka bottle
again and with a vengeance; he was more
than making up, it seemed, for the three
months' partial abstinence. In the meanwhile our hard earned savings were rapidly dwindling, and worse yet, "father" had
begun to waver; he was continually inventing new excuses to the effect that it
might be best for all concerned to send
his oldest son first. Fortunately, I carried
the deciding vote, for sixty percent of
the savings belonged, by previous agreement, to me, and this amount was snugly
and safely tucked away in my innermost
clothes.

After a week or so of indecision,
"father" began to itch for my share of
the money and tried all sorts of ways to
relieve me of it. Failing in this, he came
out in the open and one day demanded
its immediate surrender. Thereupon I
handed him a verbal ultimatum with a
twenty-four hour interval to decide whether he would rather lose me with my sixty
percent of the cash or start for America.

18

This brought him to terms within as many minutes, and on that same day a ship ticket was purchased in Vienna by telegraph, and "father" was soon off for the Land of—license, for such was then the ignorant foreigner's conception of American liberty.

Chapter III

THE VOYAGE

Three months went by but not a word of news did we hear from America. Nevertheless, we were not worried, for his last words at parting were: "My first letter home will be a ship ticket, so don't worry about my silence in the meantime."

The first week in December brought the long looked-for letter, enclosing a ship ticket and a slip of paper addressed to his oldest son. On it were the words, written probably by the agent, for he could neither read nor write: "Son! I am waiting for you at Castle Garden." A few months later a letter with like contents came for me, and on my twelfth birthday I was in Hamburg gazing across nowhere on the edge of the great Atlantic; and while thus dreaming of the land beyond, I was literally lifted off my feet by one of the crew, and a few minutes later I found myself transplanted from the cool breeze of the ocean into an atmosphere

20

of chloride of lime in a dark iron pit—the hell of the ship—the steerage. It is here that the poor, ignorant immigrant gets his first black impression of America. The poor little mite of a ship on the vast ocean he pictures as a miniature America. To be sure it is some Hyphenated American Line, but to him it is "America." He cannot reconcile it with any of the Hyphens. In this tiny floating world the people are divided not as in his world—the European world—according to blood, but are sharply classified according to wealth. At the top are the very rich, whether they be prizefighters, or railroad presidents. They occupy the choicest cabins and are served the best of food. At the bottom are the very poor. They are herded into a dungeon dirty, dark, dreary; and their food is not fit even for swine.

This was my first ride on water, and still sea-sick from a three days' journey by land, I soon succumbed to the "rocking of the boat." The second day I could no longer stand up for dizziness, and all that night I lay tossing and fretting with incipient fever. My suffering soon

21

aroused the ire of my shipmates. "Poor boy!" exclaimed one Polish young peasant in bitter tones. "It's a dirty shame the way these dogs treat us; but wait! Wait till we get to America. We'll show 'em how to treat poor devils like us."

On the third day I asked one of my neighbors to take me up to the doctor, for I could no longer even sit up; much less was I able to walk up two flights of vertical iron stairs. Realizing my helplessness, the husky young peasant gripped me as he would a fence post and carried me up to the main deck and deposited me at the doctor's door. The doctor was not in.

"Thank you very much. Don't wait for me. I shall remain here until the doctor comes."

"But who will carry you back?"

"Nobody. They shall find a better place for me. I shall not return to that filthy dungeon."

Presently the doctor arrived and asked me what I wanted. I told him that I could neither eat nor sleep nor even stand up for dizziness. He mechanically felt my forehead and my pulse and then went into his dispensary and brought out a few dark gray tablets.

"Here! Take one every four hours. Now go back to your hole, and don't come here again!"

The command to go back to the "hole" had a stimulating effect upon me, and, for the moment, I forgot I was sick.

"I don't want to go back to that hole. I am no rat, and besides I paid enough to get better quarters. I will stay right here. If I am to die I want to die like a human being and not like a rat."

My loud, sharp protest attracted the attention of a well-dressed man who was pacing the deck outside the vestibule. He stopped and gave the doctor a hostile look. At this the doctor changed his tone.

"I can't do anything for you about your quarters. You will have to see the purser about that."

"Then why did you order that poor, sick boy to go back to his hole and not to come here again? You did seem to assume that authority." This came from the man outside the vestibule. "Of course," the man continued, "it is the purser's business, but I don't think the purser can do anything at all for him, now. The boy is sick and needs medical attention, and it seems to me you might give him a bunk in the hospital; that is if there is one to spare."

Both the doctor and the man were Germans. The doctor pondered for a minute, then looked up at the man and nodded approval.

"Yes, that will be the best place for him."

The doctor's assistant then took me to the far end of the deck and down a vertical stairway ending in a box-like room about five by seven feet. To one side of this room was a low iron table about the size of a cot. Upon it the assistant threw an old cotton pad. There was no place for ventilation save through the trap door which he closed behind him when he left. The only light came from the port-hole under the water. A few minutes later the assistant returned with a large pot of tea and two slices of rye bread. Just as he was shutting the iron door above my head I happened to think of the tablets.

"The doctor said there is medicine in the tea. You don't need any tablets," and he slammed the door tight.

"A funny looking hospital," I thought. "Just an iron box with no ventilation and no light but the dim rays through the port-hole."

The few whiffs of fresh air which I had inhaled during my transfer had somewhat revived me and for the first time in three days my thumping headache had relaxed and I fell asleep. Some time later I awoke in utter darkness, and my head almost bursting with fever. Whatever I touched was scorching hot and everything about me sizzled. I thought I was being fried in a huge frying pan, and I wondered if they were roasting me to death. After a while sparks began flitting all around me, the sizzling noise intensified, and I was on fire! I tried to cry out but the sparks would rush into my mouth and burn my throat. After a long struggle I finally gave up all hope of life and lay powerless while the skin all over my body was swelling in big blisters and bursting; and as the naked flesh was bared to the fire the pain became very keen. Suddenly I felt a deathly pang, my body leaped into space and I was—in dreamland!

25

After what vaguely seemed a long time I slowly grew conscious that I was still alive, and, by the dim light from the port hole, I gradually discovered myself on the cold iron floor. With considerable effort I managed to rise and crawl back onto the iron cot. How many days I remained in this stupor I cannot tell. I remember drinking the tea which was very bitter and I also remember nibbling at the stale pieces of bread. One day I was aroused by a stream of cold air and a flood of glaring light.

"You are not dead yet, I see," and before me stood the doctor's assistant pleasantly surprised.

"The doctor said you would die soon after I lowered you down here, and so I came to fetch you up and throw you overboard to the sharks; but I see you fooled us all. I guess I had better bring you some fresh tea and more bread."

"Yes, please, and leave that trap door open," I begged. "It is so stifling in here that I can hardly breathe."

"I will, if you promise not to come near the doctor's door and make a fuss."

"No danger," I answered feebly. "I

haven't the strength to climb the stairs."

He brought me a fresh pot of tea, a piece of cake, and several slices of buttered bread. When he had climbed out he shouted down, "In two days more we shall be in New York."

One bright morning the assistant threw the door wide open and shouted down, "We are here! Hoboken! Can you climb out?"

"Yes," and I rose from my cot and began climbing the stairs—I was dressed, for I had not removed any of my clothes during the entire trip of thirteen days. When I reached the top step, he extended his hand and pulled me onto the deck. A few steps away the doctor stood grinning, his hands behind his back, like a soldier at ease.

"Well!" said he approaching us. "You came through it all right, didn't you? I certainly thought we would feed you to the sharks. Don't you ever in your life step on board a ship again, or you will surely die. You will never last through another ocean trip."

"Yes," I thought, "under such treatment anybody would die," but I did not

27

dare say it aloud for fear he would keep me from landing. As the assistant helped me onto the small boat, he whispered in my ear, "Do you know where you have been for the last ten days? In the morgue! Good-bye, and good luck to you."

CHAPTER IV

THE GREENHORN

"America at last!" I shouted, and I bounced about with great joy until I entered the place of disembarkation. Here the horror of the steerage seized me once more and with even greater force. The steerage passengers were herded, bag and baggage, into a dark corner partitioned off from the rest of humanity by stout iron bars. Fortunately for me I had no baggage, and standing there barcheaded with my hands in my pockets there evidently wasn't the slightest suggestion about me even to hint that I had just arrived, and so I escaped their attention and also the opportunity of being crammed into the prison with the rest of the dungeon folk. As I stood there deeply absorbed in watching the immigrants, freighted and billed, driven into their cell, I was suddenly startled by a familiar voice calling out my name. On the other side of the iron fence separating

29

the newcomers from the sight-seers stood
my "father." I made a rush for the gate
but to my great consternation it was
locked. A few feet away a man in uniform
was helping a brother officer pack the
immigrants into the cell to its fullest
capacity. He seemed to eye them very
keenly as they passed in line, and he kept
them from dodging the grated iron door
as one guards the passageway of a herd
of cattle driven into a freight car. A snap
of my fingers to "father" caused the
officer to face about, and catching sight
of me near the gate he walked over,
nabbed me good-naturedly by the collar,
and lifted me over the gate.

After two hours' travel, first on a ferry
boat and then on horse cars, we finally
arrived at the crossing of Ludlow and
Delancy streets, a bedlam of clattering
tongues. Each individual was shouting,
pushing and scurrying every which way
and ever struggling to force his way ahead
of all the rest. "Father" took me by the
hand: "Come on; let's get out of this
push."

"So this is America," I remarked cyn-
ically between scuffs and shoves and a

30

thousand angry, snappy looks. "I wish I had never come here. Oh, the stench is suffocating! This is a thousand times worse than the slum at home."

"Never mind, boy. Wait till you begin making money. You won't mind it then. I tell you America is a wonderful country, but you must have money. With money you can do anything you have a mind to; without money you may as well jump off the Brooklyn Bridge. And it doesn't matter how you make your money. No man criticises; no law condemns. Oh, look! See the policeman chasing that man with the push-cart? Well, he is either a fool or a poor beggar. For if he would just slip the policeman half a dollar now and then he could stay any where he wished and as long as he wished. But, as I said, he is either a miser or a beggar. Yes, here money is everything."

On Rivington street, between Ludlow and Essex, he led me into a dingy, ill-smelling hall of a tenement house, then up three flights of stairs and finally into a two-room apartment consisting of a dark kitchen and a small bedroom. This apartment was occupied

31

by a husband, wife and four children.

"Here," "father" said proudly, "is where 'brother' and I have lived since our arrival; and here is where the three of us can stay until we save up enough money to bring over the rest of the family. We pay two dollars and a half a month for sleeping, and now we shall have to pay fifty cents a month extra for you. That makes seven and a half Austrian dollars, doesn't it? Or enough to pay a year's rent at home for the whole family."

"Where is your room?" I ventured to ask, for on the only bed in the small room sat a despondent, young looking woman with a shadow of a baby at her breast.

"Our room?" he asked in tones of mockery. "Right here on the floor, and there behind the kitchen door is the mattress—it's almost new. I paid a dollar for it in a second hand furniture store."

As he pulled the door to exhibit his bargain, a regiment of roaches varying from a half to two inches in length scurried forth, executed a squad right, and disappeared into the black kitchen; and the blotches of dried blood on the mattress

32

were to me indicative of nocturnal battles with blood-thirsty bedbugs.

"So this is your bed?" I said pointedly.

"Yes, our bed," he corrected sarcastically, "until we get rich; we can then move up to Fifth Avenue, where all the rich live, and pay five dollars a bed. This is America, son, and rents are high, and greenhorns must not choose."

He then pulled out a flat pint bottle from his back pocket and offered me a drink. The liquid in it looked and smelled like the stuff at home with which we painted our slivered feet.

"What is this stuff?" I asked, turning up my nose in disgust and handing it back to him. He kissed it in the old characteristic manner and smacked his lips.

"Hm! You don't know what's good. This is perfectly good brandy; costs twenty-five cents half a quart and lasts us all day. It's a thousand times better than the vodka at home."

"I see," said I meditatively, "twenty-five cents a day. Thirty times twenty-five makes seven hundred and fifty cents, and converted into Austrian money makes— let me see—nearly twenty dollars a month for sliver paint. I see. You can afford to spend seven hundred and fifty cents a month for this stuff, but you cannot afford a private room and a clean bed. I thought you had changed, but I see you are up to your old tricks again. Well, here is where we part," and I started down the stairs. He ran after me and I waited for him outside.

"Here! Where are you running, you greenhorn? This isn't home. You are in America now."

"I know it, and I'm going to live in a clean room and sleep in a clean bed, America or no America. And remember this! I helped you to rescue your darling son from militarism, and am willing also to help you bring over the rest of your family; but unless you are decent I won't have anything to do with you."

"Well, what do you want me to do?"

"Do! Come right now to look for a room. That is what I want you to do."

After an hour's search we located a small, unfurnished second story hall room on Delancy street, a few doors below Clinton. The rent was only four dollars a month, and the cost of the furniture, comprising a couple of old chairs, an old bed with a new mattress, and a table, did not exceed ten dollars. Here the three of us lived until the rest of the family was on its way to New York.

On the following morning "father" and "brother" took me to a cheap skirt factory somewhere on East Broadway. There in a dingy third story shop, we worked by gas light from seven in the morning until six in the evening making skirts at twenty-five cents apiece. The first day I earned fifty cents, but before the week passed I turned out six and often seven skirts a day, or more than "father" and "brother" could do together. This job lasted until the end of June. Then came the slack time. Fortunately, we had more than enough money saved to tide us over July and August, the two dullest months of the year in the tailor trade; but we were not all idle. "Brother" and I would rise early in the morning, swallow our breakfast, and then start out on Delancy street until we hit Broadway; he would then walk on one side of the street and I on the other, looking for "hands wanted", which was the extent of our reading

85

knowledge. Frequently the "hands wanted" did not prove to be of the tailor variety, and this unpleasant discovery we generally made after climbing six, eight and often ten flights of iron stairs. We kept this up until the middle of September when we secured work for the fall season, for the three of us. In this house the working conditions were far superior to those on East Broadway. Every part of the building was heated, lighted and ventilated, and everybody made good wages and saved money. Even we, who were still hopelessly green, were able to earn, among the three of us, thirty-five dollars a week, while our total expense never exceeded seven dollars a week, exclusive of brandy.

In the restaurants of our neighborhood one could get three fresh rolls, a piece of butter and a cup of coffee for five cents, and lunch could be had free with a five-cent drink in any saloon. Nor did one have to drink intoxicating liquor. There were good fresh milk, rich buttermilk, and also a variety of soft drinks. This was especially true of the saloons in the manufacturing sections where one could get a

36

good, wholesome lunch with any kind of drink for five cents. In the evening when we came home from work we generally found in a Clinton street restaurant an excellent meal for ten cents; it consisted of soup, meat, vegetables and dessert. This brought up the total expense of the day to twenty cents apiece. Adding to this sum twenty-five cents a day for brandy, it barely totalled one dollar a day or, at most, thirty dollars a month, and we were earning one hundred and fifty to one hundred and sixty dollars a month! To be sure, the season was short, but in the three months of that winter season and in the two months of the following spring season we had saved about five hundred dollars. And so, after a year of hard work, sixteen to eighteen hours a day (ten hours in the shop, the rest of the time at home in our little room— everybody carried home a bundle for overtime to make up for the slack season), we were able to send for the rest of the family and prepare for their reception.

Now I was through! I felt that I had amply repaid my "father", not for my

37

bringing up, for that amounted to little, indeed, but for having taught me a trade whereby I was now able to earn a good living. Three days before the family arrived I took "French" leave of my "father" and "brother" and dropped out of sight.

CHAPTER VIII

MY SECOND BIRTH

In my frequent walks up and down Broad street I often stopped on the corner of Broad and Green to admire an artistic granite structure which was then nearing completion. From its outward appearance I guessed it to be either a modern church of the Christian Science type, or a public library. One spring morning there was assembled in front of this beautiful pile of granite a large crowd of boys celebrating something as only American boys know how to celebrate. As I stood there at a safe distance

I always kept at a safe distance from an American boys' celebration center— guessing what might be the cause of the rhythmic yelling and the thunderous applause, I was startled by a gentle slap on the back, a slap that sent me half way across the street.

"Why don't you step up and join the gang? Go ahead! Don't be a side-stepper! Ain't you one of 'em?"

69

This bit of touching encouragement came from a boy of about three score and ten.

"No!" I answered somewhat ruffled, "do I look like one of them?"

"Oh, I beg your pardon, young man! I thought you were one of the boys—the Central High School boys," he corrected himself apologetically.

"No; I am not one of them, but you seem to be—in spirit, anyway. By the way, you don't mean to say that those big fellows, too, are still in school?"

"I sure do!" he answered proudly and with such a stamp of his foot that he loosened a brick on the typical Philadelphia sidewalk. "I sure do!" he repeated with even greater emphasis. "We've spent over a million dollars for the building alone, and gladly, too. Why, it's the best investment the City has ever made! We want our boys to grow up intelligent citizens! If I had my life to live over, there's where I'd begin—and there's where you belong, too!" and he stalked away as if he had said the final word and expected me to act upon it. I had never seen the man before, and he soon passed out of my mind, but "there's where you belong, too!" would not be erased. Every time I approached Broad and Green I heard, or thought I heard, "there's where you belong, too!" until one morning I entered.

"Who do you want to see?" demanded the foreman who later proved to be Bill Huttonlock, the head janitor.

"I wish to see the manager of the school," I replied in the usual business way.

"The menagerer, eh? That's the best I ever heard yet. You'll find the menagerer right there, in that office."

Leaning back in a chair in front of a desk sat a scholarly middle-aged man of heavy physique and dressed in black. As I entered the office he wheeled about in his chair, took off his spectacles and began twirling them by the loops so rapidly that I feared they would slip off his finger and strike my face.

"What can I do for you, young man?" he asked good naturedly.

"I understand," I said smiling, "that you are turning out tolerably good citizens. I wonder if you would do as much for me."

71

For a moment he paused in apparent surprise mingled with curiosity. Then his eyebrows lowered and settled into a frown and he began to think, and, to my great relief, his spectacles ceased revolving. Finally he looked up and glanced at me quizzically.

"And how do you think I can help you?"

"Well, sir, I suppose the same way you help others."

"Yes; I see," he said thoughtfully. Then: "How much schooling have you had?"

"None to speak of, sir. Hardly enough to learn my alphabet, and that, in a foreign tongue."

"And you have made your own way, you say, ever since? And now you want to become an American citizen? And you think that I can help you?"

Reassurance smiled on the good Doctor's face. He leaned back in his chair to its maximum tilt and his spectacles resumed their gyration with such tremendous speed that they presented a solid glass ring in motion.

"Very well, then. I shall gladly show

72

you the way," and, replacing the spectacles on his nose, he took a sheet of paper
from his desk and wrote something on it.

"Take this note to the principal of the
grammar school on Seventeenth street
above Fairmount Avenue, and when you
get through there come back to us and
we will do the rest."

As I was leaving the office I ventured
to ask him how long it might take me to
prepare for high school. His curt reply
was, "It takes the average boy eight
years, but I think you can make it in a
week," and he held out his hand and
wished me success.

In my struggle to emerge from ignorance I had often listened to all sorts of
advice and encouragement with varying
feelings, but this statement of his stunned
me. I knew that to accomplish such a
task in so short a time was an impossibility. Still, it sounded interesting and
was worth investigating; but I could not
help feeling that he, like many others,
had based his judgment of my ability
upon my manner of speech. By this time
my foreign accent was pretty well worn
off, but the extent of my reading of Eng-

lish was still limited to the newspaper ads in the "Help Wanted" columns; and as to writing, I could hardly sign my name. Nevertheless, I thought that under proper guidance and with close application I might go through grammar school in two years.

It did not take me long to find the little grammar school on Seventeenth street above Fairmount Avenue. The aged principal read my note of introduction, then looked at me, then read it again as if to reassure himself of its contents. Finally he said with a pitying smile, "I'm afraid it will be somewhat embarrassing for you to be classed with little boys; but no doubt Doctor Thompson knows what he is about," and he gave me a note to the teacher in charge of the upper grades. This gentleman, believing or trying to make the pupils believe that I was a visitor, escorted me to a seat in the last row; but the boys soon caught on, and when I returned to school that afternoon they were "goin' ter hev some fun" at my expense. One little tough asked me if my mother knew I was out, while another of the gang sug-

74

gested that I get a haircut. (I did wear my hair rather long then.) When I made no reply and appeared not to understand their allusions, another youngster yelled out, pointing significantly with his finger at his own head, "Nobody home in the upper story." In short, as the aged principal had rightly feared, it soon became so embarrassing for me that Doctor Thompson's prediction came true: I did make it in one week, and it would have been a waste of time had I stayed any longer even within most peaceful surroundings; for while these boys were fed knowledge in teaspoonfuls I felt capable of absorbing it in ladlefuls. Nevertheless, in the five consecutive school days I found out what I had to know in order to pass the entrance examinations for the high school.

On the following Monday I went down to Leary's old book store and equipped myself with a set of old books, consisting of a Brown's grammar, a Butler's geography, a school physiology, an old time complete arithmetic and a history of the United States. I also procured a second hand Webster's unabridged dictionary

which served the double purpose of speller
and interpreter. Thus equipped, the next
question was, how best to acquire a thorough knowledge of the contents of the
few volumes. With good health, plenty
of clothes and with a hundred dollars in
my pocket, I decided to give myself a
fair chance to find out what I could do
with a hundred dollar vacation; but the
heart of Philadelphia is no pleasant place
to spend the summer. Those that can
possibly get away for June, July and August generally do so; those that cannot
afford to escape the depressing climate
find a great deal of comfort and pleasure,
of evenings and holidays, in the great,
beautiful Fairmount Park, the just pride
of Philadelphia. A shady nook in this
park, I thought, would be just the place
for me.

At Elm Avenue and Fifty-second street
there is one of the many natural beauty
spots of Fairmount Park. It is known as
George's Hill and rises somewhat abruptly
in places to a height of several hundred
feet above the surrounding country. At
that time (1900) this out of the way corner was seldom frequented by the pleasure

76

seeking mobs of the city. Here in the coolness of the spreading trees I found little nooks most delightful for summer study; and to make every minute count I rented a room close to the Hill and divided my day as follows: From dawn until eight I studied arithmetic, lost half an hour at breakfast, then studied grammar and English (and incidentally other languages in Webster's dictionary) till noon. From one until six I read geography and physiology and, from seven to nine, in my room, United States history. Lastly, after a two or three miles' run in the Park, I turned in for six hours sleep. This program I followed scrupulously day after day, Sundays included, from the last of May to the first of September. The result was, I accomplished in three months what I thought at the start would take me two years. Indeed, I felt quite confident that in arithmetic I was even more proficient than the smartest of Mr. Bishop's eighth graders, my erstwhile school mates; and in geography, history and physiology I was abreast of any of them. But penmanship and spelling I could not learn, and, even to this

77

day, I am far from master in these two arts. Nevertheless, I was pleased with what I had accomplished in this short time and felt quite certain of passing the entrance examination for the Central High School.

On the day of the examination I called on a neighbor's boy, and, after breakfast, we rode down to school together. On our way down he recalled some of the questions from the examination of the previous June. One of the questions was about "American Classics," of which I knew nothing. Then he recalled questions on arithmetic involving problems in banking, insurance, stocks and bonds, on all of which I was equally ignorant, and I began to tremble with anxiety, for if there was any subject that I felt absolutely sure of until that moment it was arithmetic. My companion had evidently noticed my nervousness and at once changed the subject.

"There is another question you must be able to answer," he said with forced playfulness, "before you are qualified to enter the Central High School: Why is the northwest corner of the

78

Central High School like an Irishman?"

I did not feel much like joking, but, realizing his good intentions, I merely said that I did not know. Just then the car stopped with an unusual jerk and the conductor called out: "Fifteenth and Green!"

"Here's where we fall off," said my companion. "Good luck to you. I hope you pass all right."

"Thanks. I hope so, too."

I got as far as the front entrance, when my wiser self, as on many a previous foolish move, suddenly checked me: "Don't make a fool of yourself," I heard myself saying. "You don't know as much as you thought you did. Go home and study some more. Another year's study now may save you several years of regret later." I obeyed and went back to work at my trade, but with a broken spirit. My mind was no longer on my work, and the forelady in the shop, a fine looking girl in the late twenties, noticed it even before I did and diagnosed my case as "heart ache."

"I ought to know," she winked at one of the girls. "I have been there twice, and not so very long ago."

79

Of course, I knew better. Studying stocks and bonds and annuities and compound interest until three in the morning and dreaming about them all day long at my work was not at all conducive to fancy stitching.

One morning Madam Foley, my employer, informed me that the lady, whose coat I was just finishing, would drop in before one o'clock for a special try-on, and that I was therefore to be in readiness and have both sleeves closely basted in with lining and all. To accomplish the task by one o'clock I had to work with nerve wrecking speed, and even then I had to go without lunch to do it; and when a quarter to two the coat was returned, the side-pieces were covered with chalk marks. The note of instructions concluded with, "take in on each side as marked, and have coat ready for finished try-on by five p. m."

At five o'clock the lady arrived, and right on the dot. This, indeed, was an ill omen, as she was usually late, and I hinted as much to the forelady who hated premonitions, more out of fear than belief. A few minutes later the forelady re-

80

turned, coat in hand and pale as a ghost:

"Hereafter I wish you would keep your ghastly hunches to yourself; I never did know 'em to fail! The darned thing is about three inches too tight through the waist! She can't begin to get into it."

Now, what had actually happened, as I learned later, was this: The lady dreaded getting stout and had begun dieting. That morning her flattering maid had persuaded her that she could draw in her corsets "quite a bit." Whereupon the lady was so pleased with herself that she at once called up Madam Foley and made an appointment for an extra try-on, resulting in the taking in at the side-pieces. Finding now that a few days of dieting had reduced her/ thickness by several inches she thought it safe and, indeed, desirable to have a square meal, lest she turn into a broomstick; and when after the heavy luncheon the lady began to feel rather uncomfortable and drowsy, the maid unlaced her and cozily tucked her away for an hour's rest. After her nap, when she was again laced into her corsets the lady suddenly became alarmed at the inflation about her delicate waist, and that

81

accounted for her promptness at five p.m.

The coat was three inches too tight. Unfortunately, I had to cut away the old seams to prevent wrinkling, and there was no cloth left for new side-pieces; and worse yet, the coat had to be delivered "positively" on Monday morning, and this was Saturday afternoon. When, eventually, a piece of cloth was found which matched so closely that only an expert at the game could have told the one from the other, I absolutely refused to work overtime that night, or all day on the morrow, not so much because of Sunday and because I should miss my ethical culture lecture, but just from sheer cussedness. I was hungry and sleepy and disgusted with all including myself; and when Madam Foley herself appeared and demanded an explanation of my conduct, I grabbed the jacket by the collar, threw it ceiling high and—quit. Poor Madam Foley! On the one hand, she had to deal with merciless, whimsical, self-centered creatures, who, with their millions, thought they could enslave the world! Then again, she had to contend with ignorant, sensitive tailors.

CHAPTER VIII

MY SECOND BOYHOOD

Christmas was close at hand, and with it came the slack time; and so I had nothing to do now but work at my books; and when examination time had arrived I was ready. This time, on the day of examination, I approached the school in quite a different spirit from that of several months before: I was just as confident now of passing as I had been then of failing.

Hardly had I entered the High School office when Doctor Thompson, the president, rose from his chair, stepped toward me with extended hand, and greeted me with, "I am very glad to see you."

"I hope that I am not too late to enter," I stammered, my hand gripped in his.

"That's all right. You can easily catch up. Mr. Faltermeyer, my secretary, will immediately attend to you."

The secretary received me cordially and

83

asked me to wait in his office until the "rush" was over. Deeply absorbed now in anticipation of the possible examination questions which I was about to face, I was startled by the ringing of bells.

"That's just the first bell. Did you think the building was on fire? No; we have a special bell for that," said the secretary jestingly. Soon another bell sounded, and the clattering noise which had increased to a scurrying din gradually subsided and finally died away. After a deep breath of relief the secretary said: "Now I think I can attend to you without being interrupted a hundred-and-one times." He leaned back in his chair and started to say something but apparently hesitated. Then he asked me in a casual way: "Have you seen any of the examination papers of recent date?"

"No sir, but I heard some of the questions last September."

"Hm! The best thing to do then will be to ask the men in charge of the several departments to give you a special examination; just to comply with the rules, you know, he added apologetically. "Excuse me; I will be back in a few minutes."

The secretary soon returned with the man who was to give me my first examination. As the examiner approached me his piercing black eyes set me trembling. After a curt introduction the secretary said: "Doctor Snyder, will you please start this young man with his entrance examinations. I picked you first (and he winked) because I was sure that you would be very lenient with him; he is a bit nervous, you see."

This statement, and especially the playful manner in which it was uttered set me at once at ease.

"Well," replied Professor Snyder, "I shall give him the hardest question first." Then turning to me again he asked: "Do you know complex fractions?"

"I believe so."

"Here, then! Reduce this complex fraction to its simplest form."

I have since examined many texts on arithmetic, but I have never found an example so complex. To be sure it involved but the four principles of arithmetic, yet, unless one was on to all the arithmetical tricks, it would have taken one nearly an hour merely to dismember

85

it. Had I studied a more modern arithmetic, I should never have been equal to
the task; but, thanks to the old book, I
"killed" the problem in fifteen minutes,
and I was quite sure that the answer was
correct, for the expression reduced to a
very simple fraction, which is usually the
case with such examples in the old textbooks.

Doctor Snyder watched me from the
adjoining room, and when I caught his
eye he arose and came stalking in, smiling
under his beard. Finding that I was just
nervously playing with my pencil, he began, somewhat conscience stricken, "I'm
afraid that example is a little too hard
for you. I will give you an easier one. I
suppose you would call that a "sticker."

"Oh, it wasn't so difficult," I replied.
"I rather like 'stickers' in arithmetic."

Attracted by our conversation or rather
by the silence that followed, the secretary
turned about in his chair, and focusing
his gaze upon Doctor Snyder, he awaited
curiously his next move.

"We will call it square," the professor
finally announced, and he put his O. K.
opposite "Mathematics" on my special

examination sheet. With this "O. K." at the head of the list the rest proved easy. Doctor Harley, assistant professor of history, asked me to give a historical sketch and the geographical position of my native land, about either of which I knew nothing, and he put his "O. K." in advance opposite "Geography" and "United States History." Similarly, Doctor Holt, a veterinary surgeon of the Civil War, signed his name opposite "Physiology," without asking me a single question, and said, "I hope you haven't assimilated too much of the primary school physiology so that you won't have to unlearn most of it before I can begin teaching you something." Finally, I was sent up to Professor Ernest Lacey, to be examined in English. He looked at my sheet, then asked me whence I came and when and where I had learned to speak the English language. On hearing that I had learned a great deal of it in the city of Philadelphia and in the short interval of twelve months, he exclaimed: "And you haven't acquired the 'Philadelphia lockjaw!' You pass!" (As I learned later, this remark alluded to the habit

87

many Philadelphians have of talking with their teeth shut.) He then put his "O.K." opposite the last subject on the sheet. But I knew only too well that these had been merely the preliminary examinations; for judging from my brief experience with the boys at the little grammar school on Seventeenth street above Fairmount avenue, the real test was yet to come.

When I returned the sheet with all the O.K.'s the secretary meaningly remarked: "I suppose you are glad that your first trial is over?" He evidently had overheard my thoughts. "Anyway, you have had a good start, and I shall soon enroll you as a member of the Central High School."

While the secretary was taking my nativity record a strange feeling possessed me. "How foolish!" I thought. "A creature of the world with all the bitters and sweets of a man suddenly changed into a schoolboy! Is this a dream? And is it the end or the beginning of a dream?"

"By the way," the secretary suggested as he methodically filed away my record, "if I were you I would not enter school until next fall, now that you have passed all the examinations. You see, in the freshman year the subjects extend over the entire year, and I'm afraid you will never catch up a half year's work; and even if you do, you will not do yourself full justice. I deem it advisable, however, that you attend classes for a few days as a visitor, you know; just to get a line on things; just to catch the spirit of the American boy," he hinted with a twinkle in his eye and a nod, and I understood.

That afternoon, at the friendly suggestion of Professor Snyder, my first examiner, I got an American haircut. I also bought a Central High School pin, and from that moment the "Crimson and Gold" has ever fluttered in my heart.

The secretary was right. The work was too far advanced for me, and after two days of "visiting classes" I left, but I was firmly resolved to return.

Chapter XIX

THE BACHELOR OF ARTS

Returning to school after a strenuous summer of cashiering twelve hours a day, mostly night work and with but very little day sleep in the hot city, I was hardly in a condition fit to cope with my Senior program of studies extensive in number and of greater difficulty. Yet, as the record of that year shows, I did quite well in all subjects but two; chemistry and physics. The irony of it all! Today my published work on these two subjects is a sealed book to those science teachers. Had I done as well in chemistry and physics as in the other subjects I should have ranked among those who were awarded honorary scholarships to universities. To most of the winners a scholarship meant merely prestige plus eight hundred dollars; the loss of it to me meant a stone wall in my path and a gloomy outlook on the future. I was especially distressed when I thought how disappointed Doctor Thompson probably was because of my failure to win an honorary scholarship, and I felt that I owed him an explanation; but to my astonishment the president greeted me with a welcome smile and extended hand as if I had just returned from a long and successful journey.

"I am very glad to see you," he said as he rose to meet me, and grasping my hand he continued feelingly, "and I congratulate you upon your success." I was momentarily choked, but I soon recovered and disclosed the object of my visit.

"Yes," said the doctor after a little thought, "it would have been very nice to have had the honor of winning a scholarship, but I should have been greatly disappointed if you had not declined it in favor of a more helpless boy, had you won a university scholarship. You have had your first great fight here, and you have won. And now you have a greater and much longer and harder fight before you, and you will also win because you are able and because you are fighting in a worthy cause, in the cause of Truth. Remember that the struggle for life, and

180

especially for a righteous life, is one continuous fight. Good luck to you. I expect to hear of your success in the near future."

When I entered high school in 1901 I was five feet three and weighed one hundred and twenty-five pounds; when I left in 1906 I was still five feet three, but weighed only one hundred pounds—plus a Bachelor of Arts. In this run down condition I dreaded the prospect of cashiering for Child's Dining Hall Company twelve hours at a stretch, day or night. Yet it seemed the only means whereby I could save enough during the summer to pay for at least the first term's tuition at the University of Pennsylvania; and while I was thus thinking hard of a way out of cashiering, it occurred to me that I might drum up enough summer tutoring and earn perhaps as much as in cashiering and in one-tenth of the time, for I was now charging a dollar a lesson. I immediately procured from Mr. Faltermeyer, the secretary, a list of twenty "flunkers," and after two days' drumming I secured six pupils for the entire summer and a sure income of about fifteen dollars a week. I then figured: "Subtracting seven dollars for living expenses, it will still leave me eight dollars a week clear, and in twelve weeks of vacation I can lay aside about a hundred dollars. Also, it will afford me the necessary time for recuperation," and I wasn't slow in seizing the opportunity.

Chapter XX

TWO FRIENDS IN NEED

One of the many charming and delightful suburbs of the City of Philadelphia is a bit of fairy-land in the vicinity of Ogontz, a cosy, old-fashioned village about six miles north of the city proper.

A block or so this side of the village and to the east is Elkins Park, which is not a park at all, but a sample of what fairy-land must be. It is full of green vales and mounds and winding roads. These mounds or terraces, as they are called, are studded with rich, spacious, artistic bungalows built of stone and trimmed with wood. In vain did I look at the windows and under the door-bells for the familiar sign: "Furnished Room for Rent;" or "Room and Board;" or "Private Boarding House;" or even the subtile signs of the more exclusive boarding houses of Germantown.

On the opposite side of the road and facing the entrance to Elkins Park is a rural lane named Spring Avenue. This lane leads to the great estate of the Widener family with Piere Widener, the multi-millionaire and founder of the estate, as the central figure. Although on certain days the Widener gates were thrown open to the public, only the curious ventured beyond a peep through the chinks of the tall, majestic gates. From a knoll not far away I could see the Widener mansion in all its glory. It was a large snow-white structure set among trees and shrubs of every shade of green and surrounded by innumerable beds of colorful flowers.

A little way outside and to the east of this "garden of the gods" was a tiny old two-story shingled house tucked away cosily among tall weeds. On the bit of front porch of this little house sat an aged man with smooth face and scanty, white flowing hair. The familiar G. A. R. button in his lapel at once aroused my admiration and respect for the veteran. I walked up close to him, took his hand and asked him how old he was.

"Heh? Speak louder, I'm a bit hard o' hearin'."

I repeated louder, "How old are you, Dad?" He gave a delightful little chuckle and said with a twinkle, "I hain't your dad."

"Well, then," I insisted, "Grand-dad!"

"Nor Grand-dad, neither."

"All right! How about Great grand-dad!"

"That's more like it; you want to know how old I am? Well, ef I live another week I shall be eighty-six. Out for a walk, are ye?"

"Yes, and I am looking for a place to board for the summer."

"Ye think ye'd like it way out here, eh? Mebbe my wife will take ye, ef it's only for the summer."

I had to do considerable persuading before I finally prevailed upon Mrs. Hall, his wife, to board me for the three months. She was a vivacious little woman, being then but sixty-six; twenty years her husband's junior.

After a week or so I was thoroughly at home in my new surroundings. I tutored mornings only; the afternoons I spent rambling all over the country; the evenings I devoted to reading. So the summer

185

passed, like a pleasant, fleeting dream; and when I awoke late in Spetember I had ten crackling, brand new ten-dollar bills plus some loose change, and a gain in weight of fifteen pounds. With these two items and with a reputation as first class tutor, I felt quite prepared to enter upon my new tasks at the University of Pennsylvania.

Of the hundred dollars which I had saved during the summer, seventy-five had gone to the bursar as tuition for the first semester, fifteen dollars for books and stationery, and ten dollars for a month's rent, this time at a students' rooming house in West Philadelphia near the University to save time and carfare. This left me quite poor again, but it did not worry me much, for I hoped to retain indefinitely my position as tutor with the "flunkers" of the Central High School.

During the first few weeks I had no cause to complain; but toward Thanksgiving my C. H. S. tutoring had for some unknown reason taken a slump, and I had very little for which to be thankful. Gradually things went from bad to worse, and by Christmas I had to fall back on five cent meals, and often but two a day. Worst of all, seventy-five dollars for the second semester's tuition had to be paid not later than the first of February. Failure to pay by that date would have barred me from all classes. This naturally had a depressing effect on me, and on Christmas Day I felt so blue—also hungry—that I decided to run out and see the old folks in Ogontz. I remembered that on leaving them in October Mrs. Hall had said that if I had no special place to go for Christmas dinner I should be welcome there; and so, about noon, I struck out for Ogontz and, after a two hours' hike, reached the house of welcome and good things to eat.

I found Mr. Hall a bit under the weather, but Mrs. Hall was as sprightly as ever. Fortunately, I had arrived early enough (two p.m.), so that she could "gauge her victuals accordingly," and with the help of her daughter, she prepared enough "victuals" for half a dozen ravenous appetites. In the meantime I was served a light lunch which was substantial enough to have all the effects of a full course dinner. After lunch I gladly

187

offered to help the ladies with their work. "No, sir!" said Mrs. Hall in chiding tones but with a smiling face. "You go out on a long run and fill your lungs with the fresh autumn air and get up an appetite for a good Christmas dinner. You still have three hours to wait, so out with you," and she shooed me out with her apron.

It was a gloriously clear, sunny Christmas day. As yet, there had been no snow to speak of, nor any strong winds to scatter the fallen leaves; they lay just where they had fallen, in many places a foot deep. Their fragrant perfume, their charming iridescent hues, their rustling, soothing sounds had all blended, for me, into one harmonious lullaby; and after wading for half a mile or so through these crisply toasted singing flowers, my head grew heavy, my eyes dim, my gait unsteady, and all I could now hear with every dying step was, "sleep? sleep? sleep? sleep?" I do not remember lying down, but I do recall waking with a start: there, in the dark, a cow on her way home had evidently taken me for a fallen post and, in stepping over me, had accidentally

188

stubbed her toe. Thanks to the well-meaning milcher, or I should have not only missed my Christmas dinner but should have caused great anxiety to dear old Mrs. Hall. As it was, I was just in time to hear her say, "Oh, here you are! I was just about to go out and ring you home with the cow bell."

In the few hours' glory of internal satisfaction I had forgotten all my troubles and sparkled with joy and laughter throughout the feast. It was not until the time of parting that the immediate future was again staring me in the face, and I winced with pain. Mrs. Hall saw my sudden change of mood and asked with some alarm, "Why, what is the matter? You look troubled. It did me good to see you so happy all the evening, and I don't like to see you go away sad. Cheer up and trust in the good Lord; He will help you." It wasn't so much what she said as the manner of it that, for the moment, made of me a helpless waif, and to my great shame and chagrin, I was actually weeping. Her glowing kindness had touched my very soul, and I melted away. I often wonder what a mother's

180

loving touch would feel like in time of trouble and sorrow. Perhaps that was a fair sample, for I at once opened my heart and soul to her without reserve. Mrs. Hall listened attentively and sympathetically, and at the end she exclaimed joyfully, "I know what you can do; borrow some money now, and when you get through college you will pay it back with interest."

"What! Borrow money! I couldn't think of it. I would never stoop so low. I am determined to fight my entire way to its end, or to mine."

She burst out in hearty laughter though full of compassion: "Why, my dear boy, if every business man followed your method he would go to the wall in no time. Don't you know that credit is the foundation of all business? Don't you delay this any longer! On your way home tonight think out a letter of request for a loan that shall cover all your expenses for the rest of the college year."

"You will pardon me, Mrs. Hall, but your advice is, to me, painfully funny. In the first place, such a request would not be looked upon as a business proposition,

190

but more in the light of charity, and, second, who do you suppose would lend me, say, two hundred dollars without collateral? And besides, I haven't the least idea where to begin with such a request, even if I wished to drop the thing I hold dearest—honest pride."

"Why, you won't be dropping honest pride. Indeed, you will be honestly prouder when, at the first opportunity, you begin paying back your honest loan —with interest. And I know where you can get the money at once by making a purely business request explaining the object of the loan," and bending forward as if to drive home the last words and so end all arguments she added, "and address your request to George D. Widener and you will get an answer by return mail, and a favorable answer, too."

Plodding my way home down old York road in utter darkness I had composed about twenty letters but none seemed to suit, and when I had reached Tioga I had given up all hope of being able to write a letter that would be at once business-like and telling. Indeed, I had forgotten all about the letter when I swung into Broad street aglow with glaring electric arcs, mellowed here and there by tiny, glittering, colorful bulbs on spangled trees in happy windows. "Those joys and delights are not meant for me," I murmured, and I deliberately looked straight ahead where the rows of twinkling arcs converged to a single point and then vanished – like all my hopes.

It was past midnight when I reached my room but I was neither sleepy nor tired; just sad and lonely. Again I tried to write a request, but in vain, and I remained in the chair awake, motionless; but with the first rays of sunshine my hopes were revived, and I struck off the letter with an ease and speed as from a printed page and mailed it, not to George D. Widener, but to Doctor C. P. Franklin, one of my very few friends and worthy of the name. The answer I received the next morning follows:

"Dear George:—The enclosed is so good that I did not dare alter a single word for fear of destroying its essence. Send it to Mr. Widener's office, not to his home, and remember to enclose a stamp if you want an answer. You may not

192

know it, but Mr. Widener receives many queer letters, hundreds a day. Don't be disappointed, therefore, if you don't hear from his office right away. However, I cannot help feeling that your note will get prompt attention. Send it off at once and—forget all about it.

"Yours,

"C. P. Franklin.

"P.S. –Let me hear of your success—or failure."

I dropped the letter into a post box and waited in a nearby drug store until the postman arrived an hour later and started my request on its doubtful, hopeful errand; and I did "forget all about"—everything else, but the letter. It was therefore a great relief when several days later I received a note from Mr. Daily, Mr. Widener's secretary, to call at the office at my "earliest convenience." It is rather difficult for me now to describe my mingled feelings. I was in a state of hopeful joy and sad regret. One minute I was chilled with pride and wished I had never applied, the next minute, warmed by Mr. Daily's note, I prayed for success.

193

CHAPTER XXI

TWO FRIENDS INDEED

I think it was the second day of the year, ten in the morning, that I entered the ante-rooms of the Widener offices. Mr. Daily received me kindly and apprised Mr. Widener of my presence. I tried to muster courage and look business-like and calm and cool, but my heart beats became irregular, now doubling in speed, and at times completely stopping. Then the private office door opened and I was face to face with the multi-millionaire smiling and saying cheerfully, "I am glad you have come. I wish to talk with you about helping my boy with his studies. I understand you are a first-class tutor. Is that so?"

My tongue had suddenly become pinned and I could only smile in assent.

"Could you come up tomorrow morning about this time? My son will be here then, and we may make arrangements

194

which, I hope, will prove to be mutually helpful."

The first words I remember uttering on leaving the office were, "God bless Mrs. Hall." I then walked home as if on air cushions and lived in ecstasy for the next twenty-four hours. At ten o'clock sharp on the following morning I crossed the Widener threshold and was again announced to the kind, fatherly George D. Widener. Presently the private office door opened and a rather tall, handsome young man came out, sized me up as he walked by and left the office; but I could hear him come back and re-enter his father's private office by another door. Just what followed "behind the scenes" for the next fifteen minutes I could only surmise. At the end of the argument Mr. Widener reappeared minus the smiling countenance of the day before and also minus the cheery voice. "I am very sorry, young man," he said with a frown of disappointment, "but my son says he prefers to retain his old tutor, an elderly gentleman, who has been successfully helping him with his tasks for some time. He doesn't think

195

it wise to change tutors in the middle of the school year, and especially now, the examination time."

"He is quite right, sir," I replied with forced unconcern and began to back out of his presence. At this Mr. Widener lost his frown and called out in as cheery a voice as on the day before, "Just a moment, young man," and then to his secretary, "Daily! Start an account with this young man." Then turning to me he asked, "How much did you say you needed?" "About two hundred dollars." "Well, you don't need it all at once, do you? You may lose it. Why not deposit it here with my secretary and draw it out weekly or monthly?" "That would be very pleasing to me; to draw twenty-five dollars a month; but the second semester's tuition is now due and it amounts to one hundred and twenty-five dollars. I am changing over from the College of Arts and Science to the College of Chemistry where the tuition is two hundred dollars instead of one hundred and fifty, and so I not only have to pay one hundred dollars for the second semester but twenty-five dollars additional for the privilege of transferring my last term's credits."

Mr. Widener looked puzzled. "That doesn't seem fair, does it? But I suppose it's a rule. Daily! Give this young man one hundred and fifty dollars now, and twenty-five dollars each month until— until—?"

"Until about May when I hope to secure a position for the summer and perhaps begin to pay back my debt with int ."

"No, no, no. No interest. And don't you fret and worry about paying it back so soon. When you get through college and begin to earn real money, it will be time enough."

Now that the thickest cloud had lifted, another light from an entirely unexpected corner began to shine in upon me. A new friend and a lasting one had suddenly appeared in the person of Mr. William O. Easton, Director of the Educational Department of the Central Y. M. C. A. of Philadelphia. Simultaneously with the erection of the new "home" on Arch street for the young men of Philadelphia, this clear-headed, far-sighted man was

197

laying the foundation for the "Central Y. M. C. A. Institute," a helpful school for a helpless mass of humanity of all races and colors and of all ages from boys just out of the grades to men of three score and ten. Most of the instruction was necessarily to be individual. This was especially true in the case of boys who for various reasons were unable to keep adrift mentally with their classes in the Central High School. Their parents had the money and were willing to pay for any necessary extra help in order to maintain their sons in the Central High.

Mr. Easton had first heard of me from one of my pupils, and after inquiring at the High School office about my character and ability he concluded that I was the man he wanted. My work was to be chiefly in Latin and academic mathematics, as distinguished from practical shop mathematics, and the compensation was to be seventy-five cents an hour; but these "hours" were at first rather irregular and far between, for the Y. M. C. A. was still on Fifteenth and Chestnut streets and had not yet gained the booming popularity it subsequently attained with the completion of the magnificent structure of marble and brick and with an interior equipment surpassing any institute of its kind. Nevertheless, I earned enough to get through the college year with but two more drawings from my private account generously set aside for me by the man of the kind heart and noble soul, George D. Widener.

Another important event, to me, that occurred at this time was the creating, by the University Trustees, of deferred payment scholarships. Good standing in college and real need were the only requisites, and I had both: my first year's record testified to the first, and the signatures of Doctor C. P. Franklin and of Mr. William O. Easton vouched for the second. About this time also, Professor Smith's heart began to soften a little, I thought, in spite of my sinister science record from High School. Was it because he recognized in me the potential scientist? Perhaps. At any rate, he recommended me as a worthy applicant, and I was awarded for the following year a deferred payment scholarship, under writ-

199

ten agreement that two years after graduation I was to begin making payments
on said scholarship, to all of which I
heartily agreed, and with my signature
gladly endorsed.

CHAPTER XXII

THE SCIENTIST IN EMBRYO

Before summing up my first year's work at the University of Pennsylvania I may relate briefly two incidents which are worthy of record. The first came very nearly marring my chosen life-work, the second gave it its first impetus.

During one of his fascinating lectures, Professor Smith performed a Marsch's test for arsenic. This very delicate test for the deadly element is based upon the experimental fact that the flame of burning hydrogen gas, generated in a solution containing arsenic, is colored bright lavender, but if the gas is generated in a pure solution it burns with an almost colorless flame. To be sure, Professor Smith cautioned us that before attempting to light the gas issuing from the fine glass jet we must "sweep" the entire apparatus free of air (because of the oxygen in it) with the hydrogen gas. This stage of the experiment, as every high school grad-

201

uate knows, or ought to know, is ascertained by collecting a small test tube full of the issuing gas and bringing it near a flame; if it "pops" gently the air is all out, but if it explodes with a shrieking noise, the air has not yet been swept out, and there must be no flame in the vicinity of the issuing gas! When finally the hydrogen gas has displaced all the air, the jet is lighted and the issuing gas burns with the merest suggestion of a pale blue flame. The solution suspected of containing arsenic is now introduced into the hydrogen generator by means of a water-seal glass funnel. If but a mere trace of arsenic be present in the solution just added, the almost colorless flame of the burning hydrogen will at once be tinged with the characteristic lavender color and will remain so until all the arsenic is exhausted. This test for arsenic we freshmen were not only not required to perform in the laboratory but we were actually forbidden to perform it.

That afternoon I committed a double offense. I remained hidden in the laboratory after five p. m., and I performed the forbidden experiment—part of it.

Instead of a straight stem funnel for a water-seal, I preferred a fancy thistle tube with many loops and bulbs; but this fancy thistle tube was evidently as treacherous as it was beautiful; for somewhere in its fantastic loops it had carefully and sneakingly obscured a column of air. Everything was now ready, all but the introduction of the arsenic solution into the generator through the thistle tube. The operation was successful, as the saying goes, but this time it was the operator and not the patient who died—almost. After I had recovered from the shock of the most terrific noise I had ever heard, I picked myself up, and with the left eye still half open I groped my way to the University Hospital two blocks away. A peep in the mirror with my left eye told me that the right side of my face was studded with glass and covered with blood. After a brief examination the resident young doctor announced that my right eye was punctured and would probably have to be removed, and that my nose would have to be carved to remove all the pieces of glass; but the nurse didn't seem to take his pronouncement seriously and set about picking glass out of my face and washing the wounds with some alcoholic lotion. At last she persuaded the budding M. D. to inject a little cocaine around the eye, and the pain instantly vanished. The next morning I awoke with both eyes. This was the first and, so far, the only serious explosion I have experienced at such close range. The second incident I wish to relate is about my discovery of a new compound.

From my experience in the tenderloin drug store I had learned that no kind of bug or germ can live very long in a solution of bichloride of mercury. "Therefore," I reasoned, "bichloride of mercury ought to cure all kinds of germ diseases;" but I also knew that those who had used bichloride of mercury as a germicide incautiously, or, as in the case of the tenderloin dwellers, for illegitimate purposes, had died horrible, agonizing deaths. In short, bichloride of mercury was likely to be a killing cure, and it was to overcome this killing effect that I undertook the study of mercury compounds.

One day I was dissolving mercury in strong nitric acid. Alongside was a dishful of glistening iodine flakes, and I began wondering what would happen if I introduced a handful of these into the whirling liquid mercury. It didn't take me very long to find out, but no one was seriously hurt. It did have, though, the benevolent effect of silencing the continuous roaring chatter of my classmates. When one of the fellows at the far end of the laboratory had sufficiently recovered and, after throwing the window wide open, had regained his breath, he began shouting as loud as he could, "Throw 'im out! Throw 'im out! Throw 'im out!" The cry was caught up by the dense brown clouds and was wafted across the spacious laboratory until every mother's son, and others too, had joined in the chorus, "Throw 'im out! Throw 'im out! Throw 'im out!" and I was thrown out. When I was finally allowed to come in, I found my desk perfectly clean. The porcelain dish, the pound or so of mercury, the iodine and the concentrated nitric acid had all been thrown out after me, but evidently through a different window, for I sustained no nitric acid burns either on my clothes or on my skin. After the storm had subsided and the boys had ceased showering a variety of blessings upon my head and other parts, I solemnly promised that in their presence I would never make NOO gas again.

Although I did not learn much from this rudely interrupted experiment, it nevertheless revealed to me the significant fact that the iodine did take part in the reaction, and I hoped to have better success next time.

In repeating the experiment that evening between the hours of five-thirty and seven p. m.—forbidden hours to be sure, but then, night watchmen are human—I was naturally more careful, adding the iodine flakes a few at a time. When finally the issuing of the reddish brown fumes had ceased there was revealed on the upper part of the porcelain dish, a heterogeneous mass of crystalline crusts, ranging in color from chrome yellow to vermillion red, studded here and there with unconsumed bits of deep violet crystals of iodine. The mercury had all dissolved and the solution was clear, limpid and colorless. This limpid fluid I transferred to a smaller porcelain dish, placed the dish on a steam bath, and kept it there until a drop of the liquid congealed on the end of a glass rod. At this point I transferred the hot solution into a cold crystallizing dish. Instantly a shower of pearly spangles began to fall to the bottom of the dish. The crystalline shower gradually subsided and after awhile completely stopped. Looking now through the liquid from the side of the deep crystallizing dish, I was amply rewarded for my troubles and fears and exposure to dangerously poisonous fumes. Suspended from this liquid surface was a most brilliant array of twinkling pearly scales. Indeed, the liquid itself seemed crystalline and dispersed the light into a colorful radiance never to be forgotten—but I had promised the watchman to be out before seven, so I quit gazing and finished the experiment.

On the following morning I found each scale looking as if it had been peeled from a small circle of a large, brilliant pearl; but a scientist must not be enthusiastic. He must suppress and repress every human feeling, or he is relegated to nutland; and so, assuming a very sober, scientific countenance, I ventured to approach Professor Smith with a few of the pearly scales. I described dryly the mode of their formation and then asked him what he thought they were and whether they had ever been prepared before; and while he slowly and weightily examined them, my enthusiasm broke through and I burst out upon him, "I'll bet, Professor, that this is a new compound!" The Professor looked up with a great deal of gravity and replied, "No doubt, no doubt; freshmen often do discover many new compounds." Several years later when I had become more familiar with the chemical literature I found that a substance of the same chemical composition had been discovered by a German at about the same time.

I can now best sum up my first year's work in college by a glance at the record: English Composition—Good; History of the English Language—Passed; General Inorganic Chemistry—Good; Schiller's Dramas—Distinguished; German—Distinguished; Scientific German—Good; Algebra—Distinguished; English Language Distinguished.

It is seen that in no subject did I fall

below eighty, except in the History of Ancient English in which I was not particularly interested.

*   *   *   *   *

There my story should end, for my struggles for existence would then have been over had I chosen the easiest way: got a job, married, settled down and—intellectually died. But those were not my ambitions then. I wanted to become a great scientist, and, if possible, one of the greatest. What was to hinder me in a free America? Has my dream come true? Laying aside false modesty, I can say that I have reached the summit in more than one special branch of science, and that I have a clear understanding of the fundamentals of the entire field of science. The special branches in which I have attained some distinction are Chemistry, Physics and Applied Mathematics, especially in Biology, Chemistry, Statistics, and Thermodynamics. My published work has never been adversely criticised; indeed, a considerable portion of it has already been incorporated in scientific treatises of the highest order. "Why, then," you will ask, "are you so

209

poor and so deeply buried in utter obscurity? Why, with all your knowledge and learning, aren't you holding a prominent position in some great university?" Ah! That is quite another story; I hope to tell it some other time.

Box 57, Riverside, California.

4.2   The Actual CV of G. A. Linhart in the USA

Now we might duly summarize the fruits of the above-published story by G. A. Linhart.

4.2.1   Education

He was capable of picking up English in the streets of New York and Philadelphia, occasionally working as a waiter and as a tailor, just to survive somehow. Nonetheless, he could meanwhile successfully graduate a high school in about one year, and then go to the universities for his further education.

BS in Chemistry from the University of Pennsylvania, Philadelphia, 1909;

MA and PhD in Chemistry from the Yale University, Kent Chemical Laboratory, and his supervisor was Prof. Dr. Frank Austen Gooch (1852–1929), 1909–1913.

4.2.2   Working

University of Washington, Seattle: Teaching Instructor in German and Chemistry, 1913–1914.

Simmons College, Boston: Teaching Instructor, 1915.

University of California in Berkeley:

Assistant, Chemistry, 1915–1916;

Teaching Fellow, Chemistry, 1916–1917;

Assistant, Chemistry, 1917–1918;

Drafted for the World War I, 1918 (September–December);

Assistant, Biochemistry, 1919 (Feb–May);

Instructor, Soil Chemistry and Bacteriology, 1919–1920;

Research Associate, Soil Chemistry and Bacteriology, 1920 (May–July).

Eureka Junior College, Eureka, CA: Teacher, 1920–1921 (The activity of his Junior College was officially terminated in 1921, so that Dr. Linhart had to look for some other working place).

Riverside Junior College, Riverside, CA: Teacher, 1921–1948 (till his regular retirement age).

What we can recognize from the above CV in its entirety at the first glance: A truly strange person, really from nowhere, but he could somehow reach the USA as an immigrant.

As we see, he was but a truly ambitious guy, and could consequently manage not only just his plain survival, but also the proper educational development, by landing not more and not less than in Yale University, one of the traditional elite Universities of the USA.

Still, his ambitions could not be based upon the proper 'ammunition', so to speak, so that he was just properly expelled from the academic sector to enter the pedagogic activities. That was his fully deserved niche for the most of his professional life. As a highly educated teacher of the provincial Californian schools he could be of use for the USA, his 'true native country', isn't it, dear readership? Is it a rhetoric question?

4.3   Why Should We Deal with This Guy at All?

Indeed, as I was trying in vain to publish my paper entitled "George Augustus Linhart as a widely unknown thermodynamicist" in a "widely known journal," the relevant referee's report contained the question just posed. Moreover, the referee has even allowed him-(her?)-self to be really abrupt—and to my mind even impolite—in stating, *mutatis mutandis*, that *we were all the time living and working without knowing about this guy, so our life would definitely continue without fail, even if we skip any information about him now.*

Well, he/she was definitely right in the main—and despite the clear violation of the general rules of scientific refereeing in this case, I would not like to explicitly mention here the name of this renowned journal.

Still, howbeit, we are scientific research workers, and if it comes to a memory of a colleague, whose achievements are undoubted, while he/she is not more among us, then the story definitely ought to require its pertinent continuation, so I have duly tried to do all my best to publish, what is to be published, here is the list:

1. E. B. Starikov: "Many faces of entropy or Bayesian statistical mechanics" *ChemPhysChem*, **11**, pp. 3387–3394 (2010). To be referred to as **[L1]** from here on.
2. E. B. Starikov: "George Augustus Linhart as a 'widely unknown' thermodynamicist" *World Journal of Condensed Matter Physics*, **2**, pp. 101–116 (2012). To be referred to as **[L2]** from here on.

3. E. B. Starikov: "'Entropy is anthropomorphic': Does this lead to interpretational devalorisation of entropy–enthalpy compensation?" *Monatsh. Chem.*, **144**, pp. 97–102 (2013). To be referred to as **[L3]** from here on.

4. E. B. Starikov: "Statistical mechanics in Bayesian representation: How it might work and what ought to be the probability distribution behind it" *Proceedings of the Conference: SEE-Mie, October, 2015, at the City of Tsu, county of Mie, Japan.* To be referred to as **[L4]** from here on.

To sum up, Dr. Linhart appears to be the only fruitful successor of Prof. Dr. August Friedrich Horstmann and Prof. Dr. Josiah Willard Gibbs in clarifying the actual physical sense of the Thermodynamic Entropy notion, in formally mathematically deriving the famous Boltzmann–Planck formula—which was remaining a purely ingenious guess before Linhart's initiation. Thus, the functional interdependence between the entropy and temperature could properly be established, to open the way to the true rational statistical mechanics.

The present chapter intends to present Dr. Linhart's achievements in the necessary detail, to discuss their implications—and, the last, but not the least—to provide insight into Dr. Linhart's life and destiny.

We would greatly appreciate to start our story with daring to re-publish the both main Linhart's works appeared in the *Journal of American Chemical Society* back in 1922.

To our mind, it is very important to republish them here, for these seminal works ought to demonstrate a very important point still not attracting the pertinent attention even nowadays, to our sincere regret.

In effect Dr. G. A. Linhart had in his works to 100% responded to the criticism Prof. Dr. Mohr addressed to Prof. Dr. Clausius in connection with Clausius' efforts in rationalize N. L. S. Carnot's work (we have already considered this story in Chapter 1, Pages 90–120 and the references therein, when considering the works of Peter Boas Freuchen).

Indeed, it is not necessary to derive the "true fundamental natural laws" using any kind of mathematical tools, such laws might be recognized by a purely logical analysis of their observable consequences—like Julius Robert von Mayer and other colleagues could ingeniously accomplish this. If a rule/law might be derived mathematically it is not more to be considered a truly fundamental one, whatever degree of generality it is being possessed of.

The fact of its mathematical derivation means indeed that there is/are truly fundamental natural law(s) serving as an actual principal basis for the rule mathematically derived.

With this in mind, what we see below demonstrates a clear-cut expression of the thermodynamic entropy vs. the absolute temperature. The latter functional construction enables the entropy to come to zero at the zero-absolute temperature. This way we immediately arrive at the *so-called third basic law of thermodynamics*. But, as soon as we could formally derive the latter using some mathematical toolbox, we have no more right to speak about any 'magic' basic natural law!

Indeed, some truly fundamental rule/law is in effect lurking behind Dr. Linhart's 'entropy vs. temperature' formula—and it is closely related to the experimentally well-known fact that it is practically impossible to reach the zero of the absolute temperature! One would now ask—but why it is just as it comes?

The immediate proper answer—**for the entropy would then come to zero**!

Still, any attentive reader would never be preying on my mind—and the immediate question would come then—*well, sure, but why on earth the entropy might not come to zero*???

Well, it's just exactly the point, where the significance of the actual achievements by Prof. Dr. August Friedrich Horstmann and Prof. Dr. Josiah Willard Gibbs come at the surface.

The actual thermodynamic entropy is a characteristic of all the ubiquitous hindrances/obstacles/baulks/deterrents/checks/preventives/interferences/hitches/incumbrances/handicaps/impediments on the way of any realistic natural/artificial progress—be it physical/chemical/biological/economical—or even—the last, but not the least—our own, personal, social, and professional progresses.

To sum up, any *progress* is in the *dialectic* 'unity and struggle' with the appropriate *obstacle(s)*. No progress—no obstacles—and vice versa—if there is some progress, there will always be some obstacle(s) to overcome. Thus, the entropy might never arrive at its absolute zero, as soon as some relevant progress is explicit/implicit. **Here is the fundamental law!**

140   NOTES.

the end-point for ammonia corresponds to about 0.38 volt, but the curve is steep between 0.25 volt and 0.53 volt. Evidently, therefore, bromophenol blue, changing at 0.34 volt, or resorcin blue, changing at 0.39 volt, should give a sharp end-point in the titration of ammonia.

Summary.

Hydrogen-electrode titration-curves are given for a number of reactions in the ethyl alcohol system, together with a table of indicators for use in such reactions. The possible applications of the data obtained are illustrated by the titration of a fatty acid in the presence of its glyceride.

BERKELEY, CALIFORNIA.

NOTE.

The Relation Between Entropy and Probability.[1] **The Integration of the Entropy Equation.**—The rate of increase of the specific heat with the entropy of a given element or compound depends upon the probability of the randomness of the individual particles. At the absolute zero, or at the point of zero kinetic energy we are quite certain that each particle will remain in a fixed position. The probability, therefore, will be unity. At relatively high temperatures the probability of that state prevailing is very nearly zero. Now, the mathematical expression of the above statements may be assumed to be proportional to the term, $\dfrac{C\infty - C}{C\infty}$, which at the absolute zero is unity and at relatively high temperatures approaches zero, or,

$$\frac{dC}{dS} = K \left\{ \frac{C\infty - C}{C\infty} \right\} \tag{1}$$

Assuming that when C is zero S is also zero, Equation 1 on integration gives,

$$S = \frac{C\infty}{K} \log \frac{C\infty}{C\infty - C} \tag{2}$$

The value of K may be readily obtained by substituting in Equation 1 $C dT/T$ for its equal dS and integrating. Thus,

$$\frac{dC}{dT} = \frac{K(C\infty - C)C}{C\infty \quad T}. \tag{3}$$

---

[1] The specific heats of all solid substances decrease with the temperature and approach zero as the absolute temperature approaches zero. This experimental fact led to the assumption that at the absolute zero of temperature the specific heat of all substances is actually zero. Following this announcement many attempts were made, notably by Einstein, (*Ann. Physik*, [4] **22**, 180 (1907)), Nernst and Lindemann, (*Sitzt. Akad. Wiss.* Berlin., 494, **1911**), Debye, (*Ann. Physik*, [4] **39**, 789 (1912)), and by Planck, "Theorie der Wärmestrahlung," to express the specific heat as a simple function of the absolute temperature, but with no success.

Whence,

$$\log \frac{C}{C\infty - C} = K \log T + \log k \qquad (4)$$

In the straight line equation (4) K is the slope, and $\log k$ is the intercept on the ordinate.[2] This equation may of course be written in the simpler form,

$$C = \frac{C\infty \, kT^K}{kT^K + 1}. \qquad (5)$$

Equation 4 or 5 reproduces the experimental data, within the probable error, for the specific heats of all substances thus far obtained by thermoelectric methods.

The following table giving the results for copper demonstrates this.

TABLE I

CALCULATIONS OF THE SPECIFIC HEAT OF COPPER AT DIFFERENT TEMPERATURES BY

THE FORMULA $C = \dfrac{C\infty \, kT^K}{kT^K + 1}$

1 T	2 C_p (obs.)	3 C_v Calc.	4 Diff.	1 T	2 C_p (obs.)	3 C_v Calc.	4 Diff.
Dutch (Leyden Laboratory)				English			
14.51	0.04	0.04	0.00	50.00	1.32	1.28	+0.04
15.59	0.05	0.05	0.00	90.00	3.98	3.57	+0.41
17.17	0.07	0.07	0.00	130.00	4.78	4.84	−0.06
20.19	0.11	0.11	0.00	170.00	5.23	5.39	−0.16
20.74	0.12	0.12	0.00	210.00	5.50	5.64	−0.14
25.37	0.23	0.22	+0.01	250.00	5.70	5.77	−0.07
29.73	0.38	0.34	+0.04	290.00	5.83	5.83	0.00
40.22	0.83	0.75	+0.08	390.00	6.09	....	......
50.04	1.43	1.28	+0.15	German (Nernst Laboratory)			
59.75	2.06	1.86	+0.20	23.5	0.22	0.18	+0.04
60.33	2.08	1.90	+0.18	27.7	0.32	0.28	+0.04
69.66	2.59	2.47	+0.12	33.4	0.54	0.46	+0.04
80.32	3.05	3.08	−0.05	87.0	3.33	3.43	−0.10
88.86	3.37	3.50	−0.13	88.0	3.38	3.47	−0.09
89.39	3.44	3.55	−0.11	137.0	4.57	4.98	−0.41
				234.0	5.59	5.72	−0.13
				290.0	5.79	5.83	−0.04
				323.0	5.90	5.87	+0.03

$C\infty = 3 \quad R = 5.966$

$\log k = -5.4955$

$K = 2.900$

The figures of Cols. 1 and 2 were collected by Dr. Latimer, Gilman Hall, University of California.

In the table following are given in the second and third columns, the values for K and for K_1, or $\log k$, obtained from the straight line experimental plots according to Equation 4; in the fourth column are given the

[2] Actually, the raw experimental C_p values were used in the straight line plots, as the increase in volume is negligibly small up to 100° A. and in some cases even up to 200° A.

142   NOTES.

specific heats per gram atom for constant volume, calculated from the values of K and K_1 with the aid of Equation 4 or 5. In the fifth column are given the entropies per gram atom or mol, calculated by means of Equation 2, and in the sixth column are given the values for the entropies obtained by Lewis and Gibson[3] by a graphical method.

Since Lewis and Gibson state that their calculated values are accurate to from 0.3 to 2.0 units of entropy the agreement between the two columns is entirely satisfactory. However, the values obtained by means of Equation 2 are slightly but consistently higher in most cases. This is due to the fact that for very low temperatures for which there are no available data Lewis and Gibson join to their graphical method the equation,

$$C = aT^3, \text{ whence } S = \frac{1}{3} C.$$

This is equivalent to the assumption that in Equation 1 $(C\infty - C)/C\infty$ remains unity at very low temperature and that $K = 3$ which is evidently not true, as a glance at the table will show.

TABLE II
ENTROPIES AT CONSTANT VOLUME

	K	K_1	C_v (298)	S_v (298) Calc. (Eq. 2)	S_v (298) Lewis & Gibson
Al	2.90	−5.6985	5.774	7.07	6.7
Cu	2.90	−5.4955	5.844	8.00	7.8
C, diamond	2.80	−7.4088	1.482	0.61	0.6
C, graphite	1.83	−4.7946	2.093	1.41	1.3
KCl	2.47	−4.3320	5.870	19.90	19.4
NaCl	2.94	−5.4008	5.887	17.55	17.2
PdCl₂	1.88	−3.1133	5.798	33.99	33.2
Na	2.334	−3.7157	5.916	12.23	11.7
I	1.84	−2.5074	5.911	15.20	14.8
Pb	2.18	−2.9652	5.943	15.21	15.0
S, rhombic	1.23	−2.4418	4.771	7.80	7.4
S, monocl.	1.30	−2.5627	4.882	7.83	7.6
Sn, gray	2.24	−4.0162	5.794	9.23	....
Sn, white	2.79	−4.7626	5.922	10.50	....
Cd	6.00	−10.2960	5.965	8.65	....
Ca	2.53	−4.3570	5.892	10.35	10.3
HgCl	1.60	−2.6235	5.702	23.26	23.2
Zn	2.40	−4.2204	5.880	10.54	9.7
Ag	2.36	−4.0875	5.862	10.24	9.9
Mg	2.57	−4.8409	5.790	8.18	8.1
CO(NH₂)₂	0.705	−1.7485	2.969	46.62	41.0 ±2[a]

[a] This value is taken from a paper by Gibson, Latimer and Parks, (THIS JOURNAL, **42**, 1541 (1920)). On plotting the values of C_v, calculated by means of Equation 5, and also the experimental values for C_p against log T, the difference was found to be 8(0.035), whence S_p(298) = 46.9.

EUREKA, CALIFORNIA.   GEORGE A. LINHART.
Received September 16, 1921.

[3] Lewis and Gibson, THIS JOURNAL, **39**, 2554 (1917), Tables VI, VIII.

[CONTRIBUTION FROM THE EUREKA JUNIOR COLLEGE, SCIENCE DEPARTMENT]

CORRELATION OF ENTROPY AND PROBABILITY

By GEORGE A. LINHART

Received May 3, 1922

In a recent article[1] two equations are given, one connecting entropy and probability, and the other, specific heats and temperature. These equations are

$$S = \frac{C\infty}{K} \log \frac{C\infty}{C\infty - C} \tag{2}$$

$$C = \frac{C\infty \, kT^K}{kT^K + 1} \tag{5}[2]$$

In the article just cited the validity of Equation 2 is amply substantiated by a comparison of the calculated entropies for 18 substances with the values obtained for the same substances by a graphical method. For Equation 5, however, only one set of calculated values is given, namely, those for the specific heats of copper. The purpose of the present article is to show that this equation which is based upon the laws of entropy and probability holds for the specific heats of all substances thus far obtained by thermo-electric methods. It will be seen that in all cases where check values are given (see silver chloride and aluminum) the deviations from the values calculated by means of Equation 5 are far less than the variations among the experimental values themselves. Most of these data have been obtained in Nernst's Laboratory[3] and are scattered all through the literature in fragments, each writer smoothing, readjusting and generally vitiating the experimental data to suit his particular empirical equations. In the present paper are given presumably only the actually "observed" values. These have been subjected to the treatment described in the article previously cited. The results with the 31 substances, with the exception of those for ice and silver iodide, leave nothing to be desired. The data for the specific heats of these two compounds seem to be quite erroneous.

In conclusion it may be pointed out that from the results obtained with the aid of Equations 2 and 5 the values for Sp can in every case be found as follows: Calculate $Sv(298)$ by means of Equation 2, then with the aid

[1] Linhart, THIS JOURNAL, **44**, 140 (1922).

[2] Here, as in the article cited above, C denotes the average gram atomic heat at any temperature, T; $C\infty$ equals 3 R, equals 5.966 cal.; k and K are constants. For convenience of computation and tabulation, log k is given instead of k and is denoted by K'. Where the values for K and K' are omitted in the present article they may be found in Table II of the article cited.

[3] The data for Na, Mg, Ca, Cd, TiCl₄, CCl₄, SiCl₄, SnCl₄, HCOOH, CO(NH₂)₂ were obtained in the Chemical Laboratory of the University of California and have been published in THIS JOURNAL within the past 5 years.

1882   GEORGE A. LINHART

of two or three calculated Cv values, from the point where Cp begins definitely to exceed Cv,[4] sketch Cp and Cv against log T on uniform tracing paper or cloth, cut out the segment thus obtained and weigh it. Having previously determined the relation between a unit of entropy and a square unit of the tracing cloth, the difference between Sp and Sv is very accurately obtained, provided, of course, the substance does not undergo a modification in form or a change of state for the chosen temperature interval.

TABLES

T	Cp (obs.)	Cv(calc.)	Diff.		T	Cp(obs.)	Cv(calc.)	Diff.
	Sodium					**Calcium**		
64.6	4.52	4.54	−0.02		80.6	4.42	4.44	−0.02
67.9	4.66	4.66	0.00		92.7	4.81	4.81	0.00
71.1	4.77	4.77	0.00		94.9	4.92	4.86	+0.06
74.2	4.81	4.86	−0.05		103.4	5.07	5.04	+0.03
84.6	5.08	5.11	−0.03					———[a]
94.8	5.30	5.29	+0.01		141.4	5.70	5.51	+0.19
156.8	6.02	..	....		143.3	5.68	5.52	+0.16
	Silver				145.2	5.75	5.54	+0.19
35.0	1.58	1.58	0.00		163.2	5.97	5.64	+0.16
39.1	1.90	1.90	0.00		198.5	6.36	..	....
42.9	2.26	2.19	+0.07			**Cadmium**		
45.5	2.47	2.50	−0.03		68.4	4.98	5.00	−0.02
51.4	2.81	2.81	0.00		70.7	5.20	5.15	+0.05
53.8	2.90	2.97	−0.07		72.5	5.29	5.25	+0.04
77.0	4.07	4.17	−0.10		74.8	5.25	5.36	−0.11
100.0	4.86	4.84	+0.02		75.7	5.36	5.40	−0.04
			———[a]		76.8	5.45	5.44	+0.01
200.0	5.78	5.71	+0.07		78.8	5.32	5.51	−0.19
273.0	6.00	..	....		79.1	5.56	5.52	+0.04
331.0	6.01	..	....		80.7	5.49	5.51	−0.08
535.0	6.46	..	....		81.8	5.65	5.60	+0.05
589.0	6.64	..	....		84.2	5.71	5.65	+0.06
	Magnesium				86.2	5.63	5.69	−0.06
74.9	2.90	2.90	0.00		87.0	5.67	5.70	−0.03
78.3	3.03	3.08	−0.05		88.4	5.74	5.73	+0.01
83.5	3.33	3.32	+0.01		90.0	5.78	5.75	+0.03
92.1	3.61	3.68	−0.05		92.5	5.73	5.78	−0.05
101.5	3.99	4.04	−0.05		298.0	6.35	..	....
114.5	4.47	4.41	+0.06			**Mercury**		
115.8	4.44	4.44	0.00			$K =$ 2.07		
132.7	4.77	4.80	−0.03			$K' = -2.8050$		
155.2	5.08	4.96	+0.12		31.1	3.89	3.93	−0.04
172.4	5.31	5.31	0.00		36.6	4.36	4.35	+0.01

[4] This is due to an increase in atomic or molecular volume. For the elements and compounds here presented this phenomenon begins to appear at about 100° A., except for Cu, Al, diamond, graphite, HCOOH and $CO(NH_2)_2$. For some of these substances the increase in volume does not seem to be appreciable even up to room temperature.

CORRELATION OF ENTROPY AND PROBABILITY   1883

			——a
192.6	5.55	5.46	+0.09
220.0	5.82	5.59	+0.23
237.3	5.83	5.66	+0.17
253.5	5.98	5.70	+0.28
255.2	5.86	5.71	+0.15
288.5	6.11	..	....

Calcium

67.6	3.93	3.89	+0.04
70.4	4.02	4.03	−0.01
73.3	4.16	4.16	0.00
76.2	4.28	4.28	0.00

43.0	4.70	4.71	−0.01
62.0	5.34	5.31	+0.03
65.0	5.37	5.36	+0.01
69.0	5.43	5.43	0.00
			——a
80.5	5.58	5.56	+0.02
86.0	5.66	5.61	+0.05
92.0	5.79	5.65	+0.14
164.0	6.26	..	....
168.0	6.29	..	....
201.0	6.42	..	....
207.0	6.48	..	....

a These lines indicate approximately where the increase in volume begins to be appreciable and, with the exception of a few substances, this is approximately at about 100° Å.

Mercury

T	C_p(obs.)	C_v(calc.)	Diff.
213.0	6.58	..	....
214.0	6.58	..	....
229.0	6.62	..	....
232.0	6.70	..	....

Zinc

33.1	1.25	1.26	−0.01
34.3	1.32	1.34	−0.02
36.3	1.71	1.49	+0.22
41.0	1.60	1.84	−0.24
43.7	2.17	2.04	+0.13
61.8	3.25	3.25	0.00
64.0	3.51	3.37	+0.14
68.0	3.59	3.59	0.00
75.0	3.95	3.91	+0.04
80.0	4.17	4.11	+0.06
85.0	4.24	4.29	−0.05
89.0	4.39	4.42	−0.03
89.3	4.32	4.44	−0.12
94.0	4.55	4.57	−0.02
			——a
207.0	5.83	5.70	+0.13
274.0	5.90	5.83	+0.07
290.0	6.03	..	....
366.0	6.20	..	....
395.0	6.28	..	....

Aluminum

19.1	0.07	0.06	+0.01
23.6	0.11	0.11	0.00
27.2	0.16	0.17	−0.01
33.5	0.30	0.30	0.00
37.1	0.40	0.40	0.00
41.9	0.60	0.55	+0.05
49.6	0.90	0.85	+0.05
53.4	1.11	1.02	+0.09
62.4	1.55	1.45	+0.10

Diamond

T	C_p(obs.)	C_v(calc.)	Diff.
88	(0.03)	0.064	....
92	(0.03)	0.073	....
205	0.62	0.62	0.00
209	0.66	0.65	+0.01
220	0.72	0.74	−0.02
222	0.76	0.75	+0.01
232	0.86	0.84	+0.02
243	0.95	0.94	+0.01
262	1.14	1.12	+0.02
284	1.35	1.34	+0.01
306	1.58	1.57	+0.01
331	1.84	1.83	+0.01
358	2.12	2.12	0.00
413	2.66	2.70	−0.04
1169	(5.45)	5.60	(−0.15)

Graphite

28.7	0.06	0.04	+0.02
38.1	0.07	0.07	0.00
44.1	0.10	0.10	0.00
58.8	0.14	0.16	−0.02
85.0	0.31	0.31	0.00
137	0.69	0.69	0.00
232	1.50	1.52	−0.02
284	1.92	1.98	−0.06
334	2.39	2.39	0.00
412	3.04	2.95	+0.09
622	4.00	4.03	−0.03
1095	5.45	5.10	+0.35
1250	5.60	5.26	+0.34

Tin (white)

79.8	4.68	4.64	0.00
87.3	4.87	4.88	−0.01
94.8	5.07	5.07	0.00
194.9	6.20	..	....

1884   GEORGE A. LINHART

TABLES *(Continued)*

Mercury

T	Cp(obs.)	Cv(calc.)	Diff.
73.4	2.08	2.03	+0.05
79.1	2.36	2.32	+0.04
32.4	0.25	0.27	−0.02
35.1	0.33	0.34	−0.01
83.0	2.41	2.53	−0.12
86.0	2.52	2.68	−0.16
88.3	2.62	2.79	−0.17
137.0	3.97	4.53	−0.56
235.0	5.32	5.60	−0.28
331.0	5.82	5.82	0.00
433.0	6.10	..	....
553.0	6.48	..	....

Tin (gray) (cont.)

T	Cp(obs.)	Cv(calc.)	Diff.
197.2	5.71	..	....
205.2	5.75	..	....
248.4	5.87	..	....
256.4	5.88	..	....
264.3	5.89	..	....
273.0	5.90	..	....
288.1	5.91	..	....

Lead

T	Cp(obs.)	Cv(calc.)	Diff.
23.0	2.96	3.00	−0.04
28.3	3.92	3.66	+0.26
36.8	4.40	4.40	0.00
38.1	4.45	4.49	−0.04
85.5	5.65	5.65	0.00
90.2	5.71	5.68	+0.03
200.0	6.13	..	....
290.0	6.33	..	....
332.0	6.41	..	....
409.0	6.61	..	....

Sulfur (rhombic)

T	Cp(obs.)	Cv(calc.)	Diff.
22.7	0.96	0.86	+0.10
25.9	0.99	0.99	0.00
27.5	1.04	1.05	−0.01
28.3	1.08	1.08	0.00
29.9	1.14	1.14	0.00
57.0	2.06	2.05	+0.01
69.0	2.29	2.37	−0.08
83.0	2.70	2.70	0.00
93.0	2.93	2.91	+0.02
138.0	3.63	3.63	0.00
198.0	4.72	4.22	+0.50
235.0	4.93	4.47	+0.46
297.0	5.47	4.77	+0.70

Diamond

T	Cp(obs.)	Cv(calc.)	Diff.
197.2	6.23	..	....
205.2	6.25	..	....
248.4	6.36	..	....
256.4	6.37	..	....
264.3	6.38	..	....
273.0	6.39	..	....
288.1	6.40	..	....

Tin (gray)

T	Cp(obs.)	Cv(calc.)	Diff.
79.8	3.80	3.80	0.00
87.3	4.07	4.07	0.00
94.8	4.30	4.30	0.00
194.9	5.66	..	....

Iodine

T	Cp(obs.)	Cv(calc.)	Diff.
28.3	3.78	3.54	+0.24
33.5	3.97	3.97	0.00
36.5	4.17	4.17	0.00
77.0	5.38	5.38	0.00
186.0	5.92	5.84	+0.08
235.0	6.36	..	....
298.0	6.64	..	....

Sodium chloride

T	Cp(obs.)	Cv(calc.)	Diff.
25.0	0.29	0.30	−0.01
25.5	0.31	0.31	0.00
28.0	0.40	0.40	0.00
67.5	3.06	2.90	+0.16
69.0	3.13	3.00	+0.13
81.4	3.54	3.72	−0.18
83.4	3.75	3.81	−0.06
138.0	(3.87)?	5.29	....
235.0	(5.76)?	5.81	−0.05

Potassium chloride

T	Cp(obs.)	Cv(calc.)	Diff.
22.8	0.58	0.57	+0.01
26.9	0.76	0.82	−0.06
30.1	0.98	1.01	−0.03
33.7	1.25	1.30	−0.05
39.0	1.83	1.70	+0.13
48.3	2.85	2.40	+0.45
52.8	2.80	2.72	+0.08
57.6	3.06	3.04	+0.02
63.2	3.36	3.37	−0.01
70.0	3.79	3.74	+0.05
76.6	4.11	4.04	+0.07
86.0	4.36	4.40	−0.04
137.0	5.25?	5.36	−0.11
235.0	5.89?	5.79	+0.10

CORRELATION OF ENTROPY AND PROBABILITY　　1885

Sulfur (monocl.)

83.0	2.75	2.75	0.00
87.0	2.82	2.84	−0.02
87.0	2.90	2.84	+0.06
89.0	2.90	2.88	+0.02
91.0	2.95	2.93	−0.02
96.0	2.97	3.03	−0.03
102.0	3.15	3.15	0.00
194.0	4.92	4.30	+0.62
200.0	5.00	4.35	+0.65
201.0	4.81	4.35	+0.46

Silver chloride

T	C_p(obs.)	C_v(calc.)	Diff.
23.5	1.49	1.49	0.00
26.4	1.72	1.75	−0.03
32.8	2.40	2.28	+0.12
45.6	3.63	3.18	+0.45
87.0	4.87	4.72	+0.15
116.0	5.17	5.16	+0.01
207.5	5.90	5.66	+0.24
330.0	6.51	..	....
405.0	6.80	..	....
430.0	6.86	..	....

Thallium chloride

$K = 1.90$
$K' = -2.8838$

23.1	1.89	2.02	−0.13
26.7	2.32	2.40	−0.08
29.7	2.69	2.69	0.00
32.4	2.95	2.94	+0.01
36.3	3.22	3.25	−0.03
40.9	3.53	3.59	−0.06
44.8	3.79	3.83	−0.04
50.1	4.03	4.11	−0.08
90.4	5.20	5.20	0.00
94.0	5.25	5.25	0.00
138.0	5.62	5.60	+0.02
236.0	6.02	..	....
297.0	6.34	..	....

Mercurous chloride

23.0	1.56	1.58	−0.02
25.7	1.74	1.79	−0.05
29.0	2.18	2.01	+0.17
34.5	2.54	2.43	+0.11
75.0	4.20	4.20	0.00
83.0	4.40	4.40	0.00

331.0	6.16	..	....
416.0	6.36	..	....
550.0	6.54	..	....

Silver chloride

22.5	1.39	1.40	−0.01
26.6	1.73	1.77	−0.04
31.3	1.99	2.16	−0.17
43.1	2.75	3.02	−0.27
68.0	4.20	4.20	0.00
72.2	4.32	4.34	−0.02
81.3	4.60	4.58	+0.02
91.4	4.91	4.80	+0.11

Titanium tetrachloride (cont.)

T	C_p(obs.)	C_v(calc.)	Diff.
99.3	4.54	4.54	0.00
194.6	6.19	5.45	....

Carbon tetrachloride

$K = 2.92$
$K' = -5.605$

39.1	0.60	0.59	+0.01
40.7	0.65	0.67	−0.02
63.8	2.05	1.89	+0.16
79.6	2.81	2.80	+0.01
91.0	3.40	3.38	+0.02
95.0	3.57	3.56	+0.01
99.5	3.60	3.75	−0.15
199.5	5.84	5.54	+0.30
204.1	5.90	5.56	+0.34
208.0	5.90	5.58	+0.32
229.4	6.20	..	....

Silicon tetrachloride

$K = 1.89$
$K' = -3.3882$

77.4	3.60	3.60	0.00
81.8	3.70	3.75	−0.05
86.6	3.83	3.90	−0.07
94.8	4.14	4.12	+0.02
131.3	4.80	4.80	0.00
168.6	5.68	5.18	+0.50
181.0	5.95	5.27	+0.68
185.8	6.08	..	....

Tin tetrachloride

$K = 2.25$
$K' = -3.8800$

89.0	4.59	4.55	+0.04
89.5	4.53	4.55	−0.02
95.0	4.73	4.70	+0.03

1886

GEORGE A. LINHART

TABLES (*Concluded*)

	Silver chloride					Titanium tetrachloirde (cont.)		
T	*Cp*(obs.)	*Cv*(calc.)	Diff.		*T*	*Cp*(obs.)	*Cv*(calc.)	Diff.
84.0	4.43	4.42	+0.01		95.9	4.71	4.72	−0.01
86.0	4.46	4.46	0.00					
89.0	4.52	4.52	0.00		161.2	5.89	5.51	+0.38
					200.5	6.32	..	....
198.0	5.69	5.44	+0.25			Lead chloride		
326.0	6.12	..	....		15.6	0.72	0.71	+0.01
331.0	6.24	..	....		19.8	1.03	1.04	−0.01
	Titanium tetrachloride				24.0	1.17	1.39	−0.22
	K = 1.79				27.0	1.63	1.63	0.00
	K' =−3.0725				54.9	3.49	3.52	−0.03
86.7	4.26	4.26	0.00		61.5	3.91	3.82	+0.09
92.8	4.40	4.40	0.00		84.4	4.46	4.55	−0.09
	Lead chloride					Formic acid (cont.)		
T	*Cp*(obs.)	*Cv*(calc.)	Diff.		*T*	*Cp*(obs.)	*Cv*(calc.)	Diff.
87.7	4.52	4.63	−0.11		205.0	2.73	2.70	+0.03
106.5	4.97	4.97	0.00					
					237.0	3.02	2.89	+0.13
205.5	5.80	5.64	+0.16		243.2	3.12	2.92	+0.20
330.0	6.17	..	....			Urea		
405.0	6.32	..	....		86.4	1.79	1.75	+0.04
430.0	6.42	..	....		90.0	1.76	1.78	−0.02
	Formic acid				90.0	1.83	1.78	+0.05
	K = 0.90				90.3	1.82	1.79	+0.03
	K' =−2.1650				96.5	1.89	1.84	+0.05
71.0	1.45	1.44	+0.01		97.0	1.83	1.85	−0.02
73.3	1.50	1.47	−0.03		104.0	1.87	1.91	−0.04
76.6	1.50	1.51	−0.01		107.5	1.94	1.94	0.00
77.7	1.52	1.52	0.00		128.1	2.13	2.11	+0.02
82.0	1.55	1.58	−0.03		198.5	2.57	2.54	+0.03
86.0	1.60	1.63	−0.03		199.5	2.55	2.55	0.00
89.0	1.60	1.67	−0.07		201.4	2.54	2.56	−0.02
90.0	1.66	1.68	−0.02		204.7	2.59	2.58	+0.01
90.3	1.68	1.69	−0.01		208.2	2.58	2.59	+0.01
94.0	1.73	1.73	0.00					
176.3	2.48	2.49	−0.01		223.9	2.83	2.67	+0.16
180.0	2.51	2.52	−0.01		244.2	2.87	2.76	+0.11
184.5	2.59	2.56	+0.03		274.0	3.21	2.88	+0.33
196.0	2.69	2.63	+0.06					

EUREKA, CALIFORNIA

As we might immediately recognize, the above was just the starting point for the great fundamental research project, which had but no opportunity, no chance for its regular continuation... The chapter at hand tries answering the great poser of "Why it was like this?"

2968   ADDITIONS AND CORRECTIONS

ADDITIONS AND CORRECTIONS

1921, VOLUME 43

The Nutritional Requirements of Yeast. II. The Effect of the Composition of the Medium on the Growth of Yeast, by Ellis I. Fulmer, Victor E. Nelson and F. F. Sherwood.

P. 197.   For the ordinate on the curve, instead of "Per cent. gain in weight" read "Grams gain in weight per 1000 g. of gluten."

1922, VOLUME 44

Pyrimidines from Alkylmalonic Esters and Aromatic Amidines, by Arthur W. Dox and Lester Yoder.

P. 364.   In the first column of Table I, the 5-carbon substituent on line 11, instead of "dimethyl" read "trimethylene."

The Constitution of the Secondary Product in the Sulfonation of Cinnamic Acid, by F. J. Moore and Ruth Thomas.

P. 368.   In line 14, instead of "*m*-sulfobenzamide" read "*m*-hydroxybenzoic acid."

A Simpler Method of Determining Acetyl Values, by Leon W. Cook.

P. 392.   For similar formulas developed by a slightly different method, see T. T. Cocking, *Chemist and Druggist*, **74,** 87 (1913); *Perfumery and Essential Oil Record*, 9, 37 (1918).

The Pressure of Oxygen in Equilibrium with Silver Oxide, by Frederick G. Keyes and R. Hara.

P. 479.   For the name of the second author instead of H. Hara read R. Hara.

A Simple Method of Electrometric Titration in Acidimetry and Alkalimetry, by Paul Francis Sharp with F. H. MacDougall.

P. 1195.   In the fifth and ninth lines of Table II, instead of "100–0.5–CdSO$_4$," read "10.00–0.5–CdSO$_4$."

The Molecular Rearrangement of Symmetrical Bis-triphenylmethylhydrazine, by Julius Stieglitz and Ralph L. Brown.

P. 1280.   The eighth column in the table should read 6, 17, 22, 39, 33, 37, 31, 31, 27, 29, 35, 34.

P. 1283.   Under *Analyses*, the ammonia found should read "0.0366, 0.0345."

Ion Activities in Homogeneous Catalysis. The Formation of Para-chloro-acetanilide from Acetyl-chloro-amino-benzene, by Herbert S. Harned and Harry Seltz.

P. 1478.   In Equation 1, instead of $C_6H_5NClCOCH_3 + \overset{+}{H} + \overset{-}{Cl}$ read $C_6H_5NCl\text{-}COCH_3 + \overset{+}{H} + \overset{-}{Cl}$.

P. 1480.   In the second line from the bottom, read $\log F_a' = \alpha'C - \beta'C^{m'}$.

P. 1483.   In the eleventh line of the text, read "whence E_c is found to be 2.14×10^4 cals. between 25° and 35°, and 1.93×10^4 cals.," etc.

P. 1684.   **Temperature Coefficient of Electromotive Force of Galvanic Cells and the Entropy of Reactions,** by Roscoe H. Gerke.   The author wishes to acknowledge his appreciation of the invaluable advice of Professor Gilbert N. Lewis, who directed the research.

The Structure of the Compounds Produced from Olefins and Mercury Salts: Mercurated Dihydrobenzofurans, by Roger Adams, F. L. Roman and W. N. Sperry.

P. 1791.   Line 8, for density instead of 1.507 read 1.057.

Correlation of Entropy and Probability, by George A. Linhart.

P. 1883.   The table headings at the top of the page should read "Magnesium" at the left and "Mercury" at the right.

P. 1884. The table headings at the top should read "Aluminum" at the left and "Tin (white)" at the right; a horizontal line should appear above the headings "Tin (gray)" and "Iodine."

P. 1885. The table heading at the top should read "Potassium chloride" on the right; a horizontal line should appear above the heading "Silver chloride."

P. 1886. The table headings at the top should read "Mercurous chloride" on the left and "Tin tetrachloride" on the right; a horizontal line should appear above the headings "Lead chloride" and "Formic acid (cont.)."

The System, Fe_2O_3—SO_3—H_2O, by E. Posnjak and H. E. Merwin.

Make the following changes if not so printed.

P. 1977. Line 18, "See Table III" should be "See Table IV." In Table III, "Expt." should be "No."; and $<$ should be $\angle$.

P. 1979. Line 24, III should be IV. Line 27, the last (110) and (120) should be $(1\bar{1}0)$ and $(1\bar{2}0)$.

P. 1981. Line 2, after "but" insert "on"; line 6 from bottom, 4 should be 3.

P. 1982. Line 3, omit second c; lines 3 to 7, insert, in first wide space, $>$; line 7, the second (110) should be $(1\bar{1}0)$; line 9, III should be IV.

P. 1983. Last line, 50 should be 51.

A Study of the Velocity of Hydrolysis of Ethyl Acetate, by Herbert S. Harned and Robert Pfanstiel.

P. 2201. For $\dfrac{T'_0 - T}{T_E} = x$, read $\dfrac{T'_0 - T}{T_E} A = x$.

A New Method for the Introduction of an Ethyl Group. The Reaction Between Organo-magnesium Halides and Diethyl Sulfate, by Henry Gilman and Rachel E. Hoyle.

P. 2625. The yield in the reaction between benzyl magnesium chloride and diethyl sulfate is 65% and not quantitative.

NEW BOOKS

The Chemistry of the Non-Benzenoid Hydrocarbons and their Simple Derivatives. By BENJAMIN T. BROOKS, Ph.D. The Chemical Catalog Company, Inc., 1 Madison Avenue, New York, U. S. A., 1922. 612 pp. 16 × 23.5 cm. Price $7.00.

As stated by the author in the preface "the beautiful, interesting and often facile chemistry of the benzene hydrocarbons has somewhat overshadowed the chemistry of the aliphatic open-chain and cyclic non-benzenoid hydrocarbons." In this volume we have an adequate and a very readable survey of the paraffin and cyclic (other than the aromatic) hydrocarbons. No attempt to attain completeness has been made but abundant references to the literature are found throughout the book. The author sees great opportunities for research on the theoretical side of petroleum, rubber, turpentine and essential oils. Methods of refining petroleum, for instance, are still wasteful since they are based necessarily on empirical knowledge.

Titles of the first 6 chapters are: I. The Paraffins; II. Chemical Properties of the Saturated Hydrocarbons; III. The Paraffin Hydrocarbons; IV. The Ethylene Bond; V. The Acyclic Unsaturated Hydrocarbons; VI. Polymerization of Hydrocarbons. The great importance of the ethylene bond is emphasized in a chapter of considerable length. The theories

After publishing the above two seminal communications, Dr. Linhart was stubbornly continuing his research, but, to our sincere regret, apparently he was unduly forced to leave the academic area. As we can see from his above-mentioned publications, already when publishing his seminal reports, he was nothing more than just a common junior college teacher in Eureka, California, USA, by pursuing his research on his own (This is just what he would like to point out with the term "*Eureka Junior College, Science Department*"). That is, his explicit 'academic career' was already over by the year of 1922 and thereafter could never be re-activated, as we have seen above. We have also published a summary of Dr. Linhart's professional CV in our **[L2]** paper, whereas our **[L3]** paper presents a fragment of his publications list of relevance to our current theme.

Indeed, the full publication list of Dr. Linhart is not that long. He started publishing his research work in the year 1911, when a PhD student at the Chemistry Department of the Yale University, but after publishing the both of his seminal JACS reports he encountered an ever increasing 'entropy' when trying to publish some results of his research.

Fortunately, Dr. Linhart was keeping detailed journals of his life, career, and work, which help us to reconstruct the actual events and try analyzing (together with him himself) the actual reasons for what has in fact happened to him and his work.

Moreover, he was trying to correctly publish his diaries in the form of some fiction-like, but still genuine documentary reports, in that he had obtained the copyright for the both volumes, but his efforts were in vain. The only sensible result of all those activities by him was publishing a number of copies of the both volumes at own costs and disseminating the publications as wide as possible.

This is how it could be possible for me to get the copies of the both diary volumes via interlibrary loan—and even to get a single paper copy of "Out of the melting pot" with the original autograph by its author via the 'abebooks.com' portal. And, in view of the special importance and significance of Dr. Linhart's achievements, we have decided to publish a number of separate but logically conjoint fragments of his diaries—to enable correct estimating his personal and professional development.

4.4   The Actual Stance of Linhart's Contemporaries with Respect to His Results

To sum up, in what follows, first we shall see the list of his publications (1911–1923) he composed himself in the first volume of his diaries followed by his subsequent works, the two of which exist in the preprint form only. It ought to be just the proper point, where I would greatly appreciate to express my sincere gratitude to the colleagues from the Riverside Junior College in California, USA, who have helped me to fetch these preprints and provided me with the electronic copies of them). This is why we publish here the both of these reprints, in their chronologic order.

Then, we would dare to re-publish here the correspondence between Dr. Linhart and Prof. Dr. Linus Pauling of Californian Institute of Technology—available from Pauling's archives (courtesy to the Library of Oregon State University for their kind consent to reproduce the copies of these materials in the present monograph).

Finally, we would like to try analyzing the reasons of why Linhart's seminal results were obtaining a total over-all rejection. Dr. Linhart gives a number of hints in the second volume of his diaries, which we would like to republish in the present chapter.

PUBLISHED WORK
OF
GEORGE A. LINHART
1911–23

1. The Hydrolysis of Metallic Alkyl Sulphates:

   American Journal of Science, Vol. **XXXII**, July, 1911.

   American Journal of Science, Vol. **XXXIV**, September, 1912.

   American Journal of Science, Vol. **XXXIV**, December, 1912.

   American Journal of Science, Vol. **XXXV**, March, 1913.

2. The Rate of the Reduction of Mercuric Chloride by Phosphorous Acid:

   American Journal of Science, Vol. **XXXV**, April, 1913.

3. The Instantaneous Transformation of Mercuric Compounds to Calomel, and the Application of this Process in Cases of Mercuric Poisoning:

   New York Medical Journal, June 14, 1913.

4. The Association of Mercuric Chloride in Water Solution:

   Journal of the American Chemical Society, Vol. **XXXVII**, Feb. 1915.

5. The Rate of Reduction of Mercuric Chloride by Sodium Formate:

   Journal of the American Chemical Society, Vol. **XXXVII**, Jan. 1915.

6. The Equilibria of Mercuric Chloride with Other Chlorides:

   Journal of the American Chemical Society, Vol. **XXXVIII**, July, 1916.

7. The Potential of the Mercury Electrode against
   Mercurous Ion:
   Journal of the American Chemical Society,
   Vol. XXXVIII, Nov. 1916.

8. A Comparison of the Activities of Two Typical
   Electrolytes:
   Journal of the American Chemical Society,
   Vol. XXXIX, Dec. 1917.

9. The Applicability of the Ferro-Ferricyanide
   Electrode to the Measurement of the Act-
   ivities of Electrolytes in Concentrated So-
   lutions:
   Journal of the American Chemical Society,
   Vol. XXXIX, 9pril, 1917.

10. The Reduction of Mercuric Compounds:
    Journal of the American Chemical Society,
    Vol. XXXIX, May, 1917.

11. The Preparation of the Pure Sodium Phos-
    phite as an Antidote for Mercuric Chloride
    Poisoning:
    Journal of Laboratory and Clinical Medicine,
    Vol. 11, July, 1917.

12. Equilibrium Data on the Polybromides and
    Polyiodides of Potassium:
    Journal of the American Chemical Society,
    Vol. XL, Jan. 1918.

13. The Applicability of the Precipitated Silver-
    Silver Chloride Electrode to the Measure-
    ment of the Activity of Hydrochloric Acid
    in Extremely Dilute Solutions:
    Journal of the American Chemical Society,
    Vol. XLI, Aug. 1919.

14. A Method of Purifying Water and its Significance in Laboratory Practice and Sanitary Water Analysis:
Agricultural Experiment Station, University of California, Aug. 1919.

15. The Degree of Ionization of Very Dilute Electrolytes:
Journal of the American Chemical Society, Vol. XLI, Dec. 1919.

16. The Free Energy of Biological Processes:
Journal of General Physiology, Vol. II, January 20, 1920.

17. A New and Simplified Method for the Statistical Interpretation of Biometrical Data:
University of California Publications in Agricultural Sciences, Vol. 4, No. 7, pp. 159-181, September 10, 1920.
Also Proceedings National Academy of Sciences, November, 1920.

18. The Relation Between Entropy and Probability. The Integration of the Entropy Equation:
Journal of the American Chemical Society, Vol. XLIV, Jan. 1922.

19. Unidirectional Probability. Its Manifestation in Every-day Life:
Eureka Junior College Journal, Science, Arts and Crafts, Sept. 1921.

20. Correlation of Entropy and Probability:
Journal of the American Chemical Society, Vol. XLIV, Sept. 1922.

21. Out of the Melting Pot: 1923.

To our sincere regret, the paper number 19 in the listing above seems to be lost for us, although it is definitely of tremendous significance (at least for the topics we discuss in the book at hand)—I have tried to fetch it in applying to the colleagues at Eureka, CA, USA, but in vain. Meanwhile, the notion of unidirectional probability might be fetched (although with a truly big deal of difficulty!) in the handbooks on the theory of probability and statistics, cf., e.g., the following very helpful and useful reference [1]. To get acquainted with the notion of the '*directional probability*' and its implications, please, see pp. 65–68 of the book just cited, but do not look for any reference to Dr. Linhart. And after reading this book, the question arises: Is the mentioned '*directionality*' of the probability just what Dr. Linhart was studying?

Well, sometimes, it is also possible to encounter the notion of the '*unidirectional probability flux*', see e.g., the work [2], as well as the pertinent references therein. Interestingly, the latter paper deals with the Brownian dynamics (BD) simulation of the ionic currents in protein channels of biological membranes, and is therefore related to the modern field of the computational molecular biophysics.

It is clear that the BD approach ought to be a well-tested valuable computational tool, although its physical roots are in effect questionable, for the Brownian motion, which is nothing more than just the consequence of the thermal motion commonly observable in our everyday life.

Noticeably, it is just the thermal motion that tends to be considered a fundamental basis for everything all over the world, due to misunderstanding the entropy notion and forcing as a result the Boltzmann–Planck formula, $S = k * \ln(W)$—just as it is—to be the herald of a certain "supreme truth," just a funky standpoint all the protagonists of this book were tremendously working against, with Dr. Linhart in their first row.

Remarkably in this connection, the intrinsic interrelationship between the notions of probability and symmetry, as well as the pertinent modalities of the probability inferences started attracting detailed attention only relatively recently [3–5]. Along with this the works must be underlined, which consider the probability notion in its entirety, in its philosophic details to *avoid playing tricky games with probabilities and statistics* [6–12].

But at this very point we would greatly appreciate to stop discussing this interesting and important topic and to come back to Dr. Linhart's works. And we shall without fail revert to the former as well, but only after considering the work and life of Dr. Linhart, which is of much more interest and importance, as we shall soon see.

Howbeit, in completing Dr. Linhart's publication list, we note the following works of him:

1. G. A. Linhart (1932) *J. Phys. Chem.* 36: 1908
2. G. A. Linhart (1933) *J. Chem. Phys.* 1: 795
3. G. A. Linhart (1933) *J. Phys. Chem.* 37: 645
4. G. A. Linhart (1934) *J. Phys. Chem.* 38: 1091
5. G. A. Linhart (1935) *Am. Math. Mon.* 42: 224
6. G. A. Linhart (1936) *J. Phys. Chem.* 40: 113
7. G. A. Linhart (1929) "The entropy of physical growth." *Occasional Papers of Riverside Junior College.*
8. G. A. Linhart (1929) "The relation between chronodynamic entropy and time." *Occasional Papers of Riverside Junior College.*
9. G. A. Linhart (1938) "General laws of dynamics and their application." *Occasional Papers of Riverside Junior College.*

And that was it, to our sincere regret! Hence, owing to the clear collegial reasons, we would greatly appreciate to re-publish here the points (7–9) of the above list. Then, in sincerely trying to investigate, what were the actual reasons for such an abrupt bringing Dr. Linhart's publishing efforts to a definite halt, we shall first gain insight into the correspondence between Dr. Linhart and a widely known representative of the USA academic publicity, the Nobel Prize winner (1954, 1963) Prof. Dr. Linus Pauling (1901–1994). Finally, we shall hear to Dr. Linhart himself, to learn about his ideas, thoughts and opinions.

OCCASIONAL PAPERS
OF
RIVERSIDE JUNIOR COLLEGE

Vol. 4, No. 4   MAY 9, 1929

The Entropy of Physical Growth

By

George A. Linhart

PUBLISHED BY THE JUNIOR COLLEGE
RIVERSIDE, CALIFORNIA

(CONTRIBUTION FROM THE RIVERSIDE JUNIOR COLLEGE MATHEMATICS DEPT.)

THE ENTROPY OF PHYSICAL GROWTH

BY GEORGE A. LINHART

Received May 6, 1929

INTRODUCTION

This article is written for those who are interested in the laws of physical growth, but who are not especially interested in the mathematical derivations of those laws. We shall therefore omit the theory entirely and merely give the two equations based upon thermodynamic and chronodynamic principles, one giving the relation of growth and time, the other connecting the capacity for growth and the chronodynamic entropy, which may be defined as the amount of energy rendered unavailable, in the process of a growing body, for growth purposes. The two equations are:

$$G = \frac{Gi \; k \; t^K}{1 + k \; t^K} \qquad (1)$$

$$S = \frac{2.3 \, Gi}{K} \log \frac{Gi}{Gi - G} \qquad (2)$$

In these equations G denotes the weight of the growing body at time t, Gi denotes the hypothetical maximum weight of the growing body at ti, and S denotes the entropy. It may be observed that K is the efficiency constant for a given process; for the larger the value of K, the less the entropy; while $(\log k) / K$ is the group index, which is approximately the same for closely related processes, the numerical value of which is marked on every curve by a dash across it. It may be pointed out that these curves are far more instructive and illuminating than the usual "observed" and "calculated" values. The reason is obvious. In biological processes the variability is often so large that the differences between the calculated values and the experimental data are meaningless, unless the probable error is given for each

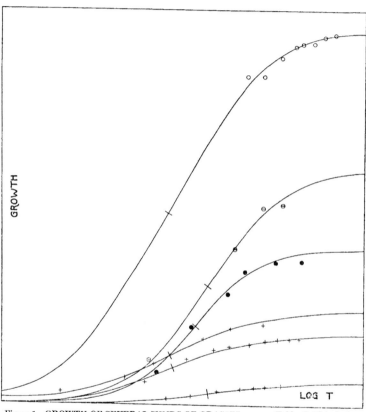

Figure 1—GROWTH OF SEVERAL KINDS OF GRASSES   Tables I–VI inclusive

set of experiments and also the number of experiments in each set, so that one could then judge of the accuracy of the mean. Finally, from the symmetry of the entropy boundary curve may be calculated the hypothetical value for Gi, which in most biological processes can not be reliably calculated from "final" weights or measurements, as in the case of a chemical reaction of definite initial concentration.

EXPERIMENTAL

Experiments on the growth of certain grasses, to be published soon by Doctors Sampson and McCarty, University of California.

1
Agropyron smithii

t days	w grams
37	34.05
79	118.95
103	150.30
122	155.27
i	180.00

II
Agropyron smithii

t days	w grams
89	249.4
103	249.9
121	264.0
136	273.0
144	274.7
160	275.5
176	280.6
188	282.3
i	290.0

III
Agropyron smithii

t days	w grams
36	37.30
52	34.54
66	41.85
76	45.14
91	46.54
105	49.74
117	50.73
127	51.13
142	51.46
i	55.00

IV
Agropyron smithii

t days	w grams
17	9.54
30	19.95
39	31.60
60	50.16
77	58.63
103	62.00
i	75.00

V
Aristida longiseta

t days	w grams
44	4.65
54	6.65
68	10.20
84	12.50
95	13.92
105	14.72
120	15.23
i	18.00

VI
Bulbilis dactyloides

t days	w grams
41	24.70
54	58.95
75	85.00
87	101.70
115	109.20
145	109.90
i	120.00

3

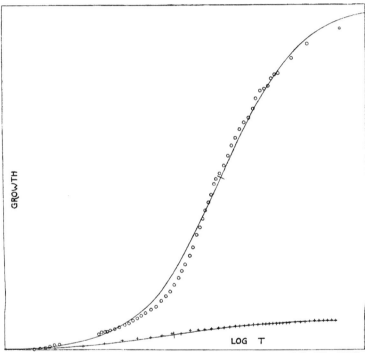

Figure II—GROWTH OF WHITE RATS AND WHITE MICE   Tables VII and VIII

Experiments on the growth of white mice and white rats, taken from T. B. Robertson's "Chemical Basis of Growth and Senescence," tables XXVII and XXVIII, and from the "The Journal of General Physiology," 8, 463, 1926, by the same writer.

VII and VIII
GROWTH OF WHITE RATS AND WHITE MICE

t days	w grams	t days	w grams	t days	w grams
13	0.040	51	29.5	114	152.3
14	0.112	53	31.8	119	160.0
15	0.168	56	34.9	124	168.8
16	0.310	59	37.8	129	177.6
17	0.548	62	42.2	134	183.8
18	1.00	65	46.3	139	191.4
19	1.58	68	50.5	146	197.3
20	2.63	71	56.7	153	202.5
21	3.98	74	62.5	160	209.7
22	4.63	77	68.5	165	218.3
33	13.3	80	73.9	172	225.4
34	14.8	83	81.7	179	227.0
35	15.3	86	89.1	186	231.4
36	15.2	89	99.3	193	235.8
37	16.5	92	106.6	200	239.4
39	17.8	95	113.8	207	239.8
41	19.5	98	121.3	238	252.9
43	21.2	101	128.2	278	265.4
45	22.9	104	135.0	387	279.0
47	25.3	107	143.8	i	300.0
49	27.4	110	148.4		

t days	w grams	t days	w grams	t days	w grams
1	0.45	15	18.34	29	23.05
2	0.85	16	19.07	30	23.05
3	1.47	17	19.59	31	23.23
4	2.99	18	20.02	32	23.41
5	5.09	19	20.45	33	23.54
6	7.99	20	20.73	35	24.04
7	9.51	21	21.21	37	24.05
8	10.7	22	21.46	39	24.43
9	12.23	23	21.70	41	25.09
10	14.05	24	21.80	43	25.36
11	15.43	25	22.09	45	25.23
12	16.52	26	22.14	47	25.69
13	16.98	27	22.73	49	26.00
14	17.59	28	22.70	51	25.93
				53	26.00
				i	27.00

5

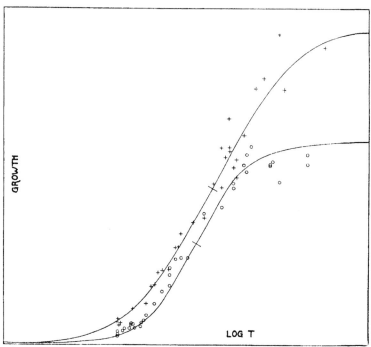

Figure III—GROWTH OF HOLSTEIN AND JERSEY CATTLE   Tables IX **and X**

Experiments on the growth of cattle, taken from a paper by Brodi and Etling, University of Missouri, Research Bulletin 89.

IX

GROWTH OF JERSEY CATTLE

t	w	t	w	t	w
days	kilos	days	kilos	days	kilos
270.5	17.6	342	45.0	840	324.0
272	21.5	360	61.0	840	336.0
273	27.0	390	85.0	930	374.0
276	29.0	420	111.0	960	395.0
293	34.0	450	122.0	960	359.0
305	34.0	450	161.0	1020	413.0
308	40.5	450	145.0	1200	374.0
310	41.5	480	177.0	1200	372.0
314	41.0	510	179.0	1330	381.0
320	34.0	540	181.0	1320	338.0
340	37.0	630	272.0	1740	374.0
350	50.0	750	336.0	1740	395.0
				i	420.0

X

GROWTH OF HOLSTEIN CATTLE

t	w	t	w	t	w
days	kilos	days	kilos	days	kilos
271	47.5	480	202.0	810	406.0
274	53.5	495	209.0	840	368.0
277	41.0	510	231.0	870	347.0
279	44.5	570	254.0	870	386.0
319	75.0	630	263.0	930	436.0
360	84.0	690	336.0	1050	533.0
375	120.0	750	327.0	1140	553.0
390	125.0	750	408.0	1320	644.0
405	150.0	780	389.0	1380	531.0
420	154.0	810	469.0	2070	617.0
450	156.0	810	413.0	i	650.0

7

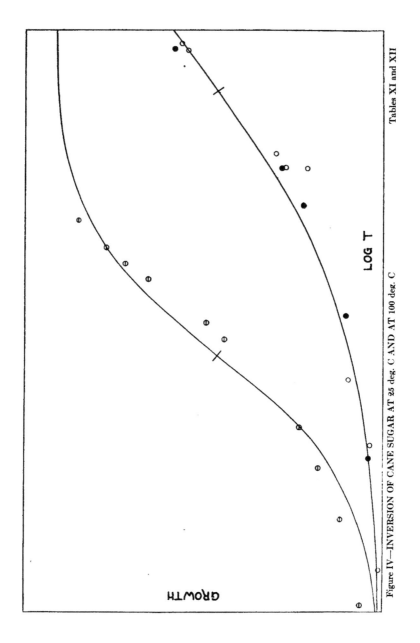

Figure IV—INVERSION OF CANE SUGAR AT 25 deg. C AND AT 100 deg. C

Tables XI and XII

Experiments on the inversion of cane sugar by invertase at 25° C. (Henri, Zeit. phys. Chem. 39, 194), and in pure water solution at 100° C. (Kullgren, Zeit. phys. Chem., 41, 408).

INVERSION OF CANE SUGAR

XI				XII	
BY INVERTASE				IN PURE WATER	
t min.	x / a	t min.	x / a	t min.	x
32	0.012	69	0.042	25	0.86
75	0.037	183	0.112	45	1.58
186	0.103	392	0.240	65	2.40
429	0.228	504	0.305	85	3.11
505	0.292	1136	0.630	155	5.90
557	0.322	i	1.000	175	6.53
1120	0.589			235	8.71
1172	0.611			260	9.58
i	1.000			290	10.31
				350	11.36
				i	12.36

Experiments on the hydrolysis of Ethyl Barium Sulphate in pure water at 60° C (Linhart, Amr J. Sc., 1911).

XIII

t hours	x (observed) equivalents	x (calculated) equivalents	t hours	x (observed) equivalents	x (calcu.) equivalents
141.5	0.0024	0.00240	498.5	0.0098	0.00955
313.0	0.0054	0.00574	862.5	0.0173	0.01730
			i	0.3188	

9

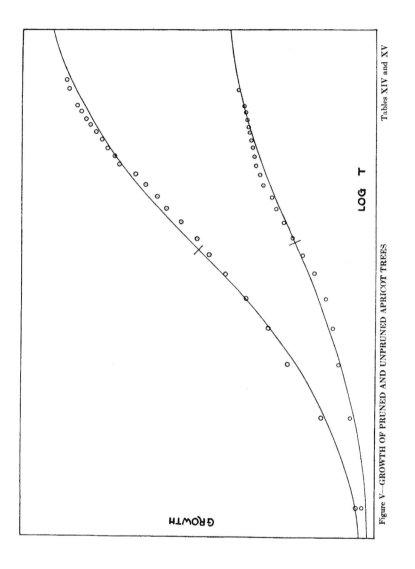

Figure V—GROWTH OF PRUNED AND UNPRUNED APRICOT TREES

Tables XIV and XV

The following three sets of experiments deal with lengths instead of weights, so that, in these cases the values for the entropies can not be calculated. However, it is not surprising that equation (1) holds also for these processes, since the correlation in many such cases of length and weight are approximately rectilinear.

Experiments on pruned and unpruned apricot trees (Reed, H. S., American Journal of Botany, 7, 327, 1920.)

XIV

PRUNED APRICOT TREES

t weeks	height cm.	t weeks	height cm.	t weeks	height cm.
1	13	10	142	19	194
2	37	11	148	20	197
3	60	12	156	21	200
4	73	13	163	22	203
5	88	14	174	23	
6	102	15	177	24	
7	113	16	182	25	208
8	121	17	186	26	
9	132	18	190	27	210
				i	240

XV

UNPRUNED APRICOT TREES

t weeks	height cm.	t weeks	height cm.	t weeks	height cm.
1	9	10	68	19	87
2	17	11	71	20	88
3	25	12	77	21	89
4	29	13	79	22	90
5	34	14	82	23	
6	42	15	83	24	
7	50	16	84	25	94
8	57	17	85	26	
9	63	18	86	27	
				i	110

11

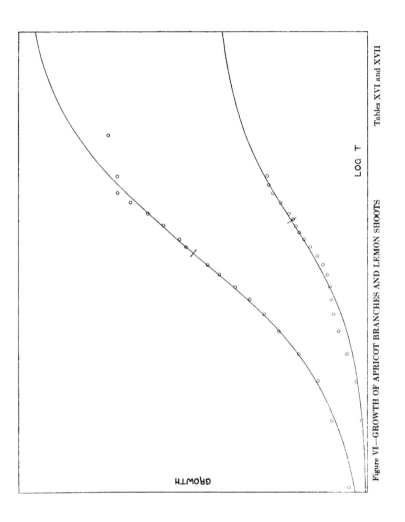

Figure VI—GROWTH OF APRICOT BRANCHES AND LEMON SHOOTS

Tables XVI and XVII

Experiments on the lengths of lemon shoots and apricot branches (Reed, H. S., Proceedings of Natural Academy of Sciences, 14, 3, 221-229; Univ. of Calif. publications in Agricultural Sciences, 5, 1, 1-55).

XVI

APRICOT BRANCHES

t weeks	x cm.	t weeks	x cm.	t weeks	x cm.
1	17.5	7	108.2	15	187.3
2	32.8	8	121.5	17	201.3
3	45.6	9	135.9	19	217.7
4	61.7	10	147.1	21	228.8
5	81.6	12	166.1	25	237.3
6	94.9	13	172.0	i	320.0

XVII

LEMON SHOOTS

t weeks	x cm.	t weeks	x cm.	t weeks	x cm.
1	1.69	8	35.19	15	65.96
2	4.81	9	37.26	16	68.49
3	10.59	10	40.89	17	72.13
4	19.26	11	46.35	19	80.30
5	26.62	12	52.72	21	87.64
6	31.33	13	58.53	23	91.28
7	33.72	14	62.90	25	92.32
				i	140.00

13

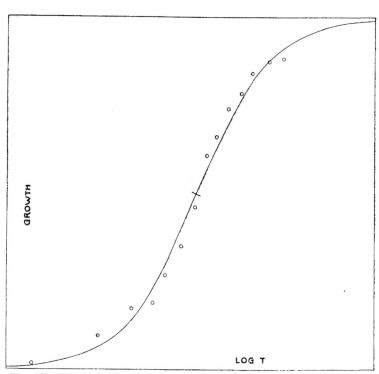

Figure VII—GROWTH OF STIPA PULCHRA GRASS   Table XVIII

Experiments on the heights of **stipa pulchra** grass, soon to be published by Doctors Sampson and McCarty, University of California.

XVIII

t days	x cm.	prob. error cm.	t days	x cm.	prob. error cm.
11	5.1	1.17	180	12.7	1.42
17	6.7	0.28	187	15.0	1.33
29	6.7	0.02	196	17.5	1.20
49	6.8	0.28	205	20.8	1.40
59	6.9	0.45	212	25.2	2.11
80	7.0	0.83	219	26.8	2.08
96	6.9	0.57	227	29.2	2.14
105	7.0	0.63	237	30.5	2.15
117	7.2	0.58	247	32.2	1.51
133	7.5	0.59	262	33.2	0.99
155	9.9	0.76	275	33.4	1.02
170	12.0	0.75	i	37.0	

CORRECTED VALUES OF THE ABOVE TABLE

t days	x cm.	t days	x cm.	t days	x cm.
4	0.0	94	5.2	143	19.8
20	-0.1	104	5.7	151	22.2
29	0.0	111	8.0	161	23.5
41	0.2	120	10.5	171	25.2
57	0.5	129	13.8	186	26.2
79	2.9	136	18.2	199	26.4
				i	30.0

15

GROWTH CYCLES

In so far as the entropy of growth is concerned, it is immaterial how many cycles a natural process goes through. All that is desired is an area within certain boundaries. If the equation for the boundary curve is known its area can often be found with the aid of calculus; if the equation is not known and the curve is highly irregular, it can be traced on uniform tracing cloth, cut out, rolled into convenient form and weighed. The values thus obtained are far more accurate than those calculated by the use of mechanical contrivances.

Attention may be called to the fact that minor cycles of growth are statistically ironed out over a long period of time. But this is not true in the case of plants that mature in one or two seasons and are susceptable to seasonal changes. Thus, the experiments of Sampson and McCarty (see table XVIII) on **Stipa Pulchra** show that this grass, starting growth on October 14, shot up to a height of approximately seven centimenters in about two weeks time and stayed at that height until about the middle of January, or ten weeks. Then the growth began to be characteristically regular, reaching its maximum height at the end of July. This check in early growth Sampson and McCarty attribute to the rather steep drop in temperature; for this period is the coldest of the year in Berkeley, California. And so, if seven centimeters are subtracted from each value, and the time counted from January 1st, the data fall closely along the entropy boundary line, as shown in figure VII.

THE GROWTH OF MAN

In the past two years the writer has examined many publications on growth and has found in most cases where the raw data were given that the probable error of any set of experiments for a given time exceeded considerably the fluctuation of its mean from the entropy boundary line, although these deviations seem to suggest the existence of cycles. However, we must admit that cycles are not merely sporadic phenomena but inherent properties of

16

growing things in general. They are, however, statistically inconsequential, except in cases like the **Stipa Pulchra**; but here the phenomenon is explained by the fact that the growth is actually checked for months because of the change of season, a sort of hibernation phenomenon. In case of man, however, who grows more or less steadily and lives, in the present generation, to an average age of nearly sixty years, it might be expected that his weight would follow the entropy boundary line rather closely. This is actually found to be the case from birth to the age of one year, and from twenty years to the end, or three score and ten. A glance at figure VIII will show, however, that at the age of one year the curve begins to digress, reaching a depression of about 33 per cent, and then steeply rises and crosses the curve at the age of fifteen and a half years. It approaches the animal and plant curve with such speed that it swings beyond to the extent of about fifteen pounds before the age of twenty is reached, and then just as rapidly settles down to the general path of growth for animals and plants. To the writer's knowledge, this enormous deviation is found only in the growth of man, and it seems plausible that the energy which has thus been diverted from the growth of the bulk of the body has been used up in the building of the more intricate human instrument, the nervous system, including the brain, which, according to modern psychology, reaches its full capacity at the age of about fifteen and a half years. This event is indicated on the graph by the intersection of the two curves.

SOURCES OF THE DATA

The weights from birth to eighteen years of age were compiled on charts for the Dry Milk Company, 15 Park Row, New York City, by Dr. B. T. Baldwin and Professor T. D. Wood. These figures represent hundreds of thousands of cases, and the charts are used extensively in hospital clinics for the growth control of children, especially of the pre-school age. The weights for the years twenty to sixty were taken from a table published by the Metropolitan Insurance Company. These figures also are based upon

17

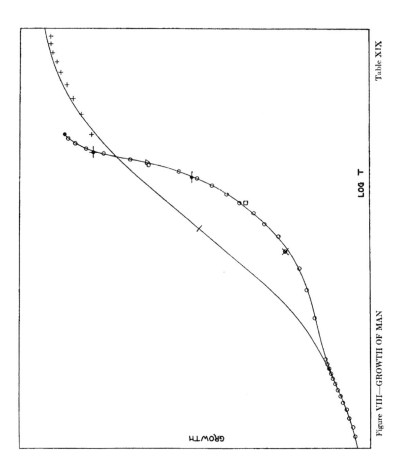

Figure VIII—GROWTH OF MAN

Table XIX

hundreds of thousands of cases and are equally reliable. Furthermore, these weights agree well with those of similar tables published by several insurance companies of the United States. However, the weights of these insurance companies represent weights in street attire and are therefore about six pounds too high. Six pounds were therefore subtracted from each of the average weights for all heights of a given age, and the values so obtained are indicated in the upper right hand corner of the graph by crosses, while the weights for the ages one to eighteen are indicated by plain circles. It may be of interest to note that a survey made of the Riverside school population of about ten thousand children completely confirms the weights on the graph for that period, namely, five to twenty years of age.

The solid circle with the forty-five degree cross represents the average weight (41 lbs.) of all the kindergarten children whose average age is five years. The square represents the average weight (60 lbs.) of the grade school children whose average age is nine years. The solid circle with the horizontal bar through it represents the average weight (86 lbs.) of a considerable group of Y. M. C. A. boys whose average age is twelve years. The triangle represents the weight (108 lbs.) of the boys of the junior high schools whose average age is 14 years. The solid circle with the ninety degree cross through it represents the average weight (134 lbs.) of the boys of the upper high school whose average age is sixteen years. The plain solid circle represents the average weight (148) lbs.) of the men of the junior college whose average age is twenty years.

19

XIX

GROWTH OF AMERICAN MALES FROM BIRTH TO SIXTY YEARS OF AGE

t mo.	w lbs.	t mo.	w lbs.	t mo.	w lbs.
9	7.5*	21	21.4	201	129.1
10	7.5-9.1	33	26.6	213	137.4
11	9.1-10.8	57	34.7	225	143.1
12	10.8-12.6	69	40.9	237	146.3
13	12.6-14.1	81	44.2	249	135
14	14.1-15.3	93	50.9	309	140
15	15.3-16.3	105	56.2	369	144
16	16.3-17.5	117	63.0	429	147
17	17.5-18.3	129	69.3	489	150
18	18.3-19.0	141	77.3	549	152
19	19.0-19.6	153	83.5	609	154
20	19.6-20.2	165	92.6	669	155
		177	107.2	729	155
		189	115.9	i	166

*The average weight at birth for 204 cases, taken from the files of the Riverside Community Hospital, is 7.6 lbs.

20

SUMMARY TABLE

Exp. No.	K	-log k	(-log k) / K
I	2.88	5.16414	1.79
II	2.52	4.18373	1.66
III	2.68	4.43793	1.66
IV	2.34	3.86088	1.65
V	2.89	5.20760	1.80
VI	3.65	6.38348	1.75
VII	2.53	5.21622	2.06
VIII	2.00	2.08360	1.03
IX	4.00	11.07080	2.79
X	3.00	8.39600	2.77
XI	1.25	3.65668	2.93
XII	2.20	4.72572	2.14
XIII	1.106	4.80150	4.34
XIV	1.40	1.20228	0.86
XV	1.40	1.23274	0.88
XVI	1.32	1.38870	1.05
XVII	1.61	1.93515	1.99
XVIII	5.46	11.52622	2.11
XIX	1.34	2.60365	1.94

It may be of interest to note that the average of the values for (-log k) / K of experiments I to VI equals 1.72. If the values of these experiments are plotted on a percentage basis they all fall along the same curve whose point of inflection comes at 1.72 which is identical with the average of the above values. This means that all these grasses for different years reach half their growth in 52 plus or minus a few days.

The numerical values of entropies will be given in a paper dealing with the theoretical part of the subject.

21

... And the theoretical paper mentioned above is just following right now...

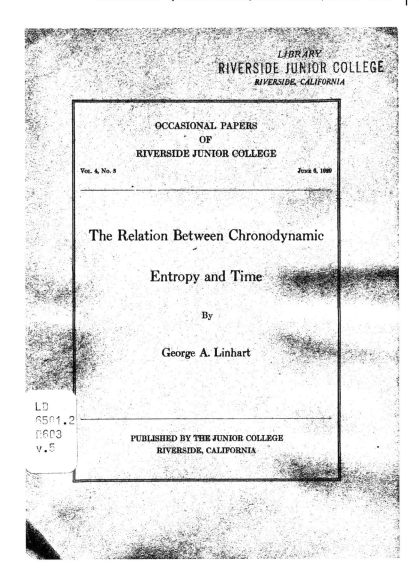

OCCASIONAL PAPERS
OF
RIVERSIDE JUNIOR COLLEGE

Vol. 4, No. 3   June 6, 1929

The Relation Between Chronodynamic

Entropy and Time

By

George A. Linhart

PUBLISHED BY THE JUNIOR COLLEGE
RIVERSIDE, CALIFORNIA

(CONTRIBUTION FROM THE RIVERSIDE JUNIOR COLLEGE MATHEMATICS DEPT.)

THE RELATION BETWEEN CHRONODYNAMIC ENTROPY AND TIME

By GEORGE A. LINHART

Received June 6, 1929

INTRODUCTION

Eddington in his delightful book on "The Nature of the Physical Universe" wonders why entropy, so intimately associated with time, should be expressed quantitatively in terms of temperature instead of time. The present writer wondered about that, too, for three consecutive years, during which time it was his good fortune to be in the very midst of multitudinous entropy calculations under the direction and guidance of G. N. Lewis. His first attempt at expressing his wonderment in mathematical form appeared in a rather humble periodical of approximately zero circulation. No wonder Eddington makes no reference to it. It was the Eureka Junior College Journal of Science, Arts and Crafts (1921).

THE ENTROPY CONCEPT

When a scientist has to define a new term through the medium of words, the faster he can get to the formula the better he feels. A young philosopher once came to the great Maxwell for advice, or rather for approval, concerning a new theory of the universe. After the young man explained his theory in detail, Maxwell impatiently exclaimed "Put it in the form of an equation, mon, put it in the form of an equation!" It is next to impossible to define anything, in words, entirely satisfactorily even to those who speak the language of the framer of the definition. On the other hand, a mathematical symbol is at once recognized by every intelligent person the world over, and sometimes even understood. For instance, one might define *entropy* by the fraction, q/T, and let it go at that; or, if one had the literary gift of an Eddington, one might write the most delightful book in the realm of philosophy solely on the definition of entropy, and especially on chrono-dynamic entropy, a term just arrived and as nervous and hopeful as a young immigrant newly arrived in New York City. It is really hard to decide which method to adopt. On the one hand, one's non-mathematical colleagues deserve some consideration, and, on the other hand, one's scientific associates may accuse one of looseness of speech. Under the circumstances, perhaps the best way to start will be with the two ideas, *progress* and *hindrance,* for it is a universally observable fact that every form of progress in an inhabited universe creates or provokes a characteristic hindrance.

By progress is meant any unidirectional phenomenon in nature, such as the growth of a plant or an animal, and by hindrance the contesting and ultimate limitation of every step of progress. In other words, progress is organized effort in a unidirectional motion, and hindrance is not so much the rendering of energy unavailable for that motion as it is the disorganization of the effort to move; it acts as a sort of stumbling block in every walk of life. It is this property of matter which the writer wishes to measure quantitatively in relation to time, and which he designates as chronodynamic entropy.

THEORETICAL

When a material body at extremely low temperature receives energy its rise in temperature is not proportional to the amount of energy received. The quantity of energy necessary to raise the temperature of the body one degree gradually increases and tends to approach a maximum. This phenomenon first suggested to the writer the notions of progress and hindrance. Thus, in the region of the absolute zero the energy flows freely through a body, and nearly all of it could theoretically be made to do work, but as the temperature rises the constituent parts of the body acquirer greate and greater agitation and hindrance to do useful

2

work begins. On the basis of this behavior the writer derived a statistical expression* which enabled him to calculate the entropies of all substances for which accurate data for the heat capacities of these substances were available at that time. Not only did the values so calculated agree with those calculated by mechanical means, but the temperature values for each element and compound investigated were reproduced to an accuracy exceeding by far the probable experimental errors.

But why should entropy be expressed in terms of temperature rather than in terms of time? Or why should it not be expressed in terms of both temperature and time? For the mass, as a whole, gains in entropy both with respect to temperature and time. This naturally suggested the search for a relation between temperature and time. But first it was necessary to establish a relation between entropy and time. Such a relation, supported by numerous experimental measurements,** is developed in this article.

When any unidirectional process takes place at constant temperature maintained by an efficient thermostat there follows an exchange of energy in the form of heat either from the reaction chamber to the thermostat or from the thermostat to the reaction chamber, depending upon whether the reaction is accompanied by an evolution of heat or by an absorbption of heat. In either case the process being unidirectional can not reverse itself of its own free will and a certain amount of organized effort has been lost. Furthermore, the extent to which this process can not reverse itself will depend upon how far above or below the optimum environment the process is made to take place. The measure of this loss of inherent organized effort due to hindrance will be identified in the present discussion with the usual thermodynamic measure, q/T, where in thermodynamics q denotes

*Linhart, Journal of the American Chemical Society, Vol. XLIV, 140, 1881, 2960, (1922).
**Linhart, Occasional Papers of Riverside Junior College, Vol. 4, No. 2, May 6, 1929.

the infinitesimal amount of heat exchange at constant temperature, and T the chosen temperature, reckoned on the absolute scale.

It is precisely at this juncture that the science of chrono-dynamics enters to take care of the element of time, so that there shall be an additional q/T, where q denotes the instantaneous exchange of an infinitesimal amount of heat under the same given conditions, and T denotes the time, reckoned not on an absolute scale but from the instant the process begins to the instant of its completion, or, for practical purposes, to 99.9 per cent, since no process in nature goes entirely to completion. It seems plausible, therefore, that the same mathematical laws, differing only in the units of measurement, should hold for both types of entropy. This is actually the case, as will be presently shown.

The idea of a process taking place at constant conditions may be carried over into the open air. Here, the sun serves as the thermostat and the weight of the atmosphere as the pressure. And while they are not actually constant over a short span of time, their variability over a single season is far less than the variability in the growth of a single species of animal or plant. This may be verified from the growth of plants in so-called green-houses, where the factors of light, heat, moisture, and so on, are all under control. It will therefore be an easy task to set up the necessary equations for the calculation of the chronodynamic entropy.

CHRONODYNAMIC ENTROPY AND THE CAPACITY FOR GROWTH

From the definition of q just given, it is evident that for an infinistesimal increase in time, $q=dE=RGdT$. Here, G denotes the mass of the growing body at time T, and R is the proportionality factor between mass and energy. It may also be recalled that q/T has been defined as the instantaneous change in entropy, which may be denoted by RdS. Whence,

4

$$RdS = RGdT / T \qquad (1)$$

From the viewpoint of thermodynamics it is immaterial whether a piece of work is accomplished in five seconds or in five hundred years. All that is required to make the desired calculations is a knowledge of the initial and final states of the substances undergoing change. Hence, the property of matter designated here as hindrance is of no concern. But from the viewpoint of chronodynamics, it is the underlying principle of all phenomena in a world of momentum. Without hindrance the "Law of Probability" becomes an empty phrase, but with hindrance the searching for a set of empty space equations to fit an inhabited universe is akin to hunting for the "Universal Solvent." But so long as such a school of philosophy is in vogue and in power the search for Hohlraum equations will continue, and any attempt to thwart it may meet with the fate of the grandfather of the famous Arrhenius. The story as told by Arrhenius is about as follows:

At a conference of the most distinguished scientists of the day there was a heated discussion concerning the properties of the universal solvent, each scientist claiming priority to certain ideas. At the very height of the discussion Arrhenius rose to his feet and in a mild inquiring manner asked the very pertinent question, "Gentlemen, judging from your animated discussion one would think that you were on the verge of discovering the universal solvent. Now, don't you think that you had better begin thinking of a container in which to store your solvent?" There followed an instantaneous explosion of human emotions, and but for the interference of a practical, wise old king, Arrhenius would have been put to death. As it was, he escaped with disgraceful banishment.

ENTROPY AND PROBABILITY

The notions of progress and hindrance suggest at once the possible relationship between entropy and probability. Such a relationship for thermodynamical processes has

5

already been established,* and the equation expressing that relationship holds equally well for chronodynamical processes, as it naturally should from the following considerations.

The question might be asked, "What is the degree of hindrance in any natural process before it occurs?" Obviously, there can be no hindrance to anything that does not exist. But at the instant of inception of the process hindrance sets in and continues to increase until ultimately it checks nearly all progress, and reduces to a minimum the chance of any further advance. At this juncture the outcome of the process is said to have approximately attained its maximum. It is clear then that the increase in mass is in the same direction as the increase in hindrance, or entropy, and the ratio of their infinitesimal increments may be expressed by the relation,

$$\frac{dG}{dS} = K\left(\frac{Gi-G}{Gi}\right) \tag{2}$$

where Gi denotes the hypothetical maximum mass of the growing body, and (Gi—G)∕Gi, the capacity for growth, which is unity at inception and statiscally zero at completion. From equation (2) it is obvious that K is the efficiency constant of the process; for the smaller the value of K, the greater the hindrance (see equation 6) and the slimmer the chance for the individual to survive. Of the many life processes thus far investigated by the writer, the value for K in each case exceeded unity.

Combining now equations (1) and (2) gives

$$\frac{dG}{dT} = \frac{K}{Gi}\left(\frac{Gi-G}{T}\right)G \tag{3}$$

which on integration gives

$$\log\frac{G}{Gi-G} = K\log T + \log k \tag{4}$$

or,

---

*Vid. loc. cit.

6

$$G = \frac{Gi\ k\ T^K}{1 + k\ T^K} \qquad (5)$$

With the value for K obtained from equation (4) for any problem, the validity of equation (2) in the integrated form, . namely,

$$S = \frac{2.3\ Gi}{K} \log \frac{Gi}{Gi - G} \qquad (6)$$

between the limits zero and 99.9 per cent, or any value in between, may then be tested by comparing its values for S with those obtained from equation (1) by mechanical means from the area under its curve. Its graph may be constructed by plotting either G against log T, or G$/$T against T. In the latter form equation (1) can now be readily integrated by combining it with equation (5), thus,

$$dS = \frac{Gi\ k\ T^K\ dT}{(1 + k\ T^K)T} \qquad (7)$$

whence,

$$S = \frac{2.3\ Gi}{K} \log\ (1 + k\ T^K) \qquad (8)$$

Two other interesting relations are those for the velocity of growth and of entropy, respectively, namely,

$$\frac{dG}{dT} = \frac{K\ k\ Gi\ T^{K-1}}{(1 + k\ T^K)^2} \qquad (9)$$

from equation (5), and,

$$\frac{dS}{dT} = \frac{Gi\ k\ T^{K-1}}{1 + k\ T^K} \qquad (10)$$

from equation (7).

CONCLUSION

From equation (9) it is evident that the velocity of growth goes through a maximum and finally approaches zero;

7

from equation (10), dS / dT approaches Gi / T, or near the limit which may never be reached absolutely,

$$dS = Gi \, d \log T \tag{11}$$

whence,

$$S = 2.3 \, Gi \log T + a \ constant \tag{12}$$

The constant of integration of equation (12) just given is equal to the entropy of the growing body from inception to the time the mass of the body had become statistically steady. Thus, the entropy of a nation whose population is fairly steady is equal to its mass times the natural logarithm of the time, reckoned from the beginning of the nation's existence minus the time the population had approximately approached the steady state. This is, of course, equally true for individuals, and for the universe as a whole, whose first law of energy may be stated as follows: The mass of the universe, and therefore its energy which is proportional to it, having presumably reached the steady state, is statistically constant.

Finally, it may be pointed out that the entropy of the universe is increasing much faster than is generally supposed. At this enormous rate of increase in entropy, will the universe run entirely down and die, or will it go on growing forever? When man builds an institution he generally provides for its maintenance. May we not then assume that the creator of the universe was at least as wise as man?

In subsequent articles will be given additional data from the fields of chemistry, physics, agriculture, commerce, and sociology, all illustrative of the phenomena of progress and hindrance.

8

Sure, the last phrase of this very interesting and important preprint demonstrates the genuine indomitable will and excellent fighting qualities of Dr. Linhart, who was uniquely determined to disseminate the important information of truly general significance he could gain in the course of his research efforts. But, this ambitious program could be fulfilled only partially—the deficiency, the truly bad luck, we are trying to amend right now.

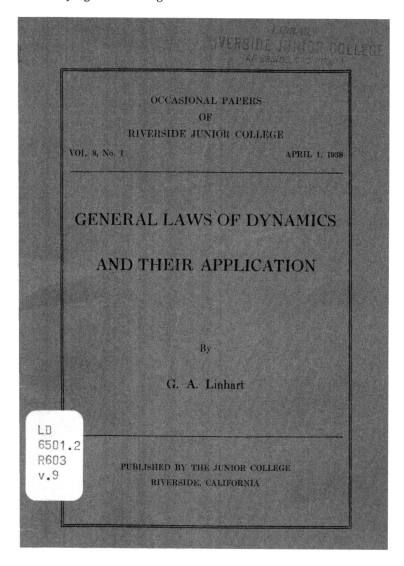

OCCASIONAL PAPERS
OF
RIVERSIDE JUNIOR COLLEGE

VOL. 8, No. 1   APRIL 1, 1938

GENERAL LAWS OF DYNAMICS

AND THEIR APPLICATION

By

G. A. Linhart

PUBLISHED BY THE JUNIOR COLLEGE
RIVERSIDE, CALIFORNIA

GENERAL LAWS OF DYNAMICS AND THEIR APPLICATIONS

G. A. LINHART

THE PENETRATION OF COSMIC RAYS INTO THE EARTH'S ATMOSPHERE

In a recent paper[1] it is shown that the penetration of cosmic rays into fresh water lakes follows the same mathematical law as the penetration of solar rays. In fact, the same law is obeyed by a great number of dynamic processes, and this leads one to surmise that all these processes have some fundamental basis in common.

When a definite amount of a perfectly insulated substance is heated at constant volume, the only visible effects are the rise in temperature and the varying amounts of energy required to increase the temperature of the substance by one degree. This quantity of energy is defined as the heat capacity of the substance and is given by the relation

(1a) $$(\partial E / \partial T)_V = C_V \ ,$$

where E denotes the quantity of energy, T the absolute temperature and C_V the heat capacity. Similarly the entropy is defined by the relation

(2a) $$(\partial S)_V = (\partial E)_V / T \ ,$$

where S denotes the entropy of the process just described, and T the absolute temperature. Another relation given in some treatises on thermo-dynamics is that between C_V and S_V, but, unlike relations (1a) and (2a), this one holds only at extremely low temperatures; for it is derived from Debye's T-cube law, namely,

$$C_V = aT^3 \ ,$$

---

[1] G. A. Linhart, Journal of Physical Chemistry, vol. 40, 1936, pp. 113-119.

1

which leads to the relation

(3a) $$C_V = 3S$$

In a recent article[2] the writer has derived a relation between C_V and S_V that holds for the entire temperature range. It is based upon relation (3a) and upon the fact that C_V approaches a limiting value, $C\infty$, as the temperature approaches infinity. This new relation is

(4a) $$(\partial C/\partial S)_V = K[1 - (C/C\infty)_V] \quad .$$

Finally, by combining (1a), (2a), and (4a), an expression is obtained which gives the relation between C_V and T, namely

(5a) $$\log[C/(C\infty - C)]_V = K \log T + \log k.$$

This relation enables us to test the validity of relation (4a); for if relation (4a) is valid, the graph obtained by plotting the left hand member of (5a) against log T should yield a straight line, and this has been found to be the fact in every case not only at constant volume[3] but even at constant pressure.[2] We shall now show that these laws are not uniquely applicable to thermodynamic processes. Let us consider, for example, the compression of a gas at constant temperature.

When a definite amount of a gas, say one mol, is being compressed at constant temperature, energy is stored among the molecules, although the temperature remains constant. This energy may be designated as ordered energy, or energy of molecular orientation. Moreover, the denser the gas becomes, the more difficult it is to orient the molecules out of their disordered state. Therefore the amount of this energy may be regarded as proportional to the density of the mol of gas, and the relation may be expressed by the equation

(1b) $$(\partial E / \partial P)_T = r C_T \quad ,$$

where C is the molal density of the gas at the pressure P and r the proportionality factor. (rC may be termed the energy capacity and defined as the amount of energy required to shrink the one mol of gas by one cubic centimeter.) Furthermore, in any such process there will generally

---

[2] G. A. Linhart, Journal of Chemical Physics, vol. 1, 1933, pp. 795-797.
[3] G. A. Linhart, Journal of American Chemical Society, vol. 140, 1922, p. 1881.

2

be a loss in the availability of the energy, due to the chaotic state of the particles. This loss in the availability of energy with change in pressure differs from that with change in temperature only in degree but not in kind. Nevertheless, we shall designate it tentatively as piezodynamic entropy and represent its relation to the other variables by the same form of functions as in thermodynamics, namely,

(2b) $$(\partial S)_T = (\partial E)_T / P \quad ,$$

(4b) $$r \, (\partial C / \partial S)_T = K \, [1 - (C / C\infty)_T] \quad ,$$

(5b) $$\log \, [C / (C\infty - C)]_T = K \log P + \log k \quad ,$$

where $C\infty$ denotes the ultimate molal density of the gas. A glance at the tables 1, 2, 3, and 4 of a recent article[1] will show how excellently concordant the values for K are, and how close the calculated values of $(d \, C / C)$ in table 2 are to the experimental errors[2] inherent in such experiments.

Of the many and various problems thus far solved the writer has not found a single exception to equation (5). However, it should be emphasized here that equation (4) is valid only if the process is (statistically) continuous. If the continuity is broken either by the process approaching a change in state of aggregation of the substance, or by some accidental or unavoidable external influence, equation (4) will not be valid; and without it, equation (5) cannot be established. Nevertheless, the values for the energy and the entropy can, in every case, be found by graphical methods. Thus, E / r can be evaluated from the areas under the respective curves of C versus T, or P, while S / r can be found from the respective areas under the curve of C versus (log T), or (log P), since by combining equations (1) and (2) there results the relation

(6) $$S = r \int C \, (d \log T) \, .$$

With this introductory discussion we are ready to take up the problem of cosmic ray penetration into the earth's atmosphere.

It has already been stated that when a substance is heated under definite conditions it requires an increasing amount of energy to raise its

---

[1] G. A. Linhart, Journal of Physical Chemistry, vol. 37, 1933, p. 645.
[2] G. A. Linhart, Journal of Physical Chemistry, vol. 38, 1934, p. 1096.

3

temperature by one degree. This means that at very low temperatures the particles of matter are more efficiently oriented to receive the quanta of energy than they are at higher temperatures; that is, the more chaotic the motions of the particles become, due to rise in temperature, the more energy it will take to orient them so as to increase their capacities. Hence, the rate of consumption of energy for orientation with respect to the absolute temperature may be set proportional to the only other increasing variable in the process, namely, the so-called heat capacity, or,

(1a) $$(\partial E / \partial T) = r C .$$

In treatises on thermodynamics, the r is generally omitted, since C itself is measured in energy units.

It has likewise been shown that in the compression of a gas, the energy for orientation is proportional to the density, or,

(1b) $$(\partial E / \partial P) = r C .$$

Similarly, in the electromagnetic induction of substances a certain amount of energy is required to orient the particles in order to render them susceptible to magnetization. The relation in this case is

(1d) $$(\partial E / \partial H) = r I ,$$

where I is the magnetization intensity; or better, the magnetization capacity, and H denotes the magnetic force.[1]

From the successful solutions of many problems, including those cited above, it seems safe to conclude that the phenomenon of orientation is characteristic of every unidirectional dynamic process, and therefore also of the process of cosmic ray ionization of material particles. Here, a certain amount of the cosmic ray energy is taken up among the particles, as energy of orientation, thus lessening the rays' intensities requisite for ionization; and the larger the number of particles per unit of space, the more energy will thus be taken up. And so, as in the case of the compression of a gas, we may say that the rate of inclusion of cosmic ray energy among the air particles, with respect to the atmospheric pressure, is proportional to the density of the particles, and therefore to the spent cosmic ray intensity. Hence,

(1c) $$(\partial E / \partial P) = r C.$$

---

[1] See article immediately following.

4

Here C is equal to the spent intensity of the cosmic rays; P denotes the atmospheric pressure, measured from the top of the atmosphere downward; and r is the proportionality factor. And here, as in the other processes described above, there is a loss in the availability of the energy, and we have, as before,

$$(2c) \qquad\qquad (\partial S) = (\partial E) / P \ ,$$

where S denotes the piezodynamic entropy. We can now readily derive equation (4c).

Looking upon the cause of entropy, which is an increasing function in all unidirectional dynamic processes, as a sort of disordered state of things in the process [1], it seems plausible that the rate of consumption of energy, for the ionization of the air particles, with respect to the entropy, will be proportional to the remaining ionization intensity $(C_\infty - C)$ of the cosmic rays; that is,

$$(4c) \qquad -r[\partial(C_\infty - C) / \partial S] = (K / C_\infty) (C_\infty - C) \ .$$

Here C_∞ is constant and denotes the ultimate value of C; it is equal to the initial ionization intensity of the cosmic rays per cc. per second. The minus sign is used because the rate is decreasing. The final equation is then obtained by combining equations (1c), (2c), and (4c); whence

$$(5c) \qquad \log [C / (C_\infty - C)] = K \log P + \log k \ .$$

Considering the highly complicated technique of obtaining and of interpreting cosmic ray results, it is extraordinarily gratifying to find how closely the experimental values of the different investigators fall along the respective straight lines when plotted according to equation (5c). This is equally true of the data from solar and from cosmic rays penetrating into fresh water lakes.[2]

Assembling and Tabulating of the Cosmic Ray Data

In the original articles numerical values are not given, except in one case, but are indicated on graphs by the usual marks. However, by drawing rectangular coordinates through the marked points the original values can be reproduced quite accurately. In this manner the values

---

[1] See E. W. Barnes: Scientific Theory. Macmillan Co 1934, p. 238, sec. 186.
[2] G. A. Linhart, Journal of Physical Chemistry, vol. 40, 1936, pp. 113-119.

for C given in the tables were obtained. It should be noted that these values for C are the complements of those given in the original graphs. The maximum value, $C\infty$, is obtained from the inflection point of the curve of C versus log P, since at this point C equals $C\infty/2$, which is obvious from the perfect symmetry of the curve about the point of inflection. This can be shown also analytically by taking the second derivative of the equation

(5c)′ $$C = C\infty\, ke^{K \log P} / (1 + ke^{K \log P})$$

with respect to log P and placing the resulting expression equal to zero. Equation (5c)′ is but another form of (5c).

6

TABLE 1 [1]

Showing the values of the ionization intensity and of the barometric height obtained from the original data as described in this paper.

Piccard and Cosyns $C\infty=3.40$; $K=3.25$ log k $=-7.3087$		Regener (1933) $C\infty=340$; $K=2.74$ log k $=-6.0630$		Millikan, Neher, Haynes $C\infty=245$; $K=3.07$ log k $=-6.9259$		Regener (1934) $C\infty=300$; $K=2.93$ log k $=-6.5798$	
C per cc per sec.	P in mn of Hg.	C per cc per sec.	P in mn of Hg.	C per cc per sec.	P in mn of Hg.	C per cc per sec.	P in mn of Hg.
20	75	1	20	6	50	8.3	13.3
28	82	15	50	18	76	11.9	18.7
34	85	30	75	48	114	—2.4	23.3
36	92	65	100	89	152	3.6	30.9
57	106	100	125	133	190	3.6	39.0
84	123	150	150	162	228	4.8	48.8
88	125	190	175	185	266	8.3	58.5
95	125	228	200	200	304	14.3	71.3
115	140	244	225	211	342	27.4	87.5
120	142	260	250	220	380	55.5	101.0
135	150	284	300	227	418	70.2	121.8
159	160	290	325	231	456	109.5	145.2
236	230	300	350	235	494	128.6	172.6
285	340	310	375	237	532	193.7	206.5
322	435	315	400	238	570	216.4	236.0
330	520	320	450	239	608	238.1	279.7
335	600	330	500	240	646	255.5	312.5
				241	684	265.7	352.9
						275.7	402.3
						280.0	434.0
						284.1	480.8
						287.9	523.8
						290.5	566.5
						290.2	611.1
						289.5	675.3
						290.0	739.0

[1] Piccard and M. Cosyns, Comptes Rendus, vol. 195, 1932, pp. 604-606.
E. Regener, Physikalische Zeitschrift, vol. 34, 1933, pp. 306-323.
R. A. Millikan, H. V. Neher and S. K. Haynes, Physical Review, vol. 50, 1936, pp. 992-998.
Regener and Pfolzer, Physikalische Zeitschrift, vol. 35, 1934, p. 782.

7

In conclusion it may be of interest to show the relation between the entropy and the height of the atmosphere and, from it, to estimate its maximum height. This is shown in the last column of table 2 where P denotes the atmospheric pressure, S the entropy, and H the height of the atmosphere. The values for the constants, $C\infty$, K, and log k, used in the calculations of the values for S are from Regener's data (1933) given in table 1. The values for the atmospheric heights and the barometric pressures are taken from the Monthly Weather Review.[1] These quantities were substituted into the equation

(7) $$S = (r\,C\infty\,/\,K)\,\log\,(1 + kP^K)\quad,$$

which is obtained from equations (1), (2) and (4), and the values for S obtained between the limits zero and P.

TABLE 2

Showing the relation between the entropy and the barometric heights.

P (mn)	S / r	H (km)	$\triangle S / r \triangle H$
20.00	0.4	25.0	— 5.7
151.80	74.0	12.0	—26.1
176.95	100.1	11.0	—32.0
206.77	132.1	10.0	—33.7
238.89	165.8	9.0	—37.2
274.98	203.0	8.0	—39.5
315.84	242.5	7.0	—40.8
361.32	283.3	6.0	—40.0
411.93	323.3	5.0	—40.8
468.22	264.1	4.0	—40.7
530.82	404.8	3.0	—41.4
564.67	425.5	2.5	—39.4
600.21	445.2	2.0	—40.0
637.81	465.2	1.5	—39.8
677.24	485.1	1.0	—39.8
718.75	505.0	0.5	—37.6
760.55	523.8	0.0	

[1] Monthly Weather Review, vol. 47, 1919, p. 161, table II.

8

Taking 40 as the ratio between $\triangle S$ and $r\triangle H$, it is obvious that if this ratio remained constant the height of the earth's atmosphere would be about 13 kilometers, or nearly 8 miles; but a study of existing facts shows that this relation cannot be expected to remain constant: If the values for H are plotted against the logarithms for the values of P all points fall practically on a straight line over the entire range for which accurate data are available; certainly up to about 40 kilometers.[1] Hence, we may express this relation by the empirical equation

(8) $H = (1 / K) \log (760 / P)$;

but equation (7) reduces to the same form if, as a first approximation, we drop the 1 inside the parenthesis $(1 + kP^K)$, so that (7) becomes

(9) $S = r C \infty \log(k'P)$.

This procedure is permissible if kP^K is large compared to unity; and this is evidently the case **only** up to a height of about 7 kilometers, which is not anywhere near the top of the atmosphere. Fortunately, reliable measurements of cosmic ray intensities have been recently obtained up to a height of about 13 mm. of Hg, which corresponds to about 28 kilometers, or nearly 17½ miles. A calculation of the corresponding value for the entropy from equation (7) yields only 0.13 entropy units, or 0.02% of the total entropy from the top of the earth's atmosphere to sea level.

Although this percentage entropy is quite small, it would be erroneous to say that it indicates a height near the top of the earth's atmosphere; for even at a pressure of only one millionth of a millimeter of mercury, there are still about 40 billion molecules within the space of one cubic centimeter of air. This is evident from the fact that at standard conditions a cubic centimeter of air contains about $3(10)^{19}$ molecules. However, since $C\infty$ denotes the ionization intensity of the cosmic rays at the top of the atmosphere, then, as a close approximation 99.99%, say, of $C\infty$ might be defined as the "fringe" of the earth's atmosphere. This value for C corresponds to a pressure of about 5.6 mm. of Hg., which, according to equation (8), (where $K = 0.0624$), corresponds to a height of about 33 kilometers, or nearly 20 miles. The only questionable point is whether the value for $C\infty$ is reasonably accurate.

The most reliable measurements up to a height of 13 mm. of Hg are

---

[1] Monthly Weather Review, vol. 47, 1919, p. 161, table II.

9

those by Regener, 1934; not that his technique or his recording instruments are more refined than those of the other investigators in this field, but simply because he chose a region over the earth where the atmospheric conditions are more favorable for cosmic rays to record their true intensities. In some parts of the earth's atmosphere the cosmic ray intensities as recorded by self-registering instruments, show a rather sharp minimum and a rapid decline at an altitude, in one case, of only about 50 mm. of Hg. However, a glance at the bottom left hand corner of figure 4 will show that the recorded values by Regener and Pfolzer, indicated on the symmetrical S-shaped curve by circles, show no tendency toward a minimum even at a height of 13 mm. of Hg; but they do show fluctuations to the extent of several per cent. That is because, this time, 1934, the author published the raw "unrounded" data, a fortunate circumstance for the present investigation. Hence the value for $C\infty$ determined from the inflection point of the curve, as described in the paragraph immediately preceding table 1, is far more reliable than would be the average of the fluctuating values at the flattening of the curve.

SUMMARY

The several outstanding features noted in this paper are: (a) the generalization of certain fundamental laws generally restricted to the science of thermodynamics are here shown to be applicable to all unidirectional dynamic processes; (b) the introduction of the concept of energy of orientation, which makes possible the generalization of the above laws; (c) the surprisingly close agreement between theory and experimental obervations in a great variety of problems in the different branches of scientific endeavor; (d) the complete and satisfactory solution of several important and interesting problems, such as the relation between entropy, heat capacity, and absolute temperature of solid substances from zero to the melting point; the pressure-volume-temperature relations of gases over the entire range of temperatures and pressures; the relation between magnetic force and magnetization intensity; and the interpretation of the behavior of solar and of cosmic rays in fresh water lakes, and of cosmic rays in the earth's atmosphere. And lastly, the estimation of the "fringe" of the earth's atmosphere to be at a height of about 20 miles above sea level.

10

In conclusion it is interesting to note that the cosmic rays have generally spent half their intensities before reaching a certain depth of the earth's

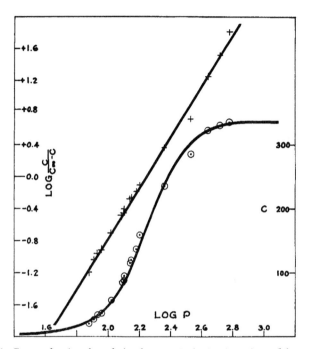

Fig. 1—Curves showing the relation between ionization intensity and barometric pressure, based upon columns 1 and 2, Table 1.

atmosphere, and this, regardless apparently of investigators, location, and, especially, of the super-refinement of measuring instruments which Milli-

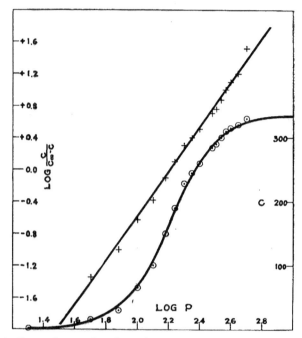

Fig. 2—Curves showing the relation between ionization intensity and barometric pressure, based upon columns 3 and 4 of Table 1.

kan and his associates specifically stress as most essential to obtain significant results. This is shown in the table below where log Pi denotes the
point of inflection on the graphs.

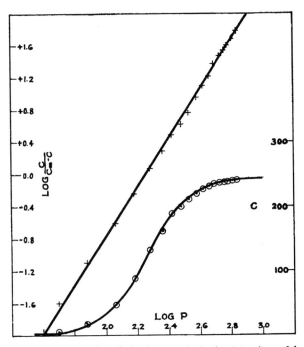

Fig. 3—Curves showing the relation between ionization intensity and baromteric
pressure, based upon columns 5 and 6, Table 1.

13

	$-\log k/K$ $= \log Pi$	Depth of atms. in mm of Hg	Date of Publication
Piccard and Cosyns	2.249	177.4	1932
Regener and Pfolzer	2.246	176.2	1934
Millikan, Neher and Haynes	2.256	180.3	1936

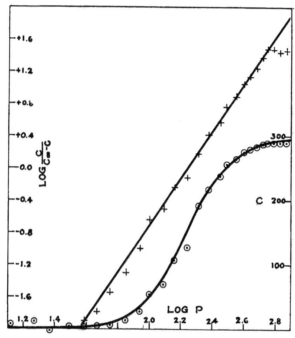

Fig. 4—Curves showing the relation between ionization intensity and barometric pressure based upon columns 7 and 8, Table 1.

14

MAGNETISM AND THE ELECTROMAGNETIC
INDUCTION

The practice of magic with the aid of natural magnets dates back to prehistoric times. In these stone bodies were supposed to reside the disembodied souls, who for certain reasons were barred alike from heaven and from hell. Even as late as 600 B. C. the phenomenon of magnetism was still so regarded by the foremost scientist of that time, Thales of Miletus. Since then, many theories concerning the nature of magnetism have flourished and gone, and after 2500 years the nature of magnetism remains more or less mysterious. However, if we introduce the concept of orientation, we can at least interpret the behavior of magnetization.

As previously stated, when a substance is being magnetized a certain amount of energy is required to orient the particles into a definite pattern. This amount of energy will be proportional to the magnetization capacity of the substance, and we have, as before,

$$(1d) \qquad (\partial E / \partial H) = r I.$$

Here, as in the compression of a gas, there will be a loss in the availability of the energy; and from the analogy between the electric force H and the gas pressure P, the loss in availability of the energy may be designated as piezodynamic entropy, and its relations to the other variables represented by the equation

$$(2d) \qquad (\partial S) = (\partial E) / H.$$

Similarly, equation (4d) is derived from the fact that the rate of increase of the magnetization capacity, I, with respect to the entropy, S, is proportional to the capacity $(I\infty - I)$ still to be attained; that is,

$$(4d) \qquad - r \left[\partial (I\infty - I) / \partial S \right] = (K / I\infty)(I\infty - I) .$$

Equation (5d) is then obtained by combining equations (1d), (2d), and (4d), giving

$$(5d) \qquad \log \left[I / (I\infty - I) \right] = K \log H + \log k .$$

15

Two typical examples will suffice to illustrate the application of these laws.

The first example[1] deals with the electromagnetic induction in a homogeneous system. In this case equation (4d) holds perfectly, and the experimental data plotted according to equation (5d) yield, of course, a straight line. In the article just cited the data were read off on the curve to an accuracy of one per cent. Here, h, the height of the liquid at any stage of the process is obviously proportional to the magnetization capacity, I; while the amperage is proportional to the magnetic force, H. The observed values for I are shown in table 1 alongside the calculated values.

TABLE 1

Showing the relation between the magnetic force and the magnetization capacity.

$I\infty = 15.8$	$K = 2.19$	$\log k = 0.0$
H	I (obs.)	I (calc.)
0.21	0.5	0.50
0.65	4.5	4.42
1.65	12.0	11.85
2.05	13.4	13.87
2.80	14.3	14.30
4.10	15.1	15.12
4.80	15.3	15.32
5.30	15.4	15.40

The second example[2] deals with the magnetization capacity of soft metallic cobalt; i. e., cobalt under weak constraints.

A glance at table 2 shows that the agreement between the observed and the calculated values for I is not as close as between those shown in table 1. This is to be expected, since the particles are still under considerable constraints, although the metal is relatively soft. A similar phenomenon is observed in plated electrodes as compared with electrodes

---

[1] Sherwood: Magnetic Susceptibility of Neodymium Nitrate. J. A. C. S., 52, 1800, (1831).

[2] Williams: Magnetic Phenomena. McGraw-Hill Book Co., (1931) p. 213.

16

in the molecular state.[1] Still, even in table 2 the agreement between the observed and the calculated values for I is within the probable experimental error. Metals under high constraints will show, of course, a marked deviation from equation (4d).

TABLE 2

I∞ = 1140	K = 1.148	log k = —1.93800
H	I (obs.)	I (calc.)
5.1	72	79
10.7	170	170
13.9	247	218
22.8	371	336
35.7	497	469
54.5	628	606
85.9	749	749
147.7	865	890
244.3	953	985
392.0	1027	1045
467.0	1047	1060
610.0	1088	1088

[1] Linhart: J. A. C. S., 41, 1775, (1919).

17

,

A STUDY OF THE FORMATION OF HYDROGEN AND
DEUTERIUM OXIDES

In Goulberg and Waage's "Law of Mass Action" it is assumed or implied that the rate of change of reacting substances in solution with respect to time is proportional to their respective concentrations. Since then, a great many modifications have been effected, and many more **ad hoc** theories proposed in order to explain the numerous deviations from the so-called mass law. These theories led to great confusion and to misunderstanding of the original idea, and we find agriculturalists interpreting the growth of trees and plants as "monomolecular reactions"; and biologists regarding life growth as "monomolecular-auto catalytic processes," in order to account for the failure of the mass law at the initial stages of certain biological phenomena. However, the theories and the mathematical formulas based upon them soon became so cumbersome that the whole structure fell of its own weight.

Nowdays, no one is seriously concerned about the failure of the mass law in reaction rates; especially in heterogeneous systems. Indeed, one is quite surprised to find that the mass law, in its original form, does hold rather well in cases where one would least expect it to hold. This frequently happens when the value for the characteristic reaction constant, K, is close to unity, as is evident from the property of the curve of P versus t; that is, when K is unity, the point of inflection is at $(0,0)$, and the curve has the shape of the simple logarithmic type, such as P versus S of equation (4e). It will now be shown that the equations developed in the previous articles apply also in the present case; that is, in the formation of deuterium and hydrogen oxides.

There are certain substances in nature which seem to possess the property of a wound-up spring. These substances contain a certain amount of latent energy which, under favorable conditions, is released at ever decreasing but characteristic rates until the process is completed. As illustrations of such typical processes we may cite the interaction of alkyl metallic sulphates and water,[1] or of mercuric chloride and phos-

---

[1] G. A. Linhart, American Journal of Science, vol. XXXIV, 289, (1912).

18

phorous acid in water solution. Some substances do not possess quite enough latent energy to carry their respective processes to completion. A typical substance in this class is a water solution of ethyl acetate. Still more interesting phenomena in this category are the growths of living things. Their driving forces do not apparently come from within, for they start from the minutest cells and frequently attain to gigantic proportions both in mass and extent; and they all approach a maximum asymptotically with respect to time, or age. This fact seems to place them all in the same class insofar as their dynamics are concerned; and hence they all obey the same general laws of progress, whether that progress denotes the transformation of substances, the growth of plants, beasts, man, or nation; or even the universe as a whole. The solutions of several such problems have already been published;[2] others will appear later. For the present we shall confine our attention to the interpretation of the processes of the formation of hydrogen oxide and of deuterium oxide.

A mixture of hydrogen and oxygen, or of deuterium and oxygen, in the proportion of 2 : 1, was allowed to react at a temperature of 557°C. The initial pressure of the reaction mixture in each case was 450 mm. of Hg. At the completion of the process the pressure had dropped to 300 mm. of Hg. Hence, the range of fall in pressure was 0 to 150 mm. of Hg. This is obvious from the chemical relation

$$(2 H_2) + (O_2) = (2 H_2O) \quad ,$$

or 2 volumes of hydrogen plus one volume of oxygen yielded two volumes of steam. Let us consider now the kinetics of the process.

As previously noted, substances such as hydrogen and oxygen possess latent energy to cause them to unite, the energy reappearing in the form of heat which is absorbed by the immediate surroundings; but not all of it reappears as heat. Some of the energy is retained by the reactance to keep them aligned for interaction. This is the so-called ordered energy or energy of orientation, without which no interaction can take place. (Catalysts, such as platinum black, seem to possess orienting properties to a very high degree). The rate of consumption of this energy with

---

[1] G. A. Linhart, American Journal of Science, vol. XXXV, 353, (1912).
[2] G. A. Linhart, Occasional Papers of Riverside Junior College, vol 4, (1929).

19

respect to the age of the process, will evidently be proportional to the amount of steam formed, namely,

(1e) $$(\partial E / \partial t) = r [H_2O] = r P \ ,$$

where P denotes the total drop in pressure at time t, and r is a proportionality factor. In other words, as the reacting molecules decrease in number, those remaining become more widely scattered among the steadily increasing number of steam molecules, so that it will take relatively more of their latent energy to align themselves for interaction. It is the phenomenon just described, together with the fact that the energy particles themselves become partly disordered in the process, or "run down," which tends to slow down the reaction. This "rundownness" of the process with the increase in its age is here designated as chronodynamic entropy, which differs from the other forms of entropy only in degree but not in kind. Its relation to the other variables in the process is expressed in the same form as in the preceding problems, namely,

(2e) $$(\partial S) = (\partial E) / t \ ,$$

where S denotes the chronodynamic entropy. And since the "rundownness," S, increases with increase in the age of the process, the rate of formation of steam with respect to S will be proportional to the remaining amount of the reactance, or, in the present case, to the pressure $(P\infty - P)$, where $P\infty$ denotes the ultimate drop in pressure at time $t\infty$, and P denotes the total drop in pressure at time t. This relation is expressed by the equation

(4e) $$-r [\partial(P\infty - P) / \partial S] = (K / P\infty)(P\infty - P) \ .$$

Combining equations (1e), (2e), and (4e), we obtain, as before,

(5e) $$\log [P / (P\infty - P)] = K \log t + \log k \ .$$

A glance at the graphs will show how well the curves, calculated with the aid of the values for K and k, agree with the observed values shown in the accompanying table. It will be noted that the values for $[D_2O]$ are more uniform, and so follow the calculated curve more closely than those for $[H_2O]$. In fact, the recorded **times** for equal drops in pressure in the two series, or duplicates, of experiments with hydrogen and oxygen are seen to fluctuate to the extent of from 7% to 14%. For this reason both series of values were plotted on the same graph to show

20

that the deviations of the experimental values from the calculated values
are within the experimental errors.

TABLE 1 [1]

Hydrogen and Oxygen				Deuterium and Oxygen	
$P\infty = 150$; $K = 2.337$; $\log k = -1.65836$				$P\infty = 150$; $K = 2.135$; $\log k = -2.16537$	
t (min.)		$\dfrac{2\,(t_1-t_2)}{(t_1+t_2)}$ % Error	P (mn)	t (min.)	P (mn)
Series 1	Series 2				
0.0	0.0		0.0	0.0	0.0
1.0			5.0	1.5	5.0
1.4	1.6	13.4	10.0	2.7	0.0
2.3	2.5	8.4	20.0	4.3	20.0
2.9	3.1	6.7	30.0	6.0	30.0
3.5	3.9	10.8	40.0	7.1	40.0
4.4	5.0	12.8	60.0	8.2	50.0
5.1	5.6	9.3	80.0	9.0	60.0
5.6	6.2	10.2	90.0	11.0	80.0
6.2	6.9	10.7	100.0	12.3	90.0
7.0	7.9	12.1	110.0	14.0	100.0
				17.0	110.0

[1] Hinshelwood, Williamson & Wolfenden: Proceedings of the Royal Society (London), **A 147**, 48, (1934).

21

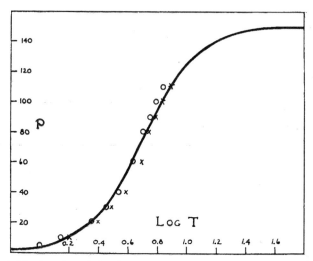

F:g. 1. Showing the relation between pressure and time in the formation of hydrogen oxide.

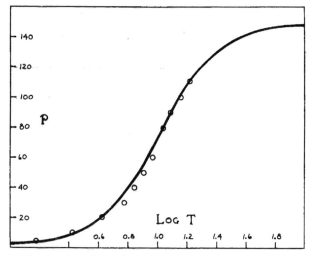

Fig. 2. Showing the relation between pressure and time in the formation of deuterium oxide.

GENERAL REMARKS

In the study of reaction velocity phenomena it is frequently essential to start with unequal initial concentrations of the reactance.[1] This makes it difficult, from the newer viewpoint, to decide which value to use for $C\infty$ in equations (4) and (5). In such cases the best method to determine the value of $C\infty$ is from the inflection point of the curve of C versus log t as described in this article. This is especially true for processes in which the value for $C\infty$ cannot be determined either at the start or reliably estimated at the completion of the process.[2]

Significance of K. In the older treatment of the subject, K is defined as the specific reaction velocity constant. This was a sound enough definition. Unfortunately K very often showed a decided increase or decrease frequently amounting to over 100% before the completion of the process, as interpreted from the "order" of the reaction, that is, from the number of molecules supposed to take immediate part in the transformation. This failure in the constancy of K generally led to the most absurd assumptions in order to render K constant; in some cases even to the extent of vitiating experimental results.[1] It was this insane tendency which eventually discredited the whole subject of reaction velocity measurements; and yet, it is the most vital phase of the physical as well as of the biological sciences, and it is hoped that the newer interpretation will restore this subject to its rightful place of preeminence. In this newer system K is defined as the **efficiency index** of the process regardless of the nature of the process. Thus in the formation of hydrogen oxide and deuterium oxide the value of K for hydrogen oxide is about 10% larger than for deuterium oxide, showing that under like experimental conditions the formation of hydrogen oxide is a more efficient process than the formation of deuterium oxide. A more marked behavior of K is observed in the compression of gases.[3] Here the value

---

[1] G. A. Linhart, American Journal of Science, vol. XXXV, 353, (1912).
[2] G. A. Linhart, Journal of Physical Chemistry, 36, 1908, (1932).
[3] G. A. Linhart, Journal of Physical Chemistry, 37, 645, (1933).
 G. A. Linhart, Journal of Physical Chemistry, 38, 1096, (1934).

23

of K ranges from unity to infinity, depending upon the temperature of the gas and upon the size of the molecule; for K increases with decrease in absolute temperature, which is in the direction of efficiency of compression. K also increases with the weight of the molecule. This, likewise, is in the direction of efficiency of compression, although there may arise some exception, owing to the configurations of the constituents of the molecule. Hence, K is correctly defined as the **efficiency index** of the process. In table 1 are shown the values for K calculated from the unity-to-infinity formula,

$$K - 1 = 1 / b\, T^a ,$$

where a and b are constants characteristic of the given substance. These values are placed alongside the observed values shown in column 2, table 1, page 1095 of the second article cited in footnote 3, page 23.

In subsequent articles we shall have occasion to show even more striking examples of the significance of K.

TABLE 1

Relation between K and T

	H_2		N_2		CO		O_2		CO_2	
a =	4.024		3.221		3.147		2.087		1.126	
log b =	—6.92668		—6.39381		—6.16822		—3.90375		—1.78448	
T	K_m	K	K_m	K	K_m	K	K_m	K	K_m	K
203	1.004	1.004	1.076	1.082	1.073	1.081				
223	1.003	1.003	1.065	1.061	1.060	1.060				
248	1.002	1.002	1.048	1.048	1.042	1.043				
273	1.001	1.001	1.038	1.035	1.033	1.032	1.066	1.066	1.110	1.110
293	1.001	1.001	1.033	1.028			1.057	1.057		
298					1.024	1.024				
323	1.000	1.001	1.023	1.020	1.022	1.019				
373	1.000	1.000	1.015	1.013	1.013	1.012	1.034	1.034		
423					1.008	1.008				
473	1.000	1.000	1.006	1.006	1.005	1.006	1.021	1.021	1.065	1.059
573	1.000	1.000	1.003	1.003						
673	1.000	1.000	1.001	1.002						

24

A PARTIAL LIST OF TOPICS TO APPEAR IN SUBSEQUENT ARTICLES

Topics from Physics and Chemistry

1. Voltage frequency relationships in action currents.
2. The scattering of high velocity electrons.
3. Motion of electrons in gases.
4. Measurements of the Townsend coefficient of ionization by collisions.
5. Diffraction of cathode rays.
6. On the specific reflections from rough surfaces.
7. Intensity factors in crystal analysis.
8. The specific surfaces of finely divided particles.
9. The turbulence of oceans.
10. The interaction of ethylene and oxygen.
11. The decomposition of amine.
12. The decomposition of ozone.
13. The kinetics of gas explosions.
14. The thermal decomposition of dimethylamine.
15. The distribution of heat in solids.
16. The reduction of potassium permanganate by oxalic acid.
17. The decomposition of silver oxide.
18. The adsorption of gases.

Topics from Biology

1. The growth of white mice.
2. The growth of rats.
3. The growth of cattle.
4. The growth of man in comparison with the growth of animals.

Topics from Bacteriology

1. The growth and populations of different groups of bacteria.

Topics from Statistics

1. The value and depreciation of physical properties in various kinds of industries.

Topics from Insurance

1. The service and the depreciation of man from the viewpoint of mortality tables.

Topics from Sociology

1. The population of every state in the union and of the United States.

Topics from Agriculture

1. The growth of grasses.
2. The pruning of fruit trees.
3. The growth of wild oats.
4. The growth of pumpkins.

25

The above truly ambitious research program has left without continuation. Why so? We shall now try to clarify at least some reasons—cf. the exchange of letters between Linhart and Pauling:

OFFICE OF THE DIRECTOR
RIVERSIDE JUNIOR COLLEGE
RIVERSIDE, CALIFORNIA

21 February 1935

Dr. Linus Pauling

Chemistry Department

California Institute of Technology

Pasadena, California

Dear Dr. Pauling:

      I have tested several sets of the data and find they fit my equation well within the experimental error. The experiments are, of course, very rough. By the way, this is the equation which I earnestly request you to help me derive. It fits a vast number of dynamic processes in many fields of research. I have submitted the equation to one editor and he rejected it without comment.

      Sincerely yours,

      , Dr. George A. Linhart
      Mathematics Department.

For the year indicated this ought to be the only Linhart's letter available in the Pauling's online archive. One might then only guess, whether there was any answer to this application and, if so, then Dr. Linhart might as well be hanged for a sheep as a lamb.

The fact is—Linhart and Pauling had absolutely no co-authored publications. Meanwhile they were in contact with each other, in lively discussing the themes interesting for the both of them, and their mail exchange proves what might be considered a productive interaction of two motivated research workers. A kind of pastoral play, isn't it?

Still, as soon as their relationship has arrived at the administrative level, the over-all tonality of the story is drastically changing.

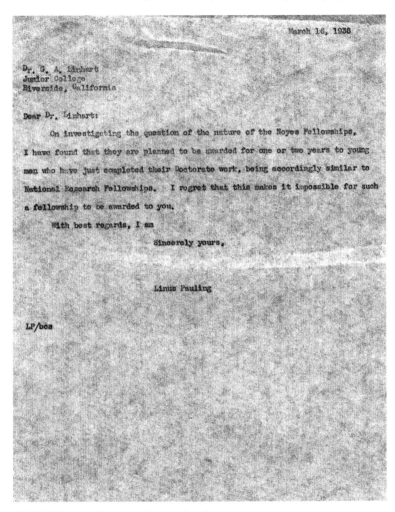

March 16, 1938

Dr. G. A. Linhart
Junior College
Riverside, California

Dear Dr. Linhart:

On investigating the question of the nature of the Noyes Fellowships, I have found that they are planned to be awarded for one or two years to young men who have just completed their Doctorate work, being accordingly similar to National Research Fellowships. I regret that this makes it impossible for such a fellowship to be awarded to you.

With best regards, I am

Sincerely yours,

Linus Pauling

LP/bca

??? ??? ??? would mean the readership.

This is a remarkable proof for the very interesting and characteristic story told by Dr. Linhart in his second diary volume, *Amid the Stars* —we dare to publish it in the following:

Chapter XIV

Staging a Comeback

> *O, never turn around when bound for home!*
> *'Tis evil luck they say for eyes to roam.*
> *If once you've lost your path to wealth or fame,*
> *You'll never find a niche to grace your name.*

Returning home from such meetings, I invariably suffered a sort of nostalgic relapse. It was not that I did not like my teaching at Sunnyside, nor my pupils, nor my new colleagues. I just felt a profound let down. It was a feeling that a race horse must have when driven off the race track and hitched to a team of "Twenty Mule Borax chips." My brain would grow dull and I had nothing on which to sharpen it. A research man among his own kind may be likened to a songster bird in a wonderland arbor, furnished with nice, clean, hard bones on which to sharpen its bill when it grows dull. Not that I regarded all my colleagues as boneheads. Far from it; but they would have served me the same purpose as the bones served the songster. On one return from Cal Tech, I felt so wretchedly isolated that I decided to do something about it.

I owned a little home that had cost me around eight thousand dollars to build and to furnish, and I knew that if I donated it to the Millikan Institute to found a graduate scholarship, I myself could become the beneficiary of around four hundred dollars a year for the rest of my natural life. Add to that a small job in the Cal Tech Chem Department, say of sixty dollars a month, and I could have given up teaching and resumed my research activities. This scheme seemed so intriguing and so promising that I took it forthwith to Millikan.

"Fine," said Dr. Millikan encouragingly, "and if you do it in the next few days, I can double your money."

117

"Dr. Millikan," I asked eagerly, "do you mean that I shall receive the overall net income from sixteen thousand dollars?"

"Oh no. I did not mean that," he chuckled. "Your *beneficiary* would receive that amount, but not you. You would get the income from the appraised value of the house. As to the small job in the Chem Department, I have no say over that. You will have to see Dr. Pauling, the chairman. By the way, you mentioned something about selling the house and giving us the money. Don't do that. We prefer the house. We find in small rentals a much better and safer investment than in huge enterprises; we own a large number of small rent-bearing houses," and he was right, judging from what followed a few years later. Dr. Millikan was evidently not only a keen scientist but likewise a shrewd realtor. No wonder Cal Tech grew by leaps and bounds soon after Dr. Millikan was installed as the head of the Institute.

Dr. Pauling, the chairman of the Chem Department, just mentioned by Millikan, was new at the game. He came to Cal Tech from a small Oregon college a few years before to take his Ph.D. Degree and found undue favor with A. A. Noyes, the director of the department. Here, as in the case of G. N. Lewis at Berkeley, Noyes felt there was no one to whom he could safely entrust the leadership and stewardship of his "dream child" chemistry department, and so broke in this young man from Oregon to take over at his death, which was rapidly approaching. Thus, I found Pauling, late in 1937, the new head of the Chem Department, but the "advisory" power behind it all was Dick Tolman's; and what could I expect from him! I therefore decided upon a different approach.

Early in the following spring, I approached Dr. Pauling with the request that he permit me to join someone on his research staff for my summer vacation, to see if I "still had what it took" in the realm of research. Pauling readily consented and hooked me up with a young man, Charles Coryell by name, who was conducting blood-curdling experiments! Charlie was around twenty-five, above medium build and height, with dark humorous eyes and a shock of black wavy hair. He was good-natured, and quite imposable. He was but a few years younger than Pauling and in many ways his superior; yet he regarded Pauling as *his* superior in a sort of hero worship manner. Pauling came from a little Oregon College, while Charlie was a B.S. of Cal Tech, 1932, and a Ph.D., 1935. Yet here he was, working for Pauling on a pittance as a glorified stockroom clerk since 1935. Hence, he had precious little time to conduct any independent research and was therefore greatly pleased to have me as a helper.

Charlie's research interests for the time being lay in a study of the magnetic properties of blue blood. I believe it was a hang-over from his thesis that had served to bolster a study of the structure of hemoglobin by Linus Pauling. Had Charlie been unencumbered by marriage, I believe he would have remained in that rut for life, despite the fact that Pauling professed profound friendship for him. Fortunately he had a sweet little slim young woman for a wife who quite likely caused him to sit up and take notice, for even slim little women like to eat and have nice things to wear and cute little homes to live in; so Charlie began thinking of a job—with pay, and soon confided in me that he had a line on a job for the coming year as assistant professor at U.C.L.A. and therefore was anxious to finish his "bloody" experiments by the end of August.

After a week of refreshing my memory of twenty years ago, I got into my accustomed stride and put in twelve to sixteen hours a day working in the laboratory. All Charlie had time to do was to go to the slaughter houses for the blood. However, Charlie had to help me frequently with the weighings, for I soon discovered that the balance was temperamental and had to be handled "just so"; while the magnet was an old piece of junk. When I called Charlie's attention to it, he admitted that the balance was "rotten" and that the horseshoe magnet was often "crazy" but that it would have to do for the present—that is, until he left for U.C.L.A.

Well, I had never liked temperamental people, much less temperamental apparatus, so I innocently invited Dr. Pauling to show me how to make a weighing. After several unsuccessful attempts to get his attention, he finally condescended to show me. He juggled and jiggled with the balance for over half an hour, and at last broke down and admitted that the balance was in "bad shape," that the magnet needed "adjustment," and that he would attend to it at once; but the "at once" was evidently forgotten, and Charlie and I kept on jiggling the instruments as best we could. Then just as Charlie and I were about to leave for our respective jobs, a young man arrived from Chicago to take over. He was one of those snooty young men who for one reason or another are favored with a Research Council Fellowship. One look at our balance and magnet and he stalked out with his nose in the air. Not that I blamed him. At the University of Chicago he had worked with the best instruments obtainable, and so naturally assumed that our results could not possibly be of any value. He was not a bit bashful in voicing his assumption in the august presence of Pauling.

A few days later, the young man from Chicago had a brand new balance of the latest and best construction and a reconditioned magnet. Pauling was taking no chances of having the glorified Fellow report to the Research Council the sort of instruments used at Cal Tech for presumably the highest class of research. Like Dr. Urey, his friend, Dr. Pauling always kept one eye peeled on more publicity and the other eye on the coveted Nobel Prize. Later Charlie assured me that our results, on the average, compared quite favorably with those by the Chicago Research Council Fellow, and that as soon as he could spare the time he would write up our results and have them published. However, he could not find the spare time at U.C.L.A., for this was not a research institution. One had to *teach* there to earn his pay. Congor Morgan, the guiding spirit of the department did not favor research, much less expenditures for costly equipment. Congor Morgan who had been Lewis' predecessor at Berkeley was now head of the Chem Department at U.C.L.A. How did this come about?

A year or so after my chief, G. N. Lewis, returned from "winning the war," he began hearing Morgan-termite tappings once more, if my meaning is clear. To prevent these tiny helmeted ants from destroying the underpinning of his own structure, Lewis, now firmly entrenched, recommended Congor Morgan for the job of presiding over the Chem Department of the Southern branch of the University of California, whose rating in 1920 was that of a junior college. In performing this strategic act, Lewis was not only applauded by Morgan's friends—thus winning them over to his side in all subsequent dealings with the Academic Senate — but also precluded a possible research rival down south for the duration of Congor Morgan's natural life, for Morgan was definitely not a research man and, in any case, only half awake . . . .

Well, Charlie kept on putting out feelers for two monotonous years, and at long last landed a job at M.I.T., my first place at genuine research under A. A. Noyes. Charlie had then forgotten all about blood-curdling experiments and began reaching out for the higher levels, and in no time at all became what the "newsboys" call a top scientist. The last time I heard from Charlie was the modest claim of having executed a successful piece of research that netted him a darling little daughter. Speaking of Professor Noyes again reminds me of another important event.

When A. A. Noyes passed away, he left his entire fortune of over $125,000.00 to the Chem Department of Cal Tech. In his last Will and Testament he specified that the income from this bequest should go to scholars who had already made their marks in the realm of chemical research, but who for lack of funds had had to quit their chosen fields of endeavor and seek to earn a living in other lines of work. That was the gist of the statement I read in the newspapers, and it seemed to speak especially to me; so before leaving for Sunny-side I went to see Pauling. I informed him of my talk with Millikan and of Millikan's advice to me to see *him*.

"Why, yes," he replied. "You certainly qualify for it. See me tomorrow."

When I saw him the next day he had quite another story to tell.

"I have taken your request before the Advisory Committee, and they seemed agreeable, when *one member* suggested that it would be wiser and fairer to split the income into smaller sums and award them to young candidates for the Ph.D. Degree, and it was so decided."

"But, Dr. Pauling, that would be contrary to the letter and the spirit of Professor Noyes' Will," I reminded him.

"Not at all"; he replied complacently, "we have found a clause which permits the change, if not according to the letter, certainly according to the spirit."

"Well," I said, "I am glad to learn of your method of treatment of sacred wills of the departed," and I pulled out my own Will with his department as my beneficiary, tore it to fragments before his face and walked away. In the three months I was there I had put in twelve to sixteen hours a day, seven days a week, of intensive lab work. And all I got in return for the twelve hundred hours of work was exactly nothing! Not even a niche in the Hall of Fame!

It was in the early summer of 1939, and the thunderous rumors of war were again crashing all around us while the warmongers already were rearing their brutish, gorillian heads. At the time I happened to be in Berkeley and thought I should visit my old chief, G. N. Lewis, to see how he was taking it. At his office I learned that he was playing after-luncheon chess at the Faculty Club. I remembered that the doors to the Faculty Club were customarily locked to those that had no business entering there, but I had not forgotten how to "crash the gates."

There sat Lewis alone bent over a table and swaying gently back and forth to the rhythm of his thoughts like an ancient seer over his cabala. It was an old habit of his when deep in thought. He happened to be playing a solitary game of four-dimensional chess, while all around him, but at a respectful distance, were standing several young men in silence and in reverent awe. As I started to advance to the edge of the magic circle, a young man checked me with a dire disapproving glance. At that instant, my chief looked up and a smile lit up his face.

"Hello, Dr. Linhart. You wish to see me about something?"

We had not met for years, but it was the same old question, as if I had just dropped in for a chat about some research problem.

"Meet me at my lab on the third floor of Gilman Hall in fifteen minutes."

In fifteen minutes on the dot Lewis came up. He inquired about my work, my new environment, and my financial status.

"You must have laid aside quite a bit of money in the past twenty years. What do you plan doing with it?"

I told him that if and when I had laid away any money, I would endow a scholarship at Stanford University.

"At Stanford! Why at Stanford? Why not here? Though we are a tax-supported institution, we are always glad to get a little private help—for our graduate chem department."

"Well, Professor Lewis," I told him, "I am still on depression wages but I shall certainly think about it. As a matter of fact, I was not aware that state-supported colleges accepted endowments. By the way, how about my coming up here summers to take part in the newer research lines?"

His prompt reply was, "Why, yes, we shall all be happy to have you again—at least I can say so for myself. I shall be released from administration duties next year and so shall have more time for research. We can work together during summer vacations. I am no longer interested in wars, and you never have been—except in those of your own making," he concluded with a twinkle in his eyes.

So G. N. Lewis at sixty-five was to be deposed, and his place was to be filled by a person so far inferior in every way that it made my heart ache for my chief. There were many new names and faces on his staff but the old crowd was still there, and I was again reminded of the ancient fable telling of a master giant and his little dwarves

AMID THE STARS

who would frequently climb to his mighty shoulders, rise to their dwarfish heights and each in turn would shout to the world, "See how big I am !"

In the twenty years just gone by, Lewis had unduly aged. A few years later, on an early morning, my chief was discovered in his private lab dead of a "broken" heart.

The above story told by Dr. Linhart might also be supported by the further available correspondence between him and Prof. Pauling.

RIVERSIDE CITY SCHOOLS

RIVERSIDE, CALIFORNIA

IRA C. LANDIS, superintendent

JUNIOR COLLEGE
OFFICE OF DIRECTOR

October 31, 1938

Professor Linus Pauling

Crellen Chemical Laboratory

California Institute of Techonology

Pasadena, California

Dear Dr. Pauling:

    I hope you will find time to read over the inclosed article, and then if you approve of it will you kindly submit it to the editor of the Proceedings of the National Academy of Sciences, for which I shall be very grateful to you.

                        Sincerely yours,

December 9, 1938

Dr. G. A. Linhart
Riverside Junior College
Riverside, California

Dear Dr. Linhart:

I have now found time to read your manuscript, which
I am returning with this letter together with the accompanying
reprints.

I am somewhat dubious about the appropriateness of the
manuscript for the Proceedings of the National Academy. Moreover,
there are some points about it which cause me concern. I would
interpret your general equation 3 as an equation valid for a gas
over a range of temperature and pressure, with V_∞ and K
constant under these varied conditions. Now it seems to me from
your table that K is not constant but varies widely and in a
systematic way for a given substance both with change in pressure and
with change in temperature. Moreover, I would interpret your statement
on page 3 that K = unity at low pressures and high temperatures as
incompatible with equation 3 in which K is a constant and in general
not equal to unity. I would be glad to discuss these questions with
you further.

Sincerely yours,

Lp/jr

A handwritten note by Prof. Pauling follows:

Looking at the table, I don't think the fit is very good. In general $K-1$ varies for T const, and also with T.

What are values of V_{oo} used in Table?

p. 3. Why does K equal unity? Isn't K a constant?

RIVERSIDE CITY SCHOOLS

RIVERSIDE, CALIFORNIA

IRA C. LANDIS, SUPERINTENDENT

JUNIOR COLLEGE
OFFICE OF DIRECTOR

December 13, 1938

Professor Linus Pauling
Director Gates and Crellin
Laboratories of Chemistry
California Institute of Technology
Pasadena, California

Dear Dr. Pauling:

Many thanks for the review of my article.
The objections concerning equation (3) and the
values of K are well taken; but I believe that
I can account for them to your satisfaction.
I would greatly appreciate a half hours' con-
ference any time you can spare it during the
vacation, weather permitting my driving to
Pasadena.

Looking forward to seeing you soon, I
remain

Yours very sincerely,

G. A. Linhart

Now, following is also the genuine note by Prof. Dr. Charles DuBois Coryell (1912–1971) mentioned in Dr. Linhart's report:

CALIFORNIA INSTITUTE OF TECHNOLOGY
PASADENA

GATES CHEMICAL LABORATORY

Nov. 29, 1938.
U. C. L. A.

Dear Dr. Pauling:

I have been planning for some time to write up the main results of my investigations on ferriheme hydroxide and those I did with Linhart, to get the results in one place and particularly to make them available to you and to Davies. There was more material than I had expected, and I have expanded it at several points dealing with technic and preparations to help out in future work. The final result is enclosed here in manuscript form, and I would appreciate it if you would have it typed with three carbon copies, for Linhart would like one and probably Davies too. The job is not a marvel, but I hope that all of the main observations on hemin that I have carried out are given in detail enough to make this work as useful as possible in further studies, for the systems already explored are quite complex.

I am enclosing also our two manuscripts sent in to the J. Gen. Physiol., together with a copy of Osterhout's letter to me. The papers are in a form that is used by several other journals, among which I would prefer the J. Phys. Chem. to the J. Biol. Chem. I am going to Deep Springs Thursday, however, so I cannot get over to Tech for about ten days to talk the situation over with you. You may send them in anywhere that you think advisable before I come over, if you care to. I would like to have the magnetic titration paper published before Taylor's if possible, but the ferrihemoglobin one need not necessarily be published at the same time.
Taylor wrote me to say that he will be down for a few days around Christmas, and will se about the manuscript then. He says also that the ferrimyoglobin data are very messy, although I still would like to see them.

Best regards to the gang,

Yours,

Charles C.

December 14, 1938

Dr. G. A. Linhart
Riverside Junior College
Riverside, California

Dear Dr. Linhart:

I will be around most of the time and will be glad to see you when convenient. Enclosed you will find a copy of the work by Coryell.

Sincerely yours,

Linus Pauling

There were also further crashes Dr. Linhart's papers have experienced during the period in question.

Remarkably, Prof. Pauling had also been involved into them—although implicitly—here comes the proof:

JOURNAL OF THE AMERICAN CHEMICAL SOCIETY

Arthur B. Lamb, *Editor*

CHEMICAL LABORATORY OF HARVARD UNIVERSITY
12 OXFORD STREET
CAMBRIDGE, MASSACHUSETTS

October 18, 1938

Dr. Linus Pauling,
California Institute of Technology,
Pasadena, California.

Dear Dr. Pauling:

        The enclosed manuscript entitled "Energy
of Orientation in Dielectric Phenomena" by Dr. G. A.
Linhart was sent in succession to Drs. Cohn and Kirkwood. Thereafter, in spite of the unfavorable comments
of Dr. Kirkwood, I sent it to Dr. E. Q. Adams for a
further opinion, since I knew Dr. Adams was not only
a qualified Referee for this type of paper but also was
a friend of Dr. Linhart, and I was anxious that the
manuscript should receive sympathetic consideration.
I have now received a report from Dr. Adams and I am
enclosing a copy of it, along with copies of the earlier
correspondence.

        It is clear that the manuscript cannot be
accepted in its present form. I am wondering, however,
whether we should even go to the extent of asking Dr.
Linhart to revise it, since that implies that its faults
can indeed be remedied. I would be grateful for your
opinion in this matter.

Cordially yours,

Arthur B. Lamb

Enclosures

The stance of Prof. Dr. Lamb is not especially remarkable. He as an editor of an important Journal has obtained two referees' reports concerning the MS by Dr. Linhart. The one by Dr. Kirkwood (John Gamble Kirkwood (1907–1959), an outstanding specialist in the field of physical chemistry, especially known for his contribution to the statistical–mechanical theory of solutions, cf. the Kirkwood-Buff theory) has been unfavorable. Another one, by Dr. E. Q. Adams (Elliot Quincy Adams (1888–1971), an outstanding specialist in the field of chemical engineering, especially known for his contribution to physical–chemical backgrounds of colorimetric effects) has instead been favorable. Prof. Dr. Lamb was aware of friendly relationship between Drs. Adams and Linhart (indeed, the both were working at the Chemical Department in Berkley under the guidance of Prof. Dr. G. N. Lewis).

The answer by Prof. Dr. Pauling is instead truly remarkable.

A series of posers arises when reading this entire story.

The mail exchange between Prof. Pauling and Dr. Linhart, which is available to us, is clearly demonstrating some normal interaction between two research workers having common areas of interest. The one of them has an administrative responsibility and has to take some decisions in this important field, whatever might be our estimate of the actual value of those decisions. In his position he definitely had much brighter horizon, so that we might well consider his decisions unsatisfactory from any permissible standpoint, but his decisions remain his decisions and nothing could be changed.

The above-mentioned story with the refereeing activity by Prof. Pauling conveys a clear taste of ambiguity.

Indeed, two colleagues are vividly discussing some research topic. One of them has some definitely interesting results. Let us assume that some inferences or conclusions by him are wrong. We would expect from his dialog partner that he notifies the author directly.

What we see is a quite different story? This looks like a thoughtfully sly action, so that it is up to us now to guess what could be the actual reasons for the latter. This is definitely not to be assigned to some "bad advice by Dick Tolman" or even a bad influence by whomever. A serious specialist well known worldwide is labeling a unique theoretical result just 'an empirical equation of no theoretical significance'. Wasn't this an intentional distortion? If so, then why?

Meanwhile, what was the reaction of Dr. Linhart? He had described this particular story—and even much more of them—in his second diary volume "Amid the Stars" as follows:

Chapter X

Ostracized

Fata Profunda:

"MICA, MICA, PARVA STELLA,
You won't go far, you little fella!
For soon the stars will pounce on you
And make you realize who is who."

I succumbed at last to Colonel Marsh's advice and settled down to the regular routine of public school teaching in an out of the way place, but the urge for research lingered on. With no facilities for experimentation, I directed my attention to the interpretation of the experimental results of others. However, such tasks were generally regarded as the prerogatives of top scientists, such as G. N. Lewis in America and Debye in Germany. Well, perchance I too was a top scientist. One never knew until one tried, and as my first task I undertook the integration of Lewis' entropy equation.

For years Lewis and Gibson had been resorting to counting squares on semi-logarithmic paper whenever they sought to calculate the entropies of chemical substances. Later, Lewis showed that from the similarity of the shape of the space under the c-logT curves, a general template could be constructed whereby approximate values for entropies could be obtained. For accurate work both methods were incorrect since they involved extrapolation to minus infinity, rather a dubious destination.

Having previously succeeded in integrating Lewis' free energy freezing point equation, I tried to evolve a similar method for the integration of his entropy equation and thus dispense with the crude procedure of counting squares or with the juggling of the umbrageous template; and I had hardly settled down to my work at Eureka Junior College when I succeeded in deriving an integrating factor. This

101

expression turned out to be much more than an integrating factor in the mathematical sense of the phrase. It likewise proved to be an important equation in its own right, for in its integral form it shows that the entropy is proportional to the logarithm of the thermodynamic probability whose value ranges from zero at the absolute zero temperature to plus infinity at infinite temperature. It thus resembles Boltzman's equation, but in a more applicable form. In addition, it enables one to express heat capacity as a simple algebraic function of the absolute temperature, and the entropy as a simple logarithmic function of the absolute temperature; all three functions reproducing the experimental results!

Hence, I at once prepared a three-page article and, modestly enough, submitted it as a mere note to Editor A. B. Lamb. In this note I made no blazoned claims; all I aimed at was the establishment of the priority of my equations.

About a month later, the note came back accompanied by three sets of criticisms which, in effect, were the crudest sort of insults. These referees had evidently gone berserk, like a prizefighter losing his temper! Poor Lamb! We were driving him insane. First, Bray warns him never to publish anything submitted by me lest it contain some coded message to the "enemy." Then Lewis returns from the war and peremptorily tells me to submit my paper to Lamb for prompt publication. Then follows a paper by Lewis and Linhart, which Lewis previously had commended Lamb, in Washington, for rejecting. And now this note comes like a bolt out of the blue, from me at Eureka Junior College. Lamb was fairly new at the editorial game and was at a loss to know how to react. He did scribble at the end of the referees' insults: "If, in view of these adverse criticisms, you still feel that your paper should be published, you might try elsewhere." The criticisms came of course from my chief's maniples. One of them had been dabbling in this field since 1913, and all he had attained thus far was the trick of counting squares on ruled paper and manipulating Lewis' template. Lewis himself was totally unaware of these petty jealousies, for he was at that time deeply absorbed in dictating his book.

I was about to tear up my paper, when Jensen walked in. "Why the disgusted look on your otherwise smiling face?" he wanted to know.

I showed him the three sets of "criticisms."

"These are not criticisms. These are the outbursts of persons afflicted with stomach ulcers; so have compassion upon your former colleagues and turn not your ire against them. You know," he continued, contemplatively, "I have often wished for a medium to express my thoughts without inflicting them upon harassed magazine editors; so why not start a magazine of our own? You could be the editor-in-chief and I your associate."

No sooner said than done, and within a week's time we created *The Eureka Junior College Journal of Science Arts and Crafts*, dedicated (on the outside cover) to Self-Expression, etc. Jensen, being associate editor, took the liberty without first consulting me, his superior officer, of sending a copy of this, our new *Journal*, to Editor Lamb. It was our first issue and contained my rejected article. That was in September, 1921. In January, 1922, my article appeared in Lamb's *Journal!* I could have sued him for plagiarism, but our *Journal* was not copyrighted, so I let it go—just as a token from one editor-in-chief to another. Then I showed it to Jensen.

"You mean he published your paper secondhand after all that fuss?" and he grinned. "You know," he confessed, "I sent him a copy of our *Journal* just to show him that out here in the Great Northwest we can roll our own when the occasion arises. Well, that winds up our *Journal*. How much has it set you back?"

I was downright dumfounded. "Mr. Jensen," I asked disappointedly, "You mean the whole set-up was a trick?"

"Of course. *You* could not afford to shell out twenty to forty dollars every time you felt an urge to tell the world what you knew. Anyway, it worked, didn't it? From now on Mr. Lamb will think several times before rejecting a contribution coming from the Eureka Junior College."

Emboldened by Jensen's ruse, I prepared a much larger paper showing the validity of my equations to the heat capacities and entropies of more than thirty substances, which I had assembled from the International Tables, and submitted the article again to Editor A. B. Lamb. This time, instead of a rejection with insults, I received galley proofs. I asked Mrs. Jensen to help me check the numerous tables—Mrs. Jensen held a Ph.D. Degree from the Physics Department at Berkeley—and returned the proof promptly to Lamb's office. Soon the article appeared in Editor Lamb's *Journal*, but its appearance was like that of a prizefighter's after a multi-round battle—it was completely battered and hopelessly pied. Headings were misplaced, large chunks of tables interchanged and legends confused. Well, it

103

was a trick for a trick, but they should have taken it out on Jensen. *I* received the portion of the "innocent bystander." I wrote the editor a mildly pointed note suggesting that he reprint the whole article. He refused to do that, stating that a "note" explaining the accident in a subsequent issue of the *Journal* should be quite sufficient. After that, I completely lost my appetite for that sort of "pie" and gave up writing scientific papers.

Sure, Dr. Linhart's story sounds to 100% clear, taking all the documentary proofs into account. But what was the actual reason for such an attitude on the part of Dr. Linhart's surrounding?

An interesting a truly far-reaching question—it ought to be somewhat rhetoric, in that the answer is more or less ready.

But still let us try analyzing the situation. There is a clear parallel to the situation we have considered in the previous chapter devoted to the life and work of Max Bernhard Weinstein. The readership might note at this point that the time periods and the countries of the both stories are quite different. Moreover, wouldn't this be an exaggeration to involve the USA in such stories—the USA is well known as a 'unique country of the open opportunities for any person willing to realize his/her dreams'.

To our mind the both stories are fully independent of any historical/national/political context—most probably this is about the *basic nature of the human beings*.

Was Dr. Linhart a highly ambitious, but in fact a fully incompetent person?

We do not think so. Why?

First, he was quite successful in all of his educational professional endeavors: The highest working position he could at last manage to get with his Yale [*sic*] MS and PhD diplomas was but nothing more and *surely nothing less* than just a school teacher in a number of junior colleges in the Californian province.

Was he somehow unsuccessful in his working?

By far not! He had told us his professional story and even duly documented it in the second volume of his diaries "Amid the Stars." We shall have a look at this part of the story later on.

Was he an unqualified, not enough motivated researcher?

By far not! He could indeed manage a breakthrough in thermodynamics and statistical physics, as we have seen from his works.

Were his colleagues so stupid and incompetent?

No-no-no, by far not! We speak here about Prof. Dr. Gilbert Newton Lewis, Prof. Dr. Linus Pauling—truly outstanding personalities and professionals—no doubts as for them.

The true answer ought to be—Dr. Linhart was all his life long remaining **alien**—practically to the whole world—be it his birthplace Austria or the country of 'open opportunities for everybody', the USA—sounds strange but ought to be true. Along with this, he could accomplish solving a quite general, quite fundamental problem—of interest and importance to the whole world. And he was aware of this.

After summing-up the entire story we have read above you would definitely ask me:

Hey, stop a bit—but why do you think, this guy was **truly alien**—at least to all of his immediate surroundings in the USA?

In fact, the answer has been given by Dr. Linhart himself—as the product of his life-long thinking over all these matters, and we shall find it in the second volume of his diaries, "Amid the Stars."

We would like to discuss them later on, but first let's try to summarize the facts of his curriculum vitae—the schematic representation of his professional way might be found above in this chapter and in the work **[L2]**, and here we just present a number of facts from his everyday life—with the due documentary proofs, where possible (all the documents have been retrieved from the Genealogy Internet site—MyFamily.com).

We start here with presenting a copy of his naturalization certificate he obtained in New Haven, Connecticut, USA:

LINHART, George Augustus   Certificate#213031
  744 Yale Station, New Haven, Connecticut

Born: May 3, 1885 - Near Vienna, Austria

Naturalized: February 24, 1912
      U.S. District Court, New Haven, Connecticut
      Petition #542, Vol.4, Pg. 92

Thus, we see from his story that he ought to have arrived at New York—most probably without any valid documents—and was accepted by the competent officials at the Ellis Island.

We go to the Internet representation of the archives due to the relevant official structure, http://www.libertyellisfoundation.org/, and find the record that on September, 3, 1896, a single guy from Austria, whose name was Josef Linhart, had in fact arrived on the ship 'Havel' from Bremen, Germany, bound to New York, USA.

We note immediately that his age at the moment of his arrival was recorded as "19 years old." Well, in the first volume of his diaries, "Out of the Melting Pot," Dr. Linhart ascribes this apparent official promiscuity with his being terribly ill—for in effect he ought to be about 11 to 12 years old at the time point of his arrival at the Ellis island—and the story told by him ought to sound throughout faith worthy. But nonetheless he decides to start the second volume of his diaries, "Amid the Stars" with the following synopsis:

Amid the stars
SYNOPSIS OF PART ONE*

A YOUNG MAN LADIES' TAILOR from Europe arrives in New York City in the Spring of 1896. He is an orphan and just eighteen years old. For four long years he struggles along among strangers in sweat shops, tailor shops, ready-made clothes shops, and finally in high-class ladies' tailoring establishments. In 1900 he goes to Philadelphia; and a year later, though still ignorant, totally illiterate and alone, he somehow finds his way into High School, graduates in 1906, at age twenty-eight, enters the University of Pennsylvania, and completes his first year of college in 1907.

Out of the Melting Pot, by GEORGE LINHART, 1923.

Whatever the actual reasons for all the promiscuity with his actual age, the true story is that an underage orphan could somehow manage his rescue from a true purgatory—and thereafter even find his strength not just to survive, but to pick up a fully foreign language and perfectly master it in the due time, then to go through the regular educational route—and at last to land at one of the highly esteemed universities of the country, to get there both MS and PhD in Chemistry, that is, in the field of knowledge of his own, fully conscious and so truly hard-won choice.

Sure, he had a definite personal aim and he could perfectly exploit all of his truly rich natural capabilities to finally arrive at the rightest position for the perfect fruition of all of his coveted dreams.

Howbeit, a natural question arises: Did Dr. Linhart have his own family in the USA? We wouldn't like to take into account his Austrian step-family, his 'so-to-speak-relatives', who, as we already know from Dr. Linhart's story, had followed him to the USA—but soon it was absolutely clear that there could be no more relationship with them.

In the Californian Alameda's census record of 1920 we might find George A. Linhart as a lone boarder in the house belonging to a family Hechturau (Albert J., the father, Corie C., the mother, and Belle, the daughter). But in the Californian Riverside census record of 1930 we might find George A. Linhart already as a head of family living in his own house together with his wife, Catherine B. Linhart, born in the state of Louisiana. The profession of G. A. Linhart is shown as a "teacher in the public school," whereas his wife bears no professional record. There are no children in this particular household. And, finally, the letter 'H' (denoting a home-maker) is inserted near the record 'Wife' for Mrs. C. B. Linhart might have a two-fold sense:

1. A person who manages the household of his or her own family, especially as a principal occupation.
2. A person employed to manage a household and do household chores for others, as for the sick or elderly.

Hence, just this particular census record alone isn't sufficient to decide, whether Dr. Linhart had really his own family, but the absence of any recollections as for the family life in Dr. Linhart's diaries, the total absence of any marriage certificate(s) etc., and of eventual children records might indicate that Catherine B. Linhart wasn't in fact an official spouse to Dr. Linhart—but solely an employed home-maker to render his lone life a bit more comfortable.

Howbeit, the main proof that Dr. Linhart had no children ought to be his bequests. The one is mentioned in his diaries, whereas another two are mentioned in the official bulletins of the Yale and Berkley universities:

Frank Austin Gooch Lectures (1953). Bequest of George A. Linhart, M.A. 1911, Ph.D. 1913, in memory of his former teacher, Frank Austin Gooch, M.A. Hon. 1887, Professor of Chemistry and Director of the Kent Chemical Laboratory 1885–1918.

Gilbert N. Lewis Lectureship: Dr. George A. Linhart by his will (1953) bequeathed one-third of the residue of his estate to support an annual lectureship at the University in memory of Prof. Lewis, who had been his teacher.

Remarkably, that was just Prof. Dr. Frank Austen Gooch (1852–1929) who could correctly foretell the destiny of Dr. Linhart—and we might read this in the diary volume "Amid the Stars."

The comments by Prof. Gooch in regard to Dr. Linhart open an apparently hidden door to the couloirs of the academic society. In effect, the described atmosphere ought to be characteristic to any kind of human professional society (Prof. Gooch mentions indeed only business/industry/entrepreneurship, but his insights ought to be valid for any creative professional society—be it letters, theater, painting, statuary, music, religion, etc.).

Is it a warning to the address of the young readership of this book? Sure, to some extent—yes, it is.

It is a pity, but any creativity seems to be connected with the **alienation** of those of us who are not only in principle capable of the **true achievements**, but do achieve their aims and substantiate their dreams.

The life and work of Dr. Nils Engelbrektsson's we have gained insight into in Chapter 2 here ought to be another bright example of this seemingly 'sturdy' rule.

But—fear not, my dear reader, for there is apparently another not less 'sturdy' rule: **The Manuscripts Do Not Burn** (Михаил Афанасьевич Булгаков (Mikhail Afanasievich Bulgakov), 1891–1940).

With this in mind let us now turn to the words by Prof. Dr. Gooch addressed to Dr. Linhart. We dare to place below a photocopy of the relevant part of the diary volume "Amid the Stars."

Dr. Linhart could perfectly clarify the interconnection between the notions of entropy and probability—which is of truly fundamental philosophic impact.

In effect, he could moreover demonstrate that the notion of heat capacity ought to be correctly mathematically expressed by the so-called 'logistic curve' (see his seminal papers) correspondent to nothing more and nothing else than a 'possibility distribution function/membership function for some fuzzy variable'.

At this point *an immediate question poses itself*: What are the fuzzy variables when speaking of thermodynamics?

A plausible answer by Dr. Linhart: If the atomistic picture of the matter is adopted, then it is not immediately clear what happens to all those atomic/molecular constituents. Hence, one might either produce some plausible theories, or consider the paradoxical, fuzzy nature of the very central notion: *'Number of Particles'*, that is, the so-called 'heap paradox', by surpassing any explicit evaluation of the 'heap'.

Right? We shall discuss this important point in more detail in the next chapter.

But let us now hear to Dr. Linhart's story. His following story could be the key to understanding the difficulties of all the protagonists of the book at hand. Interestingly, the main line of the different thermodynamics could be drawn as follows: P. B. Freuchen—N. Engelbrektsson and K. A. Franzén—M. B. Weinstein—G. A. Linhart. There had been even much more colleagues, which were not directly, but substantially following the drawn line. Noteworthy, both the main line and the branches wear their veils of oblivion. However, why at all?

Yale was now closed but Kent Hall was still open, and Professor Gooch was getting ready to leave for his summer home on Squirrel Island, Maine. He called me into his office and inquired what luck I had had on my visit to Boston, and after hearing me out, his comment was, "It will do for a beginning." Then he added thoughtfully, "It will be interesting to watch your future—a future extremely difficult to foretell. One thing is certain. You will never yield to mediocrity, and therein lie your stumbling blocks. You have set up a very high professional standard for yourself, and you expect the same of others. Well, Sir, you will meet with many disappointments—bitter, heartrending disappointments. In the field you have chosen you will find dishonesty, hypocrisy, jealousy, spite, and intrigue; and to a more vicious extent than in the world of business. For in business these human traits are indulged in more or less in the open, but in your field they are carried on *sub tela*.

"You have already met with two instances and, as a consequence, you have made several enemies, Kreman in Austria, and Garner, Wilson and Foglesong in America; but these are relatively insignificant

32

AMID THE STARS

compared to the ones you are about to run into. You see, you are not really a research man in the accepted meaning of the phrase; you are more of the detective sort. The usual run of research man chooses a line of work, dabbles at it for the rest of his active life, and eventually becomes an 'authority' in his chosen field; but I cannot visualize your doing that—not when there is so much falsehood afloat. I can safely say that you will pry into many 'finished' research projects, the results of which have already found their way into the lofty scientific literature, and find, as I have often found in my own restricted field, that a good many of the published results are falsehoods, and the experimentation sloppy, if not entirely fictitious.

"As to the theorists, they swing like pendulums; now they are at one end of the arc, now at the other; and when they finally come to rest, they have nothing but 'tentative' first approximations, and at once start swinging anew. I mention these facts to you to prepare you for the crushing impacts you are bound to come up against, and I sincerely hope you can stand up to them; and, barring malnutrition or ill health, I know that you will. How you have survived on your pittance here for three and a half years will remain for me a lifelong puzzle. You are evidently made of stronger fiber than your physique reveals. Well, here is to your success wherever you may be."

If Professor Gooch's words were prophetic, they held no fear for me, for I had nothing to lose either way. Besides, in times of sorrow, I had learned to temper my tears with laughter.

4.6   Conclusion

To our sincere regret, Prof. Dr. Gooch was in effect subliminally right—Dr. Linhart had been clearly ahead of his time, not only by employing Bayesian approach, when it was definitely 'out-of-vogue', but also by arriving closely at the problem of 'fuzzy computing' (in vogue nowadays), when the founder of the latter, Prof. Lotfi Asker Zadeh was just a newborn baby—he had to come to Berkeley much later on, as we now know.

Sure, Dr. Linhart's opponents were not blockheads, they did perfectly realize that Dr. Linhart was just going some other way than all of theirs—but going a different way doesn't immediately mean doing wrong.

Howbeit, Dr. Linhart's coveted dreams could now be manifested in effect—sure, at least with some retardation, so that he cannot feel satisfaction himself any more—but anyway!

What remains largely unanswered is the poser about Dr. Linhart's roots.

The story he tells himself on the theme sounds like just the perfect ready-made plot for the Hollywood movie. Indeed: A recent school graduate meets her first true love and this love brings the fruit. It would seem in effect that nothing might prevent the birth of a new family. Meanwhile, the notorious conflict of creeds comes into play, and the young groom gets a bullet from his father-in-law. The young bride does not endure such a development of events, and immediately deprives herself of life as well—in front of her father and faithful servant. Her son goes to an orphanage. As soon as he had reached his third birthday, some eerie family adopted him, being interested solely in an additional workforce to promote replenishing their purse, whereas the money they would like to spend mainly on alcohol.

Had Dr. Linhart invented the entire story—or something like that could be the case in reality? The poser is truly far from the thermodynamics and its foundations. Nonetheless, of definite human interest ought to be the roots of a colleague, who could not only formally derive the notorious Boltzmann–Planck formula, but also duly clarify the actual physical sense of this formula's actually notorious part: The probability under the logarithm sign. With this in mind, the present author has undertaken an investigation of the Austrian print media of the pertinent time.

The first difficulty during such an investigation consisted in defining what should actually be the "pertinent time" involved.

Specifically, in his first unpublished diary Dr. Linhart communicates us that he was just eleven years old, as he arrived at the Ellis Island in the year of 1896. His naturalization application and the resulting naturalization certificate give us his birthday as 03-05-1885, with the place of birth "near Vienna." Still, in his second diary he describes himself as an "18 years old foreigner arrived at New York," and his death certificate states that his birth date was on 12 May 1879, in "other Country," while his death date was on 14 August 1951 in Los Angeles. As he had no family, his death was most probably lone and not interesting to anybody. This is why the Los Angeles Social Security has not carefully checked his entire dossier. Thus, we could choose the first version of his birth date, as a more reliable result, for while being naturalized he had to duly prove all of his statements.

To sum up, we start from the premise that in fact he was born near Vienna, Austria, on 03 May 1885. With this in mind we have checked the printed media issued in Vienna in the year of 1885, which are available online in their digital form, as a courtesy to the Austrian National Library.

Fortunately, on the Page 9 of the newspaper *Die Presse* issued in Wien on 23rd of August 1885, we find the following two reports, we cite them in German, and then place the English translation of them:

> "**Ein vermißter Student**: Nach einer dem Polizeikommissariate Währung erstatteten Anzeige wird der Student der Pharmacie, Michael Wenrich, 23 Jahre alt, aus Schäßburg in Siebenbürgen gebürtig, von seiner Quartiergeberin Marie Mayer, Schulgasse Nr. 8, vermißt. Wenrich ist groß, mager, mit schwarzen Haaren und schwarzem kleinen Schnurrbarte.

> **Opfer der Donau**: Am 26 v. M. wurde bei Preßburg die Leiche eines ungefähr 25jährigen Mannes aus den Wellen der Donau gezogen. Die Leiche dürfte ungefähr sechs bis acht Tage im Wasser gelegen sein. – Zwei Tage später wurde bei Preßburg der Leichnam einer beiläufig 25jährigen Frauenperson, wahrscheinlich einer Magd, aus den Wellen der Donau gezogen. Die Identität der beiden Ertrunkenen, an denen keine Spuren einer erlittenen Gewaltthat ersichtlich sind, ist bisher nicht festgestellt worden."

"**A Missing Student**: *According to a complaint filed yesterday with the police inspectorate of the Vienna district Währing, a student of pharmacology Mr. Michael Wenrich, 23 y. o., a native of Schäßburg in Transylvania, had been reported to be missing by his accommodation hirer, Mrs. Marie Mayer, living in Schulgasse No. 8. Mr. Wenrich is tall, skinny, with black hair and a black little mustache.*

Victims of the Danube: *On the 26th of the last month near Preßburg (now Bratislava, Slovakia) the corpse of a man of about twenty-five y. o. was dragged out of the waves of the Danube. The body should have remained in the water for about six to eight days. Two days later, near Preßburg as well, the corpse of a ca. 25-year-old woman, probably a maid, was dragged out of the waves of the Danube. The identity of the two drowned persons, on whom no traces of any violence could be evident, has not yet been established.*"

This same communication about "a missing student," but without any reference to the "victims of Danube" has also been published in the "*Morgen-Post, Wien*" on the same day, although that newspaper had published no information about the "*drowned persons.*" Still, on 23rd of July 1885 the newspaper "*Morgen-Post, Wien*" has published an information about another missing student, namely:

"**Ein vermißter Student.** *Nach einer dem Polizei-Kommissariate Josefstadt gestern erstatteten Anzeige, wird der Mediciner Adolf Fein, zu Szarovas in Ungarn gebürtig, 24 Jahre alt, seit 20. dieses Monats vermißt. Die Quartiergeber des Abgängigen befürchten, daß der junge Mann einen Selbstmord ausgeführt haben könnte. Fein ist von kleiner, schwächlicher Statur, hat schwarze Haare, eine Adlernase, mageres, bartloses Gesicht, und ist mit dem melirtem Sacco, lichter Hose und weißer Wäsche bekleidet.*"

"**A Missing Student**: *According to a complaint filed yesterday with the police inspectorate of the Vienna district Josefstadt, the physician Adolf Fein, born in Szarovas, Hungary, 24 years old, has been missing since the 20th of this month. The accommodation hirer of the missing student fear that the young man might have committed suicide. Mr. Fein is of small, weak stature, has black hair, an aquiline nose, a thin, beardless face, and is dressed in the mottled Sacco, light trousers and white linen.*"

Now let us compare those facts with Dr. Linhart's information. He told us about "a Bohemian student" as a potential father of him, who fell victim to the "conflict of creeds," whereas he introduces his mother as "a Polish girl just out of a convent school." We might well assume that the family of his mother was a Roman Catholic one. Moreover, as soon as the girl had had her own butler, her family was a relatively wealthy one.

With the above in mind, of the both missed students we have introduced above, we would go for the candidacy of Mr. Wenrich. Indeed, although Mr. Fein might well fit into the issue of the "conflict of creeds," for at least partially he was apparently of Jewish or Gipsy origin, while his Bohemian origin ought to be highly improbable. The last but not the least, his description does not ideally fit the role of "Romeo." Still, he might indeed have committed suicide by drowning himself in the Danube somewhere in the proximity to Vienna—in absolutely independent way of the maiden—but approximately at the same time—and in some one week the both corpses could be brought to the vicinity of Bratislava by the Danube flux.

Instead, Mr. Wenrich's candidacy does seem to be at least somewhat more plausible. Specifically, his family name is a characteristic German one, which one could frequently meet in the Southern Transylvania [13], his roots might well be from Bohemia [14]. Moreover, the Transylvanian Saxons are traditionally mostly Lutherans, as well as the most of other nationalities over there, except for Romanians.

Meanwhile, when we look at Dr. Linhart's portrayal we notice that he had no black hair, no aquiline nose, and, finally, he was by far not athletic and/or tall (see him on the group photo of the Chemistry Department in Berkeley of the pertinent time [15]).

Therefore, there are serious doubts, whether we were discussing the candidacies of his actual father. So, how about his mother, maybe he had inherited her stature?

In looking about some noticeable "*young Polish girl*" in Austria of the time in question, we come to "*Else von Kolschitzky, a girl from an impoverished Polish noble family*", who had been the "*first serious relationship*" of Arthur Schitzler (1862–1931) just in July 1884:

> "*...Es ist dies das letzte Mal, daß Schnitzler Else von Kolschitzky im Tagebuch erwähnt, ihr weiteres Schicksal liegt im Dunkeln...*" [16].

> "*This is just the last time Schnitzler mentioned Else von Kolschitzky in his diary, her further fate is remaining untold...*" [16].

We sincerely hope that Austrian colleagues studying the history of science could try solving this interesting puzzle, in view of Dr. Linhart's tremendous professional achievement.

4.7   References

1. Richard Lowry (**1989**): *The Architecture of Chance. An Introduction to the Logic and Arithmetic of Probability*. Oxford University Press, New York, Oxford.

2. A. Singer and Z. Schuss (**2005**): Brownian simulations and unidirectional flux in diffusion. *Phys. Rev. E*, **71**, 026115.

3. Michael Strevens (**1998**): Inferring probabilities from symmetries. *Noûs*, **32**, pp. 231–246.

4. Michael Strevens (**2003**): *Bigger than Chaos. Understanding Complexity through Probability*. Harvard University Press, Cambridge, Massachusetts, USA; London, Great Britain.

5. Michael Strevens (**2013**): *Tychomancy: Inferring Probability from Causal Structure*. Harvard University Press, Cambridge, Massachusetts, USA; London, Great Britain.

6. David Bohm (**1984**): *Causality and Chance in Modern Physics*. Routledge, London, Great Britain.

7. Lawrence Sklar (**1993**): *Physics and Chance: Philosophical Issues in the Foundations of Statistical Mechanics*. Cambridge University Press, Cambridge, New York, Oakleigh.

8. J. Bricmont, D. Dürr, M. C. Galavotti, G. Ghirardi, F. Petruccione, and Nino Zanghi (**2001**): *Chance in Physics: Foundations and Perspectives*. Springer-Verlag, Berlin, Heidelberg, Germany.

9. Stephen Thomas Ziliak and Deirdre McCloskey (**2008**): *The Cult of Statistical Significance: How the Standard Error Costs Us Jobs, Justice, and Lives*. The University of Michigan Press, Ann Arbor, Michigan, USA.

10. Gerhard Ernst and Andreas Hüttemann (**2010**): *Time, Chance and Reduction*. Cambridge University Press, Cambridge, New York, Melbourne, Madrid, Cape Town, Singapore, São Paulo, Delhi, Dubai, Tokyo.

11. Caroline Strevens, Richard Grimes, and Edward Phillips (**2014**): *Legal Education. Simulation in Theory and Practice*. Ashgate, Farnham, Surrey, Great Britain.

12. William Briggs (**2016**): *Uncertainty: The Soul of Modeling, Probability & Statistics*. Springer International Publishing, Switzerland.

13. Fritz Keintzel-Schön (**1976**): *Die Siebenbürgisch-Sächsischen Familien-Namen*, Editura Academiei Bucuresti (Böhlau Verlag, Köln).

14. Klaus-Peter Koch (**2000**): *Böhmische Musiker in Siebenbürgen und im Banat*, in: *Musikgeschichte in Mittel- und Osteuropa*. Mitteilungen der internationalen Arbeitsgemeinschaft an der Technischen Universität Chemnitz Heft 7, hrsg. von Helmut Loos und Eberhard Möller, Chemnitz 2000, S. 141–161.

15. Melvin Calvin and Glenn T. Seaborg (**1984**): The college of chemistry in the G. N. Lewis era: 1912–1946. *J. Chem. Educ.*, **61**, pp. 11–13.

16. Johannes Sachslehner (**2015**): *Alle, alle will ich: Arthur Schnitzler und seine süßen Wiener Mädel*. Styria Premium, Wien–Graz–Klagenfurt, Austria, p. 30.

Chapter 5

The Different Thermodynamics: What Ought to Be the Proper Mathematical Instrument?

I would like to call up for discussions and debates about the scientific development ways, about what is truly important and interesting in the science, while keeping the atmosphere of mutual tolerance and goodwill, but never with bruxism.

Vitaly Lazarevich Ginzburg (1916–2009),
theoretical physicist, Nobel Laureate, 2003

We have just discussed the lives and works of the thermodynamicists 'widely unknown' for their truly valuable contributions, and at the same time we could get a feeling of some different thermodynamics being developed somehow in parallel to the conventional state of art.

Immediately recognizable from our above discussions would be the conclusion that the development of the thermodynamics was in the meantime and is still proceeding along a very specific, a rather peculiar trajectory.

Indeed, what is 100% accepted and common nowadays could be put as the two kinds of thermodynamics: The equilibrium and non-equilibrium one. Meanwhile, with the former to be widely used as a common identifier for the conventional phenomenological

A Different Thermodynamics and Its True Heroes
Evgeni B. Starikov
Copyright © 2019 Pan Stanford Publishing Pte. Ltd.
ISBN 978-981-4774-91-8 (Hardcover), 978-0-429-50650-5 (eBook)
www.panstanford.com

thermodynamics, that is, the one available long before the seminal contributions by Lars Onsager, Ilya Romanovich Prigogine, Hermann Haken, and their school. So, let us first have a closer look at this former part.

5.1 What Were and Are the Unsatisfactory Aspects of the Conventional Thermodynamics?

Prof. Dr. Clifford Ambrose Truesdell (1919–2000) had thoroughly analyzed the roots, logics and methodology of the conventional Equilibrium thermodynamics and found the whole story to be **tragicomic** [1–6]. *Why?*

To Prof. Truesdell's mind, the early history of thermodynamics was a succession of a number of annoying inconsistencies coming, first of all, from an insufficient consideration of the highly relevant works by other colleagues: For example, he blames N. L. S. Carnot for not taking into account the results by J.-B. J. Fourier, S. D. Poisson and J.-B. Biot, for otherwise Carnot could himself build up the different thermodynamics. Moreover, he anyway blames Carnot for his insufficient mathematical skills.

Howbeit, especially after reading Peter Boas Freuchen's book, we recognize that most probably, these were not the actual main points of the whole story. If Carnot would have a bit more lifetime than it was predestined to him, he would surely re-analyze his preliminary results in the proper physically/mathematically consistent manner.

Furthermore, the true main point was the actual suggestion of Carnot, namely—that any realistic process ought to be possessed of (*to put it with the proper care*—**at least**!) two basic (**groups of**!) contributions. During his active working time period he had a snowball's chance in hell to encounter the currently well-known Energy notion, but he could nonetheless put his finding at least in the proper words.

To our sincere regret, he was not fated to continue his studies, although now it is well known that he was successfully prepared for this. Indeed, his relatives had fortunately managed to publish the notes he could have handwritten some day before his final departure.

Meanwhile, that was anyway just the very start of the conventional thermodynamics, because using the proper and straightforward mechanical analogy with a kind of 'waterfall' based upon his ingenious cyclic gadget, Carnot was indeed heading towards the formulation of a unique general law governing the energy transformations.

Finally, to demonstrate the interrelationship between/among the two (or more?) ubiquitous components of every process, Carnot suggested his ingenious cyclic gadget. To our sincere regret his time upon Earth was but suddenly over.

Of interest is also Prof. Truesdell's blaming Carnot for his *"rendering thermodynamic timeless,"* for introducing *"fuzziness as a distinctive feature of the modern thermodynamics,"* for *"limiting all the general considerations to the perfect gas."* **The accusations are to 100% true! But was it at all really okay to blame Carnot of all other colleagues for all of these points?**

Meanwhile, further attentive following Prof. Truesdell's research notes reveals contributions by R. Clausius to the *"puppet theater,"* according to Truesdell's apt estimate of the whole thermodynamics development process.

Indeed, hugely inspired by graphic and conceptual re-working of Carnot's ideas due to B. P. E. Clapeyron, Clausius started to actively study in detail the cyclic gadget by Carnot by applying the proper mathematical tools and could in effect formally arrive at the sense of Carnot's suggestions. Remarkably, in the meantime the notion of Energy was born, as well as understanding the basic properties of this natural variable: namely, the conservation of the total energy and mutual transformation of its relevant parts.

With this in mind Clausius started to properly reformulate Carnot's suggestions in energetic terms—but apparently he tended to ascribe some basic, intrinsic physical properties to Carnot's cyclic gadget filled with the ideal gas.

Here we would just like to note that the cyclic gadget in question might be filled with any possible kind of working substance, according to the theorem formulated and proven by Carnot. Further, filling the cyclic gadget with ideal gas drastically facilitates the mathematical analysis of the latter. To mathematically analyze the gadget was absolutely correct decision of Clausius, for he could realize that the gadget is truly helpful in grasping a truly fundamental natural law.

Furthermore, Rudolf Clausius could absolutely correctly find the fundamental interrelationship between the basic law in question and atomistic structure of the matter. With this in mind he started his work on 'inferring the mathematically rigorous formula for this basic law based upon the mechanical atomistic theory of the matter'. In parallel to him Ludwig Boltzmann was stubbornly heading for the same direction.

To sum up in this connection, Carnot had in fact stated nothing more than the idea that his model is just independent of the "working body driving the heat machine." At the same time, the perfect/ideal gas wasn't invented by him himself, but by his followers—based upon the universality of Carnot's model—and with the clear and reasonable aim to simplify all the accompanying mathematics to the maximum. To name the *responsible persons*—R. Clausius and L. Boltzmann.

Hence, Carnot's followers had just gone a step further in theoretical idealization of the entire story—now they had not only the "ideal device," but the ideal "working body" as well! Sure, grasping the main basic features of the story to emerge could be definitely facilitated this way.

There is definitely no failure. The actual misconception *while concentrating on it* ought to be losing the necessary physical connection.

Specifically, by definition, the 'ideal gas' consists of tiny rigid balls capable of collisions among each other. There are no more interactions in the system, that is, there is no true *potential energy* in it. It is exactly here that is the stop for any realistic physics. Being deprived of the actual potential energy we must start looking for it, for we are still performing a piece of physical research. And we easily find the true source of it—it is entropy. The reason is very simple, because otherwise it is not clear what this damned entropy is for. Thus, we do solve two serious fundamental problems: On the one hand, we find the general source of all the driving sources and, on the other hand, we find the physical sense for this damned entropy (*see the details of this story in the Note 21 to Chapter 1*). Finally, we get another basic result: we could find the true physics hidden in the Carnot's cyclic gadget. So, how about the realistic aggregate states?

Interestingly, in speaking about realistic aggregate states of the matter, Dr. Nils Engelbrektsson has mathematically derived the truly universal thermodynamic equation of state, whereas Karl Alexius Franzén could thoroughly check the correctness of Nils' train of thoughts. The latter was triggered by the methodology introduced by their famous compatriot Carl Nilsson von Linné (1707–1778). *Was that physics at all?* We have already discussed this theme in Chapter 2 of the present book.

Reverting to the equilibrium thermodynamics, we note that **the cyclic gadget by Carnot doesn't require in itself any time or space variable**, for it facilitates arriving at the *important and universal mechanistic conclusions concerning the energy transformations*, to put it in our modern terms.

Sure, any exact and detailed study on how the energy transformations ought to take place in different particular cases would surely require the proper account of the time and space variables. Indeed, for such a purpose quite different mathematics ought to be necessary than that employed by Clausius, Lord Kelvin and their followers. It is truly very interesting and important topic as it is, so we shall come to discuss this in more detail later on in this chapter.

What had in fact rendered the conventional thermodynamics "timeless" ought to be the drastic overestimation of the physical meaning of Carnot's cyclic gadget, or we would like to denote this by the term '**physicalisation**'—namely, *the natural-philosophic canonization of a purely theoretical gadget helping indeed to recognize the natural laws, but in itself containing not even a tiny particle of physics.* And this is just what C. Carathéodory has demonstrated in his seminal work.

Indeed, it is the phantom of the "*physical model by Carnot*" that is the actual source of miraculous misnomers like "*quasi-static processes,*" "*adiabatic processes,*" "*isentropic processes*"—and so on, so forth—whereas **no realistic process goes without changing entropy**—and this is just the very physical sense of Carathéodory's second axiom.

As a result, those were Clausius, Boltzmann and their numerous followers, who had actually contributed to "*rendering thermodynamics timeless.*" Clausius had introduced the *fuzzy* notion of '*reversibility*', which was in fact nothing more than just a mechanically cyclic property of the process under his study. Those were Clausius, Boltzmann and their numerous followers, who had forcefully separated the two basic properties of Energy (**its conservation in total** and **mutual transformations of its sorts**).

Consequently, that was Rudolf Clausius, who did introduce two separate basic laws to where in effect a single basic law is completely enough. Those were Clausius, his contemporaries and immediate followers, who were highly fascinated by such a truly '*Multifaceted Legislative Codex of the Father God and Mother Nature*', and had thus become an overweening afflatus to produce further basic laws over and over again. Still, the latter stream seems to be truly popular and exciting [7].

To this end, see the work by Prof. Dr. Klaus Friedrich Mohr we have analyzed in detail in Chapter 1 of this book. Prof. Mohr was an outstanding German specialist in what we would nowadays denote as 'physical chemistry'. He published a clear and trenchant criticism of the original Clausius approach, by blaming Clausius for overestimating the significance of mathematical analysis, while underestimating the value of the clear-cut physical facts. Moreover, Prof. Mohr could show us the plausible ways of how we might skillfully overcome the difficulties inevitably produced by stubbornly following the original Clausius' methodology.

Howbeit, "**Truly I say to you, no prophet is accepted in his own country**" (Luke 4:24).

Bearing this entirety in mind we would like to continue reading Prof. Truesdell's work. His criticism is truly pellucid. In reading the Epilogue to his book [2], we note a number of very interesting fragments, we cite (*most of the highlighting below is ours*):

This existence of entropy for the processes **Clausius** *called reversible could have been read off from mere rephrasing of equations obtained in CLAUSIUS' second paper. Such, however, was not the course of history.*

In this third fundamental paper, published in 1862, Clausius threw in the red herring of probability's having some necessary role in thermodynamics. *He did not state what it was that had to be probable or improbable, let alone specify any mathematical probability. The passage seems to be no more than an excuse for refusing to face the physical phenomena that lift thermodynamics above triviality: the conduction of heat, deformation, internal friction of bodies, and diffusion in a mixture.*

Clausius introduced a function he called the "disgregation," the value of which he regarded as a measure of "the degree in which the molecules of the body are dispersed," but no molecules, let alone any "dispersion" of them, entered his mathematical theory, so this was merely the second red herring. *Clausius then explained what he meant by "reversible" and "irreversible" processes:*

"When an alteration of arrangement takes place so that the force and counterforce are equal, the alteration can take place in the reverse direction also under the influence of the same forces." But if it occurs so that the overcoming force is greater than that which is to overcome, the transformation cannot take place in the opposite direction under the influence of the same forces. We may then say that the transformation has occurred in the first case in a 'reversible' manner and in the Clausius *second case—in an irreversible manner.*

Since "force and counterforce are equal" in all mechanical problems, except those involving shock waves, Clausius' idea of an irreversible process must have been different from the interpretation given later to his work. For example, accelerationless laminar shear flow of a viscous fluid is "reversible" according to Clausius' definition, but nobody today thinks the entropy of a fluid mass remains constant in such a flow, which requires the continual degradation of energy.

Splendid and necessary as was the achievement of **Gibbs***, it was not thermodynamics. It was not what* **Carnot** *had called for: 'a complete theory' in which 'all cases are foreseen, all imaginable movements are referred to general principles applicable in all circumstances', so as 'to make known beforehand all the effects of heat acting in a determined manner on anybody'.*

Prof. Truesdell's conclusion sounds somewhat much too downbeat, gloomy and dreary, whereas the whole actual story was in effect never to 100% so—and, furthermore; it was not even much too tragicomic.

Sure, we might readily recognize several essential elements of a tragedy—but along with this no definite drollness at all.

Meanwhile, we owe Prof. Truesdell a very clear and apposite formulation of what are the main problems in the conventional equilibrium thermodynamics, in summing up we might cast them in the following form:

1. Clausius had no clear idea about what is the actual physical sense of the entropy notion he introduced, but was actively striving for the pertinent solution to this important problem.
2. We know that he had no more time to carefully think over the problem posed, for he had many other important activities and serious situations to deal with (In more detail we have discussed the Clausius' story in Chapter 1). His logically necessary move was **throwing the red herrings**, as Prof. Truesdell had aptly noted. Meanwhile, the latter apparent drawback ought to be widely recognized. Indeed, see for example the authoritative handbook entitled *Dictionary of Scientific Biography* and edited by Charles Coulston Gillispie (1918–2015), an outstanding American historian of science. The latter book contains Rudolf Clausius' biography by Prof. Edward Eugene Daub (1924–2015), who was an outstanding American engineer [8] lively interested in Clausius' ideas [9–12]. Studying the works of Prof. Dr. Stephen G. Brush [13–15] is conveying a similar standpoint. Specifically, more or less unanimously these authors come to the conclusion that **probability entered only when purely mechanical explanations of entropy encountered insoluble problems**. As we could see in Chapter 1, Peter Boas Freuchen had summarized the situation in stating that the then 'kinetic theory of gases' was to 100% for very dilute gases, but not for other—realistic—aggregate states of the matter, and extending that theory to the latter cases was the actual basic theoretical *complication* (in effect, *an inherently mathematical* one, but not a fundamentally physical one!).
3. The colleagues were duly and severely criticizing Rudolph Clausius, as we have already seen in Chapter 1, but to our sincere regret he had no more time for properly reacting to that criticism. To this end, of interest should also be the early criticism published already in 1868 by Theodor A. Wand, *Consistorial Assessor in Speyer* in the then professional periodical [16]. To our sincere regret, Mister Th. A. Wand could not survive in any of the scientific-historical chronicles, hence the only most probable conclusion about him and his position: He was a schoolteacher and a layman (not clergyman!) member of the local Evangelical-Lutheran Church Committee

in Speyer, Palatinate, in the Southern Germany. Howbeit, his estimate of Rudolf Clausius' work on the second basic law of thermodynamics undoubtedly deserves its citation here (we present the relevant English translation, highlighting below is ours):

"Clausius justifies his introduction of the function $\int \dfrac{dL}{T}$ *using the following statement, which he presents as an axiom without any pertinent discussion, inference or proof:*

'The mechanical work performable by heat at any change in system's arrangement/disposition ought to be proportional to the absolute temperature at which the change in question takes place'.

I must confess that I could not grasp the actual sense of the above statement, despite so many efforts I have exerted in trying to understand this. Indeed, if the heat by its nature ought to be nothing more than just a summary of molecular vibrations, then the work performable by heat ought to be nothing more than just the result of destroying/transforming the livening/driving force due to these vibrations. **Leastwise, this is just the physical sense derivable for the very words by Clausius combined with the common passable mechanical principles. Meanwhile, Clausius seems to have found some different sense of the story in question."**

Remarkably, this might just be the very point, where the physical revolutionists could pick up the paradigm and the methodology of Rudolf Clausius, namely, the priority of mathematical treatment in regard to the physical/chemical/biological reasoning. Indeed, to complete the physical reasoning we need to **create the new physics**, for the conventional physics does not allow us to clarify the points.

Noteworthy, the novel physics could appear and could be successfully developing—but at the expense of thermodynamics. Was it really a fair price for the achieved success?

4. Meanwhile, apart from Clausius' 'juggling' with the 'red herrings', there was another outstanding colleague, who did approximately the same job, but meanwhile Prof. Truesdell has not mentioned him in the above-cited treatise. We speak here of Ludwig Boltzmann. He was but massively 'juggling' mainly with one of the '*red herrings*' mentioned above: namely,

with the 'probability notion'. To properly illustrate our latter statement, we would like to place here an original statement by Albert Einstein to the theme, as cited in [17]:

"The equation 'S = k log (W) + const' appears without an elementary theory – or however one wants to say it - devoid of any meaning from a phenomenological point of view.

A. Einstein (1910)"

5. Noteworthy, the two 'read herrings' both Boltzmann and Clausius were 'juggling' with had apparently occurred to be immensely comfortable for the novel, revolutionary ideas based upon the atomistic structure of the matter. It was a definite conceptual stop for thermodynamics. Was it a tragedy? Was it a comedy? Was it the both at the same time? How to estimate the ultimate truly revolutionary result in a fully sober, correct way?

Here, in Chapter 1 we have already started discussing in detail the actual physical insolvency of the '*reversibility*' notion behind the conventional equilibrium thermodynamics. The literature references [9–15, 18–21] offer a lot of further material to think it over, as we could already see from the above short discussion. Still, of immense interest and importance would also be to consider the insolvency of the second '*red herring*', namely the '*probability*'.

This is not a straightforwardly simple task, for Boltzmann's ideas, and especially Gibbs' results were, are and are remaining by no way misleading! Indeed, Boltzmann's guess $S = k*\ln(W)$ is ingenious in its pertinence and correctness. Gibbs' analyses on the same theme were a definite development of Boltzmann's ideas, for Boltzmann had no more time for developing them himself. That Gibbs could manage solely to consider the case of 'statistical independence' does not mean that he was heading for the false direction! Dr. G. A. Linhart demonstrated the fundamental correctness of such an approach, as we could see in Chapter 4. But why then we are speaking of the 'probability' as a 'red herring' and of any insolvency in this case at all?

The actual point here is the *fuzziness* Prof. Truesdell was criticizing, but he had not reported on this topic in more detail. In effect, the *fuzziness* starts with the application to a 'system consistent of a huge number of atoms/molecules' or other minuscule particles inaccessible to a direct observation. The main point here is connected to the term **'huge'**, which is but fuzzy in itself, as it is, due to the well-known *sorites* paradox (the paradox of the heap) [18–21].

In using the arguments based upon such fuzzy terms it is extremely difficult to formulate a theory of clearly general significance. To this end, a truly outstanding achievement of Dr. G. A. Linhart has been to show the way of escaping from such fuzzy grounds by properly working with the Bayesian statistical arguments, for example (see Chapter 4). This is why Dr. Linhart's results do pave the way for the proper development of statistical mechanics.

To sum up, the main difficulty all the mentioned colleagues were experiencing during their time ought to be the finiteness of the human lives. Sure, there is nothing to laugh at, but the second difficulty in the tightest connection with the first one might still be possessed of some definitely comic element: "Нельзя объять необъятное" (Козьма Прутков), or, "*You cannot grasp the immensity*" (Kozma Petrovich Prutkov), to put the straightforward meaning of this famous Russian saying in English. But the result ought to remain as follows.

If we just put aside all the political perversity in the following connection, it would anyway be instructive to recall that Friedrich Engels (1820–1895), being a serious and thoughtful philosopher among his numerous skills, noticed in his unfinished book *The Dialectics of Nature*:

> "*Rudolf Clausius could clearly pose the problem, but he had suggested no handy solution for it.*"

Remarkably, in parallel to discussing Prof. Truesdell's conclusions, it might also be of interest to take into account that Prof. Dr. Bernard H. Lavenda had posed several striking and—at the first glance seemingly rhetoric—questions during one of our personal discussions on the theme. His very first question sounds as follows:

> 1) "*Why only one philosophy has become so ingrained in the literature: Namely that of Clausius? Would this be the actual root of confusion all around Tait's inverting the physical sense of the Clausius formulation of the Basic Second Law inequality, as well as around the final thermodynamic state—be it the minimum dissipation of energy by W. Thomson (Lord Kelvin)—or Clausius' 'Heat Death'?*"

Actually, we have started answering to this interesting poser in the Notes to Chapter 1. The book by Peter Boas Freuchen poses basically the same question, but in a somewhat less implicit form. The correct answer to it ought to be as follows. The Clausius' paradigm was indeed resulting mostly from severe mathematical complications encountered by the kinetic theory of gases, and as a result had only very-very slight connection to the actual physical posers.

Interestingly, the physical revolutionists could impose both Clausius' paradigm and Clausius' methodology on us, for these both were truly convenient for their own theoretical works.

As to the *'red herring of probability'*, that is, the second one, it was truly difficult to debunk—for the functional interrelationship between the probability and entropy was and remains to be to 100% correct.

This is why it could win a general recognition as something truly fundamental. The true authors of this 'red herring', who had also introduced the methods of treating it in a lofty manner, were definitely authoritative, active, proactive researchers. To our sincere regret, all of them had immense personal problems, including our natural human mortality. Otherwise, they would definitely analyze and re-analyze this 'red herring' over and over again in the pertinent way, until finally arriving at the physically consistent results.

Meanwhile, the very first of the 'red herrings', the 'reversibility/irreversibility' was only 'locally poisonous', that is, it was dangerous for the thermodynamics alone. Still, together with the second 'red herring' just discussed it was truly fatal for the latter. An outstanding American specialist in the thermodynamics field and pedagogue, Mark Waldo Zemansky (1900–1981) has depicted the result in his work [22]. Indeed, we are dealing with a 'conceptual corpse burst into three parts' instead of having a unique intact branch of knowledge. Interestingly enough, according to Prof. Zemansky, among those 'three parts' the chemical thermodynamics somehow seems to be functioning in the most consistent way.

Of interest for our present discussion ought to be trying to answer the poser: **How could it come** to the point that the 'red herrings' could get so mighty that they had at last not only just killed the thermodynamics, but also decomposed the remaining conceptual corpse?

In starting to analyze the situation we would like to attract the attention of the readership to a very important psychological problem, namely to our emotions. In fact, emotions constitute truly powerful weaponry, as psychology does tell us [23]. We dare to add here that our emotions might even provide us with a powerful methodology in solving whatever problems to solve.

To this end, our present story is a definite illustration of the latter statement. Since longer time many authors were noticing that the basic laws of thermodynamics look like truly **prohibitive**, especially in their formulation presented by Rudolf Clausius. In effect, they do forbid any possibility to invent *perpetuum mobile* of any kind. On the other hand, the history of mankind is full of the brightest examples of perpetuum mobile inventors, who were every time failing to achieve their unique goals.

This proves some definite attraction imposed by the perpetuum mobile idea. Indeed, all the time there were, are and will be people, who sincerely disbelieve those **prohibitions**, irrespectively of the authority surrounding the authors of the latter—be they royals, higher officials, reverends—and/or renowned scientists. This is just the point, where we come to the emotion notion! To our mind, it is just this severe prohibition of the perpetuum mobile of any kind that could have driven all of us to categorically accept the 'physicalization' of the ingenious cyclic gadget by Carnot. Indeed, in its 'physicalized' form, Carnot's cycle is nothing more and nothing less than just a perfect perpetuum mobile—forever devoid of any time and any space. It is this perpetuum mobile that is building-up the only conceptual basis of the so-called 'equilibrium thermodynamics'.

To this end, let us have a more detailed look at the truth of fact.

5.2 A Smart Balance among the Emotions and the Methodology of Scientific Research

An outstanding Russian/Soviet physical-chemist and philosopher of science Alexandr Nikolayevich Shchukariov (Александр Николаевич Щукарёв) (1864–1936), a professor at the Kharkov University of Technology (formerly, Kharkov Polytechnic Institute, in the Russian Empire and the USSR), has performed a truly apt analysis of the state of art in the thermodynamics field [24].

He concludes that the '*red herrings*' thrown in by Rudolf Clausius have triggered a truly world-wide, steady and exceedingly emotional reaction among the leading specialists of the Clausius' time as well as thereafter.

To this group of 'emotional physicists' one might surely assign Ludwig Boltzmann, Max Planck, as we could also see during the discussion of Peter Boas Freuchen's book in Chapter 1 here. To learn about more details, let us now open the book by Shchukariov [24].

This portrayal of Prof. Dr. Shchukariov has been taken from
http://kharkov.vbelous.net/politex1/shchukar.htm

When reporting about the well-known Clausius' aphorisms/ maxims concerning the energy and entropy of the universe, Prof. Shchukariov notes as follows (we present here our English translation of the original Russian text):

"...We are not entitled to apply our personal measure or mood to the natural phenomena and laws. But such an attitude is frequently being involuntarily applied. Sure, we are glad to realize that the Nature is constant in itself, which is finely expressed in the energy conservation law, but we still feel funereal, when it comes to the entropy increase law. Indeed, whereas the total energy remains always constant, it is nonetheless being spread, and its 'intensity' is decreasing. This is why; the systems' capability of transitions is also going down....

...Naturally, in considering such a scenario we get doleful – and some doubt creeps unwittingly. Is this law of the general entropy Increase indeed an absolute natural law??? ... Are there, might be, some (probably minuscule) natural phenomena, probably rare and/or persisting in some remote corners of our universe, which still do not obey this cruel general

law??? The distributed, devaluated energy may perhaps still be coming together and increasing its intensity?

Such hypotheses are rather numerous, while many of them belong to the respected scholars..."

The truly brightest examples of the above attitudes alleged by Prof. Shchukariov are listed below:

1. Rankine's *border of the universal ether*;
2. Maxwell's *demon*;
3. Svante Arrhenius' *pieces of planets as Maxwell's demons*;
4. Boltzmann's belief that *maximum entropy states are the most probable ones.*

By his throughout detailed considering the bold Boltzmann's idea, Prof. Shchukariov is forced to draw the following logical conclusion, which is trivial at the first glance but nonetheless leading to a paradoxical situation. We cite: "**From the standpoint of the probability theory, such events are not quite impossible—they are indeed the less possible, the more balls are present in the group.**"

Indeed, to exactly define, what is it in fact '*more*' or '*less*' is practically impossible due to the well-known *sorites* paradox, that is, the paradox of heap [18–21].

From such an (at the first glance) unconventional standpoint it ought to be of interest to consider the popular reports of more or less the same time period by Max Planck, Albert Einstein, as well as those by E. Warburg, F. Henning, L. Holborn, and F. Hasenöhrl (in [25]), in combination with somewhat later reports by Werner Heisenberg, Peter Debye—Josef Mayerhöfer, Hans Thirring, Karl Menger, Hermann Mark, Georg Nöbeling, and Hans Hahn, (published in the popular series [26]). Noteworthy, both Max Planck and Albert Einstein were fully aware that they had entered a kind of 'wonderland', as Einstein had 100% properly estimated it (we have just cited him in this chapter, see Page 608), the 'wonderland' having absolutely no immediate phenomenological basis, but instead, freely flying in our emotions. At this point we might also remember the call 'Surf's Ahead!' by Max Bernhard Weinstein in 1914 (see Chapter 1, Pages 348–ff).

The glimpses of that emotional ocean (it is difficult to choose another word for this phenomenon) were (and most probably even are!) throughout present—see for example a paper by an outstanding British Thermodynamicist, Prof. Dr. Peter Theodore Landsberg (1922–2010), where he had told us the following very wise and illustrative words [27]:

This portrayal of Prof. Dr. Landsberg has been taken from http://www. telegraph.co.uk/news/obituaries/7741783/Professor-Peter-Landsberg.html

"… At present, if there are 'two physicists', as Truesdell said, or even more scientists, basically agreeing on the contents on this paper. That would be a tremendous success. Some differences might still be left in cognition.

The framework of applicability for 'the impossibility to obtain the second kind of perpetual machines' is for any macroscopic systems, including the impossibility to obtain from equilibrium systems and from non-equilibrium system, the impossibility to obtain via reversible processes and via irreversible processes, and the impossibility for simple systems with only spontaneous processes and the impossibility for complex coupling systems with multiple processes.

The second law of thermodynamics is not summarized from experiences on microscopic scales or experiences for the whole universe, so it should not be extended into the microscopic range and the whole of universe. The framework of applicability of the second law of thermodynamics is any macroscopic systems, just the same as that of the thermodynamics discipline.

As a science of development, the development of thermodynamics itself had been in stagnation for such a long time. That was rarely found in other fields of natural science. The thinking on development in thermodynamics had been solidified for such a long time. The mainstream of development should be the thinking of continuous 'evolution', but it had been substituted by the thinking of degradation. That might be a pity, but might also be an internal rule in science of development.

The best answer for any difference in cognition may be that go forward based on the nature of the second law of thermodynamics, i.e., along the direction of 'arrow of time', the time proof might be the best evidence in history."

Our immediate comment: Prof. Landsberg is mentioning here the 'perpetuum mobile'.

Indeed, it is noteworthy that the actual activity of all the well-recognized founders of thermodynamics—and the actual tremendous efforts of their revolutionary followers were as a matter of fact devoted to demonstrating that the actual physical sense of the thermodynamics' basic laws ought to imply the sheer impossibility of any conceivable kind of 'perpetuum mobile'.

As a result, the considerable number of thermodynamics' handbooks all over the world teach us that the first basic law implies the impossibility of the 'perpetuum mobile' of the first kind, whereas the second basic law implies the impossibility of the 'perpetuum mobile' of the second kind.

Hence, the students must get the following impression: The peers of thermodynamics could have duly demonstrated that any trying to invent 'perpetuum mobile' has absolutely nothing to do with the basics of physics.

Meanwhile, as a result, the conventional equilibrium thermodynamics does render the ingenious cyclic model by N. L. Carnot to a kind of 'perpetuum mobile', which is acting without any apparent driving force (or, at least, whose driving force is unclear). Carnot cycle's action is infinitely slow, this is why, in fact, we do not need the time variable for analyzing its work.

Still, the realistic driving forces as well as physical time itself must somehow be generated, and this is achieved by 'negating the entropy' or in other words by producing negentropy—so that the actual holy aim of the scientific/engineering research as it is ought to be stubbornly looking for all the possible rules and means of 'generating negentropy'.

To our mind, it is the latter so-to-speak 'insight' that builds up the actual conceptual basis of the conventional equilibrium and non-equilibrium thermodynamics, and this is a result from what we would like to denote here as the 'physicalization' of the ingenious theoretical cyclic model by N. L. S. Carnot.

The interested readership could find the professionally detailed philosophic analysis of the above-mentioned standpoints (in conjunction with the *operational approach in the scientific research*, for details see Note 3 to Chapter 1) in the following sources [28, 29]. Moreover, the indication of Prof. Shchukariov ought to shed light upon the activity of the 'team' Ludwig Boltzmann, Max Planck, and Albert Einstein in the field of thermodynamics and statistical physics we have analyzed in Chapter 1 of the book at hand. Without a sole attempt to somehow diminish the truly seminal contributions of these three definitely outstanding colleagues, we would still have to boil down to the snug commonplace: "*Errare Humanum Est.*"

Meanwhile, Prof. Shchukariov concludes his above-discussed book by the following remarkable paragraph [24]:

> "*We are never contented with any kind of finish, but the universe ought to be unlimited, and its energy stocks ought to be unlimited as well, so that nothing might pose difficulties in connection to the entropy growth. So then, wouldn't it be possible to look at the latter growth as a kind of quantitative symbol of the universal life, as a continuous symbol of the time current/current time, and to apply the following enthusiastic words to this problem:* **I am the alpha and the omega, the first and the last, the beginning and the end.** "

Interestingly and most probably, it is just the latter suggestion by Prof. Shchukariov that ought to be the very emotion obsessing such highly respected researchers as Peter Guthrie Tait, as well as his undoubted followers Lars Onsager and Ilya Romanovich Prigogine, who suggested the entropy to be the '*very driving force of practically everything*'. We have started discussing this in the Note 21 to Chapter 1 and shall finish it up here.

This same pertains most probably, if we would start discussing the ingenious suggestions of the '*negentropy*' (etc.) by such highly respected colleagues from different creative fields, as

Erwin Schrödinger, Léon Brillouin, Albert Szent-Györgyi, Richard Buckminster "Bucky" Fuller, and Luigi Fantappiè.

The emotional intensity around this entire story was (and still remains to be!) so huge that even just a miniscule, but in effect ponderous, arithmetic typo by Leó Szilárd (1898–1964) is still causing the truly international research activity based upon the Maxwell demon. This is why we feel obliged to come back to the actual professional standing of Leó Szilárd later on in this chapter.

The readership willing to continue discussing and analyzing the bright and provocative idea of Maxwell demon we would like to send to the truly deep and stimulating works by Prof. Dr. Stephen Jay Kline (1922–1997), an outstanding American engineer, pedagogue, and philosopher of science [30–33].

This portrayal of Prof. Dr. Kline has been taken from
http://www.nae.edu/29074.aspx

Howbeit, it is throughout clear that to correctly answer all the questions arising in this field, one has to reveal the exact physical sense of the entropy notion. We have already touched this point in the Note 21 to Chapter 1. As we see, the point is that neither Rudolf Clausius, nor his immediate revolutionary followers (such prominent researchers as Ludwig Boltzmann, Max Planck, and their numerous colleagues) could clearly formulate the actual physical sense of the second basic law of thermodynamics, although everybody had accepted its immense fundamental importance. This apparent shortcoming was the very source of all the thermodynamic 'fuzziness'; it is exactly therefrom that the '*red herrings*' of the '*reversibility*' and '*probability*' could efficiently borrow their huge driving force.

Remarkably, this is exactly the point where Prof. Shchukariov finishes his writing—and where the famous conversation begins between Claude Shannon and John von Neumann, regarding what kind of name might be given to the attenuation in phone-line signals. **So, what is then the entropy?'** '**What is the second basic law of thermodynamics?** These both posers attracted and still attract enormously huge attention [34–39], and in duly following all this literature and the references therein we really go through the immense multitude of standpoints, slants, and perspectives without finding even a trace of at least some kind of consensus. But does this entirety have at least something to do with thermodynamics? A truly good poser.

Of course, in this chapter we are all the time reiterating: '*emotions, emotions—and only emotions*' Isn't this much too much of the stubborn non-constructive malevolence? Nowadays, we have but still another explanation, namely the one that we are "taking thermodynamics too seriously" (see the work [40] and the references therein). Sure, we agree that emotions might frequently be non-constructive, so that any scientific research worker should be as free as possible from externally or internally induced emotions, at least with respect to his/her field of professional endeavor. However, would 'taking too seriously' be much more constructive? Another one truly good poser. Still, the story isn't that murky, as it might be looking like at the first glance!

Thanks to Peter Boas Freuchen, in Chapter 1 of this book we could monitor the actual line of thermodynamics' development and would now like to sum up the story.

Indeed, a colleague of Rudolf Clausius, K. F. Mohr, timely and competently criticized Clausius. However, that was not Clausius himself, who could take into account that deserved and constructive criticism. Instead, his apprentice, August Friedrich Horstmann has managed to carry out this job. Sure, he couldn't properly finalize fulfilling this task due to fully objective reasons, but he could still convey the very principle to Josiah Willard Gibbs, who was then quite alone in pushing the whole story forward. Meanwhile, J. W. Gibbs had no infinite lifetime, to our sincere regret. Still, fortunately there was George Augustus Linhart to accept the banner dropped—and to drive the story to its further chapters (see Chapter 4).

In parallel to this, while Peter Boas Freuchen in Denmark (see Chapter 1) and Max Bernhard Weinstein in Germany (see Chapter 3), were independent of each other, but productively thinking over the actual foundations of thermodynamics, Nils Engelbrektsson and Karl Alexius Franzén in Sweden could have independently made some suggestions in the field and thus delivered the valid universal thermodynamic equation of state (see Chapter 2). All of these colleagues have duly communicated their results, hence the only principal point here ought to be our attentive monitoring of the natural-scientific development. To sum up, there seems to be definitely no space for any kind of a global tragicomic development, not to speak of any kind of a global gloomy pessimism!

Now, returning to the Russian Empire/USSR, we can immediately recognize that Prof. Shchukariov has not explicitly pointed out the regretful situation in that country. Still, the very fact of his effort to publish his above-mentioned book ought to be viewed as his definite move to communicate some extremely important information to the young interested readership.

Indeed, the leading physicists in the second half of the XIX-th and the first half of the XX-th centuries—who had paid their invaluable contributions to the formation of the Soviet School of Physics—the outstanding experimentalist, Prof. Dr. Abram Fyodorovich Ioffe (Абрам Фёдорович Иоффе) (1880–1960) and the renowned theoretician, Prof. Dr. Orest Daniilovich Chwolson (Орест Даниилович Хвольсон) (1852–1934), most probably were totally absorbed by the emotional wave in question. Meanwhile, Prof. Chwolson was stubbornly trying to cope with the latter (see Chapter 1 and Chapter 3).

Howbeit, the Russian engineers—at least at the same time—were clearly exhibiting quite fine understanding of the thermodynamics foundations. This is in fact a clear trend, for these are the engineers, who always have to take care of the usefulness of any theory, for otherwise they would not be capable of inventing and constructing any useful device and/or process. In the field of thermodynamics the undoubted pioneer of such a professional stance was Nicolas Léonard Sadi Carnot, and fortunately there were and there are many-many of his followers all over the World.

Sure, it is for this reason that the *'engineering part'* of the conventional thermodynamics *'conceptual corpse'* [22] does not cause serious objections: Motor-driven mechanical devices, fridges, as well as the entire chemical, pharmaceutical industry, etc. etc. are mostly working properly all over the world.

Very nice illustrations of the above conclusion might be, for example, the books by Prof. Dr. Alexandr Andreevich Brandt (Александр Андреевич Брандт, 1855–1932), an outstanding specialist in the fields of applied mechanics and thermodynamics.

Prof. Brandt had to emigrate from Russia due to the revolution there in 1917. This way he had fully eliminated his roots and was remaining an alien for a big number of his compatriots for a longer time. What he had written and was writing in his mother tongue was not understandable and therefore not very interesting for the rest of the world. Only recently he and his writings (either professional or journalism) could yet become available in his country.

After reading his legacy it becomes clear that it is of a truly general significance.

This portrayal of Prof. Dr. Brandt has been taken from Wikipedia.

The following book of him might be readily recommended to the interested versed readership (see [41]).

Moreover, at this same time, Prof. Dr. Nikolai Alexeevich Bykov (Николай Алексеевич Быков, 1862–1955), an outstanding specialist in the fields of internal combustion engines and thermodynamics has published the following very interesting and important book in the field [42].

It is this book by Prof. Dr. Bykov that ought to be of a separate value for our present discussion, for it contains a unique explicit chapter entitled: '*The Physical Sense of Entropy*'. Remarkably, only a few books ever published all over the world do contain such a separate chapter, where the authors do boil this important topic down to its essence.

Although both Prof. Brandt and Prof. Bykov are duly juggling the conventional misnomers, like '*irreversible/reversible/adiabatic*', etc., they allow the readership to follow the actual physical logics, to give a rather clear picture of the field they describe.

Meanwhile, Prof. Bykov was representing a different group of the Russian Empire population, as compared to Prof. Brandt.

He was quite sympathetic with revolutionary movement of the then Russian studentship and even actively participating himself in the relevant everyday activities. This is why he could sincerely appreciate the revolutionary events of the year 1917 in his Homeland, so that any kind of emigration had been out of the question for him.

With this in mind, he was duly continuing his professional service at the higher educational organizations and working in the field of thermodynamics.

This portrayal of Prof. Dr. Bykov has been taken from Wikipedia.

To be mentioned in this row are definitely also Prof. Dr. Vasilij Vasilievich Sharvin (Василий Васильевич Шарвин, 1870–1930), a renowned Russian/Soviet chemist, researcher and pedagogue, who had published a very insightful book in the field of thermodynamics [43].

This portrayal of Prof. Dr. Sharvin has been taken from
http://persons-info.com/persons/SHARVIN_Vasilii_Vasilevich

To be definitely mentioned ought to be Prof. Dr. Alexandr Alexandrovich Satkevich (Александр Александрович Саткевич, 1869–1938), an outstanding Russian military engineer, a specialist in the field of general and applied thermodynamics, who was a victim of the Stalinist regime.

This portrayal of Prof. Dr. Satkevich has been taken from http://pskovgrad.ru/
war/pervaya-mirovaya-vojna/29239-satkevich-aleksandr-aleksandrovich.html

Of special interest for our present discussion ought to be his work, demonstrating a cool professional insight of its author (see [44] and the references therein).

Contrariwise, we just get a quite different picture, as soon as we confer the lectures on the thermodynamics' foundations by Prof. Dr. Ioffe, as well as the truly encyclopedic textbook by Prof. Chwolson, where truly all the aspects (both emotional and productive!) of the then revolution in the natural sciences do remind of themselves over and over again.

To this end, a very good poser arises: Could any productive combination of the emotional and the cool professional attitudes be possible at all? The answer is throughout positive!

In principle, the positive feature of the 'emotional–operational' approach in the scientific research is obvious. Undoubtedly, it is throughout possible to solve particular basic problems of interest, high importance and impact without dwelling on the fundamental principles at all. Meanwhile, the only objection comes to mind here. Of course, without systematic dwelling on the fundamental principles we shall never extend the applicability areas of our theories. This is why, the both approaches are valuable, and the main point ought to be their proper timely combination.

A truly bright example of a specific and successful mixture between the 'emotional' and throughout practical standpoints at the beginning of the XX-th century could also be the professional activity of an outstanding Polish (first, Russian) specialist in the field of physical chemistry, Prof. Dr. Bohdan (A.) von Szyszkowski (1873–1931).

This portrayal of Prof. Dr. Bohdan Szyszkowski has been taken from Wikipedia.

He had obtained his higher education in Germany (Prof. Dr. Wilhelm Ostwald), Great Britain (Prof. Dr. William Ramsay), and then he was first a freelance lecturer (*Privatdozent*), to get soon thereafter an ordinary professorship in chemistry at the Kiev University of St. Vladimir, Russian Empire. Immediately after the Russian revolution 1917 he had moved to Poland to continue his professional work.

Prof. Dr. Szyszkowski was, on the one hand, capable of divining an empirical relationship among the solvents and solutions surface tension, as well as the concentration of the pertinent solution, whose functional form might be inferred in a straightforward way by using the J. W. Gibbs' thermodynamic approach (as could be demonstrated by I. Langmuir) and is actively used until nowadays [45].

Indeed, the empirical equation by Prof. Szyszkowski could be cast as follows: $\Delta\sigma = B\ln(Ac + 1)$, where A, B are constants to be determined experimentally, c is the concentration of the solution in question, and $\Delta\sigma$ stands for the difference between the surface tensions of the pure relevant solvent and of the solution under study, respectively. The B is constant for the homological rows, whereas there may be significant differences between the values of A for the homological row's nearest neighbors, as compared to each other, the one being up to 3–3.5 times as large as another one, which is in full accordance with the Traube's rule.

As I. Langmuir could demonstrate at his time, the purely empirical equation by Szyszkowski might well be derived theoretically. Indeed, we ought to substitute the Γ value from the Langmuir isotherm, namely: $\Gamma = \Gamma_\infty \cdot \dfrac{bc}{1 + bc}$ or $\Gamma = \Gamma_\infty \cdot \dfrac{c}{c + \alpha}$, where $b = \dfrac{1}{\alpha}$, just directly into the Gibbs equation as it is expressed $\dfrac{d\sigma}{dc} = -RT \cdot \dfrac{\Gamma}{c}$; $\dfrac{d\sigma}{dc} = -RT \cdot \Gamma_\infty \cdot \dfrac{dc}{c + \alpha}$, and integrate the latter from 0 to c, to get at last the following expression directly comparable with the empirical one: $\sigma = \sigma_0 - RT \cdot \Gamma_\infty \cdot \ln\left(\dfrac{c}{\alpha} + 1\right)$.

Remarkably, the latter story is in principle reminiscent of the inferences by Dr. G. A. Linhart (see Chapter 4). Meanwhile, this inference is mentioned in any consistent textbook on physical and colloidal chemistry, but without mentioning Linhart's name. In Russian it is definitely a book [46]. And subsequent successful uses of the Szyszkowski equation till nowadays might be found in [47–53].

Nevertheless, on the other hand, Prof. Szyszkowski was a truly enthusiastic adept of the 'emotional branch' in thermodynamics—see his inaugural lecture in connection with his getting a professorship at the Kiev University in the year of 1909 [54]. The main idea of the latter lecture seemed to be promoting the Clausius/Boltzmann's 'emotional red herring' of probability at any price. We might even recognize puffing the tiny fluffy 'herring' up (and thus, in effect, inflating it) to something like a '*Triune Whales/(or even Triune Elephants!)*' upon which our Mother Earth rests, (just like it was conventional in the ancient cosmography!), by declaring some basic interrelationship between the entropy and probability. Was the above-mentioned 'orthodox attitude' merely a pure emotional belief—or just a pure careerism—times an unrestricted desire for the immediate fame?

To our sincere regret, there are no more pertinent respondents to the poser, so that the interested readership might come up with own versions, while an urgent need remains to duly re-think over and consistently re-build the conventional thermodynamics and statistical physics.

[Please see Supplementary Note 1, Pages 670–671]

The message we would greatly appreciate to disseminate with the help of the present book:

> '*Anyway, eradicating the human emotions ought to be possible solely together with the whole mankind. So that emotions are not representing something really dramatic to be immediately avoided.*
>
> *Still, the main objective of the research ought to be the ultimate knowledge, but not emotions and/or human relations.*'

To revert to the 'red herring' of probability, it is worthwhile mentioning that there were also a lot of philosophic endeavors all around the latter—see, e.g., the works of the outstanding Austrian physicist, pedagogue and philosopher, Prof. Dr. Franz-Serafin Exner (1849–1926), and the references therein [55, 56].

Prof. Exner was a direct colleague of such a renowned Austrian theoretical physicist as Ludwig Boltzmann, a teacher of many outstanding physicists and among them—one of the most active revolutionary physicists, Prof. Dr. Wilhelm Theofil Smoluchowski von Smolan (1872–1917), who could rightly be considered one of the progenitors of the stochastic physics.

Indeed, Prof. Exner was consistently emphasizing that all the apparently deterministic laws are in fact the macroscopic limits of indeterministic basic laws valid for the huge numbers of single particles or events.

In particular, it was Prof. Exner's striving to produce a synthesis between E. Mach and L. Boltzmann's approaches that did pave the way to accepting genuine indeterminism in physics, even without any reference to quantum mechanics, which looks like much more radical standpoint than that of Max Planck.

Interestingly, Prof. Exner's firm reliance upon the second basic law of thermodynamics in its Boltzmann–Planck's formulation did not even halt at the boundaries of physics proper. By the end of his life he had completed working out a comprehensive indeterministic theory of human culture [56], which has remained conventionally unpublished (being now published in Internet) and gives vivid testimony of the cultural discussions in the large circle around Prof. Exner [57–59].

Meanwhile it is noteworthy in this connection that the most recent investigations hint that the hotly debated alleged 'passionate struggle of Prof. Dr. Ernst Mach against the atomists and indeterminists' is nothing more than just a later intentional falsification [60].

This portrayal of Prof. Dr. Exner has been taken from
http://austria-forum.org/af/AEIOU/Exner,_Franz_Serafin_(1849–1926)

Nonetheless, as we already know from Chapter 4, the ingenious Boltzmann–Planck formula is nothing more than just the consequence of some basic modalities valid at every imaginable level of consideration—from the ultra-microscopic via nano- and mesoscopic levels—up to the conventional macroscopic one. This is an unambiguous achievement of Dr. Linhart, which is still remaining unrecognized. Meanwhile, Dr. Linhart's formal mathematical inference of the 'basic' Boltzmann–Planck formula using Bayesian statistical approach is a clear and powerful conceptual thwack against the entire *'basic indeterminism inherent and rooting in microcosm and thus embracing the observable macrocosm'.*

[Please see Supplementary Note 2, Pages 671–797]

Meanwhile, in parallel to all the above-mentioned vicissitudes, the Second Basic Law of thermodynamics was finally properly formulated by August Friedrich Horstmann, a talented, but seriously ill student of the outstanding physicist Clausius and the prominent engineer Zeuner. Josiah Willard Gibbs has borrowed Horstmann's ideas. The modalities of energy transformation have in effect nothing to do with any kind of 'heat death', but with W. Thomson's (Lord Kelvin's) dissipation of the useful energy, as clearly presented by George Downing Liveing. We have already met both Prof. Dr. Horstmann and Prof. Dr. Liveing in Chapter 1 of this monograph.

That was just the 'different thermodynamics' Dr. Linhart was trying to develop—primarily along the Gibbs' line.

Noteworthy, the results by both Horstmann and Liveing have practically disappeared, while, as Prof. Truesdell could express it in a picturesque way (here we'd dare to suggest a pertinent paraphrase of it!): '*the question "What is entropy?"* —is still vibrating—*ceaselessly without damping or dissipation*'. Hence, the actual poser should be formulated as follows: "*Why such an unsatisfactory situation could come into effect?*"

[Please see Supplementary Note 3, Pages 797–812]

In our opinion, to track down the actual cause of the 'confusion all around Tait's inverting the physical sense of the Clausius' formulation of the basic second law inequality' (please see Supplementary Note 28 to Chapter 1 for the detailed discussion on this theme) it is enough to recollect the mathematical difficulties triggered by the entropy's exact definition, as clearly described by Prof. Truesdell. In fact, Prof. Dr. Moïse-Emmanuel Carvallo (1856–1945) has already long time ago noticed this very important point and tried to turn out attention to it, as we have seen above here (please see Supplementary Note 21 to Chapter 1 for details).

In addition, the actual failure was most probably the full concentration on the mathematical side of the problem—at the same time

- Without re-thinking the basic physics of the story;
- Without looking around to locate some competent authors, who would have expressed the physical truth—just in sober words, even without lots of mathematics, like N. L. S. Carnot could successfully accomplish this at his time.

After avoiding the above restrictions, the proper mathematical tools would definitely not make us wait for their advent, and even playing with the '*tiny red herrings*' would not be necessary.

With this in mind, the main point is probably that *we should not bloat the latter 'tiny red herrings'*—**for the love of God (!)**

Otherwise, our cognizance would every time remain at a very high risk of the unavoidable 'bloat and torsion'.

In the above connection, it's only at the beginning of this century that we could finally start revealing the names of Prof. Dr. August Friedrich Horstmann (1842–1929) from the Heidelberg University, Germany and of Prof. Dr. George Downing Liveing (1827–1924) from the Cambridge University, Great Britain—see the most recent works [61, 62].

To this end, several important notes ought to be due in addition to the historical investigations by Prof. Dr. W. B. Jensen [61, 62] and to summarize our corresponding discussions in Chapter 1, in bearing in mind the discussions in Chapters 2–4:

1. Prof. Dr. Horstmann started his career as an apprentice of such outstanding thermodynamicists as Prof. Dr. Rudolf Clausius and Prof. Dr. Gustav Zeuner, while attending their lectures on the subject in Zürich, and he started seriously working on the applications of thermodynamics to chemical problems practically immediately after his graduation. The first stage of his studies is clearly documented in his second PhD thesis (Habilitationsarbeit);

2. To our mind, it is absolutely not correct, to draw the following conclusion from the work by Prof. Dr. Horstmann: *"One important reason for this neglect is that Horstmann did little to propagate his explicit entropy approach to chemical equilibrium"*—contrary to what Prof. Dr. Jensen states in his above-mentioned work on this theme, we shall present here the English translation of Prof. Horstmann's brilliant publication attempting to promote his standpoint.

3. Meanwhile, with the following statement by Prof. Jensen we might more or less agree: *"However, a far more important reason for this neglect was the fact that the 19th- and early 20th-century chemical community was extremely uncomfortable with the entropy concept, which is precisely why Horstmann's approach was unique."* Howbeit, the actual reason of the mentioned 'uncomfortable situation' was the 'revolution in the physics' characterized (among other trends) by the forceful promotion of the 'interconnection between the entropy and probability notions' as a 'basic natural law'—plus the vigorous world-wide priority quarrels we have mentioned in the Supplementary Note 2 to this chapter, which resulted in 'elimination of any nonconformity'. And this is the actual reason of why the only correct formulation of the law of energy transformation by Prof. Dr. Horstmann could be downright forgotten.

4. Besides, in Chapter 1 we have already cited and discussed the address by Prof. Dr. Jacobus Henricus van't Hoff (1852–1911) (see Ref. 14 to Chapter 1 and the pertinent discussion in the latter chapter) he added to the collection of Prof. Horstmann's works edited and re-published by him in 1903, as Prof. Horstmann was increasingly suffering from progressive blindness and had to stop his active research work henceforth. Well, howbeit, the readership might nowadays feel some tiny wisp of hypocrisy in Prof. van't Hoff's statement, especially if we take into account that thermodynamics as it is was in fact not really within Prof. Dr. van't Hoff's scope of professional interests, see the following reference on this theme [63].

5. And—the last, but not the least: Prof. Dr. Gibbs was not only communicating with Gustav Robert Kirchhoff and Hermann Ludwig Ferdinand von Helmholtz, while staying in Heidelberg

for one year [64]—he was also systematically attending the lectures by Prof. Horstmann (see Chapter 1 for the details). And here is just the actual reason for the success of the thermodynamic works of Prof. Gibbs—for in his work he was consistently using Prof. Horstmann's rational representation of thermodynamics, by developing the relevant mathematical apparatus.

6. This is why we cannot agree with Prof. Truesdell's estimating the Gibbs' result as a kind of "thermostatics." Quite contrariwise, Prof. Gibbs could actually enter the hall of the different thermodynamics, so that his next step ought to be "switching on time" to justify the term "dynamics." But, to our sincere regret, he had no more 'time to spend upon earth' for bridging the existing gap. Still, now we know (see Chapter 4) that Dr. Georg(e) Augustus Linhart has managed to pick up the banner that Gibbs dropped.

7. Here we shall come to this point later on, but now it is important to demonstrate the crucial significance of Prof. Horstmann's results.

One of the main aims of the monograph at hand ought to be ensuring that Prof. Horstmann's seminal work would never disappear again. This is why we present here our English translation of one of his most important publications. Thus, we reconcile just one fully unmerited step—not mentioning the latter in the most recent historical reviews. Highlighted below is ours.

5.3 The Self-Consistent Formulation of the Second Basic Law of Thermodynamics

Prof. Dr. August Friedrich Horstmann

About the second law of thermodynamics and its application to some decomposition phenomena

(Annalen der Chemie und Pharmacie. VIII. Supplementary volume, pp. 112–133 1872)

"**Certain decomposition phenomena studied in greater detail in the recent time allow for the unique application of the second law of thermodynamics, especially in cases where heat might**

decompose chemical compounds according to some consistent rules, in that the heat performs work against the natural trend called chemical affinity*. It looks like, and it has already been often pointed out, that the doctrines summing up the theory of heat in the sentence of the latter type might also be of great interest to chemists.*

Remarkably, physicists mostly share obtuse trains of thoughts only. This ought to result [113] from the fairly abstract considerations of the main topics in the field—using rather restrictive mathematical tools till nowadays—as well as from the absence of their really popular representation. Before considering the practical applications, I would greatly appreciate presenting a concise statement of the relevant results within the theory of heat, which might hopefully contribute to their popularization. With this in mind, my current presentation pretends to be neither original, nor rigorous; in effect, it should boil down to nothing more than just familiarizing the interested readership with the actual meaning of the relevant equations and the proper ways of using the known hypotheses.

1. *On the one hand, heat can be transformed into work and, on the other hand, heat can be generated from motion. Hence, we conclude that heat is itself a movement, because movements themselves can be converted into work. **The quantity of work, which the movement might in principle produce, is determined by the vis viva (livening force) of this movement—or, in other terms—by the kinetic energy of the movement**, i.e., half the product of the squared velocity and the mass of the moving body. The same applies to the heat: A given quantity of heat corresponds to a certain amount of kinetic energy, which is, in turn, equal to the work to be produced from the mentioned quantity of heat. This is content of the first law of thermodynamics confirmed by the experience: Work and heat are equivalent.*

2. *Further, **the experience teaches that not just any amount of heat can be implemented in work to the full extent, and the very fact, that in the caloric machine heat is always applied at a higher temperature, indicates that it is just the temperature that determines how much from a given amount of heat can be converted into work. We immediately rationalize the latter statement, if we take***

into account that while producing work out of heat the working body always expands under the influence of heat and has usually to overcome some resistant obstacles/ hindrances/interferences/impediments (... etc.) during its expansion. The extent [114] of the latter effects is clearly dependent on the actual type of the working body employed, but all of these resistant obstacles are indeed the greater the higher the applied temperature, which is in particular very clearly observable when choosing gases and vapors as the working bodies. Therefore, the work performable by the heat in such operations increases with the temperature; meanwhile, the law governing the latter increase can't always be deduced from direct observations, because in general, we have to work not only against the ubiquitous external resisting factors (observable obstacles/hindrances/impediments, etc.), but also against various internal resisting forces (cohesion, affinity, and/or further snags), which are in principle expected, but might in fact be inaccessible to the conventional measurements. The following considerations should first clarify the nature of the law in question; and we shall then use the same train of thoughts to draw our conclusions on the amount of work to be performed against the mentioned snags (i.e., resisting internal forces).

3. Temperature of a body is the measure of the actual amount of heat present in this body, i.e., of the entire livening force/kinetic energy of the thermal motion. Importantly, at the zero points of the ordinary thermometric scales the body still ought to contain heat. Thus, in the following we shall speak only of the so-called absolute temperature, with the initial point of this scale being the one in which all the thermal motion really ceases. This zero point of the absolute temperature scale is in effect equal to -273°C, according to an assumption we shall consider in more detail later. Then, the entire livening force/kinetic energy of the thermal motion should be essentially proportional to the absolute temperature.

4. If we imagine that the thermal motion is in effect carried out by the smallest particles of the body under study, the livening force/mean kinetic energy of such particle motions should then

also be proportional to the temperature; Thus, the particles move with certain speeds determined by the temperature, in following certain [115] trajectories dependent on the actual physical state of the other body. For example, the trajectories of the oxygen and hydrogen atoms will be exhibiting different designs, if they are pertinent to solid, liquid or gaseous water or just to a gas mixture of chemically unbound oxygen and hydrogen. As a consequence, the different physical–chemical properties exhibited by the same chemical substance in these various physical–chemical/aggregate states are due to the different shapes of the trajectories in question; conversely, the shape of the atomic trajectories ought to fully reflect the manner in which the atoms are chemically bonded, and thus determine what is the actual physical state of the body to which they belong, but not the speed of these atoms.

5. *With this in mind, the unanswered question still remaining is about the interrelationship among the average speeds of different atoms in the various physical–chemical states, but at one and the same temperature; or, in other words, whether the amount of heat present in the body is dependent not only on the temperature, but also on the physical–chemical state of the body in question. That the latter dependence is not the case should certainly be the simplest and the most plausible assumption in this respect; indeed, if we bring different systems containing hydrogen and oxygen in their various physical–chemical states, e.g., as a water vapor and as a gas mixture without chemical bonding, into contact with each other at the same temperature then there will be no change of the total temperature. The average livening force/kinetic energy of all the atoms involved will also not change as a result, for it was already the same before the contact. Such an assumption is also in pretty good accordance with some earlier observations on the specific heat of gases—the fact only to be noted here.**

Therefore, we might safely assume, that the average livening force/kinetic energy of the atoms and, consequently, their

---

**See on this subject: Clausius, Abhandlungen I, 266 ff. as well as my own remarks, Berichte der Deutschen Chemischen Gesellschaft II, 723 ff.*

average speed [116] is dependent on the temperature only and never on the shape of the trajectories followed by the atoms during their movements.

6. Still, the velocities of the atoms ought to be in a certain relationship with the forces exerted on them, which is of importance for our present purposes. The forces partly resulting from the inter-atomic interactions, partly from some external influences, should then determine the shape of the trajectories corresponding to the given physical state. The actual nature of these forces, as well as the shape of the resulting trajectories, are practically unknown to us, but, as to the amount of the forces involved, we might nonetheless plausibly assume that at any given moment of the time and at any point of a particular atomic trajectory they ought to boil down to the centrifugal force generated by the movement, for otherwise the atom would just have to leave its actual trajectory. Then, for a circular movement the centrifugal force is $\dfrac{mv^2}{R}$, where **m** stands for the mass, **v** for the velocity of the atom, and **R** for the radius of our imaginary circle. The atomic trajectories are in reality not necessarily circular; but whatever the actual shape they could acquire, we might always take the actual atomic orbit into such small sections that each of the latter in its entire length might be coincident with a circle (called the curvature circle), whose radius is equal to **r**. Then, at this particular point of the trajectory, the centrifugal force is $\dfrac{mv^2}{r}$. If the velocity of the atom gets greater due to the increase in temperature, e.g., it's now about $v_1 > v$, then so does the centrifugal force, and—therefore—all the other forces ought to grow as well, if they still have to hold the atom within its actual orbit. If we denote such forces by **k** and k_1, we arrive at the following expressions:

$$k = \frac{mv^2}{r}; \quad \text{and} \quad k_1 = \frac{mv_1^2}{r}; \qquad (1)$$

And finally we get:

$$\frac{k}{k_1} = \frac{mv^2}{mv_1^2}; \qquad (2)$$

[117] that is, the interrelationship among such forces ought to safely mimic the one among the actual livening forces/kinetic energies of the atoms. This applies to each atom at each point of its trajectory. Since only the average livening force/kinetic energy of the atoms should be proportional to the temperature, we should also conclude that the actual inter-atomic forces must increase in proportion to the temperature as well.

7. *Summing up all the above, we stress that the temperature increase ought to trigger the change of physical state, whereas the atoms ought to leave their previous orbits, if there would be no simultaneous increase in the resistant factors/forces tending to preserve the atoms in their current trajectories. And, as soon as the resistance in question stems from the inter-atomic forces only, there is no way to change anything. Still, according to the experience in many cases, the external pressure supports the internal forces holding together the atoms in the system in their resistance against the heat-induced forces driving the atoms apart from each other, and the magnitude of the external pressure might be arbitrarily variable. The latter fact, e.g., allows preventing the evaporation of liquids by increasing the external pressure despite an elevated temperature. And, as the experiments triggered the present communication clearly show, the same holds for heat-induced desorption of absorbed gases, escape of the crystalline water, decomposition of chalk and of the ammonium-chloride compounds, etc. It is for the present all the same, how exactly the external pressure acts on the orbits of individual atoms; but, if the latter effect takes place and if it is possible to prevent a change of physical state by increasing the external pressure along with the temperature growth, then we know from the foregoing that the pressure-induced resistance to this particular change of physical state must anyway grow in the same proportion as the temperature.*

8. *Any change of state [118] preventable by external pressure might be arbitrarily observable at various temperatures, but one solely has to let the pressure be so high that the resistance to the heat-induced change of the physical state might still be overcome at this particular temperature.*

With all this in mind, the work to be performed in the latter case is definitely growing along with the resistance; because the other factor, namely the way, which the atoms have to lay back, remains just the same.

Then, it follows from what we know about the resistance that the whole amount of work done by the heat at any change of the body's physical state ought to be proportional to absolute temperature at which this particular change occurs.

And this is just the law we are seeking for, which must anyway be carrying the name of Clausius, who was undoubtedly pioneering its actual formulation.

It should be mentioned here, how to formally get to the law by Rudolf Clausius starting from the ordinary form of the second basic law of thermodynamics. The analytical expression of the latter conveys that for a reversible cyclical process the following is true $\frac{dQ}{dT} = 0$; *then, the quantity of heat supplied to the body ought to be a function of two variables, one of which is the absolute temperature* **(T)** *itself, and the other* **(x)**, *which should determine the form of the atomic paths. It is then possible to write dQ = Xdx + CdT and to satisfy the latter equation according to the rules of differential calculus. We get the following expression as a result:*

$$\frac{d}{dT}\left(\frac{X}{T}\right) = \frac{d}{dx}\left(\frac{C}{T}\right);$$ (3)

or after consequently performing the operation of differentiation:

$$T\frac{dX}{dT} - X = T\frac{dC}{dx};$$ (4)

The above equation should, among other things, take into account that the following statements are simultaneously true:

$$T\frac{dX}{dT} - X = 0; \quad \text{and} \quad T\frac{dC}{dx} = 0.$$ (5)

The second of the expressions (5) corresponds to the condition given by us—namely that the amount of heat actually existing in the body *[119]* is not dependent on the atomic/molecular arrangement, while the first of the expressions *(5)* gives by integration *X = T Const*, which is nothing more and nothing less than just the Clausius' law, because **X** is the amount of heat transformed into the work during the change of **x** to **dx**. *(see e.g., Clausius, Abhandlungen I, 264). This is exactly the context I have pointed out on another occasion (Berichte der Deutschen chemischen Gesellschaft, II, 726).*

9. *To express [119] the law in an equation, let us assume that the total work to ensure a particular change of the physical–chemical state (e.g., the evaporation of a quantity of water) at the temperature **T** is equal to **W**, so that **dW** would be the increase in this work when the temperature increases up to **T + dT**. Then, according to the law under study, we get:*

$$(1) \quad \frac{W}{W+\delta W} = \frac{T}{T+\delta T} \quad \text{oder}: \quad (2) \quad W = T\frac{\delta W}{\delta T}. \qquad (6)$$

*Meanwhile, the total work **W** ought to consist of two parts, which might be viewed separately. The amount of work against internal forces is, say, **J**, and its amount isn't known to us for the present, but it is plausible to assume that such a work amount should be conserved for one and the same change of the physical state at different temperatures; Indeed, the way/trajectories the atoms have to cover/follow remain the same, and the internal forces are not temperature-dependent, if the spacing between the atoms remains the same. Strictly speaking, the forces under study might surely change in effect, because it's not possible to completely prevent any inter-atomic arrangement alteration by applying the external pressure, but it is nonetheless possible to arrange for the temperature change δT to be always so small that the inner work remains noticeably the same. The increase in the total work might then arise solely out of the increase in the external work. Bearing this in mind, we denote here **E** to be the work against the external pressure, so that δE would then be its increase resulting from the temperature increase δT—and finally we get:*

$$W = E + J \text{ and } \delta W = \delta E \qquad (7)$$

As a result, the Eq. (2) in (6) [120] turns into:

$$W = E + J = T\frac{\delta E}{\delta T}. \qquad (8)$$

10. *From now on we shall deal with such changes of physical state that are accompanied by some gas formation. Let us then imagine some body enclosed in a cylinder with a movable punch, so that the external pressure should be brought about by some weights placed on the upper surface of this punch. In generating the external pressure to be just overcome at the temperature* **T**, *we have to charge the punch with* **p** *units of weight; hence, if the upper surface of the punch is equal to* **f**, *then an increase of the gas volume inside the cylinder would lift the weight* **pf** *to a certain height* **h**. *Therefore, the external work to perform the latter movement could be cast as follows:*

$$E = pfh = p\delta v, \qquad (9)$$

where δv stands for the increase in volume during the change of the physical state. If we denote **p** + δp *the external pressure corresponding to the temperature* **T** + δT, *then we arrive at the following statement:*

$$E + \delta E = (p + \delta p)\delta v, \qquad (10)$$

And consequently:

$$\delta E = \delta p \cdot \delta v, \qquad (11)$$

11. *Using the above results for* **E** *and* δE *the equation (8) might be recast as follows:*

$$W = p\delta v + J = T\frac{\delta p}{\delta T}\delta v. \qquad (12)$$

And for the inner work we therefore get:

$$J = \left(T\frac{\delta p}{\delta T} - p\right)\delta v. \qquad (13)$$

Finally, to mathematically formulate the equation under study in entirely strict manner we have to replace the expression $\dfrac{\delta p}{\delta T}$ by the true derivative of the pressure **p** *at the temperature* **T**, *that is, by $\dfrac{dp}{dT}$ —because, as already noted, the smaller*

δT, the exacter would be the equation for the infinitely small temperature changes.

*Remarkably, any measurement of the inner work could be possible solely via determining the amount of heat spent for carrying out the work of interest, and [121] therefore we have to bring the equations just derived into a usable form by multiplying them with the caloric equivalent of work = **A**. Thus, we obtain as a result:*

$$Q = AT \frac{\delta p}{\delta T} \delta v. \tag{14}$$

$$Q_J = A\left(T \frac{\delta p}{\delta T} - p \right)\delta v. \tag{15}$$

Q *stands here for the amount of heat to be transformed into work during the physical state change. But the latter represents also the total amount of heat that has ever been acquired—if, according to our assumption, the livening force (kinetic energy of thermal motion) contained in the body was not altered during the change of physical state.*

Here we have just arrived at the most remarkable result of the mechanical theory of heat, which in words conveys the following facts: **If we know the amount of volume increase brought about by the heat-induced change of the physical state in a body, as well as the pressure to be used in overcoming the change in physical state at different temperatures, then the external work is completely determined, and hence the internal work as well as the amount of heat consumed during the change of physical state might also be calculated using Clausius' law in a straightforward way** **.*

12. *Meanwhile, if we initially apply the above-considered law to permanent gases, then the reason for which the zero point of the absolute temperature scale should be set to the value of –273° C immediately follows. Indeed, according to our experience, if no internal forces resist the expansion of a gas, then the inner work is equal to 0, and for a small increase in volume, by which the pressure does not change markedly, it is in accordance with Eq. (13):*

---

*The derivation of this result using the ordinary form of the second basic law could be found in Kirchhoff, Pogg. Ann. *108, p. 177.*

$$J = \left(T\frac{\delta p}{\delta T} - p \right)\delta v = 0, \qquad (16)$$

[122] i.e., it should come as follows:

$$T\frac{\delta p}{\delta T} = p \qquad (17)$$

*The gas expansion should be carried out at 0°. Then, if **p** stands for the initial pressure, the pressure at a temperature higher by about δT ought to be increased according to Mariotte's law:*

$$\frac{\delta p}{\delta T} = p\alpha, \qquad (18)$$

where α stands for the gas expansion coefficient. It is equal to 0.003665 or (1/273). With this in mind we calculate:

$$Tp\alpha = p, \qquad (19)$$

*And thus **T**α = 1. Hence, in taking into account the value of α, we get **T** = 273, that is, the absolute temperature corresponding to 0° lies at –273°C. Therefore, the zero point at –273°C is the zero of the absolute temperature scale."*

The further deliberations in this paper by Prof. Dr. A. F. Horstmann are dealing with the particular applications of the above-sketched thermodynamic theory to different physical–chemical phenomena and detailed numerical work of crucial interest for physical chemists.

What are the most important and general points touched by Prof. Horstmann in his above communication? We have highlighted them in bold and shall now summarize them as follows.

Every realistic natural process is always possessed of the two basic groups of the main factors:

  (a) The *driving force* (or, as the ancient tradition would require, *vis viva*, in other words, the livening/driving force);
  (b) The ubiquitous, omnipresent hindrances/obstacles/ obstructions/hurdles/vaults/...etc.

Have we just communicated something exceptional? Not at all, the above statement is in fact trivial! But the real problem ought to be: How to correctly describe and analyze such situations—in terms of physics/chemistry/... etc.?

By the way, there is a pertinent citation by The Great Helmsman, Mao Tse-tung (1893–1976), if we get rid of the conventional political husk around his name [65]:

"What is a problem? A problem is the contradiction in a thing. Where one has an unresolved contradiction, there one has a problem. Since there is a problem, you have to be for one side and against the other, and you have to pose the problem. To pose the problem, you must first make a preliminary investigation and study of the two basic aspects of the problem or contradiction before you can understand the nature of the contradiction. This is the process of discovering the problem. Preliminary investigation and study can discover the problem, can pose the problem, but cannot as yet solve it. In order to solve the problem it is necessary to make a systematic and thorough investigation and study. This is the process of analysis. In posing the problem too, analysis is needed; otherwise, faced with a chaotic and bewildering mass of phenomena, you will not be able to discern where the problem or contradiction lies. But here, by the process of analysis we mean a process of systematic and thorough analysis. It often happens that although a problem has been posed it cannot be solved because the internal relations of things have not yet been revealed, because this process of systematic and thorough analysis has not yet been carried out; consequently we still cannot see the contours of the problem clearly, cannot make a synthesis and so cannot solve the problem well. If an article or speech is important and meant to give guidance, it ought to pose a particular problem, then analyze it and then make a synthesis pointing to the nature of the problem and providing the method for solving it; in all this, formalist methods are useless."

With this entirety in mind, a good idea in our case would first of all be to use the Language of Energy (it would be handy to denote the latter the '*energetics*'!). Further, we have to ascribe some portion of energy to the livening/driving force and another one to the hurdles. The actual energetic balance would then determine the course of the process under study.

Apart from the general methodological indications provided by The Great Helmsman, to succeed in energetically describing the livening/driving force we ought to follow the suggestions by Herr Gottfried Wilhelm Leibniz (1646–1716), as duly analyzed and proven in detail by Mme Gabrielle Émilie Le Tonnelier de Breteuil, Marquise du Châtelet-Laumont (1706–1749). This is why we might readily accept that the livening/driving force ought to be delivered by the kinetic energy. It is important to note here that the livening/driving force cannot arise by itself from nothing, for there always ought to be the relevant prerequisites. The latter could be energetically analyzed in terms of the potential energy, in other words, the energy supply. If such a supply isn't dwindling, it is in principle possible to somehow convert the potential energy to its kinetic counterpart.

And now it's just the very time to revert to the results by Hon. Sir Isaac Newton, who wasn't accepting the ideas of Gottfried Wilhelm Leibniz, but had built up the logical basis of the classical mechanics, so that we would now greatly appreciate applying Newton's third basic law, according to which *"For every action, there is an equal and opposite reaction."*

The latter point was/is/(remains to be) extremely important for any process evolving in time. Whatever the actual nature of the process in question there always ought to be two main energy contributions governing the course of the process, using our 'energetic language': The kinetic energy (vis viva), as well as some energy contribution from the omnipresent hurdles/obstacles/... & so on, so forth.

And this is just what Nicolas Léonard Sadi Carnot was most probably intending to demonstrate us using his ingenious cyclic gadget. Then, the great contributions of Rudolf Clausius and Hon. William Thomson, 1st Baron Kelvin, was to stress that the energy contribution due to the obstacles ought to consist in the *dissipation/ devaluation* of the so-called *'useful energy'*, that is, of the energy promoting the processes which are capable of carrying out some useful work.

The next step was to work out some valid mathematical methods for describing and properly analyzing the processes of interest, to formulate the relevant theory for the designing the required devices adequate to performing some useful work.

Here the pioneering works by Hermann Ludwig Ferdinand von Helmholtz and Gustav Robert Kirchhoff in Germany ought to be mentioned in the first line.

Noteworthy, August Friedrich Horstmann was but the only colleague in Germany, who was capable of correctly summarizing all the achievements in the thermodynamics of his time. Still, to our sincere regret, he had but no more time to duly continue his important work.

In the United Kingdom at the same time such outstanding colleagues must be named as the physicist Prof. Dr. Hugh Longbourne Callendar (1863–1930) and the physical chemist Prof. Dr. George Downing Liveing (1827–1924), who were intensively trying to contend for the correct thermodynamics, but their voices had somehow gotten lost in the over-all 'emotional shuffle'.

In Russia/USSR such renowned colleagues in the field must be mentioned as Vladimir Alexandrovich Michelson (1860–1927), Nikolai Nikolayevich Schiller (1848–1910), Boris Borisovich Gallitzin (1862–1916), Nikolai Alexeevich Umov (1846–1915), Alexandr Nikolayevich Shchukariov (1864–1936), Tatiana Alexeyevna Afanassjewa-Ehrenfest (1876–1964)—as well as many-many others, with some of them being already mentioned in this volume, whose complete list being truly long. The personal destinies of all of them were/are in effect very different from each other—for all of them could experience the whole spectrum of possible outcomes—from more or less quiet careers in the proper academic environment till the full scale of emigration or just the Stalinist terror.

In Japan, an outstanding biophysicist Prof. Dr. Motoyosi Sugita (1905–1990) should be named in connection with correctly understanding and duly employing the thermodynamics foundations.

Sure, there is definitely much more to the story about the *'different thermodynamics'*!

[Please see Supplementary Note 4, Pages 812–852]

5.4 What Is the Correct Meaning of the Well-Known Free Energy Notion?

To sum up, the works of all the colleagues mentioned in the Supplementary Note 4 here ought to be considered a serious collective contribution to the basement of the different thermodynamics and definitely deserve a separate detailed and thorough attention.

Now we have to move to the point, where Prof. Gibbs had to stop his important studies, owing to his inevitable departure. One of his most important findings was the thermodynamic potential known also as a Gibbs free energy/Gibbs energy, the sense of which could **apologetically** be put as follows [66]:

> "*When a system changes from a well-defined initial state to a well-defined final state, the Gibbs free energy change equals the work exchanged by the system with its surroundings, minus the work of the pressure forces,*

during a reversible transformation of the system from the initial state to the final state. Gibbs energy is also the chemical potential that is minimized when a system reaches equilibrium at constant pressure and temperature."

Sure, to neophyte's ears this ought to sound like a description of some special particular case: Indeed, the constant pressure and temperature are absolutely essentially defining some **unique static, i.e., unsurmountable situation**. Hence, any further consideration ought to stop here. Still, to avoid such a sad outcome, one has to add a couple of important words to this end, to break the logical cage inevitably formed after reading the above (and the like) deliberations: It is but throughout possible to control the both mentioned parameters of state from outside!

Sure, this way, it is possible to voluntarily choose both the actual temperature value and the actual pressure value. *As a result, these both would not be any more just the 'lugubrious' **'state variables'**, but play instead a role of the **parameters of the state** to be voluntarily set.*

The crucial difference here is that the **variables** could also be changed by some external/internal reasons beyond the researchers' control, whereas we are capable of voluntarily changing the state **parameters**. Thus, in fact, the apparently blind alley turns out to be a true esplanade/mall, for what is changeable in principle should anyway be changing, whatever the driving force be responsible for the change of interest.

The latter interpretation seems to lie in parallel with what Profs. Helmholtz and Gibbs were just thinking about, when speaking of the **available energy**.

Meanwhile, if we would like to follow the original conclusion by Prof. Gibbs himself, then the latter should in fact be nothing more and nothing less than [67]:

"The greatest amount of mechanical work which can be obtained from a given quantity of a certain substance in a given initial state, without increasing its total volume or allowing heat to pass to or from external bodies, except such as at the close of the processes are left in their initial condition."

Remarkably, it is just in parallel to Prof. Dr. Gibbs that Prof. Dr. von Helmholtz was also working on some valid description of the quite general situation described by the above Gibbs' citation. Meanwhile, his work boils down to the following **apologetic** conclusions reflecting the conventional terminological hotchpotch [68]:

> "*In 1882, the German physicist and physiologist Hermann von Helmholtz coined the phrase 'free energy' for the expression E – TS, (where E stands for the internal energy, T is the temperature, and S is the entropy) in which the change in F (or G) determines the amount of energy 'free' for work under the given conditions.*
>
> *Thus, in traditional use, the term "free" was attached to Gibbs free energy, i.e., for systems at constant pressure and temperature, or to Helmholtz free energy, i.e., for systems at constant volume and temperature, to mean 'available in the form of useful work'* [66]. *With reference to the Gibbs free energy, we add the qualification that it is the energy free for non-volume work* [69]."

Why we do consider the above citations to be **apologetic** will be clarified later on in this chapter (i.e., *apologetic* with respect to *what?*). See Note 2 to this chapter for the details.

Still, first of all, we do remain with a feeling that the Helmholtz' result is nothing more than just another special particular case (now, for the constant values of temperature and volume), essentially like the result by Gibbs.

Meanwhile, from any serious textbook on thermodynamics we learn that '*there is an intrinsic interconnection between the Helmholtz' function and the statistical–mechanical partition function*', with the latter bearing the well-recognized universal physical sense, and immediately come across the pertinent well-known mathematical formula. The questions immediately arise:

Are the results of Gibbs and Helmholtz just two parts of one coin, or two rather different particular cases?

Aren't we getting into some blind alley again?

On the other hand, the engineering thermodynamics introduces some fine difference between '*flow*' and '*non-flow*' processes, so that the former ones occur in the systems having open boundary and thus permitting mass interaction across the system boundary, whereas the latter are the ones in which there is no mass interaction across the system boundaries during the occurrence of process (see, e.g., [70] and the references therein).

Even using such a classification there is a need for both the Helmholtz and Gibbs functions, on the condition that these both are different expressions for the '*energy availability*' or '*exergy*'—just in the sense both Helmholtz and Gibbs were investigating the problem.

This entirety but makes one to assume that there ought to be some essential, intrinsic interrelationship between the both. Is it really so?

To try answering the latter poser, let us just consider very attentively and in detail the conventional equations of thermodynamics. What we would need is not exceeding the conventional school mathematics.

Is this idea briskly new indeed? Surely, not at all!

Howbeit, we know only one publication, namely the small book by a physical–chemical engineer Dr. Robert Pauli /Friedrich Robert Philipp Pauli (1866–1931), who could express a valid criticism of the then standpoints, as well as clearly formulate the problem to be solved. This seems to be a truly unexpected result, for all around the world there were really lots of colleagues actively working and publishing on the theme in the time of XIX-to-XX centuries [71].

This portrayal of Friedrich Robert Philipp Pauli has been taken from https://www.geni.com/people/FriedrichPauli/6000000018805122846

Remarkably, the then academic world has accepted this work with a clear and evil protest, closely reminding the reaction in connection with Peter Boas Freuchen's book (for the details, see Chapter 1). Meanwhile, unlike in Peter Boas' case, the 'academic' reactions were truly evil—both in Germany and in the USA [72, 73].

To our mind, such a community's reaction could be explained by a striking dissonance between the over-all, ubiquitous emotional atmosphere of that time we have already mentioned and the clear-cut suggestions published in Dr. Pauli's book, which in effect consisted in the appeal to attentively, thoroughly reconsider the two basic equations of thermodynamics—and Dr. Pauli had presented some starting points of such a re-consideration.

5.4.1 Who Was Dr. Robert Pauli at All?

To our sincere regret, Dr. Pauli's work could find no continuation. Why?

This poser is difficult, for Dr. Robert Pauli had really disappeared in the unknown direction. Meanwhile, it is known that he stemmed from a wealthy family of Palatinate, Southern Germany.

He was a son of an outstanding chemist and businessman in the region, Dr. Philipp Victor Pauli (1836–1920), a founder of his own chemical factory at Oggersheim/Ludwigshafen and then a director/CEO of the chemical factory 'Meister, Lucius and Brüning', presently well known as the 'Höchst AG'. Being thus a hereditary chemist, Robert Pauli had graduated the Leipzig University and gotten his PhD in chemistry under the academic leadership of Prof. Dr. Wilhelm Ostwald.

Sure, it is just under the influence of such an outstanding colleague that Dr. Pauli could be lively interested in the problems of thermodynamics. According to his own statements in the preface to his book, he devoted several years to attentively study the problems.

Still, he hadn't joined the academic world, but became a successful chemical engineer (at the time of his book's publication he was working at the electrolytic department of the 'Deutsche Solvay Werke' in Magdeburg), who could duly patent his achievements.

His further life could only be monitored from the following publications in the USA media (here are just two examples of a number of similar reports across the USA):

> **"German, wed here, guilty of bigamy**
>
> **Dr. Pauli contended that marriage in Boston in 1893 was not legal.**
>
> **forgot second ceremony**
>
> said he was intoxicated at the time, but court believes bride, a young Russian woman.
>
> special cable to the New York Times
>
> Berlin, Nov. 1909

Robert Pauli, a young university graduate of a prominent and wealthy Berlin family, was sentenced today at Leipzig to eight months' imprisonment on a charge of bigamy, following an American marriage. On his graduation in Germany in 1892, Pauli wanted to marry the idol of his affections, a young German girl named Munkelt, but his parents objected, whereupon the couple tied to the United States. Pauli took up a postgraduate engineering course at the Massachusetts Institute of Technology, while the lady became an opera singer in New York. Although they were married in Boston in October 1893, and Pauli claims the bond was lived up to until 1907, when the wife and two children deserted the head of the family, he says the marriage was illegal.

In December 1907, Pauli turned up in London with a young Russian named Lydia Chotow. One day, according to the fantastic tale related by Pauli on the witness stand, somebody packed the couple into a cab at the Savoy Hotel, took them to a saloon, intoxicated the groom, and then drove them to a remote part of the metropolis, where a man clad in a gabardine performed a serio-comic wedding ceremony, Pauli says this function was a joke, because he was too drunk to know what was happening. Later, however, a real marriage took place in London at the registrar's office. Pauli says he was not present at this function, and was impersonated by another man.

The bride told the court a different story, and as the signature at the Registrar's office tallied with Pauli's, the court said he was guilty and sent the absent-minded bridegroom to prison for eight months.

Published: November 9, 1909

Copyright the New York Times"

"The Salt Lake City Tribune, Salt Lake City, Utah, USA, Monday morning, November 29, 1909.

A strange story brought out at bigamy hearing

Berlin, Germany, November 28. *A sentence of eight-month imprisonment was passed on Dr. Robert Pauli, a young Berlin university man of a prominent and wealthy family, at Leipsic this week.*

Dr. Pauli married at Boston, Mass., in October 1893, a German woman, but he denies the bigamy of this marriage, which he says was nullified by the action of his wife and children in deserting him.

Fourteen years later, on December 23, 1907, he was married to Lydia Chotow, a Russian woman, in London. But Dr. Pauli considers that this ceremony was a joke, because, he says, it took place while he was drunk and against his will. On the morning of the wedding somebody put him and bride into a cab at the Savoy hotel and took them to a restaurant, where he became intoxicated.

Pauli declared that he was then taken to a mysterious dim-lit chamber in a remote part of the metropolis where a man 'clad in a somber gabardine' performed a serio-comic marriage. Later, the defendant says, an actual ceremony took place at a registrar's office, but there a certain person had impersonated him. The Chotow woman told a different story, and as the signature of the registrar's records was identified as Pauli's he was found guilty."

Indeed, Dr. Robert Pauli being 27 y. o., had reached USA on 29.08.1893 by a ship 'Waesland' going from Antwerp, and was registered as a *'Doctor of Medicine'* on the Ellis Island in New York. In fact he was heading both to spend his postgraduate time at the MIT and in October 1893 to marry the *'the idol of his affections'*, Miss Martha Adeline Katharina Munkelt (Pauli), who was a daughter of a musician in Weißenfels, Saxony-Anhalt (Eastern Germany) and had become an opera singer in New York in 1893.

This portrayal of Martha Adeline Katharina Munkelt (Pauli) has been taken from https://www.geni.com/people/Martha-Pauli/6000000018805260584

The couple had two sons, the elder, Lothar Romeo Pauli (1894–1959), who had later become a concert pianist, and the younger one, Roland Pauli, about whom nothing is known (he was born before the year 1907 and unexpectedly died being most probably still a schoolboy at the most).   This entirety of events had most probably caused the family drama lasting for several years, having a serio-comic end described above and distracting Robert Pauli from his seminal work.

Finally, there is one more remarkable point in all this story: There ought to be absolutely no parallel between Dr. Robert Pauli and Prof. Dr. Wolfgang Pauli (1900–1958), a renowned researcher of the world-wide class.

The latter colleague doesn't need any separate introduction, for he had paid a very significant contribution to quantum physics and quantum chemistry—while fully sharing the conventional emotional standpoint as concerns thermodynamics, statistical physics and being a noticeable representative of the revolutionary physicists [74, 75].

5.5 Entropy–Enthalpy Compensation: Its True Physical Sense

Meanwhile, it is of considerable interest for us here to try developing here the *'Energetic Theory of a Chemical Molecule'* initiated by Dr. Robert Pauli in 1896—but, to our sincere regret, never finalized.

For this purpose we shall start with writing the both basic equations of thermodynamics in their well-known conventional form:

$$dS = \frac{\delta Q}{T}; \quad dU = \delta Q - pdV, \tag{20}$$

With the first of the above equations being the well-known definition of entropy by Clausius, to be taken as the valid expression of the second basic law of thermodynamics, whereas the second one is the not less known Clausius–Gibbs equation to express the first basic law of thermodynamics. Here we choose the following designations: Q stands for heat, U for internal energy, p for pressure, V for volume, and δ means the inexact (path-dependent) differential, as opposed to the exact (path-independent) differential, d.

Our first step is to get rid of the inexact differential operator by substituting the first expression of Eq. 20 into the second one. As a result, we get the following ordinary differential equation:

$$dU = TdS - pdV. \tag{21}$$

To fruitfully integrate Eq. 21 we have first to define the variables of integration. Our choice here would be to take the conventional isobaric–isothermal scheme, while mentioning, as we have already discussed above that in our 'thought experiment' we take over controlling both pressure and temperature of our system of interest. This is how we remain with the pressure and temperature as two state parameters, to separate the volume as a realistic system's state variable. Sure, there might be lots of other variables and parameters, but for our present consideration our choice would in fact be fully enough.

After formally integrating Eq. 21, we arrive at the following result:

$$U(V,...) - U_{const} = TS(V,...) - TS_{const} - pV(...),$$
$$U(V,...) + pV(...) = TS(V,...) - TS_{const} + U_{const}.$$
(22)

It is straightforward to interpret the anti-derivatives of Eq. 22, if we recall that $U + pV \equiv H$ is the conventional definition of enthalpy (heat content). Then, if we recast the right-hand constant in the second expression of Eq. 22 as $a \equiv TS_{const} + U_{const}$ and, by designating the temperature in the latter constant as $T \equiv T_c$, we arrive at the expression of the so-called 'enthalpy–entropy compensation' (EEC):

$$H = T_c S + a.$$
(21)

It is important to mention here that the EEC is still a hotly debatable phenomenon, whereas it is stubbornly called an 'artifact of the experimental data processing', and, even if there are no objections as for the data processing procedures, it is stubbornly rejected as an 'extra-thermodynamic relationship'. Meanwhile, a careful analysis demonstrates that, in case we have to do with the valid EEC, it is always straightforward to find an imaginary Carnot cycle behind it. It is even possible to consider and analyze a non-linear EEC. In case the relevant experimental physical–chemical data are thoughtfully collected and carefully processed, the EEC becomes a very useful interpretation tool, irrespective of the level of consideration (from macro- through meso- till nano- and microscopic events, just like Dr. Robert Pauli had indicated it—*it is valid for all the imaginable types of molecules*).

To sum up, here are the references to our works on the EEC [76–90] (please, see also the references therein, which are of tremendous importance for rendering the correct picture).

In our present discussion it is but very important to mention one more point demonstrated by Eq. 21. Specifically, the latter expression delivers an intrinsic interrelationship between the Helmholtz and

Gibbs functions. From the conventional emotional standpoint this ought to be a quite unexpected result. Howbeit, as we could see, this is a clear consequence of the two basic equations of thermodynamics, and we could even connect this with an imaginary Carnot cycle.

Using the term 'imaginary' is in effect an excessive step, for Carnot's cycle ought to be a kind of purely theoretical gadget, kind of ingenious 'thinking experiment', but in recalling here the ideas of N. L. S. Carnot we would like to *emotionally* underline the intrinsic interrelationship between the EEC and the different thermodynamics. Otherwise, *one might ask in despair*, so what more ought to be done to conclusively defeat the emotions strangulating thermodynamics since more than hundred years.

With all this in mind, it becomes clear that the EEC presented by Eq. 21 is nothing more and nothing less than just a common consequence of the 1st and the 2nd basic laws. The latter law in its correct formulation (by Prof. Dr. Horstmann, see above) is anyway indicating some kind of *compensation*:

> *The Livening Force is every time being consequently fought and compensated by all the ubiquitous obstacles/hurdles.*

Remarkably, the latter ones are really 'stubborn': their influence is increasing the more, the higher levels the livening force is reaching. But at some point the resistance, the opposition ought to reach its maximum—and we know this thanks to the ingenious works of two colleagues independent of each other: Rudolph Clausius and William Thomson—Hon. Lord Kelvin.

Therefore, any realistic process would successfully come to its end, if, to put it in the merchant's terms, *"the available stock of the livening/driving force"* behind the process in question would be enough to fulfill two (logically tightly interconnected) tasks:

(a) To compensate all the relevant hurdles/obstacles at their peak, at their maximum performance;

(b) To perform all the necessary useful work.

For some process under study, where we have no preliminary information about the intrinsic mechanisms, the actual outcome might be successfully described in terms of probability theory, as Dr. Georg(e) Augustus Linhart had clearly indicated in his unpublished reprints (we have re-published them in Chapter 4 in this book), for it is then not clear beforehand, whether the 'livening force's stock' would be enough to fulfill the two above-mentioned tasks.

Hence, the fact of reaching the successful end of the process, the fact of fulfilling the both above-mentioned tasks ought to correspond to the *equilibrium* between the livening force and the entropy (that is, hurdles/obstacles) ... thus, here is the true sense of the equilibrium thermodynamics!

5.6 How Many Basic Laws Thermodynamics Has?

To sum up the entire above story we might state that there is really no place for the conventional separation of the two basic laws of thermodynamics.

There is one and only one basic law: The law of conservation and transformation of energy.

And its latter formulation already reveals that it has in effect two basic sides: energy conservation and energy transformation. Remarkably, the relationship between these both sides is dialectic—this has a definite connection to the 'unity and struggle of opposites'—indeed, the amount of the total energy is always constant, whereas some specific portions of energy might cancel/ exceed each other. Basically, the latter conclusion ought to have some intrinsic relationship to the Western concept of 'Janus Bifrons', as well as to the Eastern concept of 'Yin-Yang'.

Now, before starting to deal with the non-equilibrium thermodynamics we would first like to consider the problem of numerous basic laws in the conventional equilibrium thermodynamics.

The conventional literature always speaks of **four basic laws** of thermodynamics—see for example the most recent commentary [7].

Except for the two basic laws—the energy conservation law and the energy of entropy growth, there is also the third basic law—that the entropy at zero absolute energy is equal to zero, and that the zero absolute energy is never accessible (experimentally feasible) in fact. Moreover, there is the fourth—or—the zero-th basic law, whose conventional formulation could be found in any conventional textbook.

Meanwhile, here we would just like to cite the following commentary book with the contents of which we agree to 100%—

and might perhaps add just a couple of ideas at several points (see the work [91]).

The original publication [91] is in German, but there are also English translations of some chapters to be found under the URL address [92].

And as for the *zero-th basic law*, the Paper/Internet book [91, 92] contains the following instructive story:

[Please see Supplementary Note 5, Pages 852–855]

Many sincere thanks are due at this very point to Prof. Dr. Gottfried Falk (1922–1991), Prof. Dr. Friedrich Hermann and Prof. Dr. Georg Job for their above insightful comments!

When discussing the results by Dr. G. A. Linhart in Chapter 4 of this book, we have already seen, what the exact physical representation of the entropy notion is in effect. Indeed, the actual thermodynamical entropy might be expressed as a handy logarithmic-based function of the absolute temperature.

In the present chapter we have also seen what the exact formulation of the second basic law is in fact, and that there ought to be no conceptual gap between the latter and the first basic law. Indeed, these both ought to be the two dialectically coupled faces of the basic energy notion. Therefore, there is in effect *the only one basic natural law—namely—the energy conservation and transformation law.*

Undoubtedly, the latter statement sounds like an Odious Heresy in the ears of the most educated physicists.

There is but a different physics as well.

The developers of the KPC

Author: Friedrich Herrmann, Karlsruhe Institute of Technology
Numerous basic ideas stem from books and other publications of, and from personal contacts with Gottfried Falk (until his death at the University of Karlsruhe) and Georg Job (University of Hamburg).

The following PhD students and other collaborators at the Institut für Didaktik der Physik participated in the development of the KPC:
Karen Haas-Albrecht (Nuclear physics)
Holger Hauptmann (Electrodynamics, oscillations and waves, rotational mechanics, thermodynamics)
Matthias Laukenmann (Atomic and solid state physics)
Lorenzo Mingirulli (Mechanics)
Petra Morawietz (Thermodynamics)
Dieter Plappert (Energy and energy carriers)
Peter Schmälzle (Electricity, data physics)
Michael Pohlig (Relativity)

Gottfried Falk

Friedrich Herrmann and Georg Job

These portrayals and credentials of the above co-authors have been taken from http://www.physikdidaktik.uni-karlsruhe.de/Strategien/Autoren_englisch.html

5.7   What Should Be the Proper Mathematical Toolbox to Describe Approaching Thermodynamic Equilibrium?

Now we are to 100% ready to come to the final poser of this book:

What ought to be the correct mathematical toolbox for the non-equilibrium thermodynamics?

Here a Clio's smile surfaces again: We are just about to encounter Prof. Dr. John von Neumann (1903–1957), who was allegedly discussing with Prof. Dr. Claude Shannon what kind of name might be given to the attenuation in phone-line signals, according to the legend widely known in the immediate academic circles [34].

Meanwhile that was just Prof. Dr. Neumann, who could deliver the fully pertinent mathematical tool to describe the magic entropy notion: *The game theory*.

Sure, we might also name lots of other colleagues who were, are, and will be working on this mathematical theory and its widespread applications, but here we would just like to point out the very basic idea.

We have seen that entropy is nothing more than a proper description for all the ubiquitous hindrances/obstacles/handicaps/impediments/encumbrances/baulks/deterrents/hitches/preventives etc. This notion does in no way pertain to the objects of natural sciences only—this is indeed concerning all the spheres of the life in the universe. And here is in fact absolutely no emotional stress at all—in fact, it ought to be a very straightforward and ubiquitous story:

To trigger any kind of realistic process we need some driving force and the latter ought to be guaranteed by the available energy. Normally, by now it is way a trivial task to *employ the available energy*—but here we are interested in the very principle!

Basically, it is throughout important to point out the difference between the energy stocks and the actual available energy—in the conventional mechanic terms these both ought to correspond to the potential and kinetic energy, respectively. In fact the latter ought to deliver the very basis for the livening force—the vis viva—or, in more modern terms, the driving force for the process of interest.

Moreover, it is also very important to note that—if there is no driving force, there will also be *no obstacles*, and, *no process* at all.

As soon as we start triggering something at the expense of the non-zero driving force, we would immediately encounter some *obstacles* to be surmounted. The higher the intensity of the driving force, the higher should be the intensity of the *obstacles*. Finally, the latter ought to arrive at their maximum value at some particular time point. This is just the very essence of the second basic law, i.e., of the entropy increase law, i.e., of the energy transformation law.

Now, it is crucial to ensure that we would have enough livening/ driving force, which would be spent *not solely* to *overcome* the *obstacles*, but also to *fulfill* our *entire actual tasks*. It is just this way that we shall arrive at the *equilibrium among the driving forces* and *the obstacles*. Thereafter it is possible for us to *fulfill* our *entire actual tasks without any further obstacles*.

If we would now put this in the original Clausius formulation, then after equilibrating all the negative processes, we might count on the positive processes only. A simple example of the above story from our everyday lives has been alleged in the Note 19 to Chapter 1. of the volume at hand, where we were considering the possibilities of buying a fridge.

The next important poser ought to be:

How to describe the entire story mathematically?

The answer comes immediately to mind: Let us take the differential game theory, for example—let us imagine that there are two energetic players, who play the so-called *Attrition and Attack* game—a kind of an antagonistic non-cooperative game.

The pertinent mathematic tools to describe the latter 'game situations' had been suggested by an outstanding British engineer Frederick William Lanchester (1868–1946), who could write down the differential equations of state for the 'warfare games' during the First World War [93, 94].

It is already after the Second World War that Prof. Dr. Rufus Philip Isaacs (1914–1977) had properly modified them [95].

The first poser to arise here is: Why at all do we suggest using here the notion the 'differential game'?

The basic idea here is that we are dealing with continuous variables and functions to describe realistic processes of changing realistic physical states in the continuous time, but first we would like to give a short 'technical' introduction into this mathematical area.

With this entirety in mind the exact definition of a differential game is as follows:

1. The game element might be represented as a vector $x = (x_1, x_2, ..., x_n)$, which describes some state variables (x is dependent on time t);

2. At every moment t the first player chooses the so-called 'controlling vector', u, $u = (u_1, u_2, ..., u_n)$, which is called 'a pure strategy' and the components of which are in general also some functions of time and ought be restricted to some definite stretches of their values, for example, $a_i \leq u_i \leq b_i$; $i = (1,2,..., n)$; the same way the second player chooses his own 'controlling vector', w, $w = (w_1, w_1, ..., w_m)$, the components of which are in general also some functions of time and ought be restricted to some definite stretches of their values, for example, $c_i \leq w_i \leq d_i$; $i = (1,2,..., m)$;

3. The vectors x, u, w, i.e., the states of the game, are obeying some differential equations in the form $\dfrac{dx_i}{dt} = f_i(x,u,w)$; $i = (1,2,...,n)$. Here $\dfrac{dx_i}{dt}$ is the first derivative of x_i by t, whereas f_i are some definite functions of x, u and w;

4. The choice of the control actions u, w at any time moment t leads to the choice of the vector x, which obeys the above differential equation, whereas the game is running during some time lapse of $0 \leq t_i \leq T$ up to the moment till the point x would reach the border C of some closed set A; hence, the mentioned border might be denoted 'the terminal/border surface';

5. The actual payoff of the first player (his win, to put it in the simple everyday language) is defined using the following sum:

$$M = \int_0^T K(x_1,...,x_n)dt + G(x^*).$$

Here $K(x_1,..., x_n)$ is the 'payoff density' for the first player depending on the game state x, whereas the form $G(x^*)$ ought to describe the salvage value function, that is, the payoff of the first player, if the latter manages to reach the terminal surface C at the point x^* of this surface.

There is a lot of applications dictating the pertinent interpretation of the above-sketched mathematical scheme, so what should be the proper story in our particular case?

In addition, we have mentioned continuous games of the *Attrition and Attack* type, so the questions would be: How this would correspond to our current problem, and what should be the proper equations of the game state?

Indeed, in our case, the driving force/enthalpy is 'playing' against the obstacles/entropy, in that the both are attacking each other—and at the same time 'running rings around'—plus 'making mincemeat out of'—each other.

Furthermore, the mathematical (game theoretical) model of such a situation should properly depict two rivaling forces fighting against each other, so that the total sum of the entire disbursals and wins should to be equal to zero. In other words, the story under consideration should perfectly correspond to the so-called *zero-sum game*, which is in our case just the mathematical consequence of the energy conservation law.

Indeed, the total energy must always remain constant, whatever driving forces and obstacles are involved and whatever the degree of interaction between them.

At this very point we would like to gratefully acknowledge the useful advice of Dr. Stefan Wrzaczek, a game theory specialist from Vienna University, Austria, who was extremely helpful in constructing the following simplest mathematical example.

This portrayal of Dr. Stefan Wrzaczek has been taken from
http://uttb.sportunion.at/start.php?contentID=54522

Let us then consider some system possessed of the available energy stocks, and of the possibility to generate the driving force out of the latter. Furthermore, there is a process of interest, which is somehow developing in time.

Hence, in any case, there must be the process driving and the process thwarting forces.

Now, for the further mathematical treatment the following information should not be so important, but this would be still a physically significant statement. Namely, the process-driving forces come from some particular energy storage (one might say, the so-called useful energy is used in this or that way for the generation of the driving forces). On the other hand, the thwarting forces are among the most ubiquitous counter effects (such as in the Newtonian mechanics, e.g., any kind of action causes the corresponding reaction). To sum up, we would denote the latter kind of permanently thwarting agents the entropy factors.

With the above objective in mind, we would suggest to build the following mathematical model:

$x[t]$: describes the development of the driving forces in the course of time (the dynamics of the useful energy stocks),

$y[t]$: describes the corresponding development of the entropy factors (that is the dynamics of the devaluated, useless energy amount).

The over-all objective of the whole game might be described as follows—the driving forces should be exploited to achieve the minimum of the corresponding potential energy (just as efficiently as it would be possible)—and thus the goal of the whole process. The entropy factors have to permanently increase in response to increasing of the driving forces, they would this way neutralize the driving forces (just as efficiently as it would be possible), and must eventually reach their maximum value, by trying to completely kill the drivers (according to the second law of thermodynamics). Thus, the energy–entropy equilibrium ought to be reached at last, or, in other words, the desired goal of the entire process.

The latter result might be thought of as a unique outcome of a kind of war, that is, an intrinsic conflict between the energy and entropy. Hence, to model the very approach of the system under study to the above-mentioned equilibrium, we might think of some two-dimensional system of kinetic equations (mathematically, differential equations). Of interest in this regard ought to be the model of 'war of attrition and attack', as described in detail in the well-known book by Prof. Dr. Rufus Isaacs [95].

Thus, in our present case we would like to describe the 'war of attrition and attack' delivered by some (natural or artificial) driving force x and entropic factor y, with the former and the latter being expressed using the energy dimensions. Bearing all this in mind, we ought to carefully re-consider the situation described by Prof. Dr. Isaacs.

5.8 Enthalpy–Entropy Compensation as an Intrinsic 'Conflict', as a Kind of 'War'

Now, we take into account the second basic law of thermodynamics in the form suggested by August Friedrich Horstmann. We then assume that there are two basic factors, namely, a driving force (kinetic energy), E, and an opposing force, entropy, S, engaged in a protracted war. And now, several questions come.

The conventional armies involved into the war are representing military forces of some rivaling states/nations, so that one might speak of weapon supplies of the hostile forces, which are produced by the respective industries and destroyed in the course of military actions.

With this in mind, Prof. Dr. Isaacs suggested the following train of thoughts [95]:

> *The both enemies have respective supplies of a vital weapon, xx and x2, at time t. Each has at all times the choice of how to allocate his stock between "attrition," that is, depleting his enemy's rate of weapon supply, and "attack," that is, entering them in the major conflict. It is the accumulation of the latter entries that count; each player seeks more than his opponent, and the excess will be the payoff.*

> *Thus, the basic decisions here are between the long-range policy of attrition and the short-range one of direct attack on the essential targets.*

But in the case of our interest we have a quite different situation: There are definitely energy supplies, energy stocks—this is namely the potential energy. The latter ought to serve as a source of the kinetic energy, the vis viva—that is, the basis for the driving force. here we won't dwell on the ways of getting the kinetic energy from the potential one—to simplify our consideration, we assume the Hamiltonian dynamical picture, where our system ought to win the kinetic energy from the potential without hindrances. Sure, that this is in effect nothing more than a drastic idealization. But in our simple case we assume that the following functional relationship takes place, namely that $x(t) \equiv -z(t)$, where $x(t)$ is a kinetic and $z(t)$ a potential energy.

Further, we have to assume that there is *no entropy* supply, *no entropy* stocks, and if there are no driving forces, there will be no obstacles as well. The obstacles appear as soon as the corresponding driving forces become non-zero, and the more intensive the former, the higher the latter.

Thus, physically, we have to take into account that the obstacle and the driving force are tightly correlated with each other, and there might be some *physically definite interaction* that would lead to the entropy increase at the expense of the energy devaluation.

In other words, the potential energy should decrease to its minimum to enable the maximum of the kinetic energy. Then, the interaction between the livening force and the obstacles ought to cause the increase in the entropy and decrease in the kinetic energy.

Such a train of thoughts helps arrive at the following system of the kinetic equations:

$$\begin{cases} \dot{x}(t) = w_{11}(u_1,u_2)\cdot x(t) - w_{12}(u_1,u_2)\cdot x(t)\cdot y(t); \\ \dot{y}(t) = w_{21}(u_1,u_2)\cdot x(t)\cdot y(t); \end{cases} \quad (5.1)$$

Further, to define the pertinent game-theoretical framework, we would like to introduce here the objective functions (pay-offs) for the driving forces and for the entropic factors.

$$\begin{cases} \max_{u_1(t)\in U_1} \int_0^T e^{-r_1 t}\left(1-u_1(t)\right)\cdot x(t)\cdot dt + e^{-r_1 T}S_1\left(x(T),y(T)\right); \\ \max_{u_2(t)\in U_2} \int_0^T e^{-r_2 t}\left(1-u_2(t)\right)\cdot y(t)\cdot dt + e^{-r_2 T}S_2\left(x(T),y(T)\right); \end{cases} \quad (5.2)$$

The first basic law, that is, the energy conservation law, ought to impose the following constraints:

$$\begin{cases} x+y=\Gamma\geq 0; \\ 0\leq x,y\leq\Gamma. \end{cases} \quad (5.3)$$

The main question of the physicist here would be—shall we thus mimic the increase of the entropy to its maximum? In other words, may we this way mimic the second basic law?

Meanwhile, to carry out the proper game-theoretical investigation, we would need to answer somewhat more questions of mathematical nature, the background of which ought to be physical/chemical/etc.

After discussing the details of the pertinent game-theoretical model we shall not immediately go in for the relevant calculations/computations in some realistic particular cases, for this is beyond the scope of the present book.

Instead, here we shall just try to mimic the second basic law in a simple numeric example, to prove the over-all validity of our above deliberations.

5.9 The Details of the Game-Theoretical Model for the Non-equilibrium Thermodynamics

In the above expressions, r_i stands for the discount rate of the player number i. It is not necessary in the model under study to have $r_i > 0$, but we could include it. If we would assume an infinite time horizon (i.e., $T \equiv \infty$), we would need that; otherwise we would get no solution (infinite integral). It is better to keep r_i in the analytical derivation of the conditions. But if we decide to have no discount rate, we just ought to choose $r_i = 0$. In our present case the most mathematically skillful method would probably be to go for introducing both the discount rate and the infinite time horizon—or, instead, for some short finite time horizon, but then with $r_i = 0$.

Further, the term $S_i(...)$ denotes the salvage value function for the player i—i.e., the value of the respective variable at the end of the process. This is often very important in engineering models, because in the latter type of the models it is rather important to drive the system under study as closer as possible to some predefined state, but at minimal cost. It is definitely a good question, whether such a representation would make sense for a general physical model. Howbeit, mathematically seen, we have to be careful, because the salvage function values might sometimes considerably influence the choice of optimum strategies (especially, if the time horizon is short). To sum up, introducing the salvage functions having such a sense ought to be of importance in our case as well, for engineering models are in fact particular kinds of a physical model.

Moreover, of special importance ought to be the poser of how the functions $w_{ij}(...)$ should mathematically look like. And this is definitely one of the main problems of the whole study! My train of thoughts could be described as follows. Physically, if the $x(t)$ stands for the "livening energy" then $z(t)$ stands for the "bank of the livening energy," or, to put it in modern terms—using the mechanical analogy—the kinetic energy—and then the bank of the kinetic energy, kinetic energy stock, that is, the potential energy.

The sum of the former and the latter is the total energy, which is conserved, and what we put as Γ, that is: $x(t) + z(t) = \Gamma$.

Please, note that here we haven't introduced $y(t)$ as yet, for it will come later on, as soon as we are ready with the potential and the kinetic energy. Hence, by maximizing $x(t)$, we minimize $z(t)$, and all this is contained in the Γ to 100%, so that we don't need any additional mathematical action here. To sum up, the useful work done, W, could finally be viewed as the pertinent loss in the $z(t)$. In accordance with this, the latter ought to be correspondent to the appropriate win in $x(t)$.

Thus, the dynamics of our model ought to show its actual power balance, for the derivative of energy $x(t)$ with respect to time t is nothing more than the corresponding power, whereas the coefficient $w_{11}(u_1;u_2)$ should have the form $w_{11} \cdot u_1 \cdot u_2$, where w_{11} stands for some constant with the reciprocal energy dimension (to cancel the dimension of the $x(t)$ itself), u_1 for the force and u_2 for the velocity. Thus, in the simplest case we might take the linear or even quadratic time dependencies for both the u_1 and the u_2.

Also of interest could possibly be adding some impulse-like and/or random-noise-like disorders—both in the force and in the velocity terms, but this ought to be some next step.

And it is exactly at this point that we just come to discussing the physical essence of the energy variable $y(t)$ and the corresponding coefficient $w_{12}(u_1;u_2)$. This variable ought to describe the losses in the useful energy with time (or entropy growth with time). Thus, the coefficient $w_{12}(u_1;u_2)$ is also dependent on some force, u_1, and just the same velocity, u_2.

Here, to come nearer to thermodynamics, we might first of all take the mechanical example and consider the dynamics in the presence of the sliding friction force. Let us take such two solid bodies that one of them lies on the ground and another one is sliding over a surface of the latter. Then, some amount of heat will develop owing to the friction between the both surfaces, so that the amount of this heat will be proportional to the product of the friction force and the velocity.

The over-all force acting on the system is in general case a vector in the 3D space, so that there are in fact three components of the force: the two are parallel to the sliding surfaces, F_x, F_y and another one which is perpendicular to the surfaces in question, it is called 'normal force', or 'load' F_z. With this in mind, the friction force is proportional to the latter one and is possessed of the opposite sign.

The velocity would in effect be also a 3D-vector, but here we are interested in its absolute value—mathematically, in its norm—and have therefore to deal with a scalar variable.

Most recently, there have been many attempts to re-consider realistic physical/chemical situations from the thermodynamic standpoint (see, e.g., the most recent reviews on the theme, where the 'non-equilibrium' picture in its modern form is being presented [96–99]).

In fact, the above-mentioned works try to spread the 'logical cage' imposed by the 'equilibrium thermodynamics', and they do succeed in fulfilling this task indeed. But there still remain some steps to go, as we have seen in all the stories of the book at hand.

In line with all the above deliberations we might now draw the conclusion that the following scalar form represents the coefficient of energy change

$$w_{11}(u_1 ; u_2) = w_{11} \cdot u_1 \cdot u_2 = w_{11} \cdot \sqrt{(\|v\| \cdot \|F\|)};$$

$$(\|v\| \cdot \|F\|) = \left\{ (v_x \cdot v_x) + (v_y \cdot v_y) + (v_z \cdot v_z) \right\} \cdot \left\{ (F_x \cdot F_x) + (F_y \cdot F_y) + (F_z \cdot F_z) \right\}.$$

$$(5.4)$$

Herewith we would like to introduce the vectors of velocity v and force F, as well as their scalar products. In principle the time dependencies of every velocity and force component could be quite different from each other.

Furthermore, possible additive disorders (impulse-like or random-noise like) might enter into some particular vector component only—or into a definite number of them.

To conclude let us introduce the 'entropic loss' coefficient, w_{12}, and the 'entropy increase coefficient', w_{21}, with their magnitudes to be (possibly!!!) different, but their intrinsic signs to be just the same as that for the coefficient of the energy change introduced above—in order to preserve the general appearance of the basic equation of motion (or, in other words, the kinetic equation).

$$w_{12}(u_1 ; u_2) = w_{12} \cdot u_1 \cdot u_2 = w_{12} \cdot \sqrt{(\|v\| \cdot \|F\|)};$$

$$(\|v\| \cdot \|F\|) = \left\{ (v_x \cdot v_x) + (v_y \cdot v_y) + (v_z \cdot v_z) \right\} \cdot \left\{ (F_x \cdot F_x) + (F_y \cdot F_y) + (F_z \cdot F_z) \right\}.$$

$$(5.5)$$

$$w_{21}(u_1;u_2) = w_{21} \cdot u_1 \cdot u_2 = w_{21} \cdot \sqrt{\left(\|v\| \cdot \|F\| \right)};$$

$$\left(\|v\| \cdot \|F\| \right) = \left\{ (v_x \cdot v_x) + (v_y \cdot v_y) + (v_z \cdot v_z) \right\} \cdot \left\{ (F_x \cdot F_x) + (F_y \cdot F_y) + (F_z \cdot F_z) \right\}.$$

$$(5.6)$$

Finally, the physical dimension of the constants w_{12} and w_{12} ought to be the square of the reciprocal energy, to incorporate the energies of both rivalling contributions. Importantly, whereas the velocities in Eqs (5.4)–(5.6) ought to be the same, the forces in them might well be different from each other (these might in principle be both internal and external forces).

5.10 A Possible Simplification

Due to the constraint $x + y = \Gamma$ we can reduce the system to a one-state differential game. We just define $y: = \Gamma - x$. This way we have an objective function

$$
\left\{
\begin{aligned}
&\text{Player 1}: \max_{u_1(t) \in U_1} \int_0^T e^{-r_1 t}(1 - u_1(t)) \cdot x(t) \cdot dt + e^{-r_1 T} S_1\left(x(T), y(T) \right); \\
&\text{Player 2}: \max_{u_2(t) \in U_2} \int_0^T e^{-r_2 t}\left(1 - u_2(t)\right) \cdot y(t) \cdot dt + e^{-r_2 T} S_2\left(x(T), y(T) \right);
\end{aligned}
\right.
$$

$$
\text{s. t.} \left\{
\begin{aligned}
&\dot{x}(t) = w_{11}(u_1, u_2) \cdot x(t) - w_{12}(u_1, u_2) \cdot x(t) \cdot \left(\Gamma - x(t) \right), \\
&x(0) = x_0 \geq 0.
\end{aligned}
\right\}
$$

The following constraint has to be imposed in our case:

$$x(t) \leq \Gamma. \qquad (5.7)$$

Meanwhile, the constraint $x(t) \geq 0$ is not necessary, for it ought to be fulfilled automatically.

Remark 1: Note that in such a way we get rid of the function $w_{21}(u_1;u_2)$. At the first glance this looks like a bit strange, but as soon as we think over the situation, it still turns out to be completely natural, due to the constraint $x(t) + y(t) = \Gamma$.

This implies in turn that the function w_{21} is not independent of w_{11}, w_{12}. Whenever the latter functions are defined w_{21} is predefined, since the constraint has to hold.

Remark 2: Such a simplification ought to be fully okay from the physics' standpoint, for here we witness that the first basic law produces a mathematical constraint to become crucial for the mathematical formulation of the relevant kinetic equations.

With the above in mind we have qualitatively numerically analyzed the resulting set of differential equations for both $x(t)$ and $y(t)$. To accomplish the latter task we have been using the options of the software package 'Mathematica'.

Thermodynamics dictates that during some process with the time we should expect a steady increase of the latter up to some definite level and then it should stop changing anyhow, whereas along with this, the former should be steadily decreasing up to some stable value.

As a consequence, it is the compensation that should be expected between the both variables in the course of time (we shall present the relevant illustrations in the next paragraph). Remarkably, it is just the latter qualitative picture that we would greatly appreciate to achieve in our preliminary study. If so, this ought to open the way for further detailed simulations based upon diverse realistic situations. The paragraph below is just to demonstrate this result on a very simple, purely idealistic model. Still, we guess the latter ought to have definite physically reasonable roots.

Indeed, here we might think of some purely 'artificial example', like 'buying a fridge' in the Supplementary Note 19 to Chapter 1.

5.11 Solving and Analyzing a Simplified Particular Case of Eq. (5.1)

Posing the problem in *Mathematica*:

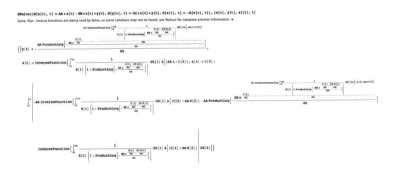

```
sol = NDSolve[{x'[t] = 2*x[t] - 3*x[t]*y[t], y'[t] = 3*x[t]*y[t], x[t]+y[t]+z[t] = 5, z[0] = 0, y[0] = 1*10^-16}, {x, y, z}, {t, 10}]
Plot[Evaluate[x[t] /. sol], {t, 0, 10}, PlotRange → All]
Plot[Evaluate[y[t] /. sol], {t, 0, 10}, PlotRange → All]
Plot[Evaluate[z[t] /. sol], {t, 0, 10}, PlotRange → All]
ParametricPlot[Evaluate[{x[t], y[t]} /. sol], {t, 0, 10}, PlotRange → All]

{{x → InterpolatingFunction[{{0., 10.}}, <>], y → InterpolatingFunction[{{0., 10.}}, <>], z → InterpolatingFunction[{{0., 10.}}, <>]}}
```

Now, the results come (the *time* is always abscissa on the three first graphs below, whereas the *energy* is the ordinate):

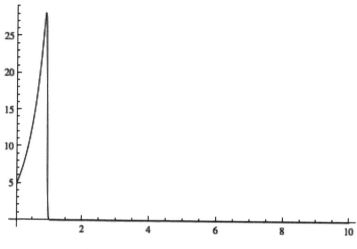

The plot of *x*(t), the driving force vs. time

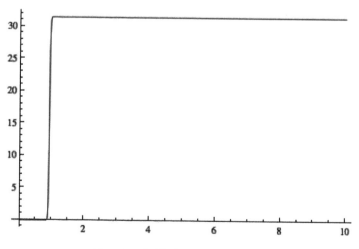

The plot of *y*(t), the entropy vs. time

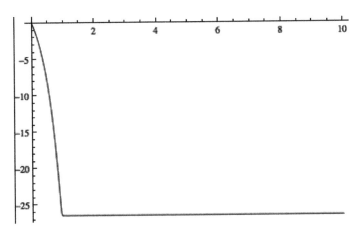

The plot of $z(t)$, the potential energy vs. time

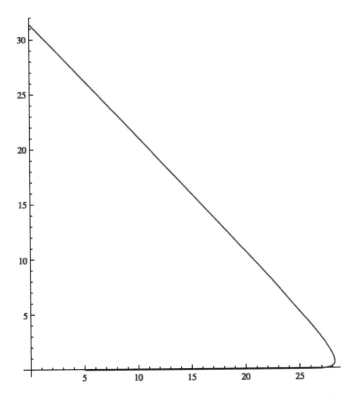

Parametric plot of $x(t)$ vs. $y(t)$, depicting the 'driving force–entropy'
(enthalpy–entropy) compensation.

5.12 Conclusion

And here comes the final point of our travel via the jungle of thermodynamics.

Indeed, here we have learned a number of true thermodynamicists, who were for a longer period of time remaining 'widely unknown', notwithstanding that they did have paid great contributions to the field. We have cast a look 'behind the curtains' of thermodynamics, to try visualize the real reasons for such a strange state of affair.

Now we know that there is only one basic law of thermodynamics, namely, the energy conservation and transformation law, a 'coin having two opposite sides'—the energy conservation and the energy transformation. That is, any realistic process ought to have the two opposite sides: The *Driving Force* enabling the actual progress and the *Ubiquitous Obstacles* to the latter. These both 'coin sides' represent a dialectical unity, whereby they are struggling against each other forever. All other rules and laws of thermodynamics ought to be just logical consequences from this basic law. With this in mind the equilibrium thermodynamics ought to consider the possibilities of equilibria between the driving forces and entropic hindrances.

Finally, we know that one has to be extremely careful with statistical interpretations of thermodynamics at the microscopic level—to avoid paradoxes, like that of the sorites (heap paradox)— this is why the Bayesian approach ought to be of primary use. And to correctly consider the time and, consequently, the kinetics in the non-equilibrium situations the differential game theory ought to be of serious advantage as the proper mathematical toolbox.

From here on, *dear young enthusiasts*—we sincerely wish you: Every success!

5.13 Supplementary Notes

[Note 1]: In this connection, there was a popular cinema piece in the time of the USSR, entitled 'Трест, который лопнул'/'A Trust that Failed', on the basis of the stories set 'The Gentle Grafter' by the famous American story-teller, William Sydney Porter (O. Henry), 1862–1910, where the following verse song appears, which is in effect of direct relevance to our present discussion (here we present the original text in Russian and then its authorized English translation):

"Как плоская истертая монета,
на трех китах покоилась планета.
И жгли ученых умников в кострах.
Тех, что твердили: «Дело не в китах».
Есть три кита, есть три кита,
есть три кита и больше ни черта.
Летит Земля в пространстве утлой
лодочкой. Но молим мы, ловцы и рыбаки:
Сглотни наживку, попадись на удочку,
О! Чудо-юдо, Чудо Рыба-кит ...
... Лишь три кита, лишь три кита,
лишь три кита. Другое - суета."

"Our Mother-Earth, just like a flat worn coin,
Is duly resting on Three Pillars/Whales.
The clever scientists were burned at the stake,
Yeah-yeah, all those who were protesting:
"Hey, look, but these are NOT the Whales!!!"
Still, we should always duly speak of Whales,
Be they Three Whales, or A Triune Whale, It's up
To you: Not Whales? Put it: Pillars, Elephants,
But nothing more than these, God damn it...
... Still, our Mother-Earth – it's like a fragile boat.
Where all of us – the trappers, fishermen, or else ...
Have nothing more to do than simply praying:
"Do swallow the bait, while getting caught on it!!!"
I mean YOU, the Whale, or YOU! The Elephant! ...
... YEAH-YEAH, Three Whales, or just A Triune Whale –
Or – simply Vanity – in case it's something else ..."

Meanwhile, the 'trust' of the 'emotional thermodynamics' hasn't failed at all. Instead, it was, is and still remains at its perfect self-reproduction, as could be clearly seen from the wealth of the thermodynamics textbooks available in any of the World's language.

[Note 2]: Here it seems to be the proper place for trying to answer several interesting posers of relevance to our topic here. Prof. B. H. Lavenda has put them in our personal communications and they are concerning the actual significance of the work by George Augustus

Linhart, who was several years long fruitfully associated with the Berkeley school of Gilbert Newton Lewis (see Chapter 4 in the present book for the wealth of details). Moreover, the main poser in this row concerns the stances and attitudes of such truly outstanding colleagues of Dr. Linhart as Prof. Dr. Gilbert Newton Lewis (1875–1946) and Prof. Dr. Richard Chace Tolman (1881–1948). We shall cite them below and successively present our answers, respectively.

2.1 What are the real backgrounds of the puzzle posed by Gilbert Newton Lewis together with Richard Chace Tolman—who were both on the verge of using Lobachevsky's geometry in treating the relativity, but had suddenly dropped this activity to move to California and turn from theoretical physicists into practical chemists?

It ought to be surely difficult to try reconstructing the ideas and hopes of those who are not more with us since longer time. Still, we might try reconstructing the actual portrayals of those persons according to the memoirs of their contemporaries. For the detailed description of the Lewis' school history and estimates of Prof. Lewis' achievements—in Berkeley, as well as everywhere else—we might check the following sources [2.1, 2.2], not to forget the copyrighted, but at last still unpublished diaries of G. A. Linhart. In combining all these sources we might get a more or less objective picture of what was taking place at the time as Linhart was in Berkeley and how colleagues were treating Linhart's work in particular.

We have already presented a detailed account on this topic in Chapter 4 of this book, so here we would like to restrict ourselves by some short summary only.

As we have already seen, G. N. Lewis appears in G. A. Linhart's memoirs as a true teacher, which is in full consistence with the entire official representation. Specifically, G. A. Linhart is sure that it was just G. N. Lewis, who could trigger his sincere and productive interest in the problem of the entropy notion. Moreover, both in the official sources and in Linhart's memoirs G. N. Lewis is portrayed as the true chief of the Chemistry Department in Berkeley, thus G. N. Lewis was undoubtedly possessed of both professional and administrative authority and talent.

This is why it is to 100% clear that G. N. Lewis could immediately recognize the ultimate significance of G. A. Linhart's results.

And the most of the other Berkeley 'inhabitants' of the Linhart's time appear in Linhart's memoirs as nothing more than 'greedy rats', as soon as it has come to the attitude toward G. A. Linhart's personality and his works.

As for Richard Chace Tolman, in Dr. Linhart's memoirs he appears to be just one of the 'most active rats', whose move to the Californian Institute of Technology had appeared to be a great personal relief for G. A. Linhart, but, regretfully, not for longer time.

Sure, George Linhart himself was never a divinity, but just a rank and file citizen, like all other in his entire surrounding, including G. N. Lewis. But we remember the wise note by Prof. Dr. F. A. Gooch as to this delicate theme (see Chapter 4 of this book).

With this in mind we might finally state: What seems in fact to be rather strange would be the attitude of G. N. Lewis. On the one hand, he had moved no finger to stop—or, at least, to tune down to some reasonable extent—the massive "rats' hunt" for G. A. Linhart.

Then, remarkably, Prof. Lewis' own deliberations concerning the entropy notion [2.3] are fully consistent with the tiny 'Red Herring' literally bloated up to an infinite 'triune whale' by the 'orthodox-revolutionary thermodynamicists' (see the Note 1 to this chapter).

And, the last, but not the least, was G. N. Lewis' application to G. A. Linhart with the suggestion to re-initiate their professional cooperation (see Chapter 4 here), shortly before Lewis' strange and sudden departure to elsewhere.

Moreover, as concerns Prof. Dr. Tolman, we would like to cite the clever words by Prof. Dr. Otto Redlich (1896–1978). The details of his biography could be borrowed from the following most recent paper [2.4].

Although Prof. Dr. Redlich was originally of Austrian origin and due to Holocaust had to spend a lot of time in Berkeley and Californian Institute of Technology; although in tight personal and professional contacts with Prof. Lewis and his school; although he was lively interested and productively working in the field of thermodynamics, he had absolutely no information about Dr. G. A. Linhart. The following citation of Prof. Redlich comes from his work [2.5]:

This portrayal of Prof. Dr. Redlich has been taken from http://photos.lbl.gov/ viewphoto.php?albumId=126055&imageId=4647150

In the last 100 years, several authors noticed that the variables now called generalized coordinates and forces belong to two classes, differing from each other and from all other variables. But repeated attempts to characterize these classes were unsuccessful. Consequently, the concept of work, defined as the integral of a generalized force with respect to its generalized coordinate, never had a solid foundation. On the basis of the present discussion, work can be introduced in a clear and unambiguous manner. Temperature can be defined in the conventional way. Carathéodory has shown the path to the first and the second law. Serious mistakes have been the curse of thermodynamics. They have caused the uncertainty and uneasiness that often have emerged. They have been repressed by the habit-forming procedures of applying thermodynamics. Wrong definitions of extensive and intensive quantities, of generalized coordinates and forces, and defective definitions of work have been quite common.

A historical accident—namely, Tolman's unawareness of a prior definition—has greatly contributed to a widespread confusion of extensive and intensive quantities with coordinates and forces. The so-

called zero-th law going back to Carathéodory is not only unnecessary but also actually misleading. The problems of the relationship between the physical sciences and mathematics become more manifest in thermodynamics than in any other part of science. The wonderful efficiency and elegance of so many mathematical tools, starting from the concept of the limit, should not induce us to see in them more than tools. The theoretician as well as the experimenter must be the master of his tools, not their slave.

It has been the purpose of the present discussion to show that thermo-dynamics can be developed in a clear and consistent conceptual structure. We may conclude that thermodynamics can be understood.

After reading this fragment, we greatly appreciate Prof. Dr. Redlich's move to clarify the thermodynamics and to hint that Prof. Dr. Tolman was never a true specialist in the field of thermodynamics.

Reverting to Prof. Dr. Lewis, he was clearly a much more profound specialist in thermodynamics than Prof. Tolman, irrespective of his unambiguous support of the revolutionary physical ideas. The further remarkable fact having nothing to do with Prof. Lewis' professional qualifications, but rather with his personality: All the 35 Nobel Prize nominations ended up with nothing for him. This might definitely speak in favor of the widespread official suspicion that he could have simply done away with himself.

Most probably, the fact of *getting* or *not getting* the Nobel Prize was of tremendous importance for him—like for Richard Chace Tolman, as well.

Meanwhile the actual potentials of G. N. Lewis and R. C. Tolman as research workers were practically incomparable—with the clearest favor going to the former colleague. Interestingly, R. C. Tolman wasn't even an apt enough thermodynamicist, see Prof. Redlich's remark above. But this is certainly not to diminish his over-all professional contributions and accomplishments!

To sum up answering this poser of Prof. B. H. Lavenda and in recalling his poser as to the stubborn stability of Clausius' philosophy: To our mind, it is rather difficult even to formulate a kind of specific "Clausius' philosophy," for this sounds approximately like Prof. Truesdell's accusations to the address of Carnot for his alleged 'ineptness at mathematics'.

Indeed, like N. L. S. Carnot was not a mathematician, but an engineer with extremely wide horizon, R. Clausius was not a philosopher, but a physicist with extremely wide horizon. Then, the clear description of the problem and its possible solution solely in sober words by Carnot could trigger the proper graphic and mathematical analysis by Benoît Paul Émile Clapeyron. Similarly, just correctly posing the problem—even without truly solving it—by Clausius could still be throughout fruitful. The actual achievements by Gilbert Newton Lewis, one of the true originators of the modern physical chemistry, and Richard Chace Tolman, an outstanding specialist in the physical chemistry, statistical mechanics and relativity theory were definitely fruitful. All of them were never 'know-it-alls'; all of them did illustrate the truism *'Errare Human Est'* and they had, they have and they will have true followers.

Hence, the main point here is that scientific research goes on irrespective of wars, revolutions, and human life's finiteness and sometimes truly dirty taste of the human relationships—as well as of other known and/or unknown cataclysms.

2.2 Why Linhart was so fascinated by the isotherm of Langmuir?

Dr. Irving Langmuir (1881–1957) is well known as a winner of the Nobel Prize in chemistry, 1932, for his ingenious studies on the physical chemistry of surface. To the best of our knowledge, Dr. Linhart was fascinated neither with his name, nor with the isotherm presently bearing Langmuir's name. The actual point of Linhart's fascination was just the general and abstract functional form of the isotherm equations suggested by I. Langmuir—and even earlier by A. V. Hill, which are nowadays very well recognized as powerful interpretational tools, but still remain to be largely semi-empirical [2.6–2.10].

Remarkably, Dr. Archibald Vivian Hill (1886–1977) was also a Nobel Prize winner for physiology/medicine in 1922. In effect, the Langmuir's expression is mathematically just a particular case of the Hill's one. See the most recent work by Andres Ortiz for the detailed analysis of this issue, where but the Linhart's contribution isn't even mentioned explicitly [2.11].

Interestingly, the Hill's expression could be theoretically derived in the straightforward way using G. A. Linhart's thermodynamic approach. Dr. Linhart wasn't dealing with these issues himself, but our recent work demonstrates how to employ his approach in building up the true thermodynamic basement for the general functional form introduced by Hill/Langmuir [2.12].

Remarkably, the Hill/Langmuir functional form in question is mathematically nothing more than just the well-known 'logistic curve'. Meanwhile, the latter ought to have a deep philosophic sense of the *possibility distribution function/membership function for some fuzzy variable* as well.

Dr. Linhart could rigorously demonstrate this important property of the heat capacity. Moreover, it is possible to show that the Hill/Langmuir equations are also possessed of the same property. A good poser would then be: *What actual physics ought to be behind such a property*?

It is not the main goal of the volume at hand to dwell on this very interesting and important theme, so here we just restrict ourselves by placing the references to a number of most recent pertinent reviews on this theme [2.13, 2.14].

2.3 Why Linhart states that Planck, Einstein etc. could not write down an expression for the heat capacity as a function of temperature?

It is a truly good question! So, let us try to properly answer it. This would require a more detailed analysis of what the named colleagues were actually doing at their time. They did not seem to spend their main time in looking for the most skillful mathematical expressions, for they were in struggle against the 'energetics', which to their minds was a 'reactionary paradigm'. The 'progressive paradigm' could be boiled down to a basic indeterminism, as we could see in this chapter (see the works [57–60] for the pertinent detailed analysis). We could already catch the objective sketch of the relevant vicissitudes from reading the unbiased Peter Boas book about this struggle (see Chapter 1 of this book), so let us have a look at it in somewhat more details.

2.3.1 The acrimonious struggle of atomistics against energetics

Here we start with presenting the English translation of the relevant paper by Max Planck (*all the highlighting below is ours*).

The full citation below ought to help grasping the very idea by Planck as to the apparent drawbacks of energetics, in trying to find the ways of their amendment.

Against the newer energetics, by Max Planck, *Annalen Der Physik*, **Volume 293, Issue 1, 1896, Pages: 72–78.**

Soon after the discovery of the mechanical heat equivalents by R. Meyer and J. G. Joule, in 1853, an English physicist, W. J. M. Rankine* published his essay 'On the general law of the transformation of energy', in which he had made an attempt, by classifying the energy into its various forms and partitioning it into two factors, to extend the Carnot's principle to a general law, which ought to bestride all the natural phenomena. And later he had published several other works on the same subject, but neither he nor any other physicist could be capable of achieving any result worth mentioning with this method, so that they could kindle no general attention, and finally fell quite into oblivion. Thereafter this direction, called "energetics" by Rankine, remained entirely silent for some time.

Meanwhile Clausius could formulate his 'second law of the mechanical theory of heat' and discover the notion of entropy in the course of his further studies, so then a long series of fruitful applications of this law began by Clausius himself, Lord Kelvin, Gibbs, Helmholtz and others; those were just the applications that could have shed new light on various fields of physics and chemistry, as concerns the mutual dependence of the phenomena, and all of them had been confirmed by the experience without any exception.

But after a number of excellent works have created the firmest possible basis for the sound development of the thermodynamic principles, the energetics had appeared on the scene again in recent times, regardless of Rankine's work, but using essentially the same ideas. Specifically, this newer energetics tends to consider the thermodynamics laws from a quite different, seemingly 'more universal' standpoint, by inferring them by 'much simpler means', so that their physical sense could be obvious without any further trouble.

---

*Rankine, Phil. Mag. (4) v. 6, p. 106, 1853.

Then, based upon its 'more generalized approaches', it lifts itself up to the highest possible position over the theories developed up to date in different fields of physics and promises to solve problems other than the earlier theories could ever manage to; it tries even to tackle the mechanical worldview, which is nowadays already well known to lead the natural sciences to their most noticeable successes.

I do not intend at this point to defend the placement of the mechanical worldview, for this would require profound and partly very difficult investigations. Here I would like to consider much simpler things, namely whether the newer energetics is possessed of any mathematic basement.

The point is that every specialist in the field might ineluctably draw the conclusion about the absence of any such basement, that the 'simple inferences', exactly where they could be of importance, are in effect bogus, that the energetics has nothing to do even with posing the actual questions, let alone their actual answering.

To illustrate the above criticisms, let us now consider one of the characteristic notions of the energetics, namely the 'energy of the volume'. The latter term is so frequently discussed in the energetic literature that even those physicists and chemists, who clearly stand apart from the field of energetics, could get accustomed with it and wouldn't check its actual sources. Still such a check would nonetheless clearly demonstrate that the 'energy of the volume', to put it in a nutshell ought to be a mathematical absurdity, that is a variable which might never become one such. Indeed, we have to require that any physical variable, not to speak of the energy, which ought to be the notion $\kappa \alpha \tau' \varepsilon \xi o \chi \eta \nu$ (which is quite intrinsic) for energetics, should be determined by the chemical and physical state of the pertinent system. Moreover, any interrelationship among the variables and the pertinent system—especially that between the latter and its energy— should take into account that, if the system after some changes comes back to its initial state, the variables must adopt their initial values. Without such a rule the energy principle ought to be senseless.

Now let us ask about the 'volume energy' of some particular gas with given mass, temperature and density. The energetics has an immediate answer: This energy should be equal to $\int pdv$, where p stands for the pressure and v for the volume of the gas in question. And the energetics represents this integral as the change in the 'volume energy' resulting from the change in the state of gas, so that *this energy change ought to be determined by the state change. Let us now assume that the gas would now undertake a continuous series of state changes, for example, along the cyclic process of Carnot, to return back to its initial state. Then, according to the above-mentioned principle, the 'volume energy' ought to arrive at its initial value, so that its over-all change should be equal to zero. Nonetheless, it is well known that generally this is not the case, instead—during the above-mentioned cyclic process—the gas delivers either positive or negative work, in accordance with the actual conditions. With this in mind, it is absolutely senseless to represent any gas 'volume energy' as a kind of physical variable to be taken into account.*

There ought to be absolutely no problem, if we would always speak not about the absolute value, but about differences of the 'volume energies'. Meanwhile, there ought to be only one difference, so that the actual difference between the 'volume energies' in two various states should unlike the above-mentioned integral be independent upon the way along which the change of the gas state takes place.

Clausius considered this circumstance to be so important that he published the following mathematical introduction to the collection of his works 'Gesammelte Abhandlungen über mechanische Wärmetheorie'*: "About the treatment of differential equations which are not integrable in the usual sense," and justified such an addition by the following words in his introduction:

> *"A further drawback, as I have heard many times, hitherto encroached upon the proper use of my publications, was the difficulty they offered to their understanding at some places. The mechanical theory of heat has introduced new ideas into science, which clearly differ from the formerly widespread standpoints, and hence they did require some peculiar mathematical considerations as well. To be mentioned especially is a certain kind of differential equations, which I have applied in my studies. They are quite different in one essential point from those usually occurring otherwise, which might lead to misunderstandings, if this difference is not followed precisely. The importance and the treatment modus of these differential equations are in fact already noted by Monge, but do not seem to be known generally enough, and, in fact, it was an incorrect view of these equations that caused a violent attack on my theory."*

**Clausius: The First Volume, Braunschweig 1864.*

[***Our immediate comment****: Now we know that solely Nils Engelbrektsson was not only keeping the relevant mathematical results by Monge under close scrutiny, but in properly using them could also duly produce the correct mathematical basement for the theory of thermodynamics, see Chapter 2 in this book.*]

And the contents of his mathematical introduction deal just with the differential equations not integrable in the general case, like the 'pdv' statement mentioned above, and therefore not capable of being treated as a full differential of the variable dependent upon the physical state.

Also in the second edition of his works collection (1876) Clausius had decidedly stressed this same point (§3 of his mathematical introduction), but, as we see to our sincere regret, this hadn't attracted the necessary attention.

The present examination, which has revealed the basic unsustainability of one of the main concepts of energetics, might also be performed with a number of other similar ideas and with the same result; but here I can forego the discussion of the further points, especially as in the next few years I shall have an occasion to summarize all the concepts and results of thermodynamics in a separate detailed presentation.

On the other hand is clearly undeniable that the energetics is possessed of a certain healthy core, by virtue of its relationship with the principle of conservation of energy. And this is also suited to explain the special influence the energetics' viewpoint exerts on a number of natural scientists.

But anyone practiced in such investigations might immediately recognize that the energetics, after being dispensed from its instantaneous excesses, would then find its applicability to be limited to an area, whose circumference would just boil down to extremely modest exempts from the highest claims, with which it currently appears on the scene.

Most probably, energetics expects a cruel twist of fate to get persuaded that its applicability area is just the one to which it glances down with a certain contempt, considering it a single special case: namely the mechanics, if we would disregard the friction, diffusion, imperfect elasticity and related phenomena, and also the electrodynamics, including magnetism, if we would disregard Joule's heat, as well as the magnetic hysteresis, etc., then the optics, if we would disregard the absorption, dispersion, etc.—or, to sum it up in short: in case we would disregard all those phenomena taking place in finite times and having the property that they can run not only forward, but also in the opposite direction, being thus the ideal abstractions of the actual processes.

And it is exactly here that the correct energetics, mind you, might well contribute much to illustrate the processes of nature. It is just in this case that the partition of various energy forms into the capacity and intensity factors, and everything connected with the latter, ought to be undoubtedly valid. Thus, it is on this basis that a communion of all the valid examples could be found, through which the actual sense of the energetics might be brought closer to the understanding. Still, energetics would in fact hardly offer something really new in this field, because it is precisely for this purpose that the science has since long ago a method to answer all the posers asked in terms of the measurable progression of phenomena, to answer much clearer and much more completely than, most probably, the energetics would ever be able to: The Hamilton's principle of least action.

Meanwhile, the pertinent notions of energetics are not valid in the field of thermodynamics, for rationalizing the concept of chemical affinity, for the electrochemistry, as well as for all the phenomena listed above. In such cases the energetics might achieve the simplicity of its inferences only through mislaying the latter to the pertinent definitions, whereas the contents of the laws to be inferred ought to be already well known. What happens to such definitions we could already see above using the example of the 'Volume Energy'. Such shortsighted efforts cannot brighten up any actual interrelation among the phenomena under study they ought solely to obnubilate them.

In particular, the energetics has inflicted the obfuscation of the basic antinomy between the reversible and irreversible processes, whereas, in my opinion, the elaboration and further deepening of this antinomy are connected to every advance of thermodynamics and related fields. The energetics don't express at all, or only peripherally, the fact that the adjustment of the level for a heavy liquid in two communicating tubes ought to be fundamentally different from the temperature balance between the two bodies being in the direct heat exchange with each other. Therefore, it will never be capable of drawing any new characteristic conclusions from such important facts.

*And in fact: Energetics has brought no, absolutely no positive achievements up to this day, although time and opportunities were offered to it in abundance since Rankine's publications. "But," you might reply, "it has also never resulted in any contradiction with the experience as yet." That's right—for the simple reason that due to uncertainty of its concepts and notions the energetics is not at all capable to produce a new result, which can be checked in the course of some experiment. And it is just this particular accusation that must be considered as the most serious at all. Indeed, any theory, which relies on avoiding the real problems to preserve its existence, is no longer rooted in the realm of science, but stands on the metaphysical ground, where the weapons of empiricism are unable to harm it. And that is why, I consider it to be my duty to clearly express my fiercest protest against the further expansion of energetics in its recently chosen direction, which is opposite to the previous results of theoretical research and constitutes a sensitive step backwards. **So, the only practical success the energetics might achieve ought to be: Just cheering up younger researchers to entering hacky speculations, instead of thorough deepening into the study of the available masterpieces, and thereby to lay a broad and fertile field of theoretical physics broke for the years to come.***

Berlin, December 1895

Remarkably, the highlighted most important points of M. Planck's work appear in fact to be highly emotional statements. We learn from the above-cited fragment that in effect its author was forcefully trying to clip any development of the thermodynamics foundations, by drastically overestimating the results of Rudolf Clausius—in declaring the latter to be the ultimate general basis for any development.

Along with finding some terminology inconsistencies in the opponent branches of research, he tries to invent and intermingle a practically non-existent 'antinomy' among the reversible and irreversible processes. As we have already seen above (see Chapter 1 here and the Notes in addition thereto), this so-called 'antinomy' ought to be nothing more and nothing less than a result from an occasional combination of mathematical difficulties.

There seems to be no other way than just to call the mentioned result '*throwing a red herring*', according to Prof. Truesdell. We must but point out here that the very pioneer of '*throwing a red herring*' was in fact Rudolf Clausius, while his followers were just stubbornly and forcefully marketing/promoting the very approach, the methodology of R. Clausius.

To sum up, what M. Planck is in effect calling for ought to be trying to clarify the actual physical sense of the ingenious gadget by N. L. S. Carnot. Still, the only achievement of the gadget's author involved was in fact nothing more than its theoretical perfection in visualizing the generality of the second basic law. In other words, N. L. S. Carnot's undoubted achievement was to clearly demonstrate the latter basic law is coming up to surface even in such a highly idealized case as his cyclic process. Indeed, he could demonstrate, yet purely logically, without any explicit mathematics: Even some perfectly idealized process proceeding without any apparent counteractions, if it is nonetheless a realistic one, should be incapable of delivering the maximum working efficiency, as a result. Some inherent counteractions must be active anyway and anyhow. It is the latter decrease of the process' efficiency that demonstrates the desired basic law.

To this end, Rudolph Clausius' undoubted achievement was to formally mathematically demonstrate the existence of some basic natural law in addition to the law of energy conservation, on the basis of Carnot's deliberations, making use of B. P. É. Clapeyron's graphic representation of the latter. That the novel law had also something to do with the energy notion was immediately clear to Clausius, as well as the clear fundamental generality of the discovered law. As a result, Clausius' immediate decision was just to start looking for the pertinent mathematical toolbox to express this law, as well as to try inferring the latter from some more general principles. Meanwhile, this was just the field where Ludwig Boltzmann, a young researcher at that time, was also very active, so his reaction was to deliver a priority quarrel with Rudolf Clausius [2.15, 2.16].

Moreover, Ludwig Boltzmann was also a persuaded and fierce objector of energetics, see, e.g., his own statements about the topic, which are clearly supporting the standpoint of Max Planck and at least clarifying, who were the actual opponents of M. Planck. Meanwhile, Boltzmann's viewpoint looks at least like the fully factual statements of some competent but still a sideway observer without any irrelevant emotion [2.17, 2.18].

We shall not consider here the full English translations of the pertinent Boltzmann's papers, but most probably several points in the latter might nonetheless shed light on the actual roots of Max Planck's emotional anti-energetics statements. Below comes our English translation of the relevant paragraph from the Ludwig Boltzmann's paper [2.17]:

§ 24. Let me conclude.

It is throughout clear that first of all we would have to set the description of nature as free from hypotheses as possible; this has been most clearly accomplished in the works by Kirchhoff, Clausius (in his general theory of heat), v. Helmholtz, Gibbs, Hertz etc. Still, the language of energetics has so far been proved to be ill suited for this purpose.

Similarly, the educational value of energetics, at least in its present form, must be in contest, and therefore its further development in such a form ought to be sheer disastrous for precise apprehension of the nature. For example, the general part of a large textbook of chemistry contains numerous sites confusing to students, owing to the prevalence of the energetics' parlance.

In addition to this general theoretical physics the images of mechanical physics should also be preserved and even fostered nowadays, because these are extremely useful both to find something new and to arrange the already available ideas, to clearly display and keep them in memory. The very possibility of a mechanical explanation of the nature as a whole is not yet proved, and whether we might ever perfectly achieve this goal is hardly conceivable.

Nor is it proven that we may perhaps achieve even more significant progress and get even more new benefits thereof. No one may be further away than the representatives of today's theoretical physics from arguing that we certainly know the already formed ideas to eternally appear as the most appropriate ones. No one may equally be further away from trying to block the other, newer trains of thoughts, or from labeling the latter as amiss from the outset.

And it is equally not admissible, before gaining real achievements, to polemically oppose the tried and true ways of thinking or to view the latter as something only slightly different from the utter nonsense. The parlance of the general theoretical physics still remains to be the most appropriate and practical, and the ancient images of mechanical physics have by no means become superfluous.

No one knows whether the above-mentioned situation would always be the case, but it may perhaps be completely pointless to puzzle our heads over the question of which ideas will become the most appropriate after the centuries. In this sense, I am pretty far away from denying the possibility that the further development of energetics would bring tremendous benefits for sciences. But equally should not happen something like it was attempted in the recent times by some researchers (not rightly, to my mind) to represent themselves as the only true successors of Gibbs.

Vienna, November 2, 1895

In addition, there are also some relevant fragments from the Boltzmann's paper [2.18]:

A discussion similar to the present one concerning the energetics was not undertaken to demonstrate that the one is right, while the other wrong, but to finally clarify the views. Therefore, I declare that I can be fully satisfied with the clear achievement, as concerns the energetics to the mechanics. The last article by Mr. Helm* seems to clarify everything.

Messrs. Planck and Helm (it is now obvious that they both could accomplish this at the same time) could demonstrate that the ordinary equations of motion for a system of material points can be inferred from the Energy Principle, by just assuming that the latter ought to be applied for each of the points under study in each direction of the coordinate axes – or, according to Mr. Helm – to be separately applicable for any arbitrary direction. ... "

And then a sober, constructive criticism of the Helm's work[†] follows.

To summarize the above forays into the forefront of physics at the end of XIX-th century, the attitudes of both Max Planck and Ludwig Boltzmann concerning the energetics are remaining rather unclear from the viewpoint of a pure researcher. A careful analysis reveals the main objections by Boltzmann and Planck to be rather of *linguistic* than of *basically methodologic nature.*

2.3.2 What was the actual energetics and who were the 'dissidents'?

Hence, what ought to be of clear interest for the specialists in the history of the natural sciences is the *actual* standpoint of the both outstanding colleagues with respect to the general energetics field.

For our discussion here it is but important to have a closer look at the latter, for this way we might recognize what was actually

---

* *Helm, Wied. Ann.* 61. p. 646. 1896.
[†]Helm, *Math. Chemie* pp. 45, 46, 47, 60.

the 'different thermodynamics' and who were duly working on its development.

Our poser would sound as follows: Were there at least some resistance to the over-all physicalisation of Carnot's cycle? The immediate answer: Sure, of course!

Indeed, there is a well-known course book by a renowned German specialist in the field of engineering and chemical thermodynamics, Prof. Dr. Ernst Heinrich Wilhelm Schmidt (1892–1975) [2.19].

Noteworthy, there is also an English translation of this book carried out by a renowned USA specialist in the field of engineering thermodynamics, Prof. Dr. Joseph Kestin (1913–1993), in collaboration with its author, Prof. Dr. Schmidt [2.20]. Prof. Kestin himself was actively and fruitfully carrying out a thorough and detailed research of the thermodynamics foundations. Of special interest for him was the sense of the second basic law, and he edited and partially authored a volume on this hot topic: 'Second Law of Thermodynamics'. John Wiley & Sons Inc. (February 1977).

In the book [2.19, 2.20] we find the following remarkable paragraph (it is of interest and importance to cite it here as a whole, *whereas all the highlighting below is ours*):

§30 Introduction of the absolute temperature scale and the entropy concept without the aid of cyclic processes

In the previous section we could manage to derive the absolute temperature scale free from the use of the characteristics of the perfect gas. Now we would like to derive the same results using another method suggested by M. Planck and forgoing the conventional usage of the cyclic processes.

As we had already seen above, the following expression for the heat supplied to an arbitrary body was not a complete differential

$$dQ = dU + PdV = \frac{\partial U}{\partial t}dt + \left(\frac{\partial U}{\partial V} + P\right)dV \, ,$$

whereas U and P were some functions of V, the volume, and empirical temperature t.

But still, the pure mathematics dictates that an integral denominator N (t, V) should always exist, which ought to render any incomplete derivative the complete one. We thus get

$$dS = \frac{\partial U + PdV}{N(t,V)} \, . \tag{112}$$

Bearing this in mind we introduce S (t, V) as a property of state of the body under study, which is represented by an integral denominator and completely defined up to the integration constant, if we could determine the values of two state variables, say t and V. As we have already seen, there might be a lot of valid variants of the integral denominators, because any expression of the form N (t, V)*f (S), where f (S) ought to be any arbitrary function of S, should just deliver a valid integral denominator. Therefore, the magnitude of S standing here for the entropy notion ought to be unambiguously defined, if and only if we might clearly specify the above-mentioned arbitrary function. By its nature the latter uncertainty ought to be the same as that of the empirical temperature scale, and we would now like to allow this uncertainty to be as it is to choose any arbitrary integral denominator and require that such a denominator obeys the only condition, i.e, N > 0.

Then, it follows from Eq. (112) that the adiabatic curves should obey the condition S = const and one could assign a definite value of S to every adiabatic curve, if for some particular adiabatic curve the appropriate value of S could be set up.

Now we shall consider the behavior of two bodies whose states could be described by two independent variables, namely, as V_1, t_1 and V_2, t_2, and t stands for the temperature measured using some empirical scale. These both bodies should be capable of effecting reversible changes of their respective states, so that we might think about storing up mechanical work through lifting and sinking two weights see also the figures to this paragraph: [2.19, 2.20]).

If the bodies of our interest are adiabatically separated both from each other and from the surrounding, then their respective states might be changing along some adiabatic curve to the effect that the variables S_1 and S_2 do retain their respective definite values—after we have chosen the pertinent integral divisors.

Interestingly, the above-cited fragment doesn't contain any clear reference to some particular work by Max Planck, but it nonetheless perfectly depicts his actual standpoint. Of definite interest to compare the latter standpoint to the one by Prof. Dr. Horstmann (this chapter pp. 630–640), just to achieve a clear feeling of what are the differences between the conventional equilibrium and the different thermodynamics. Howbeit, several serious authors, in particular Hermann von Helmholtz and Max Planck himself among them, have already noticed that all the conventional thermodynamic notions, like entropy and absolute temperature might well be introduced without cyclic processes, ideal gas and other purely logical-mathematical—but expressly non-realistic—auxiliaries. We know that this train of thoughts was initially triggered—by such authors, as Reech, Massieu, Poincaré and Duhem in France—by Hermann von Helmholtz and Max Planck in Germany—and along with all the latter colleagues—by Josiah Willard Gibbs in the USA (see, e.g., the works [2.21–2.30] and the references therein). Meanwhile, the Energetists confronting Max Planck and Ludwig Boltzmann could not suggest a handy counter-paradigm, as Peter Boas Freuchen could conclude in his thermodynamics book (see Chapter 1). Noteworthy to this end, the notorious 'anti-atomism' of such energetists as Prof. Dr. Ernst Mach (and his worldwide allies) seems to be unduly exaggerated [60].

Then, bearing all this in mind, the poser arises: Why such an outstanding philosopher as Prof. Mach, as well as his school, his allies taken together, were finally not capable of suggesting a handy counter-paradigm? It is a truly good poser to be investigated by the philosophers and historians of the natural sciences.

The immediate idea coming to mind in connection with the above deliberations: The marketing of the 'physicalist indeterministic paradigm' stemming from the 'two tiny red herrings' by Rudolf Clausius was truly skillful and rather aggressive, as we have already seen. The relevant counterparty members were partially

(a) in their retirement age or even going to their final business trip (Prof. E. Mach, Prof. W. Ostwald, Prof. G. F. Helm, Prof. O. E. Meyer, Prof. C. G. Neumann, Prof. O. Wiedeburg, Prof. K. von Wesendonck, etc.), or

(b) having huge personal problems, like, e.g. Prof. Dr. Max Bernhard Weinstein (see Chapter 3), or

(c) just on the brink of the 'very academic society', to be marginalized the easiest way, like e.g. Peter Boas Freuchen, Nils Engelbrektsson, Karl Alexius Franzén, Georg Augustus Linhart (see Chapters 1, 2, and 4), or

(d) just incurring the well-known worldwide sociological-political drawbacks of the epoch in question (the end of the XIX, the beginning of the XX century).

That is, Max Planck was in effect among those actively and fruitfully working in the field of energetics. And no wonder that his actual original and tremendous achievement was nothing more and nothing less than just the suggestion and promotion of the *energy quantum* concept. Meanwhile, that was not Max Planck who had pioneered introducing the term 'quantum' into the natural sciences.

Noteworthy, this notion was employed by much earlier professional authors when describing the effects obeying to the second basic law (see, e.g. [16]).

Then, a good poser is: Was the emotional outing of Max Planck connected with personal problems initiated by professional rivalries, e.g., that under participation of Prof. Dr. Georg Ferdinand Helm (1851–1923), just as mentioned by Ludwig Boltzmann in his above-cited works?

When thinking over the answer to the latter poser, not to forget the universal human striving for the recognition—yet, as soon as possible. Meanwhile, among the colleagues discussed here the revolutionary physicists could get it to the full, whereas in their counterparty solely Prof. Ostwald could have 'a stroke of luck'.

Interestingly indeed, both Ludwig Boltzmann and Max Planck were extremely active against any colleague, who was at least to some extent dissident from their thermodynamic standpoint. Among the outstanding 'dissidents' we, find Prof. Dr. Oskar Emil Meyer (1834–1909), Prof. Dr. Carl Gottfried Neumann (1832–1925), Prof. Dr. Otto Wiedeburg (1866–1901) and Prof. Dr. Karl von Wesendonck (1857–1934). At least to pay the due tribute to all of them we would like to present the references to the relevant publications. First come but the examples of reactions by L. Boltzmann and M. Planck to the works of the colleagues named [2.31–2.33].

On the other hand, a complete list of Max Planck's works together with the analysis might be found in the book [2.34]. A complete list of Ludwig Boltzmann's works together with the analyses might be checked in the following volumes [2.35].

Hence, of interest ought to be at least the listing of the dissident's works, with the commentaries and analysis to come from those interested in thermodynamics (see [2.36–2.65]). Among the colleagues whose works are just referenced, Prof. Dr. Neumann was an outstanding mathematician, whereas Profs. Meyer and Wiedeburg were physicists. Of the both latter colleagues, Prof. Meyer is more or less known—at least due to his eminent brother, Prof. Dr. Julius Lothar von Meyer (1830–1895), a chemist who invented the elementary periodical scheme independently of Dmitri Mendeleev, while Prof. Wiedeburg is definitely wearing the veil of oblivion nowadays, although he was truly actively working and publishing his results about the irreversibility notion. The only colleague working in detail with Wiedeburg ideas and results was Prof. Dr. von Wesendonck, who received a 'destroying' comment from Max Planck—and nowadays is also wearing the veil of oblivion. The list of Prof. von Wesendonck's publications is nonetheless impressive [2.52–2.65].

What we can clearly see from the above publication lists: A physical chemist in Berlin, Carl von Wesendonck was attracted by the thermodynamic approach due to a physicist in Leipzig and then in Hannover, Otto Wiedeburg. The physical chemist in Berlin had started his own thermodynamic studies, in trying to grasp the logical structure and the physical sense of the thermodynamics' basic laws.

> Then comes Max Planck together with his allies, and all this activity ultimately stops'. Is this just the only correct diagnosis? Of course, not!

Prof. Wiedeburg had to enter his last departure in 1901, whereas Prof. von Wesendonck was continuing his studies up to the year of 1913.

Most probably, the ultimate stop had to come due to the eruption of the First World War in August, 1914 plus some purely personal reasons: Prof. von Wesendonck had just immigrated to Switzerland, where he had ultimately stopped his entire scientific research as well as his academic activity—and was just living through his life, up to the inevitable final departure.

Hence, there seemed to be no special evil efforts due to Max Planck & Co., so everything could just be following its logical natural way.

What is still bothering indeed: Nowadays nobody, except the nearest descendants, could be able at least to recall the names of Otto Wiedeburg and Carl von Wesendonck, not to speak of their contributions to establishing the foundations of thermodynamics. In the book at hand we have learned a lot about the lives and works of a number of thermodynamicists all around the world, who have just obediently accepted their veils of oblivion, in spite of their truly seminal contributions and achievements.

Is the true reason for such a result some intrinsic fault in the standpoints of Wiedeburg, Wesendonck et al.—with respect to those by Boltzmann, Planck, and their followers??? Is this the result of the systematic aggressive marketing by the latter ones—or just a bizarre communion of purely occasional coincidences??? Are these posers rhetoric indeed?

In effect, these are looking like pretty good posers for the historians and philosophers of the general scientific research—while we shall continue our picking up the zests of the thermodynamics foundations.

Howbeit, according to §30 from Prof. Schmidt's handbook cited above, the actual viewpoint of Max Planck might in effect be formulated as follows:

Sure we must try to circumvent blindly using the gadget suggested by Carnot, but there is nothing more at our hands than the latter instrument. Hence, let us just skip mentioning it, even when using it in fact!

The above is just what we have estimated as a purely operational approach. And Prof. Schmidt clearly and consecutively follows the above-formulated principle. Indeed, by setting up the correct headline of his paragraph, he does clearly come back to the non-existent, clearly artificial and therefore superfluous '*adiabatic—i.e., isentropic—state changes*'.

To sum up, the only problem does remain here. If we would carefully think over the actual physics, there might be a possibility that we learn something principally new and important, which would significantly widen the horizons of our 'operational space'. Would this be a truly bad result?

2.3.3 The contribution to thermodynamics by Albert Einstein

Meanwhile, Albert Einstein, a younger still unknown researcher at that time, was faithfully accompanying his both already prominent colleagues in their 'thermodynamic struggles'.

Below we would like to present the list of Albert Einstein's reviews concerning the work on thermodynamics' foundations. His reviews are solely delivering nothing more than the shortest notes about the contents of the work reviewed. As a rule he was not analyzing in detail the latter. Moreover, Albert Einstein's own contribution to thermodynamics and statistical physics was in fact rather modest and sober, as compared to that by his elder colleagues and actual "fellow soldiers."

Instead, of clear interest and even importance for thermodynamics' foundations ought to be rather the very works reviewed by A. Einstein, as compared to his referee's reports. So, let us have a closer look at the stories.

1. Review of Giuseppe Belluzzo: "Principi di termodinamica grafica" ("Principles of Graphic Thermodynamics") *Beiblätter zu den Annalen der Physik*, v. 29, p. 235–236, 1905.

   This is about the original work by Prof. Dr. Giuseppe Belluzzo (1876–1952) [2.66, 2.67], a renowned Italian engineer, inventor and, later on, one of the active political functionaries under the government of Benito Mussolini.

   Prof. Belluzzo's work was trying to find a systematic, logically consistent presentation of the ideas by N. L. S. Carnot, in helping engineers to correctly understand and practically employ the latter, especially as concerns the difficult notions of 'irreversibility' and 'adiabaticity'.

2. Review of Albert Fliegner: "Über den Clausius'schen Entropiesatz" ("On Clausius' Law of Entropy") *Beiblätter zu den Annalen der Physik*, v. 29, p. 236–237, 1905.

   This is about the original work by Prof. Dr. Albert Fliegner (1842–1928), an outstanding Swiss engineer and pedagogue [2.68–2.69].

   Prof. Fliegner's work is analyzing in detail the same notions as in the Prof. G. Belluzzo's work—as well as, in addition, the actual sense of the entropy notion and of the second basic law of thermodynamics.

3. Review of William McFadden Orr: "On Clausius' theorem for irreversible cycles, and on the increase of entropy" *Beiblätter zu den Annalen der Physik*, v. 29, p. 237, 1905.

This is about the work by Prof. Dr. William McFadden Orr (1866–1934), a prominent British and Irish mathematician. His work contains a grounded criticism of Max Planck's deliberations about the 'reversible' and 'irreversible' processes, which Prof. Orr has found to be logically inconsistent [2.70–2.72].

The response of Max Planck to William Orr's paper is of tremendous interest and importance for our discussion here (*it has also been refereed by Albert Einstein in this same volume of the*

Beiblätter, vol. 29, p. 635), so that we dare to reproduce here Planck's full text (*highlighting below is ours*):

On Clausius' theorem for irreversible cycles, and on the increase of entropy*
Gentlemen,

Berlin, Nov. 2, 1904.

I beg leave to make a few observations on Prof. W. McF. Orr's paper[†] of the above title, in which he criticizes among other things my treatment of Thermodynamics[‡], for otherwise I fear that it may perhaps give rise to one or two misconceptions.
I will not quarrel with Prof. Orr as to whether he is right in saying (p. 509) that I use the words "reversible" and "irreversible" in an unusual sense, since these words are seldom expressly defined. Yet, I really must show by an example that the form of the definition I use is practical.
Clausius, as is well known, founded his proof of the second law of thermodynamics on the simple proposition that heat cannot of itself pass from a colder to a hotter body. Here it is not only stated, as Clausius repeatedly and expressly pointed out, that heat does not pass directly from a colder to a hotter body, but that heat can in no way whatsoever be conveyed from a colder to a hotter body without leaving behind some lasting changes (that is, without any compensation).
If I would now say, the process of heat conduction is irreversible; this statement means, according to my definition of irreversibility, just exactly the same as the Clausius' fundamental proposition. Whether this proposition is in reality true cannot be directly settled and requires a special investigation; but if once we assume its truth, then the whole import of the second law of thermodynamics can be deduced from it. And if, on the other hand, we understand by the irreversibility of a process, only that it cannot be directly reversed, then the proposition that the passage of heat from a higher to a lower temperature is irreversible is, to be sure, self evident, but it should be of no value for the derivation of the second basic law of thermodynamics, because we are not

---

*Translated and Communicated by Alexander Ogg, Ph. D.
[†]Phil. Mag. Oct. 1904, 48. 509.
[‡]Treatise on Thermodynamics, translated by Alexander Ogg, 1903.

in a position to draw therefrom any conclusion as to the other processes.

Prof. Orr states (p. 511): "Planck gives one definition of reversibility, but he uses another." As I was searching for a proof of this assertion, the only thing I could find was the following statement: "Under ordinary circumstances, however, no body can expand without producing a change of density in some other body." The actual truth of this statement, and the conclusions therefrom, I must directly contest.

There is nothing to hinder us from supposing that the gasholders together with the weights, which are acting against the gas pressure, are placed in a vacuum. If, then, the gas pressure lifts the weights, the gas expands without producing a change of density in some other body. That it is sheer impossible to obtain an absolute vacuum, that absolutely unchangeable weights do not in reality exist, and that there ought to be other difficulties opposing the realization of this process, does not of course affect the validity of the proof.

I am sorry to see from this misconception, that the note, which Dr. Ogg on my advice added to that particular part of his translation of my thermodynamics, has not been sufficient to make the point clear.

If a process, e.g., the flow of a gas into a vacuum, takes place so violently that one can no longer define temperature and density, then the usual definition of Entropy ought to be inapplicable. Of course, on this point I completely agree with Orr and Bertrand. So, my observations with regard to the entropy of a gas (p. 512), which is not in an equilibrium state, refer only to the case where we may speak of a temperature and a density of the gas in all its parts. **Still, as is well known, Herr L. Boltzmann has shown from the point of view of the kinetic theory of gases that, even in the case of a violent motion, the definition of entropy, which includes the usual one as a special case, may be deduced from the theory of probabilities. The suppositions of the kinetic theory of gases, however, were purposely excluded from my treatise on thermodynamics.**

Prof. Dr. Orr claims to give (p. 518) "A shorter proof of the principle of increase of entropy, if the substance of Planck's definition of irreversibility were adopted." And any closer examination shows that this proof starts with the assumption expressed in the Lord Kelvin's version of the second law, and that, therefore, the essential import of the thing to be proved is already assumed to be true. I can't, therefore, accept this proof as an improvement on my own. With these remarks I shall conclude. I should not, however, have again given my views on these questions at such length, had it not been that I wish to express my pleasure at having found in Prof. Orr's paper much that was stimulating and interesting to me.

I am, yours faithfully,

Dr. Max Planck

To sum up the highlighted ideas of the above fragment, the 'theory of probabilities' is used by Max Planck as a kind of apology for promoting apparently unphysical assumptions, for trying to *'physicalize'* apparent idealizations—like, e.g., the notion of 'reversibility', while clearly realizing that the true processes are irreversible by their nature.

The main idea of such an approach ought to be

- That some fundamental probabilistic laws are working at the microscopic level, unlike those in the perceptible macroscopic picture;
- That due to the huge number of micro-particles building up the macroscopic objects some barely perceptible microscopic physics ought to be perfectly expressible using the probability theory;
- And therefore the work of a physicist should be aimed at finding the (mathematical) interrelationship between the probabilistic concepts and the macroscopic observations.

Remarkably, the above standpoint is still agile, as seen from the most recent citation:

...Enormous progress has been made in statistical physics research in the last hundred years and it is now the moment not only to reflect this in the teaching of future generations of physicists, but also to acquaint a larger audience, such as students at the École Polytechnique, with the most useful and interesting concepts, methods, and results of statistical physics. The spectacular success of microscopic physics should not conceal from the students the importance of macroscopic physics, a

field, which remains very much alive and kicking. In that it enables us to relate the one to the other, statistical physics has become an essential part of our understanding of Nature; hence the desirability of teaching it at as basic a level as possible. It alone helps to unravel the meaning of thermodynamic concepts, thanks to the light it sheds on the nature of irreversibility, on the connections between information and entropy, and on the origin of the qualitative differences between microscopic and macroscopic phenomena. ... This is just a citation from the book [2.73, 2.74].

Moreover, the recent journal papers and the references therein can clearly demonstrate that the above-mentioned 'phantom' research field (reversibility + adiabaticity et al.) remains to be fully active [2.75, 2.76].

Well, but why we tend here to treat the above-cited research field as a *'phantom'* one? A good poser! So, what might be the proper answer?

Because everything in this field is okay apart from its conceptual ground, namely

- On the one hand, the ambiguity is formed by stubbornly making use of such true phantoms as *'reversible'* together with *'adiabatic'* processes;
- On the other hand, by a 'huge number' of microscopic particles producing a 'huge number' of 'complexions' composing the macroscopic system under study.

The phantom features of the former both we have already discussed above in detail, whereas the latter one drives the stories based upon it into the well-known sorites paradox (that is, the 'paradox of the heap') [18–21].

Moreover, the latter feature renders the actual situation fuzzy, in addition to the apparent fuzziness of the 'reversibility' notion.

Definitely, Ludwig Boltzmann and Josiah Willard Gibbs systematically and successfully demonstrated that it is throughout possible to successfully employ the probability theory to the physical/chemical etc. problems.

Meanwhile, the results of these both outstanding colleagues ought to be just the very starting points of the further fruitful development in the field.

It is definitely not okay to use their seminal results for building up the logical blind alleys around the palaces of the '*absolute truth*'. The latter does sound like a kind of 'metaphysics', isn't it?

Hence, to effectively prevent from the latter kind of development a grounded detailed criticism should every time be openly expressed and duly perceived—for triggering the relevant discussions to be as broad as possible.

This is just what had happened already at the lifetime of Ludwig Boltzmann, when the following book has been published:

"*Festschrift Ludwig Boltzmann, Gewidmet Zum Sechzigsten Geburtstage (Am 20 Februar 1904). Verlag Von Johann Ambrosius Barth, Leipzig, 1904.*"

(*A commemorative publication devoted to the sixtieth birthday of Ludwig Boltzmann, February, the 20-th, 1904.*)

This volume contains 117 reports by a truly international Collegium of professional authors, the specialists in different fields of physics, chemistry, and physical chemistry.

Apart from the majority of more or less neutral publications, the four of the contributions to that commemorative volume ought to be pointed out here, and namely those by:

1. Prof. Dr. George Hartley Bryan, F. R. S. (1864–1928), a professor of mathematics at the University College of North Wales, Bangor, U. K.
2. Prof. Dr. Nikolai Nikolayevich Schiller (1848–1910), an apprentice of Prof. Dr. H. von Helmholtz in Berlin, from the year 1903 on—director of Kharkov Technological Institute, Russian Empire—later on at the Soviet time: Kharkov Polytechnic Institute. Presently: National Technical University of Kharkov, Ukraine.
3. Prof. Dr. Max Planck (1858–1957)—one of the Holy Fathers of the Quantum Physics.
4. Prof. Dr. Marian Wilhelm Theofil Smoluchowski von Smolan (1872–1917), at the time—professor of physics at the University of Lemberg, Austrian-Hungarian Empire (presently: Lviv, Ukraine).

We shall combine reports about the above contributions with the referee's reports on them by Albert Einstein.

4. Review of George Hartley Bryan: The law of degradation of energy as the fundamental principle of thermodynamics, *Beiblätter zu den Annalen der Physik*, v. 29, p. 237, 1905.

Albert Einstein: *"The author starts from the energy principle and the principle of the free energy decrease. The free energy (available energy) of a system is defined as the maximum mechanical work done by the system along with the changes of the system's state compatible with the external conditions. Following is the definition of the heat supplied to the system. Then the concept of thermal equilibrium, the second basic law, the concept of the absolute temperature are derived in an elegant way from the general principles, ultimately arriving at the equations of the thermodynamic equilibrium."*

Noteworthy, George Hartley Bryan introduces this same point himself, but in the following way:

---

18. The Law of Degradation of Energy as the Fundamental Principle of Thermodynamics

By G. H. Bryan in Bangor (North Wales)

(Boltzmann Commemorative volume: pp. 123–136)

1. In most textbooks, the study of thermodynamics is approached from a historical point of view being based on the discoveries of the Mayer-Joule principle of equivalence of heat and work and Carnot's principle as modified by Clausius and Kelvin. Very few writers have attempted to present thermodynamics as a purely deductive subject or to render it independent of preconceived notions concerning heat and temperature in the same way, that rational mechanics has been rendered independent of preconceived notions of mass and force. But the study of abstract dynamics has led to such valuable results in the interpretation of physical phenomena that it appears desirable that the fundamental principles of thermodynamics should be presented in an equally formal manner. As it is unnecessary to again traverse ground that has already been covered by writers on dynamics, this is best done by examining what modifications have to be made in the properties of an ideal dynamical System in order to obtain a thermodynamic system.

It has for some time past appeared to me that the principles of Conservation and Degradation of Energy afford the best starting points for a treatment such as is here proposed. In the present paper I propose to give an outline of the results at which I have arrived in working out this method, in the hope that other workers may be induced to turn their attention in the same direction more than they have done hitherto.

And this is nothing more and nothing less than—just the 'notorious' *energetics*—just the formulation Giuseppe Belluzzo and Albert Fliegner were behind in their works as well—just the one both Ludwig Boltzmann himself and Max Planck were so forcefully, emotionally and *successfully* [*sic*] fighting against.

5. Review of Nikolay Nikolayevich Schiller (Kharkov, Russian Empire): "Einige Bedenken betreffend die Theorie der Entropievermehrung durch Diffusion der Gase bei einander gleichen Anfangsspannungen der letzteren" ("some concerns regarding the theory of entropy increase due to the mutual diffusion of gases, where the initial pressures of the latter are equal") *Beiblätter zu den Annalen der Physik*, v. 29, p. 237–238, 1905.

As the above-refereed communication was expressing some doubts in the ways the theory by Ludwig Boltzmann were further developed, Albert Einstein had presented a very short and dry, but still a somewhat more detailed report on it (for the original work see pp. 350–366 in the commemorative volume):

First of all, it is shown that it is possible to bring a homogeneous gas to a n-times smaller volume without importing work and/or heat, but just by assuming that the walls permeable to a part of the mass of a gas do exist, while being impermeable for the remaining gas amount. The author means that such an assumption doesn't contain any logical contradiction.

Then, it is demonstrated that the expression for the entropy of a system composed of some spatially separated gas at the same temperature and pressure might be cast in the following form

$$S = \left(\sum m_i R_i \right) \lg v + f(\theta);$$

The entropy of the system after diffusion may be represented by the same formula. It is then concluded that the entropy ought to be the same before and after the diffusion took place.

At the same result the author arrives through the consideration that cannot be reproduced here. The latter consideration operates with an area separating a chemically homogeneous gas into two parts in such a way that the gas pressure in the resulting two parts, while the latter both in thermal and mechanical equilibrium in regard to each other, should nonetheless have different values. In such a case, it is (implicitly) believed that, while gas is passing through this area, the latter performs no work on the gas.

In fact, Prof. Schiller himself puts the main idea he would like to express in somewhat different and actually in the much clearer form:

"Finally, when considering any chemically homogeneous amount of gas held in equilibrium by the external forces, one must admit that in such a mass system in its equilibrium state the internal molecular motions would nonetheless continue to exist, and should in fact consist in that every two adjacent gas volume fragments ought to exchange their molecules with each other. Meanwhile, the latter motions are of such a kind that they must anyway be considered a diffusion process of the like masses of gas into each other. And if one would like to insist that entropy increases during the diffusion and by means of the latter, one should also have to recognize that the above-mentioned molecular motions must lead as a result to a continued entropy increase up to the infinity."

Noteworthy, the grounded doubts expressed by Prof. Schiller in his contribution to the Boltzmann's commemorative volume are connected with the novel (at the time being) idea as for the fundamental interrelationship among the Brownian motion, the entropy notion and the second basic law of thermodynamics, expressly presented first by Prof. Smoluchowski and then by Albert Einstein himself a couple of years after the commemorative publication we discuss. Prof. Schiller hadn't presented any detailed and formal inference on the theme, but solely his clear deliberations about the apparent logical gap between the idea of the Brownian motion conceptually underlying the second basic law and the law of the entropy increase.

Here comes but one very important note: During several last years of his life Prof. Schiller was very intensively working on the physical-mathematical foundations of the second basic law and could with his seminal results even forereach the famous publications by Constantin Carathéodory, but taking into account the fact that the relevant Schiller's work was published in Russian only—as well as—the then politically stormy times all over the world, it is throughout clear, why the important ideas and findings by Prof. Schiller remained longer time 'widely unknown'.

To sum up, in Chapter 4 of the present book we could see that the pertinent and purely formal physical–mathematical inference just depriving the famous Boltzmann–Planck formula, reading as usual, $S = k \ln (W)$ of its ostensibly 'tremendous fundamental significance' had in effect been presented by Dr. Georg(e) Augustus Linhart. Moreover, the Boltzmann's commemorative volume we are discussing right now does also contain the insightful contribution by Max Bernhard Weinstein, which has been refereed in the same Beiblätter volume 29 by Prof. Dr. Siegfried Valentiner. Here we are publishing the English translations of the latter both and discuss them in Chapter 3, Pages 329–338 as well.

Interestingly in this regard, into the same Beiblätter volume 29 its editor had also placed a detailed review of the book by James Swinburne [2.77]—just after the review of the contribution by Prof. Schiller by Albert Einstein.

Furthermore, just after this review of the Swinburne's publication a clear editorial comment comes about the heated, extended, truly international debate about the actual sense of the entropy notion in the leading British engineering journals.

There is also a book by Paul J. Nahin about the life and work of Oliver Heaviside (we have already referenced this book in the Note 18 to Chapter 1) describing this debate in detail, as well as an interesting Internet review of its truly show details: http://www.eoht.info/page/What+is+entropy+debate.

In effect, one might safely conclude that by the end of the XIX-th and the beginning of the XX-th centuries practically nobody could understand the actual sense of the entropy notion.

Remarkably, when reading the materials of that debate, one of the pithiest features appears to be the standpoint of Max Planck, who was continuing to forcefully insist on the Boltzmann's way of the probabilistic interpretation for the entropy notion, by fiercely rejecting all other viewpoints, which appeared to be more or less dissident in comparison to that of him.

In this connection it is definitely of interest to recall at this very point about Max Planck's own contribution to the Boltzmann's commemorative volume at hand (see pp. 113–122, "*17. Über die mechanische Bedeutung der Temperatur und der Entropie,*" "17. As concerns the mechanical sense of the temperature and the entropy").

Here we would like to present only several fragments of the above-mentioned Prof. Planck's contribution, which appears of crucial importance for our present discussion (*the English translation below is ours*):

Of fundamental importance for any mechanical theory of thermal processes ought to be questions about the mechanical sense of the temperature concept, which is closely connected with the poser about the mechanical meaning of the entropy notion, for the latter is indeed related to the former by the well-known thermodynamic equation TdS = dQ.

Therefore, after successfully answering one of the both questions we shall be immediately done with answering the other as well.

While in earlier times the most active interest was naturally dedicated to the temperature as the directly measurable quantity, and the entropy appeared to be a more complicated concept derived from the temperature considerations, today this interest ratio seems to become rather the opposite: Now it is especially the entropy that is to mechanically explain; and then the temperature might straightforwardly be defined thereby.

The reason for the above-mentioned change in the posers' prioritization lies in the following:

All the currently available experimental attempts to interpret thermodynamics on a purely mechanical basis, like, for example, those following from the theory of monocyclic systems developed by Helmholtz, could prove again and again (what is in effect evident already from the outset), that one can only arrive at some justified on all sides mechanical definition of the temperature, if you go back to the peculiarities of the "thermal equilibrium."* Still, the latter term in its fullest physical sense might be properly understood solely from the standpoint of the irreversibility, because the heat balance could only be defined as the **final state***, which all the irreversible processes are always striving for.*

Therefore, the question as for the temperature should at all costs lead to the poser about the nature of irreversibility, and the latter is in turn exclusively based upon the existence of the entropy. Furthermore, the latter variable constitutes primary, general, the most meaningful notion for all the possible kinds of states and changes in a state, whereas the temperature follows therefrom only under the special condition of thermal equilibrium, where the entropy reaches its maximum. According to the second law of thermodynamics a body has definite entropy in each state, but a temperature might only be defined for some particular state

---

*And the condition that T ought to be nothing more than just an 'integrating denominator' of the heat differential dQ, is well known to be not enough to fully define the temperature, but just renders indefinite the most important properties of the latter.

if the latter is stationary in some definite sense. Hence, you can always specify the entropy for a gas with quite arbitrarily prescribed velocity distribution, but to define the temperature is possible, if and only if the velocity distribution is exactly or nearly exactly coincident with that derived by Maxwell.

Clausius and Maxwell didn't seem to have yet attempted arriving at a general mechanical definition of entropy. To accomplish this very step was only reserved to L. Boltzmann who, starting from the kinetic theory of gases, could conclude that entropy is generally and clearly defined by the logarithm of the probability of the mechanical state. Then, the Boltzmann's mechanical definition of entropy could be complemented by another one, and then even by the second and the third ones, based again upon the basis of probability theory, as appeared in the works by J. W. Gibbs about statistical mechanics. The Gibbs' definitions might in so far be entitled to get a more general meaning, for they are from the outset based upon no special assumptions about the nature of the considered mechanical system; they might in principle be equally applicable to the systems with many, as well as with a few degrees of freedom, to those consisting of similar or dissimilar ingredients; each of the above-mentioned individual definitions of entropy corresponds to a pertinent special definition of temperature, according to the equation $dQ = TdS$, as we already mentioned above.

For the systems possessed of a very large number of degrees of freedom, the three different entropy definitions by Gibbs lead always to the same result, as he shows in general, so that, e.g., for a collection of a huge number of molecules such as that present in any warm body, only one Gibbs' definition of the entropy would survive, which is fully correspondent to that of the conventional thermodynamics.

To sum up, the ultimate solution to the problem concerning the most common and most rational definition of entropy might solely be delivered by directly calculating the entropy under the conditions defining some state to be markedly different from the most probable one, as has been highlighted in the introduction to the present communication. Still, to my mind, Gibbs gives no general directions for the latter cases, since he treats them in so far as if they ought to be composed of stationary states. And the Boltzmann's definition is instead sufficient without more ado, as is well known, for the states with quite arbitrary prescribed positions and velocities of the particles constituting the system under study.

Therefore, I hope to be able to conclude as a result of this investigation that the universality Gibbs assigns to his different definitions of entropy seems to be impressive at the first glance, but, without taking into account the nature of the systems under consideration, it actually turns out to restrict the physical meaning of those definitions. For all the reversible processes the three Gibbs' definitions ought to be throughout useful, likewise several other entropy definitions of more formal nature, already available to us.

Meanwhile, for the irreversible processes on the contrary, it is just the Boltzmann's definition of entropy that turns out to be the most powerful and relevant one under all those known until now—for, first of all, these are the irreversible processes that deliver the actual significance to the entropy notion, and secondly, these are the irreversible processes that provide us with the only key to a complete understanding of the thermal equilibrium.

Berlin, July 24, 1903

(Received July 26, 1903)

Well, then where is here the actual point at which the conventional thermodynamics is '*to die a miserable death*' in the above report? It is extremely difficult to localize in a clear-cut manner.

Indeed, everything in Planck's deliberations is absolutely correct: Sure, both Boltzmann and Gibbs had successfully started exerting their seminal research efforts, but could never bring the work to conclusion due to purely natural reasons, to our sincere regret. Indeed, J. W. Gibbs solely had time to perform his studies for the systems consisting of the statistically independent subsystems only, whereas the ingenious guess by Ludwig Boltzmann, the $S = k*\ln(W)$, could 'hit the bull's eye', indeed!

To our sincere regret, neither Boltzmann, nor Gibbs had any more time for the thorough and in-depth analysis of the actual interrelationship between the entropy and probability notions. The latter is in fact *not* the *ultimately basic fundamental link*—for there are definitely much more general principles underlying this basic interrelationship.

We have already discussed this important point in more detail in Chapter 4 of the present book, while analyzing the breakthrough achievement accomplished by Dr. G. A. Linhart.

Meanwhile, Prof. Smoluchowski in his own contribution to the commemorative volume at hand (see pp. 626–641, "*79. Über*

Unregelmäßigkeiten in der Verteilung von Gasmolekülen und deren Einfluß auf Entropie und Zustandsgleichung," "*79*. About the irregularities in the distribution of the gas molecules and their effects on the equation of state") is expressing his grounded and detailed ideas about the apparent insufficiency of the well-known van der Waals' equation of state.

To this end, in Chapter 2 of the present book we have already seen that Dr. Nils Engelbrektsson could in effect present the fully formal and consistent physical–mathematical inference of the very general equation of state, while his colleague Karl Alexius Franzén had managed to experimentally verify the latter.

Now, just to finalize the study of all those Albert Einstein's reviews in the proper way, we notice that the most of his 'thermodynamics' reviews of that period are dedicated to physical chemistry, which was but never the main area of Albert Einstein's research interests and activities as well.

Meanwhile, Albert Einstein is widely recognized as an outstanding specialist in the field of statistical mechanics. Indeed, he is one of the actual authors of the statistics of photons/phonons—that is the Bose–Einstein statistics [2.78] of the quantum–physical quasi-particles describing the radiation phenomena.

Howbeit, Albert Einstein's refereeing work reveals a truly international effort in the fields of thermodynamics and kinetic gas theory.

To this end, we continue below the listing of works revealed by Albert Einstein.

6. Review of Jakob Johann Weyrauch's work: "*Über die spezifischen Wärmen des überhitzten Wasserdampfes*" ("On the specific heats of superheated water vapor") *Beiblätter zu den Annalen der Physik*, v. 29, p. 240, 1905.

7. Review of Jacobus Henricus van't Hoff's work: "*Einfluss der Änderung der spezifischen Wärme auf die Umwandlungsarbeit*" ("The influence of the change in specific heat on the work of conversion") *Beiblätter zu den Annalen der Physik*, v. 29, p. 240–242, 1905.

8. Review of Arturo Giammarco's work: "*Un caso di corrispondenza in termodinamica*" ("About corresponding states in thermodynamics") *Beiblätter zu den Annalen der Physik*, v. 29, p. 246–247, 1905.

9. Review of K. F. Slotte's work. *"Über die Schmelzwärme"* ("About the Melting Heat") (*Öf. Finska Vet. Soc. Förh.* 47, S. 1–8, 1904) *Beiblätter zu den Annalen der Physik*, v. 29, pp. 623–624.

10. Review of K. F. Slotte's work. *"Folgerungen aus einer thermodynamischen Gleichung"* ("Consequences from a thermodynamic equation") (*Öf. Finska Vet. Soc. Förh.* 47, S. 1–3, 1904) *Beiblätter zu den Annalen der Physik*, v. 29, p. 629.

11. Review of M. E. Mathias' work. *"Die Konstante a der rechtwinkligen Diameter und die Gesetze der übereinstimmenden Zustände"* ("The constant **a** of the orthogonal diameter and the law of the corresponding states") (*J. de Phys.* (4) 4, S. 77–91. 1905) *Beiblätter zu den Annalen der Physik*, v. 29, pp. 634–635.

12. Review of Edgar Buckingham's work. *"Über gewisse Schwierigkeiten, welche man beim Studium der Thermodynamik begegnet"* ("about some particular difficulties, which we encounter during the thermodynamic studies") (*Phil. Mag.* (9) 50, S. 208–214. 1904) Beiblätter zu den *Annalen der Physik*, v. 29, pp. 635–636.

13. Review of Paul Langevin's communication: *"Über eine fundamentale Formel der Gastheorie"* ("Concerning one fundamental formula of the kinetic gas theory"). This work has been published in: *Comptes Rendus de l'Académie des Sciences* (France), v. 140, pp. 35–38, 1905. The reference to the referee's report sounds as follows: *Beiblätter zu den Annalen der Physik*, v. 29, pp. 640–641, 1905.

14. Review of M. Ponsot's work. *"Die Wärmezufuhr bei einer Gleichgewichtsänderung eines kapillaren Systems"* ("Heat supply during an equilibrium change in a capillary system") (C. R. 140, S. 1176–1179. 1905) *Beiblätter zu den Annalen der Physik*, v. 29, p. 952.

15. Review of Karl Bohlin's work. *"Über den Stoß als Fundament der Theorien des Gasdruckes und der Gravitation"* ("About the Collisions as a Foundation for the Theories of Gas Pressure and Gravitation") (*Arkiv för matematik, astronomi och fysik* 1, S. 529–540. 1904) *Beiblätter zu den Annalen der Physik*, v. 29, pp. 952–953.

16. Review of Jakob Johann Weyrauch's book: "Grundriß der Wärmetheorie: Mit zahlreichen Beispielen und Anwendungen" ("An Outline of Heat Theory: With a Big Number of Examples and Applications") *Beiblätter zu den Annalen der Physik*, v. 29, pp. 1152–1153, 1905.

Several of the refereed works mentioned above are definitely to be estimated as very interesting contributions to the field of thermodynamics (see [2.79–2.82], as well as [2.83] and the references therein, e.g., especially [2.84], where Max Planck estimated his own actual contribution to thermodynamics).

In total, we have to be grateful to Albert Einstein for his depiction of the publication activity in the field. At the time of this activity A. Einstein was still in Bern, Switzerland.

His story is meanwhile very well known, so that we shall not go in for its details here.

Most probably, for Albert Einstein that was a preliminary work to start his own contributions to the field.

Indeed, the literature search leads us to the Reference [2.85], which is a report by Albert Einstein to the 1911 Solvay Conference entitled: *La théorie du rayonnement et les quanta: rapports et discussions de la réunion tenue à Bruxelles, du 30 octobre au 3 novembre 1911, sous les auspices de M. E. Solvay* (The theory of radiation and quanta: Reports and discussions of the meeting held in Brussels from 30 October to 3 November 1911, under the auspices of Mr. Ernest Gaston Joseph Solvay (1838–1922).

At this meeting Albert Einstein was already representing the Austrian-Hungarian Empire as a full university professor of theoretical physics at the German University of Prague).

Albert Einstein's report entitled: *L'état actuel du problème des chaleurs spécifiques* (The current state of the problem of specific heats) presents a story about how to infer the proper general formula for the specific heat (heat capacity). This story is dealing mainly with the heat radiation theory by Max Planck, as presented in his well-known book [2.86].

The system considered in the both works [2.85, 2.86] just mentioned is clearly confined to the *ideal gas at equilibrium*.

The result is therefore ought to be not a general, but a particular one, because the system considered ought to be just a clear-cut particular case of the actual systems in their various aggregate states.

It has been a perfect extension of the quantum theory just emerging at that time, but not an ultimately final contribution to the foundation of thermodynamics. What is then wrong with thermodynamics in this case?

As we have already discussed, the main points here ought to be (a) the physical insufficiency of the ideal gas model and (b) the *fuzziness* of the picture operating with '*huge number*' of the subsystems under study, that is, particles/complexions/etc.

This is definitely ***not*** the basic failure of the atomistic picture of the matter!

The reason is also throughout fundamental, but it has faintly anything to do with the physics or other natural science branch: It is just *the bent and genius of our tongue* [18–21].

Interestingly, right at the time of the revolution in physics we are discussing right now, a renowned Dutch physicist Prof. Dr. Johannes Diderik van der Waals, Jr. (1873–1971) has published a seminal communication [2.87].

Noteworthy, he was a son of the Nobel Prize winner 1910, Prof. Dr. Johannes Diderik van der Waals (1837–1923).

Meanwhile, the latter work by this outstanding researcher is still remaining in shadow, not to speak of its apparent falling on deaf ears, as revealed by the work [2.88].

This portrayal of Johannes Diderik van der Waals, Jr., has been taken from
https://nl.wikipedia.org/wiki/Johannes_Diderik_van_der_Waals_jr.

Howbeit, the paper [2.87] is very short but nonetheless contains a clear cue to make use of the Bayesian approach in working with statistical toolboxes. In fact, such a cue was considered a malicious heresy in 1911 and even later on, during many years, as a result of the revolutionary reconstruction of physics, but nowadays Bayesian approach is very well known and widely recognized as a reliable tool to solve fuzzy problems (see, e.g., [2.89–2.92] and the references therein).

Nonetheless, the 'revolutionary assertions' to the contrary, the idea of successfully using the Bayesian approach in the statistical physics was not only definitely up in the air but could also clearly prove its throughout fitness and usefulness, as we could see in Chapter 4 of the monograph at hand.

Of our interest at this very point ought to be Albert Einstein's attention to the field of thermodynamics. There is a recent publication dealing with this interesting topic, where the actual interrelationship between the thermodynamics and "the rest of physics" is clearly depicted [2.83] as well. Noteworthy, it is this publication that offers the best opportunity to clarify the reasons for our using the term *apologetic* everywhere in the monograph at hand.

The book [2.83] starts its consideration of the period between the end of the XIX-th and the beginning of the XX-th centuries. From the beginning on, after some rather detailed discussion of the deficiencies exhibited by Newton's mechanics, a clear border is drawn between the 'classical physics' (the physics before the advent of the quantum theory and the special theory of relativity) and the 'modern physics'. The immediate conclusion without any in-depth analysis: 'Okay, everything before the modern physics was erroneous, or at least largely insufficient'. [*sic*] Remarkably, what Max Planck and Ludwig Boltzmann had once foretold to be the logical consequences of following the way shown by the energetics could happen indeed, isn't it?

Further, the book mentions the two most important experimental achievements having come up at that time (the discovery of electron and of radioactivity). Meanwhile, the theoretical achievements at that time could boil down to the 'hypothesis of quantizing the radiant energy by M. Planck leading to the advent of quantum theory' and to the 'special relativity theory by A. Einstein'. Then, the author duly and correctly sums up as follows: It is just this novel entirety that *'has absolutely changed the way physicists (and many other professionals) regard the world'.* Here the warning expressed by Max Bernhard Weinstein (see Chapter 3, from the page 348 and on) comes to mind immediately.

Of extreme interest for us ought to be the following analysis [2.83] of how it could come to the novel theoretical ideas just mentioned and what/who was the *'goody'* (or the *'goody two-shoes'*?) for thermodynamics.

Indeed, from this book we learn: It was the thermodynamics (in its then state, apparently) that *'inspired Einstein to develop the theory of relativity. In fact, many brilliant scientists who played a role in the development of these theories were very familiar with developments in thermodynamics'.* The author [2.83] declares the aim of his monograph to prove that such outstanding physicists as Max Planck, Albert Einstein, and Erwin Schrödinger *'were not only strong supporters of the concepts developed by Boltzmann, but also were experts in thermodynamic theory'.* Meanwhile, it is well known that Max Planck did never viewed himself a *'specialist in thermodynamics'*, although he had spent a truly significant time in performing theoretical research in this field [2.84].

Prof. Dr. Erwin Schrödinger (1887–1961) has clearly expressed his standpoint as concerns statistical thermodynamics [2.93]:

> There is, essentially, only one problem in statistical thermodynamics: the distribution of a given amount of energy over N identical systems.

Sure, this problem has been successfully solved by a number of outstanding colleagues, including Albert Einstein. Indeed, the Bose–Einstein and Fermi–Dirac distribution functions are very well known. They represent the reliable tools of theoretical research of such important particular cases as Fermion and Boson quasiparticles. But there is still no truly general solution to the problem, for *'working with N identical systems'* causes two immediate objections:

(A) What about the non-identical systems?

(B) Which N is exactly *'a huge number'*? Or, in other words: How many systems should we take into account to reach a generally acceptable result for our theoretic efforts? Here is just the point where we arrive at the sorites paradox, the paradox of the heap.

Sure, the above point B) is but truly non-trivial, for it poses a conceptual blind corner to our train of thoughts. To the best of our knowledge, there was the only colleague, who could indeed suggest some plausible way to find the general solution: Dr. G. A. Linhart (see Chapter 4). The author of the monograph at hand had recently dared to express his humble suggestions as for the possible ways of cultivating Dr. Linhart's suggestions [2.94, 2.95].

A very good thorough and detailed analysis of the actual situation in physics and chemistry, including the pertinent human interrelationships at the time period we are just discussing, might be found in the following recent book [2.96]. In trying to sum up all the well-known facts into account, the following poser arises:

> *Whose joint evil effort was building up the so-called 'equilibrium thermodynamics'?* Still, here we start immediately hearing the voices suggesting to stop playing with the entire story: *If it ain't broke—don't fix it!*

To respond to such a poser, along with the fully grounded criticism, we would have to consider in detail the entire true story about the statistical interpretation of thermodynamics. This is but clearly beyond the volume of the monograph at hand. So, here we would greatly appreciate arriving at an objectively correct estimate of Albert Einstein's contribution to thermodynamics. We could already summarize the actual contributions claimed to belong to him.

Specifically, these ought to be a) solving one of the problems of the quantum statistics and b) introducing the relativity theory. Whereas we have already discussed the latter point a), just a short note ought to be due in connection with the latter point b).

Indeed, in considering the *Great Physical Revolution* of the XIX-th and XX-th centuries, we seem to deal with the brightest case demonstrating a sheer difference between the '*mathematico-physicists*' and '*physico-mathematicians*', as originally suggested by Lieut.-Col. Richard de Villamil (1850–1936), an outstanding British military engineer. He was lively interested, *inter alia*, in the general physics and the legacy of Hon. Isaac Newton. This is clearly reflected in all of his publications [2.97–2.103].

By the way, his monograph *"Rational Mechanics"* [2.102] represents a detailed and logically consistent introduction into the field of energetics.

Moreover, the *Great Physical Revolution* alleges the brightest example of how to effectively *'get rid of the actual ordinary time'* in the course of skillful mathematical prestidigitations, with the aim to reduce the actual time phenomenon to a solely theoretical parameter.

The latter conclusion ought to be in full accordance with the forceful warnings issued by late Prof. M. B. Weinstein (see Chapter 3), Prof. Dr. Daniel Berthelot (1865–1927), an outstanding French physicist-chemist, a son of the well-known French chemist Prof. Dr. Marcelin Berthelot. Prof. D. Berthelot was keenly interested in Einstein's work and had published his detailed thoughts on the theme [2.104–2.106]. We get more information about Prof. D. Berthelot, as well as a detailed analysis of his contributions from the recent works [2.107–2.109].

The last but not the least ought to be the opinion of another outstanding French theoretical physicist, Prof. Moïse Emmanuel Carvallo [2.110], who was, *inter alia*, also thoughtfully working on the basics of thermodynamics, as we have already seen in the Note 21 to Chapter 1. To this end, he was tightly involved into the pedagogic activity, and published a number of useful handbooks, e.g., those dealing with the general mechanics and numerical maths for theoretical physics [2.111, 2.112].

We shall revert to Albert Einstein's activity and its professional estimates in the next Note to this chapter. Noteworthy, right here is but a yet persisting and truly worldwide bias in estimating the role of Albert Einstein in the Physical Revolution of the XIX-th–XX-th centuries. In reading the books like, e.g., [2.113–2.116] we do find Albert Einstein starting out to be *'a kind of animalcule chased by a lot of cruel hunters, but then suddenly giving a powerful rebuff to his pursuers and gaining his final victory, which is valid both for his Centennial and for all the Centennials to come'*. Moreover, the 'dissident hunters/pursuers', who were understanding their actual task in trying *'to stop the high-handed adventurer'*, are all in all thrown into the historical trash dump with the label of *'Alterscience'* (a funny neologism composed of two keywords: *'alternative'* and *'science'*). Noteworthy, the actual meaning of this unique neologism sounds very interesting: '*The Alterscience is a movement challenging the results of the official science and claiming to present an alternative pathways for the latter, either for personal reasons or due to ideology.*' Thus, it looks like a very convenient social standpoint: Indeed, considerable ought to be solely the '*official science*', otherwise we deal with some kind of rebellious heresy deserving no serious analysis.

It is this unique one-sidedness, combined with the clear pretension to perform a deep historical–philosophical analysis, which urges us to condemn this kind of literature as '*apologetic*' one. The analysis performed by such authors and their numerous world-wide followers brings elusive results only—for two clear reasons: the drastic insufficiency of the materials used, as well as the stubbornly biased way of handling the latter.

The only important conclusion to draw at last ought to sound as follows: *Stop Rendering People to Astral Deities*! Both revolutionary physicists and all of their opponents were highly esteemed professionals and—the last but not the least—all of them were human beings, who are all capable both to stage breakthroughs and to commit errors.

In reverting to our main topic, we should point out that Albert Einstein's analysis of the then state-of-art works in thermodynamics was, is and remains to be extremely useful from the standpoint of thermodynamics. With the above entirety in mind, of separate interest for us ought to be the reaction by Albert Einstein to the works by the Scandinavian colleagues (K. Bohlin, K. F. Slotte) and an American colleague (E. Buckingham).

To this end, we might definitely notice that the different thermodynamics did exist well apart of the main protagonists of the book at hand (see Chapters 1–4).

Hence it is of interest to indicate some important conceptual interconnections otherwise remaining completely hidden.

2.3.3.1 Scandinavian publications on thermodynamics and statistical physics at the beginning of the XX-th century, as reviewed by Albert Einstein

First of all, many sincere thanks are due to Albert Einstein for attracting our attention to the works by Karl Petrus Teodor Bohlin (1860–1939), a Swedish experimental and theoretical astronomer (primarily a theoretician in celestial mechanics).

Another colleague fetched by Albert Einstein Prof. Dr. Karl Fredrik Slotte (1848–1914), a molecular physicist of Swedish origin from Finland (professor at the Technical University of Helsinki/ Helsingfors).

While concentrated on finding the proper mathematical tools for the research in celestial mechanics (see, e.g., his very thoughtful and insightful work [2.117], as well as the reports of his life and accomplishments in [2.118, 2.119]), Karl Bohlin had dared to criticize the way R. Clausius, L. Boltzmann and their followers treat the interactions in their many-body atomistic picture.

Indeed, it is well known that these colleagues tended to systematically boiling down diverse possibilities of inter-particle interactions to pairwise rigid collisions between the virtually neighbouring particles.

Meanwhile, from Chapter 1, we know that Peter Boas Freuchen had pointed out the latter trend as well and underlined the physical insufficiency of such a conceptual approach.

This photograph of K. P. T. Bohlin has been taken from
https://sok.riksarkivet.se/Sbl/Presentation.aspx?id=17864

Remarkably, Albert Einstein could bring to our attention Karl Bohlin's work [2.120]. This communication is of special interest for our recent discussion, for in that work he had reviewed in detail the available results of the then many-body theories. As a result, he had suggested concentrating the relevant theoretical research on trying to plausibly describe the actual complicated inter-particle and particles-vessel interaction modalities. Remarkably, Albert Einstein's reaction to Karl Bohlin's seminal suggestion is as follows (*below comes our English translation*):

> 'The author starts from the observation that both in the kinetic theory of gases and in the dynamic theory of gravity solely repulsive forces among the particles are introduced to explain the inter-particle clashes. With this in mind the author examines the possibility to avoid any introduction of repulsive forces. He tries to attribute the impact of collisions solely to the attractive forces among the colliding corpuscles constituting the body under study. He argues that it is possible to trace any attractive force to the collision effects among relatively infinitesimal corpuscles. Consequently, vice versa, any actual collision may be (kinetically) explained by the attraction among relatively infinitesimal corpuscles. Thus, he introduces corpuscles of infinitely many scales to represent the elementary properties of matter. A. E.'

We would like to confide the final estimate of the above citation to the readership. Still, to our mind, we could hardly attribute the above to Albert Einstein's weakness in French (K. Bohlin's paper is published in French). Similarly, it is tremendously difficult to lead the above back to Albert Einstein's weakness in the general physics. Meanwhile, Karl Bohlin's actual suggestion ought to consist in properly taking into account *both* repulsions *and* attractions, when considering the inter-corpuscular coupling. It is solely this way that we arrive at the physically correct picture of the inter-atomic/inter-molecular potential.

However, it is throughout clear that following the latter train of thoughts is connected with enormous mathematical complications bedeviling any consequent analytical mathematical effort (Peter Boas Freuchen has also recognized this point, see Chapter 1). Howbeit, it is clear nowadays that to correctly write down and solve such many-body equations one has to employ computer simulations (molecular dynamics, Monte Carlo simulations). The purely mathematical problem is thus solved. To get the physically sensible results from the computer simulations involved, it is necessary to physically correctly interpret their results.

Then, how to ensure the correctness of the physical interpretation needed? A good poser! Answering it is clearly beyond the scope of the present volume.

Of the two works by another Scandinavian author publishing a thermodynamic work, Prof. Slotte [2.121, 2.122], as reviewed by Albert Einstein, the first one is of purely applied physical–chemical significance: It is devoted to comparison of experimental data with the relevant theoretically derived functional relationship between the melting heat and the absolute temperature for different materials. Albert Einstein is pointing out that *a remarkable accordance of the theory and experiment in this case ought to be due to Prof. Slotte's theoretical assumption about the rectilinear harmonic oscillations of the atoms in the simple solid bodies.*

Most probably, it is the latter statement that has driven another outstanding revolutionary physicist, Prof. Dr. Max Born (1882–1970) to mention Prof. Slotte's name (meanwhile, without any detailed reference!) in his very well-known book [2.123].

Remarkably, that was just Prof. Born, who suggested the probabilistic interpretation of Schrödinger's wave function and therewith ended the determinism in physics, but provided instead a firm ground for the quantum theory.

This photograph of K. F. Slotte has been taken from
https://www.myheritage.de/research/collection1/myheritagestammbaume?te
mId=3141065211500038&action=showRecord

Meanwhile, the second paper by Prof. Slotte is clearly of general significance, for it is systematically treating the temperature dependence modalities of the state variables.

We would greatly appreciate presenting here the English translation of the latter paper originally published in German: **The reasoning starting from a thermodynamic formula.**

It is well known that for reversible changes the following equation ought to apply:

$$\left(\frac{dQ}{dv}\right)_T = T \cdot \left(\frac{dp}{dT}\right)_v. \tag{1}$$

In this equation, the left-hand term is usually referred to as expansion heat. We normally express the latter in mechanical units and relate it to a quantity of heat at some particular volume. This amount of heat is consumed during an isothermal expansion of the unit weight of a body at some particular absolute temperature T to overcome the internal and external forces. In other words, this is just the amount of heat, which might be produced along with an equally large isothermal compression.

Then, the differential coefficient in the right-hand term of Eq. 1 refers to the change in the external pressure, p, when the body is being heated or cooled at constant volume v.

Now, if we assume that Eq. 1 is valid at the zero point of the absolute temperature scale as well, the conclusions resulting therefrom seem to be truly remarkable, and we would like to discuss them here.

With this in mind, let us set T = 0 in this equation. Then, if the $\left(\frac{dp}{dT}\right)_v$ *is not infinite, the* $\left(\frac{dQ}{dv}\right)_T$ *must be equal to zero as well. Hence, in such a case at the zero absolute temperature the isothermal volume changes ought to proceed without any heat consumption prior to them, which is inconsistent with all the known laws of Nature.*

Contrariwise, lest we assume that $\left(\frac{dp}{dT}\right)_v$ *be infinite at the zero absolute temperature,* $\left(\frac{dQ}{dv}\right)_T$ *should not to be equal to zero at this point. In such a case, any slightest increase in temperature at a constant volume should lead to such a pressure increase that would clearly be out of proportion with respect to the increase in temperature.*

As a result, the curve, for which pressure would be ordinate and the absolute temperature—the abscissa, would describe a change of state at some constant volume quite correctly.

Such a curve must in principle be tangent to the ordinate at the zero absolute temperature, but at the same time it should be more and more divergent from the ordinate with increasing temperature in this case.

Further, if we would now consider homogeneous bodies only, we would arrive at the following relationship:

$$\left(\frac{dp}{dT}\right)_v = -\frac{\left(\frac{dv}{dT}\right)_p}{\left(\frac{dv}{dp}\right)_T}. \tag{2}$$

Therefore, if at $T = 0$ the following is true for the differential coefficient: $\left(\frac{dp}{dT}\right)_v = \infty$, then the right-hand side of Eq. 2 must be infinite as well. The latter would be true, when $\left(\frac{dv}{dT}\right)_p = \infty$, while $\left(\frac{dv}{dp}\right)_T$ remains finite. Otherwise, this would be true as well, when $\left(\frac{dv}{dp}\right)_T = 0$, while $\left(\frac{dv}{dT}\right)_p$ is finite.

As for the first option, namely that for $T = 0$ $\left(\frac{dv}{dT}\right)_p = \infty$, then in this case any kind of the slightest warming at the zero absolute temperature would lead to a disproportionately large increase in the body's volume under any constant pressure. The latter effect ought to be analogous to the rapid spreading, which occurs during melting some solid bodies, like sulfur and phosphorus.

To sum up, if the absolute zero temperature could be reachable at some constant pressure, the homogeneous system would undergo a rapid volume reduction, as if rapidly entering into its much denser aggregate state.

To this end, at the zero absolute temperature any curve describing a change in state at constant pressure in the state coordinate system with T as abscissa and v as the ordinate would be tangent to the v-axis. Meanwhile, as experimental observations teach us, the higher the T value, the slower would be the volume increase during such state changes, as a rule.

If, contrariwise, $\left(\dfrac{dv}{dp}\right)_T = 0$ *at T = 0, then no isothermal volume change would be possible at this point, that is, every homogeneous body ought to become incompressible at the zero absolute temperature.*

Noteworthy, if both $\left(\dfrac{dv}{dT}\right)_p = \infty$ *and* $\left(\dfrac{dv}{dp}\right)_T = 0$, *then the rapid volume reduction due to cooling a body up to the absolute zero temperature, as described by the former equation, ought to entail the body's incompressibility, in accordance with the latter equation.*

Finally, the conclusions drawn using Eq. 1 might in principle be drawn from the corresponding equation as well:

$$\left(\frac{dQ}{dv}\right)_T = T \cdot \left(\frac{dv}{dT}\right)_p .$$

The above paper is an elegant proof that actually reaching the zero absolute temperature is impossible. Hence, if we might infer some effect formally mathematically, there is no more space for hypotheses about actual arriving at some basic natural law.

Indeed, the basic equations analyzed by Prof. Slotte are in effect dealing with entropy, for $\left(\dfrac{dQ}{dv}\right)_T = T \cdot \left(\dfrac{dv}{dT}\right)_p \Leftrightarrow \left(\dfrac{dS}{dv}\right)_T = \left(\dfrac{dv}{dT}\right)_p$;

$\left(\dfrac{dQ}{dv}\right)_T = T \cdot \left(\dfrac{dp}{dT}\right)_v \Leftrightarrow \left(\dfrac{dS}{dv}\right)_T = \left(\dfrac{dp}{dT}\right)_v$; $dS = \dfrac{dQ}{T}$. Moreover, Prof.

Slotte points out just at the beginning of his above communication: "*This amount of heat is consumed during an isothermal expansion of the unit weight of a body at some particular absolute temperature T to overcome the internal and external forces.*" Hence the latter statement does clearly relate the entropy notion with 'overcoming internal and external hindrances'. This is in full accordance with the actual physical–chemical–...etc. sense of the entropy notion, as suggested by Prof. A. F. Horstmann, what we could already see in this chapter. The result presented above is killing the notorious third basic law of thermodynamics, but in a different way, as compared to the result by Dr. G. A. Linhart (see Chapter 4), for Prof. Slotte is not going over to the 'kinetic theory', 'statistics', etc., but still remaining within the frame of the conventional thermodynamics.

Noteworthy is Albert Einstein reaction to the result analyzed above, so we publish below the English translation of his referee report:

"Staring from the well-known equation $\left(\dfrac{dQ}{dv}\right)_T = T \cdot \left(\dfrac{dp}{dT}\right)_v$, *the author draws a number of conclusions about the behavior of a system near the absolute temperature zero. The latter conclusion has been drawn under the unmotivated premise that* $\left(\dfrac{dQ}{dv}\right)_T$ *remains finite at infinitely small temperature values."*

Once again, we would like to confide the final estimate of the both citations above to the interested readership. Howbeit, it was just Albert Einstein's stimulus that could assist the author of the monograph at hand in digging into the wealth of professional publications appeared at that time in the Scandinavian area. Indeed, just at that same time there was one more colleague in Sweden, who had paid his contribution to the field of our interest. Most probably, he is still remaining widely unknown, because his important communications on the theme were published partly in Swedish and partly in German [2.124–2.126]. The colleague we would like to introduce here is Prof. Dr. Henrik Petrini (1863–1957), a renowned Swedish mathematician, physicist, and writer.

This photograph of Henrik Petrini has been taken from
https://sv.wikipedia.org/wiki/Henrik_Petrini

The general information about Prof. Petrini and his work might be fetched in the review book [2.118]. Remarkably, he was well known for his critical statements concerning not purely scientific issues—he was also publishing pamphlets about the Swedish educational system. After successfully defending his PhD and starting his career as an associate professor in mechanics at the Uppsala University, he had decided to continue his pedagogical activity in the Swedish secondary schools throughout the country. Here we would like to present the original Swedish version as well as the authorized English translation of the introduction to his thermodynamics paper [2.125].

Inledning

I de vanliga framställningarna av den mekaniska värmeteorin förekomma vissa oklarheten, som göra, dels att de olika begreppen ej införas i rätt ordning, dels att åtskilliga af kropparnas grundegenskaper ej med tillräcklig skärpa framhållas, dels slutligen att vissa lärosatser, exempelvis den andra huvudsatsen, ej bliva fullt begreppsmässigt bevisade, hvadan de nödvändiga förutsättningarna för sakens giltighet ej alltid klar fattas. Den, som synes hava gått längst i försöket att mer begreppsmässigt än empiriskt framställa principerna för den mekaniska värmeteorin, är Poincaré i sin "Thermodynamique" men han står dock bland annat kvar på den ståndpunkten, att begreppet temperatur behandlas före begreppet värme. I fråga om dessa tvänne begrepp inträffar det egendomliga förhållande, att den dunkla språkkänslan har varit en säkrare ledare än det abstrakta tänkandet, i det ordet temperatur, som lämpligen kan översättas med (värme-)utjämning, synes fordra en uppfattning af temperaturen hittills tämligen oklart fattade begrepp. För att ej för mycket betunga framställningen antagas de begrepp, som förekomma i den rena mekaniken vara bekanta, ehuru medgivas måste, att även de kunde behöva klareras. En ringa antydan därom i fråga om begreppet tryck kommer att lämnas i sammanhang med frågan om den adiabatiska sammantryckningen (sida 12).

Introduction

In the usual representations of the mechanical theory of heat there is a definite uncertainty. The latter arises in part because some of the relevant concepts are not introduced in the correct way and in part owing to our reluctance in clearly bringing into prominence several properties of the physical bodies. Furthermore, the inconsistency mentioned ought to result from the fact that some laws, such as the second law of thermodynamics, have not yet become fully conceptually shaped, for the necessary conditions for their actual validity are not always clarified.

To this end, Poincaré in his "Thermodynamique" seems to have succeeded in attempting to build up rather conceptual than empirical foundations for the mechanical theory of heat. Inter alia, he is meanwhile hanging lantern on the idea that the concept of temperature ought to be treated prior to the concept of heat. Apparently, in the case of these both concepts a rather peculiar state of mind is dominating, for in fact we tend to deem right that the obscure language feeling be a safer leader than the abstract thinking. Probably, it is for this reason that we prefer considering the temperature notion in terms of the (heating) balance, in rendering temperature this way a rather unclear concept so far.

For not to discommode ourselves, we tend to take over the familiar concepts from the pure mechanics, though even the latter ones might definitely have to be clarified. A slight hint to how one might fulfill the latter task will be provided here for the concept of pressure in connection with the issue of the adiabatic compression (page 12).

First of all, we have to point out that the above citation is clearly summarizing Nils Engelbrektsson's standpoint. Noteworthy, the years of Prof. Petrini's publications are just the years of the secondary school graduation by Nils. Nils, being just a graduate interested in scientific research, could definitely fetch the works by Prof. Petrini and get his inspiration from them. Furthermore, after Dr. Petrini's PhD thesis approval and the then internship at the Uppsala University (1892–1901) he was acting as a teacher of mathematics and physics at the gymnasium in Växjö (1901–1914). Remarkably, this is just the time period, when Karl Franzén could be his pupil and get his inspiration from the teacher. Most probably, it is the set of Prof. Petrini's works that could help establishing a collegial contact between Nils Engelbrektsson and Karl Franzén.

Still, it is important to note that the different thermodynamics was not only developing in the Scandinavian area: In fact it was vivid virtually everywhere all around the World.

To this end, Albert Einstein's refereeing activity could also help us revealing a very active at his time, but nonetheless 'widely unknown', American researcher in the field—namely, Dr. Edgar Buckingham.

2.3.3.2 Forgotten American physicists/thermodynamicists at the beginning of XX-th century

The personality of Edgar Buckingham (1867–1940) is of tremendous interest irrespective of Albert Einstein's attention to his works.

This photograph of Edgar Buckingham has been taken from
https://en.wikipedia.org/wiki/Edgar_Buckingham

Unlike our protagonist Dr. G. A. Linhart (see Chapter 4), Edgar Buckingham was a descendant of a Native American family; he was born in Philadelphia, Pennsylvania, USA, and finally graduated from Harvard as a physicist. Then he had acquired additional graduate in Europe, namely chemistry, in Strasbourg, France and in Leipzig, Germany. In 1893 he had defended his PhD at the University of Leipzig under the mentorship of Wilhelm Ostwald. Most probably, it is the latter part of his educational endeavor that had stimulated his interest in thermodynamics. Being back home in the same year, Dr. Buckingham started his pedagogical activity at the private college in

Bryn Mawr, Pennsylvania by getting the position of reader in physics and physical chemistry. In parallel to this he is actively working on thermodynamics foundations, having published a monograph on this theme [2.127].

Nonetheless, after some six years of his throughout successful academic service he had suddenly left the College and was then working at a mining camp in the Arizona State for some two years (1899–1901). His second (and, actually, the last) attempt to enter the academic society was his serving as an instructor in physics at the University of Wisconsin. The latter activity lasted about one year (during 1901). Thereafter Dr. Buckingham was successfully working at the USDA Bureau of Soil (1902–1906) [2.128]—and finally from the year 1907 till his retirement in 1937—in the National Bureau of Standards (NBS). Noteworthy, at the NBS he was highly estimated as a highest-level specialist in the field of the engineering thermodynamics (see, e.g., his NBS report [2.129]): Even after his retirement due to the mandatory age, he was duly consulting the NBS in different research projects until his entering the inevitable last business trip in 1940. His fields of expertise included soil physics, gas properties, acoustics, fluid mechanics, and blackbody radiation. He is also the originator of the well-known Buckingham π theorem in the field of dimensional analysis. The only conclusion coming to mind in connection with all the latter 'info at a glance' is that thermodynamics had definitely lost a skillful theoretician.

Howbeit, the problem of our immediate interest here ought to be trying to answer the difficult poser of why Dr. Buckingham has anyway decided to quit the academic circles. At the very least, let us try to answer the poser—What were the most probable reasons for his decision in question?

Here we start with the reviews of his work appeared in the Beiblätter, vol. 29, 1905, where his work was refereed by two German physicists: by Albert Einstein, as we already know, and also by Prof. Dr. Siegfried Valentiner (1876–1971).

Interestingly, Edgar Buckingham was clearly supporting the critical address by Prof. McOrr as concerns Max Planck's thermodynamical activities at the time [2.70–2.72]. Above in this Note, we have already cited the very response by Max Planck to Prof. McOrr's criticism. Interestingly, after this response, the reaction by the latter colleague was still much softer than from the beginning on [2.71, 2.72]. Nonetheless, Dr. Buckingham was still remaining somewhat more restrictively critical in his conclusions [2.130].

Albert Einstein was monitoring this whole story very attentively, indeed. Here are his reactions (we place the English translations):

> *Beiblätter, vol. 29, page 237, Albert Einstein's reaction to Prof. McOrr's criticism:*
>
> *26. The author shows that Planck uses the terms "reversible" and "irreversible" in a somewhat different sense than he defines it in his "Lectures on Thermodynamics." Further, he introduces a series of objections expressible in regard to various representations of the fundamentals of thermodynamics; Among these objections, the one by Bertrand is particularly noteworthy, namely, that pressure, temperature, and entropy are defined only in the case where at least a small portion of the system in question can be regarded as being in equilibrium; A similar objection is raised in regard to the way the heat might be supplied.*

Our immediate comment: The '*objections by Bertrand*' mentioned above are related to the work of Prof. Dr. Joseph Bertrand (1822–1900), an outstanding French mathematician and pedagogue. Among other mathematical topics, Prof. Bertrand was dealing with both the probability theory [2.131] and the theory of thermodynamics [2.132]. Of interest for our present discussion ought to be the standpoint of French mathematicians of that time, as concerns the usage of probabilistic-theoretical inferences in natural sciences, where there is an *intrinsically menacing danger* (*in effect, we do not consider the latter choice of word an exaggeration!*) of conceptually overestimating the huge power of the pertinent mathematical toolbox.

An epigraph chosen by Prof. Bertrand for his book on the probability theory [2.131] is the following relevant Latin citation from Daniel Bernoulli (1700–1782), the worldly famous Swiss mathematician and physicist:

> *Facile videbis hunc calculum esse saepe non minus nodosum quam jucundum.* (*It is easy to see that this calculation is often not less knotty than jesting.*)

The latter phrase is just reflecting the actual train of thoughts represented by such famous colleagues of Prof. Bertrand as Henri Poincaré (in Sweden, as we have just seen, Prof. H. Petrini had accepted Poincaré's standpoint and tried to pursue the work in this direction), as well as—the last, but not the least—Prof. Dr. Antoine-Augustin Cournot (1801–1877), a fine French mathematician, economic theorist, and philosopher. The readership interested in the details of the ongoing 'fierce feud between the philosophy and theoretical physics' might fetch much more details about this very interesting and important topic in the following literature sources in different languages (see Refs. [2.133–2.140] and the references therein). The feud just mentioned was—and in fact is—still developing around the notions of probability and, consequently, of entropy in the tightest connection with the space, time, and energy notions.

Bearing the above entirety in mind, we would now like to revert to Albert Einstein's thermodynamic studies.

Beiblätter, vol. 29, page 635, Albert Einstein's reaction resulted from Max Planck's response to Prof. McOrr's criticism:

78. In response to Mr. Orr's criticism (Beiblätter, v. 29, p. 237) concerning the author's treatment of the thermodynamics foundations, the author declares that he used the terms "reversible" and "irreversible" in the same sense as Clausius. The author points out the fact that it is impossible to speak of temperature and density of the particles of a tumultuously moving gas, and thus of its entropy, if we are not taking the kinetic theory of the gas into consideration. Finally, the author finds that, in principle, the inference proposed by Mr. Orr is closely resembling that given by Lord Kelvin, but unlike the latter it contains a circular reasoning.

Our immediate comment: *No words, a truly perfect chess move. The detractor's apparent plagiarism and non-professional reasoning do prove doubters wrong. Howbeit, the main point to stress ought to be the fundamental indeterminism of the universe as a whole.*

Beiblätter, vol. 29, pages 635-636, Albert Einstein's reaction to Edgar Buckingham's criticism [2.130]:

79. The author follows the work by Mr. Orr (Beiblätter, v. 29, p. 237) and agrees with the results of the critical considerations contained in that work. He also expresses the conviction that it is impossible to derive the Clausius inequality $\int \frac{dQ}{T} < 0$ without further assumptions from the second basic law, as formulated by Lord Kelvin. The work contains some critical remarks on the treatise by Mr. Orr.

Our immediate comment: *Alack, all those detractors are definitely acting up: Now this next guy is just trying to reinvent the wheel.*

Could Edgar Buckingham have a rough luck in German academic circles? Not quite. Of interest for our discussion was also his work [2.141], where he would like to develop the work by Ludwig Boltzmann on theoretical interpretation of Stefan's law on the basis of a thermodynamical consideration. The *Beiblätter* volume 29 we are dealing with right now contains a referee's report to this work. The relevant referee was Prof. Dr. Siegfried Valentiner, whom we have already met in Chapter 3.

Prof. Valentiner's report is fully unbiased and sounds as follows (our English translation):

Beiblätter, vol. 29, pages 777-778, Siegfried Valentiner's reaction to Edgar Buckingham's work [2.141]:

Optics. *44. The author is dealing with the following process. A stamper is moving isothermally and reversibly and without any friction within a cylindrical channel, in which a radiative energy of the density φ at some uniform temperature Ξ. In his relevant inference Boltzmann deals with the direct application of the Second Basic Law to an accomplished circular process. Instead, in considering the first part of the latter the author employs the equation for the free energy ψ and internal energy ε: Eq. (1)*

$\psi = \varepsilon + \Theta \dfrac{\partial \psi}{\partial \Theta}$. *This way, the author takes into account the fact that any work produced via the reversible isothermal process ought to be equal to the loss in the free energy. As a result, it becomes possible to arrive at the relationship between φ and ψ. Specifically, Eq. (1) might then be recast*

to give $\dfrac{1}{3}\phi\delta v = -\phi\delta v + \dfrac{1}{3}\Theta\dfrac{\partial\phi}{\partial\Theta}$, *so that one gets immediately* $\phi = C\cdot\Theta^4$.

Instead of Eq. (1) above, it is throughout possible to consider any other equation deducible from the Second Basic Law, e.g., $\left(\dfrac{\partial\eta}{\partial v}\right)_\Theta = \left(\dfrac{\partial p}{\partial\Theta}\right)_v$

where η stands for the entropy. The author performs the pertinent calculations for the both sides of the latter formula by setting the pressure p over there to be equal to (φ/3), and he gets again the correct expression for the Stefan's law.

The above is just a discussion about the modalities of the black-body radiation. The latter is well known to be the electromagnetic radiation within or surrounding a body in thermodynamic equilibrium with its environment. This topic was truly explosive that time, for it was the central point of Ludwig Boltzmann's [2.142] and the Max Planck's [2.143] theoretic activities that could finally lead to formulating the principles of the quantum physics. Albert Einstein has suggested a statistical interpretation of this effect, which could serve as a basis for Max Planck's theory. As we see that was a truly fortunate occasion that Edgar Buckingham's work had not landed onto Albert Einstein's table. With respect to the then emerging quantum theory Prof. Valentiner was solely a friendly professional observer from the outside [2.144, 2.145], whereas his interest in thermodynamics was mainly concentrated on the equations of state for different aggregate states of the Matter [2.146].

'Meanwhile, in this same Beiblätter volume 29 Prof. Valentiner had placed the referee's report as for Max Planck's statement as to the mechanical definitions of temperature and entropy in Ludwig Boltzmann's commemorative volume (we have already translated the relevant original publication into English above in this Note, see pp. 702–705). Here we place the English translation of Prof. Valentiner's report to have a look at the unbiased standpoint of that time.

Beiblätter, vol. 29, pages 636-637, Siegfried Valentiner's reaction to Max Planck's ideas about the significance of statistical mechanics:

80. The mechanical definitions of the entropy in Boltzmann's and Gibbs' works are compared with each other to obtain the proper clues for answering the fundamental question as to which definition of the entropy (and thus the temperature) must finally be given priority over the other. 'By the number of similar microscopic complexions, which do comprise a certain macroscopic state Boltzmann defines the probability of the latter and, consequently, the entropy of the system in question. Such a definition is valid for the case of identical particles only and implies from the outset that the number of complexions ought to be modified, if the system is not composed of identical particles. Instead, Gibbs takes into account the entirety of the complexions obtained by letting the system adopt all the different values of coordinates and velocities, which are possible within some special limits, but without any concern as to the uniformity or diversity of the particles. In the latter case, as the definitions by Gibbs lead to the known thermodynamic formulas, we must conclude afterwards that the number of different particles must be taken into account, when determining the number of complexions. Thus, a straightforward analysis does show clearly that the expressions for the entropy derived from the both definitions in question in the simplest case of the identical particles (one-atomic gases) are the same, but this is true solely for the latter case. As a result, Boltzmann's viewpoint seems to be considerably simpler and more appropriate, because of the necessary reformulations of Gibbs' definitions for the more general case. Meanwhile, as the author points out already in his introduction, the problem of choosing the most general and most rational definition of entropy cannot be conclusively solved the above-sketched way. The ultimate solution to the latter problem ought to be dependent on the result of checking the performance of the different definitions when applying them to irreversible processes. After taking the latter point into account, it is clear that Boltzmann's approach turns out to be superior to its counterparts as well.

Our immediate comment: *First of all, many sincere thanks are due to Prof. Valentiner for his very attentive reading and proper analyzing in detail that communication by Max Planck. At the publication time of Planck's comments J. W. Gibbs had already taken his final leave of us, to our sincere regret. His seminal work was just remaining unfinished. Moreover, L. Boltzmann had at that same time severe health troubles and was to take his final leave of us as well. Hence, to our sincere regret his work was about to be unfinished as well.*

As we have already seen in Chapter 3, as well as above in this very note, the theories of the both outstanding colleagues had in no way grasped at nothing, in no way gone and in no way forlorn, for there were competent colleagues working on the pertinent development of these seminal theories. Indeed, Max Bernhard Weinstein was a truly versed detractor, and at the same time pursuer, of the seminal works by L. Boltzmann, but Prof. Dr. Weinstein's truly immanent 'Vis Viva' and therewith his lifetime had also come to their natural end untimely soon.

Finally, as we have already seen in Chapter 4, Dr. G. A. Linhart could clearly and formally prove that the thermodynamic entropy ought to be the logarithmic function of something, indeed. Moreover, as a matter fact, the notion coming out as magic 'number of complexions' turns out to be a clear-cut algebraic function of the temperature. On the other hand, the further detailed and rigorous analysis of how to skillfully connect both L. Boltzmann's truly ingenious guess and J. W. Gibbs' ingenious inferences with the actual modalities of the probability theory [2.147] clearly shows that such an interconnection is throughout possible in principle. In summing up, everything seems to be consistent. To this end, what is then the very point where thermodynamics starts 'feeling forlorn and miserable'?

Remarkably, it is exactly here that Dr. Buckingham's seminal results allow us to find the very feeble point. Indeed, it is the notorious 'blackbody radiation' that ought to scheme the whole intrigue. This kind of radiation ought to be the sensible effect in the absence of any kind of externally driven processes [2.148–2.152].

Therefore, the first truly bright idea might be stating that **this be just the activity of any aggregate state of the matter at the thermodynamic equilibrium** *[2.153]. Constructing the handy theory of the blackbody radiation [2.143] has been the very starting point for the glorious advance of the quantum physics [2.154–2.155], being nowadays an old and good physical theory without any doubt. However, at the very same time, this could promptly allow to* **'physicalize'** *the cyclic gadget by Carnot, thus giving rise to the notorious 'equilibrium thermodynamics'. Meanwhile, the latter is based upon nothing more than the 'basic natural indeterminism' having hardly anything to do with the actual physics/chemistry, and so on.*

Thus, the statistical–mechanical theory of blackbody radiation was in fact the first skillful sketch for the gem of the emerging quantum physics.

At the same time, it was the first 'gorgeous' nail into the emerging coffin of the general thermodynamics.

Otherwise, why should we substitute J. W. Gibbs' systematic and rational thermodynamics' theory by L. Boltzmann's ingenious guess? It is well known that the both have occurred to be just the ultimate first steps only.

Howbeit, Dr. E. Buckingham was in effect trying to demonstrate that the blackbody radiation is not a '*special effect demonstrating some novel basic physics*', that there is indeed a fully valid way to explain this important effect without any 'revolutionary breakthrough'. Meanwhile, the latter result could not change the destiny of Dr. Buckingham, alack! Noteworthy, his theoretical book on thermodynamics [2.127] had no success even in his homeland. A renowned American experimental physicist, Prof. Dr. Edwin Herbert Hall (1855–1938), has published the only referee's report presently known, as for the monograph by Buckingham [2.156]. Prof. Hall is the author of the Hall effect well known in the physics (the details about the life and work of Prof. Hall could be borrowed in the work

[2.157]). The report by Prof. Hall is throughout polite, but withering in fact. Very strange is the latter result because Prof. Hall was never dealing with thermodynamics or even any of the physical fields bordering to the latter.

Finally, Dr. Buckingham had decided to quit the academic areas being that time to 100% full of the turbulent 'revolutionary emotions' [2.157]. Instead, he had chosen to try his skills and experience in some areas of practical significance.

Interestingly, in that time there was another 'widely unknown' American colleague, who had experienced approximately the same destiny, as Dr. Buckingham.

That was Prof. Dr. William Walker Strong (1883–1955).

This portrayal of Prof. Dr. Strong has been taken from http://www.ancestry.com/genealogy/records/william-walker-strong_73468779

Just at the time we are interested in he was active in the area of radiation physics, one of the then topics in vogue, as we have just seen above in this note. There is an example of his pertinent publication outside the USA and in German language [2.158]. Meanwhile of urgent interest for our present discussion are his two books [2.159, 2.160]. The first book is now available in the open access through the portal archive.org, whereas the second one has been re-published most recently under the series *Books on Demand*. In his books he has presented his full CV.

This is why first we would greatly appreciate to learn something about his life and work. Further, we would like to cite here a very interesting fragment of the first book, by just noting that the second book is developing the ideas of the first one.

Of himself Prof. Dr. Strong communicates us the following information:

> The author was formerly scholar, fellow, fellow by courtesy, and assistant at the Johns Hopkins University; research assistant of the Carnegie Institution of Washington; instructor in physics at the Carnegie Institute of Pittsburgh; fellow of the Mellon Institute and professor of electrical theory of the University of the Pittsburgh; and a director of the S. I. E. M. Co.

> The author is a fellow of the A. A. A. S., a Member of the American Physical Society, a Member of the A. I. E. E., Franklin Institute, etc. consulting physicist, expert on ionization of gases, electrical theory, the precipitation of fumes, dust, smoke, mist, fogs and all kinds of vapors, research investigator on the causes of fires and explosions due to electric static sparks, expert on electrical methods of evaporation, oil clarification and deblooming, electrical smoke apparatus, grating replicas, gas methods of treating contagious diseases, fume masks, etc.

His above CV suggests that initially he was a competent and active physicist, but then decided to quit the academic field by starting his own engineering company, which was successful enough to allow him, among other things, publishing at his own costs the books on the themes he was interested in. In this respect his destiny might partially be comparable with that of Dr. Buckingham and of all the protagonists of the volume at hand.

Specifically, Prof. Dr. Strong had graduated the Johns Hopkins University and started his academic career at the physical laboratory of this university under the leadership of an outstanding American experimentalist in the field of spectroscopy, Prof. Dr. Joseph Sweetman Ames (1864–1943). The then young research associate, William Walker Strong expresses his sincere gratitude to Prof. Ames in his (most probably, the very first!) publication [2.158].

More details about the life and work of Prof. Ames could be borrowed from the book describing the correspondence of Henri Poincaré with the natural scientists of his time [2.161]. Specifically, Prof. Dr. Ames was a graduate from Johns Hopkins University in 1886 and then he had moved to Berlin, Germany to work in Hermann Helmholtz' laboratory. In 1887 he returned to Hopkins to study spectroscopy with Henry Rowland, working from 1888 to 1891, as an assistant in the latter's laboratory. He defended his thesis in 1890, and became an associate professor of physics at Hopkins in 1891. Ames spent his entire career at Hopkins; he was promoted to professor in 1899, served as provost from 1926 to 1929, and university president from 1929 to 1935. He succeeded Rowland in 1901 as director of the Physical Laboratory and was actively attending international conferences, where he could be in intensive contacts with the leading squad of the European scientists.

Noteworthy, W. W. Strong's experimental work [2.158] has received an immediate attention in connection with the then on-going discussion about the blackbody radiation (see a detailed theoretical review paper by an important Swiss theoretical physicist Prof. Dr. Walther Ritz (1878–1909) [2.161] published in that same journal volume). Remarkably, both this work and even a name of Walther Ritz are to 100% absent in the comprehensive historical treatment on the theme [2.162].

2.3.3.3 Quantum mechanics vs. thermodynamics: A true Gigantomachy

This is why, the life and work of Prof. Ritz ought to deserve a separate story, the full details of which might be borrowed from the volume containing his biography and collected works, which was edited and published by the Swiss physical society after his death [2.163]. Here we would just like to mention a number of points of importance for our present discussion.

Walther Ritz was studying both in Zürich and Göttingen. He is most famous for his work in the field of theoretical spectroscopy performed together with the eminent Swedish physicist Prof. Dr. Johannes Robert Rydberg (1854–1919). This collaboration resulted in formulating the well-known Rydberg–Ritz combination principle [2.164, 2.165]. On the other hand, Ritz is also known for his working-out the variational method, in collaboration with Prof. Dr. John William Strutt, Third Baron Rayleigh (1842–1919). This method is widely known in theoretical physics as the Rayleigh-Ritz method [2.166, 2.167].

To our sincere regret, Prof. Ritz had left us very early. He died in 1909, at the age of 31. First, he caught tuberculosis in 1900, which had caused the subsequent pleurisy and finally led to his death in 1909.

This portrayal of Prof. Dr. Walther Ritz has been taken from
https://de.wikipedia.org/wiki/Walter_Ritz

Of extreme interest for our current discussion is the fact that Prof. Ritz's seminal theoretical results and approaches could finally become the important mathematical tools of the quantum physics [2.165, 2.167]. At the same time, he was a powerful detractor of Maxwell–Lorentz electromagnetic theory [2.162]. He contended that the connection of the latter theory with the so-called 'luminescent ether' renders it *"essentially inappropriate to express the comprehensive laws for the electrodynamics"* [2.168]. The latter notion was introduced in the ether theory suggested by Prof. Dr. Hendrik Antoon Lorentz (1853–1928).

Specifically, in his review papers [2.168, 2.169] Prof. Ritz has summarized the main feeble points of the Maxwell–Lorentz electromagnetic field equations as follows:

- Electric and magnetic forces really express relations about space and time and should be replaced with non-instantaneous elementary actions.
- Advanced potentials don't exist (and their erroneous use leads to the Rayleigh–Jeans ultraviolet catastrophe).
- Localization of energy in the ether is vague.
- It is impossible to reduce gravity to the same notions.
- *The unacceptable inequality of action and reaction is brought about by the concept of absolute motion with respect to the ether.*
- Apparent relativistic mass increase is amenable to different interpretations.
- The use of absolute coordinates, if independent of all motions of matter, requires throwing away the time-honored use of Galilean relativity and our traditional notions of rigid ponderable bodies.

The basic point of Maxwell–Lorentz electromagnetic theory was the notion of the *'luminescent ether'*, which was unacceptable from the standpoint of the natural philosophy. Remarkably, this notion was one of the reasons for Wilhelm Ostwald to start fighting against the 'scientific materialism' (see a short discussion on this theme in Chapter 3, pages 343 ff., as well as his work in Italian [2.170]). Remarkably, the mentioned paper by Prof. Dr. W. Ostwald presents a clear and instructive introduction into the modern energetics. Moreover, on the latter theme in this same journal there is also a work in English [2.171] by Prof. Dr. G. H. Bryan, whom we have already met in this Note and whose work is even much closer related to the thermodynamics' foundations than that by Prof. Ostwald.

To our mind the statement underlined and highlighted in the above list of Prof. Ritz' criticisms ought to be the sharpest feeble point of the Maxwell-Lorentz' theory. It is thus of tremendous importance for our present discussion. Indeed, it points out the violation of the third basic law in the Newton's mechanics, which states that any action causes the equal counteraction. Moreover, it is just this basic law that represents the actual physical/mechanistic sense of the thermodynamic entropy notion, as we have learned in Chapter 4 and the present chapter.

Interestingly, Max Planck has immediately taken this truly very serious demur into account when participating in the 1908 Meeting of the German Natural Scientists in Köln [2.172]. Meanwhile, in his latter report Planck is not referring to Walther Ritz at all, but instead to Henri Poincaré as the main detractor on the theme.

Max Planck ensures the readership that the investigations by Max Abraham "*do demonstrate that the mentioned Newton's Law, as well as the Energy Conservation Law, should both remain intact, if we would introduce the new 'motional quantity' in addition to all the well-known ones, namely the 'electromagnetic' motional quantity.*" Further on in this report, Max Planck clearly states: "*introducing such a novel quantity ought to be throughout possible, if one is based upon the relativity theory by Albert Einstein.*"

Remarkably, this report by Max Planck has caused the following discussion between him and Hermann Minkowski (here we place its English translation):

Discussion

Minkowski: *To my mind, the laws concerning the motional quantity could be derived directly from the energy law. In fact, according to the Lorentz' theory, the latter law ought to be dependent on the coordinate system to choose for the space and time. If we now write down the energy law for every coordinate system possible, we would arrive at a number of equations, which would all contain the laws for the motional quantity.*

Planck: *Sure, of course. Meanwhile I am not considering the independence of the coordinate system a solid physical result, but rather a hypothesis, which, to my mind, definitely looks like promising, but not yet proven. It is still necessary to examine whether such relationships are really present in the Nature. However, we might only answer this poser by experimental means, and hopefully the time is not far off the beaten truck, when we would finally learn it.* See Hendrik Antoon Lorentz (**1895**): *Versuch einer Theorie der elektrischen und optischen Erscheinungen in bewegten Körpern.* J. Brill, Leiden, the Netherlands. Page 28.

Our immediate comment: *The statement by Hermann Minkowski was actually expressing a sheer surprise concerning the actual interrelationship between the theories advertised and the actual physics. Max Planck did act here as a 'Viola da Gamba virtuoso' in his field. Remarkably, by the time in question, the above-mentioned book by H. A. Lorentz has been republished in Germany in the year 1906 [2.173].*

In fact that Max Planck has done nothing more than just avoiding the correct answer in a virtuoso way, and we might draw this conclusion after reading both the very insightful works by Max Abraham [2.174–2.176] and the referee's report in regard to them, by Prof. Dr. Edwin Bidwell Wilson [2.177, 2.178]. Max Abraham was clearly aiming at building up a '*Novel Mechanics*'.

To this end it is very important to note that the story of Prof. Dr. Max Abraham (1875–1922) is truly very sad. A promising young theoretician, after graduating his University and then doctoral studies in Berlin under the leadership of Max Planck in 1897, he was in vain trying to get a stable academic position—first in Germany (Göttingen), then in the USA (University of Illinois), and only his professional contacts with Prof. Dr. Tullio Levi-Civita (1873–1941) ensured that Max could get the professorship in the rational mechanics at the University of Milano, Italy. The First World War had driven him back home, to Germany, where he could get successive professorships in Stuttgart, and then in Aachen. But finally he became seriously ill—he had an intracranial neoplasm—and had to have left us preterm. Remarkably in this connection, Prof. Levi-Civita was himself a powerful proponent of the 'novel mechanics' and a close ally to Albert Einstein in the field of mathematics [2.179–2.182].

Still, the final help to the revolutionary physicists is very well known to have finally come in 1924 from France, namely from the then young researcher Louis-Victor Pierre Raymond de Broglie (1892–1987) [2.183]. After thorough analysis of the equivalence of mass and energy suggested by Albert Einstein and expressed in his meanwhile famous formula $E = mc^2$, de Broglie comes to the conclusion that energy, as it is, like mass, ought to be localized in the form of particles in small spatial regions.

At the first glance this sounds crazy from the then conventional physical standpoint, but this way it does turn out to deliver a consistent explanation of the observed atomic/molecular spectra.

Indeed, the only logical closure to the conceptual gap would be finding the correspondence between every mass *m* and the frequency.

According to the relationship $E = h\nu$, as postulated by Max Planck, the frequency in question might then be immediately found from Albert Einstein's formula, to be equal to $\nu = mc^2/h$.

This way we arrive at the fundamental conclusion that the structure of matter ought to be of essentially *quantum nature*, as fully supported by the wealth of spectroscopic experiments. The latter conclusion seems to perfectly eliminate all the demurs in connection with the conventional Newtonian mechanics. Indeed, the basic elements of the matter ought to be of the dual nature: they are to be simultaneously viewed as both particles and waves! Successfully developing this standpoint further on, as concerns optics, was a renowned German physical theorist Prof. Dr. Gustav Adolf Feodor Wilhelm Ludwig Mie (1868–1957) [2.184].

Our immediate comment: It is important to note here that Prof. Dr. Mie was *not* following the theories by Antoon Hendrik Lorentz. He did go quite a separate way, which was independently followed by an outstanding Danish specialist in the optics field, Prof. Dr. Ludvig Valentin Lorenz (1829–1891). This means that the success of Prof. Mie's theory—or better to put it Mie–Lorenz theory—is not due to following the revolutionary breakthrough in the physics, but just to the contrary. Meanwhile, this fact should not anyway cast shadow on Hendrik Antoon Lorentz, who was undoubtedly a serious specialist in optics. Indeed, a very important optical relationship, namely, the mathematical formula connecting the refractive index and the density of a medium had been published by Lorenz in 1869 and by Hendrik Lorentz in 1878, independently of each other. This is why we now refer to this as Lorentz–Lorenz equation [2.185–2.187]. To this end, of definite interest and importance for our present discussion would be to point out that Prof. Dr. de Broglie, who could deliver the clinching contribution to the quantum physics, was exerting powerful efforts to clarify the actual physical sense of the entropy notion [2.188]. This whole story excites only one comment: *A phantom limb pains of a great thinker and fine specialist.*

It is throughout clear that in the year 1908 Prof. Ritz, who had soon to leave for the 'last ultimate business trip', could not be aware of such a handy solution to the conceptual problem we are discussing right now. He was but fully correctly indicating that the realistic light should *not* be considered as *propagating* (in a medium), but in effect, it should be *projected* in the latter.

This was just the point of the powerful controversy between Walther Ritz and Albert Einstein. Specifically, Albert Einstein was just consequently building up his relativity theories on the Maxwell–Lorentz theoretical basis mentioned above, and this is why Walther Ritz has heavily criticized Einstein's efforts [2.189, 2.190].

Noteworthy, Prof. Dr. Max Bernhard Weinstein was another outstanding colleague, who was criticizing Albert Einstein's activity in the professionally consistent way (see Chapter 3, pages 348 ff., for the relevant English translation).

Meanwhile, of especial significance for our current discussion is the common statement by Walther Ritz and Albert Einstein published in the volume X of '*Physikalische Zeitschrift*' [2.191].

Here we place our authorized English translation of this extremely very important and eye-opening publication:

On the current state of the radiation problem

By W. Ritz and A. Einstein

In order to elucidate the disagreements, which have appeared in our previous publications,* we would like to note the following:

In the special cases, in which an electromagnetic process is confined to a finite space, the representation of the process is throughout possible in the form

$$f = f_1 = \frac{1}{4\pi} \int \frac{\varphi\left(x', y', z', t - \frac{r}{c}\right)}{r} dx' dy' dz'.$$

It is but equally possible to represent the process in question in the following form as well

$$f = f_2 = \frac{1}{4\pi} \int \frac{\varphi\left(x', y', z', t + \frac{r}{c}\right)}{r} dx' dy' dz'.$$

To sum up, any other plausible forms of the above expression should also be feasible.

---

*W. Ritz, *Physikalische Zeitschrift*, 9, 903–907, 1908 and A. Einstein, *Physikalische Zeitschrift*, 10, 185–193, 1909.

While Einstein believes that one might successfully stick to the latter case without substantially restricting the generality of the consideration, Ritz argues that such a limitation is not permitted in principle. Ritz suggests that it is the experience that makes it necessary to consider the representation with the aid of the retarded potentials as the only possible one, if we consider that already the fundamental natural laws are dictating the fact of the radiation processes' non-reversibility. Ritz considers the restriction to the form of the retarded potentials as one of the roots of the second basic law of thermodynamics, while Einstein believes that the non-reversibility is only due to the basic indeterminism, which could be duly described by the probability theory.

Zurich, April 1909.

(Received April 3, 1909)

By the way, the combination of the above-cited works by Walther Ritz and Albert Einstein, as well as their common publication [2.162, 2.168, 2.169 and 2.189–2.191] might still help clarify the situation with thermodynamics, among other useful things.

To start with, what we clearly reveal from reading the above citation is the proof of a *stubborn voluntarism* shown by the revolutionary physicists. Specifically, the revolutionary physicists group Albert Einstein could finally penetrate by the time we are considering now, had the only hotly intended aim.

The latter was—to formulate quantum physics, as we know. Sure, everybody is free to go his/her own way to reach his/her own aims.

A good poser arises in this connection: Should the end every time justify the means?

Another bright and eye-opening example of the '*revolutionary voluntarism*' might bring up the story taken place at about the same time in connection with the publications by some unknown *Mr. F. Cohen* in the '*Physikalische Zeitschrift*', vol. X, 1909 [2.192]. *Mr. F. Cohen* presents a clear proof that the well-known H-theorem by Ludwig Boltzmann is based on debatable physical grounds. The English translation of the two communications by Mr. F. Cohen do definitely require their separate publication.

Of primary interest for our present discussion is but the clearly furious reaction by Max Planck in response to the both insightful communications by Mr. F. Cohen in this same journal volume [2.193]. We place here our English translation of his rebuttal:

Concerning the Kinetic Gas Theory

A Critical Investigation

By Max Planck

The above-entitled communication by Mr. F. Cohen is an attempt to prove that the well-known H-Theorem by Boltzmann ought to be devoid of any justification. In fact, Mr. Cohen manages to prove this way that it is pestilent, if we deal with mathematic symbols without grasping their actual meaning. To demonstrate the latter point let us consider a simple differential equation

$$\frac{d\varphi}{dt} = -\varphi^2.$$

Let φ here adopt some positive value at t = 0. With this in mind we might draw the immediate conclusion that the above equation would require φ to decrease continually with the growth of t and to asymptotically approach its zero value. According to Mr. F. Cohen, we have to draw a quite different conclusion, for example, the following one. 'Let us start from the value φ = 0 and consider a small change of the latter, so that we arrive at some φ = δ. Then, for the variation of $\frac{d\varphi}{dt}$ *should hold*

$$\delta \frac{d\varphi}{dt} = -\delta^2.$$

This is in turn an essentially negative value. As soon as $\frac{d\varphi}{dt}$ *is negligibly low for φ = 0, then the latter should never be correspondent to the minimum value of φ, for any lowering the latter (δ < 0) would render* $\frac{d\varphi}{dt}$ *even more negative. This way φ is free to approach its further minimum.' The inadmissible point of the latter deliberation consists in that the dependent variable φ must be varied in an unrestricted manner. Equally, it is inadmissible to assume, together with Mr. F. Cohen, that Maxwell's velocity distribution function might experience some perturbations leading both to the increase and to decrease of the H-function, which is dependent on the distribution function mentioned. Specifically, the H-function's mathematical definition clearly demonstrates: It is for Maxwell's distribution—and only for the latter one—that the H-function would reach its absolute minimum.*

The further demurs by Mr. F. Cohen against the H-theorem are based upon the tacit assumption that the dimensions of Boltzmann's 'velocity cells' dω might be time-dependent und especially that the latter might be rendered arbitrarily small. Such an assumption is inadmissible. On the contrary, it is well known that among other important conditions, it is essential for the H-theorem that the dimensions of the 'velocity cells' remain constant. Moreover, it is very important that the number of the molecules contained in every particular 'velocity cell' might be set proportional to its dimensions. The latter condition ought to be true if and only if the number of the molecules in question is either very high, or appreciably low.

As one might conclude according to the introductory statements contained in Mr. F. Cohen's communication, he would greatly appreciate continuing to publish his 'critical investigations'. In such a case, the above notes ought to serve as a guideline for the degree of caution, which should be offered against such criticism.

Berlin, February 22, 1909.

Our immediate comment*: Truly impressive is the perfect expeditiousness of Max Planck. Indeed, the first submission by Mr. F. Cohen has been received on 12.12.1908. It has been published in the Number 4 of the Volume X, that is in April 1909. His second submission has been received on 22.01.1909 and published in the same Number of the Volume X, being placed right after Max Planck's rebuttal. The latter being dated by 22.02.1909 has been received on 23.02.1909.*

This clearly demonstrates how serious Mr. F. Cohen's criticism was for Max Planck's own work, and he knew very well who Mr. F. Cohen was in fact.

To our sincere regret we cannot pose this interesting question to Max Planck personally, but we have material enough to solve this instructive puzzle.

Indeed, the surname Cohen was very well known in the German academic circles of that time. First of all, we mention Prof. Dr. Hermann Cohen (1842–1918), the University of Marburg, who was one of the most important Jewish philosophers in the XX-th century, but along with this had nothing to do with the natural sciences.

Further, there was Friedrich Cohen, a publisher and distributer of the scientific literature in different areas of knowledge. Friedrich

Cohen's company was situated in Bonn since 1828, short time after the grounding of the Bonn University, and was very well known. Now it is also well known in Bonn as the Bouvier publishing house and University bookstore. This entirety but has definitely nothing to do with the statistical mechanics stemming from the kinetic gas theory.

Meanwhile, above in Chapter 3 we have learned about Prof. Dr. Max Bernhard Weinstein, who was in Germany one of the outstanding specialists in the field. In 1904, the commemorative volume devoted to the sixtieth birthday of Ludwig Boltzmann contains two interesting communications, among others, as we have seen.

In the one of them Max Planck has told us about the advantages of Ludwig Boltzmann's approach to the kinetic gas theory as compared to that of Josiah Willard Gibbs, whereas in the second one Max Bernhard Weinstein has presented a suggestion of how to properly re-cast the Clausius–Boltzmann–Szily formula for entropy to infer the handy formula for the internal friction.

As a result we see from the above that Max Planck found it fully adequate to 'like or lump' the probabilistic interpretation of entropy, whereas Max Bernhard Weinstein was sincerely trying to find an interconnection between the abstract inferences by Ludwig Boltzmann and the actual physics. It is this way that Max Bernhard Weinstein was trying to find the rational thread between the thermodynamics and statistical physics.

Most probably, that was just Max Bernhard Weinstein who dared to act as the only serious opponent. Max Bernhard new Planck very well. What we can definitely see from Max Planck's rebuttal: There is no logical ground for preferring the Boltzmann's H-theorem in its original elegant formulation to some other formulation or even to something else. Moreover, the last part of his rebuttal is a clear call to his allies to get rid of the criticisms undermining the Boltzmann's theory. Knowing Max Planck personally, having learned his standpoint in personal discussions and bearing the latter in mind, Max Bernhard Weinstein could have decided to choose a handy pseudonym for publishing his insightful ideas.

Here is but the final serious point in this instructive story: We have seen that Max Planck was categorically against including time variable into the statistical–mechanical inferences. The actual reasons are remaining unclear.

Meanwhile, the actual contribution by Albert Einstein—the 'Einstein's cud', as Nils Engelbrektsson has duly dubbed it (see Chapter 2)—was to 'kill the time', to get rid of such a physical variable.

One might ask the author at this point: Does the above story demonstrate the inconsistency of the statistical mechanics?

The immediate answer is: *definitely not*! Sure, the works by the actual founders of this field of knowledge—Ludwig Boltzmann and Josiah Willard Gibbs are seminal without fail! The point is that they were the *starting successful steps* requiring the proper further development. Max Bernhard Weinstein in his works has demonstrated that the latter is definitely possible, as we have already seen in Chapter 3.

Moreover, an outstanding British physicist Prof. Dr. Sydney Chapman (1888–1970) and renowned Swedish physicist Prof. Dr. David Enskog (1884–1947), independently of each other and of Prof. Weinstein, have presented a very interesting development of the field in question in their clearly seminal works [2.194–2.197]. To this end, a truly non-linear line of Prof. Enskog's destiny is definitely remarkable. Indeed, reading his biography [2.198] ought to catch in a quandary thereover.

To this end, in reverting but to Albert Einstein's activity in all the fields we are discussing now, in Chapter 3, pages 348 ff., we have learned about the clear statement by Max Bernhard Weinstein to the theme. Here we would like to add the relevant conclusion by Prof. Dr. Eberhard Zschimmer (1873–1940), a renowned German engineer, manager, pedagogue and philosopher of technics. He has published his impression shortly after the first seminal publication by Albert Einstein has appeared [2.199].

Indeed, Prof. Zschimmer could assign Albert Einstein's activity to a kind of '*photoshop*' usage, to put it in the modern terms. That was definitely not a personal attack against Albert Einstein, but rather a general philosophic/methodological observation. Specifically, in having the main aim **not** to reveal/interpret some important natural interrelationships, **but** to more or less correctly describe the latter, we would tend to skip some natural features. The features to skip should be, to our mind, not truly essential in the sense of producing the rightest '*visual image*' we are looking for. This way, in adopting the relativistic standpoint, we are free to skip whatever we wish. When building up a physical theory, we might first of all skip the space without fail, for in itself it does contain no matter we are willing to describe. In going farther and far worse, it seems to be not a problem to skip both time and energy, for these both physical notions ought to be intrinsically related to each other. To sum up, a theorist should play the role of an external observer, who furthermore acts as a skillful reporter—and nothing more or less than the roles named.

At this point an attentive reader would probably throw the book at Prof. Zschimmer and the author of the monograph at hand, in charging the both with accepting all the quirky philosophic standpoints. With this in mind we concede the final conclusion to the readership. Howbeit, here we finish up our short philosophical/methodological foray with noting that the book by Prof. Zschimmer [2.199] and the both books by Prof. Strong [2.159, 2.160] do help arrive at the fully objective picture of the then physical revolution, in that these both colleagues could help embracing the actual manifold of the ultimately revolutionary results.

To this end, the work by W. W. Strong [2.158] helps us to reveal that a considerable number of publications in the volume number IX of *Physikalische Zeitschrift* are devoted to the topic of our current interest. First of all, we would like to refer here to the detailed German translation of the statements by Lord Kelvin concerning the atomistic standpoint [2.200]. Moreover, of particular interest for us here ought to be the discussion between Prof. Dr. Max von Laue (1879–1960), the well-known proponent of quantum physics, and an outstanding Italian colleague, Prof. Dr. Orso Mario Corbino (1876–1937) [2.201–2.203]. The last but not the least, of interest ought to be a short comment on the same theme by Prof. Dr. Hendrik Antoon Lorentz [2.204], one of the main proponents of the debatable 'luminescent ether theory'. To be mentioned here is also a communication by the renowned British physicist, astronomer, and mathematician, James Hopwood Jeans (1877–1946) [2.205], one of the well-known proponents of the quantum physics emerging at those days. Remarkably, most of the colleagues just named were actively exerting their research efforts just in the directions criticized by Prof. Ritz.

In fact, Prof. Corbino was wondering about the connection between theoretical inferences concerning the blackbody radiation and the second basic law of thermodynamics, likewise Prof. Ritz. Prof. von Laue could just skip the detailed answer to this poser in a virtuoso manner.

Meanwhile, the thermodynamic works seemed to have nothing to do with the turbulent activity just mentioned (see e.g., a work by another renowned Italian experimental physical chemist, Prof. Dr. Angelo Battelli (1862–1916), who has reported on specific heats of the liquids adopting solid states at lower temperatures [2.206]).

The volume X of *Physikalische Zeitschrift* published in 1909, is presenting a number of further works the careful analysis of which leads to a truly complicated picture of the then development in the general physics field. Applied thermodynamic analysis was definitely exciting live interest, as we might see e.g., from the insightful work by Edmund Altenkirch on the useful effect of thermopiles, that is, of thermal-to-electric energy converters [2.207].

To this end, of separate interest for us should be the work by Mr. Josef Weiß (University of Freiburg in Breisgau) [2.208]. Over there, he was that time a young research and teaching assistant at the physics faculty working on his PhD thesis devoted to checking experimentally the electron theories and defended just one year later [2.209]. Remarkably, Mr. Weiß presented his work [2.208] as a 'preliminary communication' and deals in it with the thorough analysis of Max Plank's theory of blackbody radiation. Specifically, Josef Weiß has presented a detailed enquiry into why there is no perfect accordance among Planck's theory and the experimental data on thermoelectric effects.

Meanwhile, his further publication in *Annalen der Physik* [2.210] together with Prof. Dr. Johannes Königsberger (1874–1946), one of his PhD advisors, deals exclusively with discussion on his PhD thesis results. Remarkably, neither the text of the PhD thesis in question [2.209], nor his later journal publication [2.210] do ever mention the criticism published in the work [2.208]. Moreover, there is practically no information about the further CV and work of Dr. Weiß, except that most probably he was stemming from the traditional family of visual artists in the region of Bodensee, Southern Germany [2.211].

Nonetheless, the above 'initial kick' to study electric phenomena from thermodynamic standpoint has not died away in an irreversibly traceless way. In particular, later on in Berlin there were such colleagues as Dr. Wilfried Meyer und Mr. Hans Neldel, who could experimentally fetch the definite signs of the entropy–enthalpy compensation during electric conduction phenomena (for the detailed historical-methodological analysis of that work, see [2.212]). Meyer–Neldel effect is well known and widely used in the modern research activities [2.213–2.215], although there are colleagues who try to skip it, for whatever reasons [2.216, 2.217].

To sum up, it is well known that the actual birth of the quantum physics was definitely not a 'gorgeous triumphal parade'.

Still, even using a truly peculiar belief system, the revolutionary physicists could definitely have won, while thermodynamics, which is in fact nothing more and nothing less than energetics, had clearly lost this Gigantomachy!

Everybody knows about the quantum victory nowadays, whereas the basic sketch of the belief system resulting from this victory might be formulated the following way, we cite from the book [2.218]:

Werner Heisenberg is best known in quantum physics for his discovery of the uncertainty principle, which has the consequence that to make measurements of very short distances—such as those required by string theory—very high energies are required.

The uncertainty principle states that the more precisely you measure one quantity, the less precisely you can know another associated quantity. The quantities sometimes come in set pairs that can't both be completely measured.

What Heisenberg found was that the observation of a system in quantum mechanics disturbs the system enough that you can't know everything about the system. The more precisely you measure the position of a particle, for example, the less it's possible to precisely measure the particle's momentum.

The degree of this uncertainty was related directly to Planck's constant—the same value that Max Planck had calculated in 1900 in his original quantum calculations of thermal energy. Heisenberg found that certain complementary quantities in quantum physics ought to be linked by this sort of uncertainty:

- *Position and momentum (momentum is mass times velocity);*
- *Energy and time;*

*[Our immediate comment]: **This uncertainty is a very odd and unexpected result from quantum physics. Until that time, no one had ever made any sort of prediction that insights/awareness might somehow be inaccessible even at some fundamental***

level. Sure, there were technological limitations to how well measurements were made, but Heisenberg's uncertainty principle did go even much further, in asseverating that The Mother Nature itself does not allow us to make measurements of the both physical quantities beyond a certain level of precision. Such a statement might thus be dictated solely by the fascination with the Principle of the "basic indeterminism" dominating the fantasies of the revolutionary physicists, who were stubbornly and successfully trying to forcefully introduce such a kind of "make-believe" into the publicity—as wide as possible.

Instead, the defeat of the energetics was, is—and remains to be—by far not so spectacular.

Not everyone could come to terms with such a state of art! This is just the case of practically all the protagonists of the monograph at hand, including Edgar Buckingham and William Walker Strong. With the above entirety in mind, we might in particular immediately recognize that it is from his young years on that Prof. Strong was very tightly involved into the highly turbulent, truly international research activity. The latter has been in direct connection with the blackbody radiation and all the relevant topics of our current interest. Prof. Strong was definitely aware of all the developments we have just discussed. This same pertains to Dr. Buckingham. The final decision of the both colleagues was to stop any relationship to the academic sector. What was then their actual reason for such a decision?

Nobody could now be asked personally and directly, but some thoughts do come to mind. What was in embryo state at their time could finally find its logical result in the above citation from the book [2.218], together with our immediate comment in this connection, highlighted thereafter.

This is just what we 'the dummies' have to consume and be to 100% happy with this. Indeed, serious physical literature is still euphorically discussing not only the '*quantum–mechanical uncertainty*', but also the '*thermodynamic uncertainty*' as well as related stuff, for example, cf. the works [2.119–2.225] and the references therein.

Fortunately, the latter publications flow could not fully eliminate serious analytic works in the both fields, as well as in different languages: for example, like the works [2.226–2.241].

From the literature of the type [2.118–2.225] we might borrow quite a wrong impression that the renowned German physicist Prof. Dr. Werner Heisenberg (1901–1976) was in effect just 'a true emperor of the basic indeterminism'. Meanwhile, there is a book about him published in German in 1957, when Prof. Heisenberg was still among us [2.226], as well as the most recent book published in English [2.239] and depicting the true situation in the atomic physics in the years 1913–1925. These books are unambiguously describing Prof. Heisenberg's life, achievements and his actual role. The conclusion we borrow from these serious books is that Heisenberg's uncertainty relations are describing by far not the 'basic uncertainty', but just a sheer impossibility to experimentally investigate some pairwise sets of physical variables at that time.

If to consider the practical sense of such a truly fundamental conclusion without any emotional stress, just from 'a pedestrian' standpoint, Prof. Heisenberg's seminal result is nothing more and nothing less than a stimulus to reconsider the methods of the experimental data processing. We shall revert to this important discussion in the Supplementary Note 4 to the chapter at hand.

As long as there is still a considerable number of the publications like [2.118–2.225], it is of tremendous interest for us to learn about Prof. Strong's early impressions concerning the then emerging trend, as published in his book [2.159]. Below we would appreciate to cite Prof. Strong.

2.3.4 'Disappearance of energy'—the story told by William Walker Strong

5. The Disappearance and Conservation of Energy, Electrical Charge and Mass

It has long been assumed that energy disappears or changes its form and one can consider phenomena as largely if not altogether marked by energy changes. To us the conception of energy is most satisfying when we consider kinetic energy or one-half mv^2. Potential energy is in an intangible condition or form for how can we conceive of this energy as existing except in the medium or the ether. Since we apparently know so little of the ether we can almost speak of the energy as being lost. The classical illustration of the transformations of kinetic and potential energy is that of the vibrating pendulum. In the *middle of its swing the energy is entirely kinetic, while at the ends of the swing the energy is entirely potential. Our laboratory work consists of experiments with the energy in the kinetic form. To be studied energy must first be converted into the kinetic form adapted to the problem at hand. A piece of uranium or a piece of coal can be used as regards their radioactive or chemical energy content only when radioactive energy or when heat is developed.*

The law of the conservation of energy is found to hold when we consider the existence of the two forms of energy and we feel entirely justified in treating potential energy as "existing" in the same way as kinetic energy because it can be converted into the kinetic form.

Our ordinary world as we see it every day is one in which kinetic energy is usually being converted back and forth into potential or "heat" energy. And it requires much of our engineering effort to maintain our kinetic condition. Friction is ever working against the moving elements. Most of us are kept busy to keep things going. Even in the astronomical world the philosophy of many is that the suns are becoming cold and dead and that the revolving planets and moons are slowing down in their motion. Yet this death of kinetic energy that is everywhere about us and which we will designate as a condition of our type of physical world is counteracted by more or less unknown elements so that conservation may always apply.

The disappearance of energy

The energy world is then ruled by these conditions—energy is conserved—the energy changes of our every day life is marked by the natural tendency of kinetic energy to disappear—and the second law of thermodynamics which may be stated in a number of forms such as the ever increasing value of the entropy content of any isolated system, the impossibility of perpetual motion apparatus or the ever tendency of the universe to run down as regards many energy changes (in the same way as a clock, the potential energy of the spring being converted into the kinetic energy of the pendulum and this kinetic energy changed gradually into the kinetic energy of heat motions).

The continual flow of energy through so many transformations leads one to suppose that when it ultimately is found to consist of units, particles, quanta, entities, atoms or whatever term we may call them, these ultimate elements may be all alike because all quantities of energy, like electrical charges, depend upon the total amount of energy present and are independent of the nature of the elements of energy combined to form the whole.

The appearance of electrical charges bears some analogies to the appearance of energy. The commercial problems relating to electricity have to do with the generation, the transmission and the absorption of charges of electricity, the phenomena of magnetism being assumed as being due to the motion of electrical charges. The constitution of electrical charges is very simple in that all electrical charges appear to be built of elementary parts all having the magnitude of $4.7 \ (10)^{-10}$ e. s. units, all these units being identical except as they may be positive or negative in character.

Like kinetic energy electrical charges never appear except in connection with masses of matter thus making the trio of kinetic energy, mass and electrical charge an inseparable ensemble as far as experimental knowledge extends. As kinetic energy is characterized by its tendency to disappear so free electrical charges always disappear unless special precautions are taken to prevent this change. The electrical elements possess the unique property however of neutralizing themselves in that the approach of equal charges of positive and negative electrical discharges causes the more or less complete disappearance of the charges.

The smallest element of negative electricity, the electron, is associated with a mass of matter about the one-eighteen hundredth part of that of the smallest element of matter, the hydrogen atom. Since the ratio of mass to charge is found to vary with the velocity of the electron it has been assumed that this mass is variable rather than the electrical charge. Experimentally the magnitude that is measured is the ratio of the charge to the mass so that for the rapidly moving electron we have a condition arising where electrical charge, mass and of course the kinetic energy appears as a variable ensemble for which no experiment has been devised that will permit of the determination of the varying elements.

As the philosophy that assumes that natural phenomena are extremely simple has led to many of the greatest discoveries, so the assumption that the magnitude of the electrical charge is constant rather than the mass of the electron, makes the treatment of the ensemble of energy, charge and mass more simple and apparently just as accurate as would any other assumption. The law of the conservation of electrical charge can thus be considered to be absolute. The second law and the general condition that applies to electrical charges is that all the natural changes in a closed electrical system are such as to make the magnitude of the free charge a minimum.

No positive electrical charge has been found except as being associated with atoms of matter and the view commonly held is that the positive charges are an essential part of the nuclei of the atoms. The generation of free charges of electricity always results in the disappearance of energy. The disappearance of free electrical charges always takes place in matter and the picture of the process assumes that the positive and negative charges neutralize each other inside the molecules and atoms. The energy that is lost in the generation of free charges is considered as the energy of the electrical field between the positive and negative charges and that on the localization of the electrical field inside of atoms or molecules some of its potential energy is changed into kinetic energy.

As in energy changes the disappearance of kinetic energy was considered as accompanied by the appearance of an equal amount of potential energy in the ether so the disappearance of negative charges into molecules and atoms could be associated with the appearance of certain characteristics of matter, which we ordinarily associate with the neutral state. In other words the characteristic properties of matter may be those associated with what we consider as matter after it has lost all of its electrons. Such matter we will speak of as "natural" or "free" matter. "Ordinary" matter will be called "neutral" matter. All electrical charges then become aggregates of electrons and all masses become aggregates of electrons and the "atoms" of "natural" matter.

The constitution of matter is found to be atomic, all matter being built out of some 92 elementary atoms with accumulating evidence that these atoms may themselves be composed of much simpler systems such as hydrogen or helium. The fundamental property of matter is usually believed to be its mass and the science of chemistry is based on the assumption that mass is conserved. We have seen that the mass of the electron is considered as a variable. Whether this kind of mass is acted upon by gravitational forces remains for future experiments to determine and whether the mass of the various atoms varies, as does that of the electron at high speeds also remains for future investigations to tell. There may be an electrical and a material mass, or a kinetic and a potential mass corresponding to the energy terms. The law of conservation might be considered as applying to the two types of mass. "Natural" matter would always naturally disappear into ordinary or neutral matter and the law corresponding to the second law of thermodynamics would be that the complex atomic systems are mining down or in other words the radioactivity of the elementary atoms always results in simpler atomic systems through the disintegration of the more complex atoms.

The law of conservation and the disappearance of kinetic energy, electric charge and natural matter appear to be the same. These elements naturally partition themselves between the ether and neutral matter. Neutral matter is then the vehicle of kinetic energy, electrical charge and natural matter. The disappearance of kinetic energy, of free electric charges or of natural matter is accompanied by potential energy changes for which the ether is the vehicle. The ether about an ensemble of energy, charge and natural matter may be considered to be in a state or to be approaching a state of equilibrium with the ensemble. Thus the ether about ordinary matter could be considered as "ordinary" ether and might be the same or different from the ether at a very great distance from any of these elements or from ensembles of one or more of these elements. Experimental work of the most fundamental character remains to be done as to the relations between the free ether, energy, electric charge, and natural matter and the partitioning of these elements between ordinary ether and matter. The law of conservation applies by definition to electrical charges absolutely and to energy and mass only as elements in a closed ensemble of ether, energy, matter and charge.

The writer proposes the ether to be an analogue of matter possessing a much more fine-grained structure. The elements of electrons and natural atoms of matter become the 'electroethons' of the ether. The potential or hidden energy becomes the kinetic energy of the 'electroethons'. The vibrations and sound waves of neutral matter are paralleled by the electromagnetic waves of the ether. As sound waves are pictured as the ordered collisions of particles of matter so the electromagnetic waves are the ordered collisions of 'electroethon' particles. For purposes of illustration rather than as an exact analogue there can be pictured an atmosphere of mobile electroethons in a lattice work of immobile electroethons somewhat similar to the existence of an atmosphere of electrons in a framework of the natural atoms of matter in a metal. An electric field is a region of the ether where the neutral condition has been disturbed in much the same way that an electromotive force disturbs the condition of equilibrium of the electron atmosphere of a metal.

In treating the second law of thermodynamics Maxwell considered that there could be "demons" that could reverse the changes whose direction was given by the law. Our credo as optimists leads us to believe that there are 'angels' that can direct the transformation of any natural phenomena. For example the direction of the radioactivity of matter is that of the disintegration of the more complex elements. We believe that experimental discoveries will be made that will indicate how "angels" (and perhaps ourselves) can build the more complex atoms from the simpler conserved elements. These "angels" might possibly work in the center of the sun, in very intense electric fields or by "breaking down" the ether under electrical stresses so that even electrical charges or natural atoms would result. And the Director of these "angels" we might define as the "Creator" or "Ruler" of the Universe, and this form of philosophy as "monotheism."

As regards kinetic energy we know that its magnitude is relative depending on how (½) mv² is defined and measured. Presumably potential energy and possibly mass and electrical charge may be viewed in the same way for the latter two quantities are measured by "forces" and accordingly if the system upon which measurements were being made was subjected to acceleration the measurements would be subject to modification. The condition of relativity thus appears amongst the conserved elements. An example of such relativity is that of electrical charge and mass. No electrical charge has ever been separated entirely from a mass, the electron possessing a certain charge and a certain mass defined with reference to the system with reference to which its motion is measured.

These problems as to the way the other is modified by its energy content are very important and can be studied by the effect produced upon the velocity of the electromagnetic radiations (the writer is now engaged in experiments of this kind). Indeed apparently no more fundamental and isolated experiments can be performed because these are the simplest ensembles at present open to the new science.

The marked effects of electric charges and natural matter upon the ether (the electric and magnetic fields), the apparent absolute conservation of these elements and their disappearance by their "mutual neutralization" have few analogues in the coarser grained entity systems unless we consider phenomena like the formation of salts from acids and bases, the phenomena of permanent magnetism and some life phenomena.

Experiments have already indicated that the structure of electrical charge is more "fine grained" than that of ordinary neutral matter though the atoms of natural matter, the positive nuclei of the ordinary atoms, possess about the same size as the electron. The "fineness" of structure of natural matter is thus made to depend upon the magnitude of the nuclei of the atoms. The ultimate and individual elementary units of energy and of the ether have not been discovered experimentally and it is reasonable to assume that these elements are much finer grained than mass or electrical charge.

From the Zeeman, Stark and similar effects and the modification of the velocity of electromagnetic radiations by matter it seems certain that electrical charges remain as such in the neutral atomic and molecular systems. The disappearance of energy and mass in the ether without changing its properties to any great extent would lead us to believe that the ether may be such a "dense" energy and mass medium-that the increment due to these disappeared elements affects the "density" but little. The older elastic ether theory developed from these disappearance phenomena.

The possible conditions of relativity in any ensemble are so numerous that it is always highly important to reduce the number of elementary units in the ensemble to the minimum. The new science aims to obtain the ultimate units or atoms of all the entity systems of the universe and then consider if any of the properties of these elementary atoms can be defined as "absolute" in the sense of "non-disappearing." Other elements will then be described relative to the "absolute" elements. If the elements that appear to be definable in an "absolute" manner do not determine the physics uniquely, then the new science will develop all the consistent systems of physics that appear to equally simulate phenomena and employ the system that is simplest and most convenient.

After going through the above fragment, the readership might get a feeling of the author's truly radical stance—or may perhaps that we are dealing with some evil piece of science fiction. Meanwhile, this is just a fully consistent view of the physics the '*Revolutionary Passionaries*' were stubbornly suggesting (and had finally imposed on!) us. The following facts duly illustrate Prof. Strong's clear-cut warning.

2.3.5 Should the first basic law of thermodynamics be indeterministic as well?

The expression $S = k \ln (W)$, the well-recognized Boltzmann–Planck equation, opens the well-known story about the bold guess by Ludwig Boltzmann as to the intrinsic and basic interconnection between the notions of thermodynamic entropy and probability. The purpose of this expression ought to be the statistical representation of the second basic law of thermodynamics. Meanwhile, the true story about the attempts to introduce the statistical interpretation of its first basic law is somehow much less discussed.

Howbeit, the latter story seems to be of considerable interest and importance for our present discussion. Indeed, the powerful attempts to establish a throughout statistical interpretation of thermodynamics ought to start with the work by such prominent physicists as Niels Bohr (1885–1962), Hendrik Anthony Kramers (1894–1952), and John Clarke Slater (1900–1976) [2.242].

They had employed the well-known result by Albert Einstein, who could in particular give a very simple derivation of Max Planck's heat radiation law under the assumption that the physical behavior of an atom in a given stationary state is being guided by probability laws. And they had come to the conclusion that at the first basic law of thermodynamics, the energy conservation and transformation law, might at some particular circumstances get obsolete at the microscopic level.

In other words, one might in principle imagine some microscopic situations when the latter law becomes obsolete, but such situations ought to be extremely improbable.

Specifically, by the time when the above-mentioned Bohr–Kramers–Slater publication was being prepared, the seminal atomic theory of Niels Bohr was already well recognized, but certain unsolved theoretical problems were still remaining, like, for example, the difficulties when explaining radiation intensities and studying in detail the atoms possessed of a more complicated electron structure than the hydrogen atom. Moreover, in its then state the Bohr's theory could not be successful at explaining such features essentially related to wave processes as wave interference and wave diffraction.

The classical wave theory was pretty successful at explaining them, whereas the novel quantum approach had thus come to a kind of standstill. This is why the ingenious idea by Bohr, Kramers, and Slater was to build a conceptual bridge between the classical and quantum approaches as follows. The classical theory dictates that atoms should lose some amount of energy by transferring it to the radiation energy, while the latter one could then be absorbed by some other atom.

With this in mind, the interaction among the atoms ought to be based upon some kind of "virtual radiation field" not capable of carrying any portion of energy, according to the suggestion by Bohr, Kramers, and Slater. Hence, the energy released by the radiating atom might in principle be lost, transferred to nothing. *Vice versa*, if some atom gets energetically excited, it is then not absorbing the energy coming from elsewhere, say, from outside, but it instead, creates the energy from virtually nothing.

The entire story mentioned above is instructive for our current discussion owing to the two following points.

1. The above suggestion by Bohr, Kramers, and Slater was carefully checked during several detailed experiments in independent laboratories all over the world, but at last it could not get any experimental support, so that the work on the atomic theory was continuing along other directions (see the works [2.243–2.260] and the references therein).

   Interestingly, the above clear and instructive situation seems to be remaining widely unknown, although the experimental stories are instead very well known and being cited. See, e.g., one of the most recent truly huge projects on the philosophy of scientific research [2.261].

2. And of extreme, of special interest for us now is the reaction of Max Planck to the fact of such a suggestion, see his work

[2.262] (remarkably, this work of Max Planck is possessed of a clear philosophical, methodological scent [*sic*]). Indeed, Max Planck has commented the relevant situation as follows (we translate):

The deep seriousness of the difficulties described above casts a revealing light on the fact that most recently a suggestion has even come from the utterly competent side to sacrifice the assumption of the exact validity of the energy conservation principle—a definite way out, which probably might with some justification be considered a desperate move, but could soon be proven to be inaccessible by special experiments.

On the other hand, in his work about the second basic law published practically at the same time Max Planck asserted forcefully contrariwise [2.263], namely as follows (we translate):

Every natural process is taking place in the sense that the sum of the entropies of all bodies involved in the process is increasing. In the limiting case—for a reversible process—this sum remains unchanged. Or, in putting it more succinctly: The entropy ought to be a measure of the probability (§ 3). Herewith the contents of the second basic law of thermodynamics could be exhaustively characterized and, at the same time, the transition to the statistical definition of entropy could be made possible.

Remarkably, in the introduction part of this same communication—before all the paragraphs including the mentioned (§ 3)—Planck analyzed and criticized the work by Constantin Carathéodory. It is of considerable interest for our present discussion to cast a detailed look at the logics of Max Planck. Here it would definitely be appropriate to recall the conclusion by Prof. Dr. Mark Waldo Zemansky (1900–1981) as for the hard digestibility of the Kelvin–Planck's statement of the second basic law for students [2.264]. Along with this, Prof. Zemansky does complain of the mathematical over-complications of the original Carathéodory's approach. To this end, we recall the assertion by Prof. Peter Fong in his book [2.265] that Carathéodory 'had stopped halfway'.

Finally, it seems to be extremely important to try answering the poser of why Max Planck considered the statistical interpretation of the first basic law 'a desperate move', whereas the statistical interpretation of the second basic law—'an exhaustive solution'. Constantin Carathéodory, an outstanding world-class mathematician, has published only two works on the basics of the conventional thermodynamics [2.266, 2.267]. To analyze the logical nuclei of the latter and provide it with the clear rationale on par with the classical mechanics was in effect Carathéodory's very aim. Indeed, he could achieve this crucial goal in a generally comprehensive and abstract way. Carathéodory's effort was met by most of the colleagues enthusiastically and triggered some further refinements and revisions [2.268–2.271].

It is important to underline here that Carathéodory was by far not alone in his endeavor. In fact, already several years earlier, an outstanding Hungarian mathematician and physicist, Farkas Gyula (Julius Farkas, in German) had published substantially similar reports [2.272].

This portrayal of Gyula (Julius) Farkas has been taken from
https://de.wikipedia.org/wiki/Gyula_Farkas

Independently of Prof. Farkas, a renowned Russian physical chemist, Nikolai Nikolayevich Schiller, whose contributions we have already mentioned above, was trying to clarify the actual formulation and the second basic law of thermodynamics [2.273–2.294]

This portrayal of Prof. Dr. N. N. Schiller has been taken from
http://thebestartt.com/nikolay-nikolaevich-shiller

N. N. Schiller had in fact started his thermodynamic studies and the consequent publication of his results in the German, Holland's, and Russian periodicals from the year 1897 on, but those earlier original publications of him seem to be inaccessible right away. Remarkably, N. Schiller's studies of thermodynamic relevance were dealing with thermo-elastic and thermomagnetic properties of Matter, and had been published already in Russian in 1879 [2.278], after the relevant work by Prof. Dr. William Thomson (Lord Kelvin) appeared in 1857 [2.295]. In fact, that was Prof. Thomson's first detailed publication on the theme, whereas his next relevant detailed paper had been [2.296], the one triggered Prof. Schiller's interest in such a research topic. Noteworthy, an outstanding German theoretical physicist Prof. Dr. Woldemar Voigt (1850–1919) was actively continuing both Lord Kelvin's and Prof. Schiller's studies [2.297, 2.298].

Strikingly, although the both of the reports came out prior to those by Carathéodory, the report by Farkas remained unnoticed until very recently [2.299].

As concerns N. N. Schiller's reports, the most relevant of them had originally been published in Russian. Meanwhile, for the first time only Tatiana Ehrenfest-Afanassjewa (in Russian: Татьяна Алексеевна Афанасьева) (1876–1964) had consequently noticed Prof. Schiller's activity. She could publish in Russian a detailed logical analysis of the Carathéodory's work vs. the so-called Kelvin-Planck's trains of thoughts [2.300]. She had also provided her readership with a succinct note about her coming across the Schiller's work and accompanied her Russian note by a concise German translation. But anyway all this wasn't helpful in disseminating that important work by Prof. Dr. Schiller throughout the world.

According to Tatiana Ehrenfest-Afanassiewa the main conceptual (both mathematical and physical) point of the conventional 'equilibrium thermodynamics' consists in what follows. Conventionally, the system in question may be considered consecutively trespassing some states quite similar to the equilibrium ones, if the pertinent real process is sufficiently slow. This is just what they usually describe by such a mathematically explicable phrase:

> *The quasi-static processes ought to be limiting cases of the realistic infinitely slow processes.*

Max Planck was also actively re-phrasing the above idea in his writings, but he didn't seem to go beyond any kind of a sole re-phrasing. Instead, his above-mentioned rancorous fighting against all the interested mathematicians and physical theorists with the only aim to eradicate any kind of referencing the energetics might be (and is indeed) considered a fight for the novel (at that time) branch of physics—the quantum physics. **Our immediate comment:** Well, howbeit, but one apparent methodological problem to remain after consequently eliminating the energetics ought to be the inevitable separation between the first and the second basic laws. If we insist on splitting the fundamental law of energy conservation and transformation into two laws to act separately from each other: The Energy Conservation Law and The Energy Transformation Law we alienate the actual notion of Energy.

Indeed, in trying to formulate some separate Law of Energy Transformation, Rudolf Clausius had to explain the actual mechanism of how the Energy should be transformed. He had to introduce the basic notion of Entropy, by putting it at the heart of the Energy Transformation Process. Still, that Clausius fully separated the Energy Transformation from the Energy Conservation had immediately posed a conceptual challenge of losing the connection between the Energy and Entropy notion. Stubbornly insisting on the separation between the Two Basic Energy Laws had inevitably led to the "Entropy Puzzle," which is remaining unsolved till nowadays. It is just this stubborn separation that had also led to the "physicalization" of the Carnot cycle.

With the above in mind, we now come back to the story told by Tatiana Ehrenfest-Afanassiewa.

In view of the usual massive over-interpretation of N. L. S. Carnot's widget, the term "quasi-static" is very frequently substituted by the term "reversible," by completely forgetting (or, may perhaps, intentionally capping?) the fact that Carnot's 'widget process' is indeed reversible, but only because its starting and final points are coincident—owing to Carnot's ingenious idea, but anyway not owing to some 'mysterious physics underlying it'. This is why, concludes Tatiana Ehrenfest-Afanassiewa, it ought to be much more physically appropriate from the beginning on to try keeping the identifiers 'reversible' and 'quasi-static' separate from each other. Furthermore, any realistic process ought to go through the pertinent succession of the non-equilibrium states. Bearing this in mind Tatiana Ehrenfest-Afanassiewa suggests using a handy notation "non-static processes" to describe the latter events, for the term "irreversible processes" usually employed to denote them ought to be inappropriate at all. Indeed, the very work by Carathéodory is in effect devoted to analyzing the realistic, that is, irreversible processes. Nonetheless it should be noted that Carathéodory had introduced the term "quasi-static process" himself. And it is exactly in this connection that Tatiana Ehrenfest-Afanassiewa had equitably concluded, that what he was considering in effect ought to be conceived rather as a ***quasi-process***.

This portrayal of Prof. Dr. Ehrenfest-Afanassiewa has been taken from
http://lthmath.tumblr.com/post/126004740644/smithsonianlibraries-portrait-of-tatiana

Further on, Tatiana Ehrenfest-Afanassiewa does carefully analyze the notion of the infinitesimally small heat amount, dQ, accepted by the system when it carries out some work on its surrounding, but first of all, it is placed into the expression for the first basic law: dQ = dU + dA (Eq. 1.1). Here dU and dA stand for the infinitesimal changes in the amounts of internal energy and work, respectively. Further, to mathematically analyze changes in the system's state, we introduce some state variables, x_1, ..., x_N and assume that dQ can finally be re-cast as some proper function of the latter of the as follows: dQ = $Y_1 dx_1$ + ... + $Y_N dx_N$ (Eq. 1.2), where $Y_i = Y_i$ (x_1 ... x_N), with i = 1, ..., N, stand for some functions of the system's state variables determined by the structure of the system under study.

Eqs. (1.1) and (1.2) are properly expressing the first basic law in that any dQ < 0 means that the system acquires some non-zero amount of heat from its surrounding, when the system works on the latter. The word "acquires" should be understood in the algebraic sense, that is, the system might both "accept the heat from" and "transfer the heat to" its surrounding and, therefore, both dQ > 0 and dQ < 0 ought to be equally physically possible, respectively.

In effect, Eqs. (1.1) and (1.2) constitute the physical contents of the Carathéodory's first axiom [see his original works]. Besides, the processes exhibiting dQ = 0 stand for the so-called adiabatic ones. Consequently, the adiabatically isolated systems should have no heat exchange with their respective surroundings. And, accordingly, the second axiom by Carathéodory states (see his original works cited above—here we translate):

In any close proximity to any state of a system of bodies, there are neighboring states that are not reachable from the initial state by an adiabatic way.

It is exactly this Carathéodory's axiom that was forcefully attacked by Max Planck in 1926, but nonetheless, this axiom is nowadays fully acknowledged to be the true logical basis for the definition of the entropy notion, see the detailed work [2.301].

Meanwhile, Max Planck had said in this same regard [2.263] (we present here the English translation):

... The Thomson's principle readily proves to be superior as compared to the Carathéodory's. While the problem of perpetual motion of the second

kind countless times has been treated experimentally, probably no one has ever experimented with the intention to reach all neighboring states of any particular state by an adiabatic way. This consideration might seem to be not fully conclusive, but another viewpoint is also available. Indeed, in the present case all the numerous experimental confirmations of the second law consequences result from the Carathéodory's principle just in the same manner as from the Thomson's principle.

Yes, of course, with respect to the latter fact, one might be basically inclined to attach just a formal meaning to the whole discussion on the question posed here. This would only be correct if the thermodynamics ought to be a perfectly formulated, complete in itself, isolated area within the physical science. But this is well known to be definitely not the case. On the contrary, within the structure of the current physics, which is entirely built upon the atomistic basis, the so-called general thermodynamics serves as a specific part, as a certain limiting case, which is, strictly speaking, still not built-up with the absolute perfection.

And if we would consider the problem from the above standpoint, everything gains a completely different appearance. Then, the second basic law of thermodynamics loses its fundamental importance; it appears only as a statistical record, not valid for the properties of a single body system, but only for the mean values of the properties of a very large number of macroscopically identical copies of the system under consideration. The fluctuations of the individual values around the means are all the more significant, the fewer degrees of freedom has the system.

After reading the above citation a really strong feeling arises that the mathematical difficulties experienced when working on the thermodynamics foundations, while being based solely upon the ingenious cyclic widget by N. L. S. Carnot, had driven R. Clausius and the vast majority of his followers to looking for some intrinsic physics or even some deep philosophic significance of the actual theoretical gadget. Consequently, the fruitful idea was indeed to harp on about the atomistic basis of the matter, because that time it could seemingly deliver useful ideas without any thorough and detailed experimental examination. Of course, the availability of nothing more than sole optical microscopes at the time being was definitely not enough to skillfully organize the latter.

Undoubtedly, the latter circumstance ought to point out the significance of theoretical research, but would it be okay to explicitly play with the 'very large numbers of copies', 'fewer degrees of freedom', although it pushes the whole story into the Sorites paradox?

Meanwhile, playing with Sorites ought to deprive the theoretical conclusions to be reached of the desired generality. Therefore, at any rate we should avoid the explicit Sorites, if we are intending to introduce the theories of truly generally validity.

Here we come to the very point, where the actual roots of how the energetics, the unique basis of the thermodynamics could have been defeated in favor of the quantum physics. This is just what Peter Boas Freuchen was trying to clarify (see Chapter 1).

Indeed, at the time of Peter Boas Freuchen's book publication the story was looking like as follows (we dare to summarize here the points underlined by Peter in taking into account the whole breadth of the international situation):

1. Max Planck, most probably, the only hopeful theorist-thermodynamicist had no idea about what is the actual physical sense of the second basic law of thermodynamics, and how to correctly deal with the latter. In that truly revolutionary time any principal complications in the course of research works had caused the respective specialists to make desperate conceptual moves.

2. The Bohr–Kramers–Slater work discussed above was a desperate move to overcome theoretical difficulties of atomic physics by introducing the statistical interpretation of the first basic law. Fortunately, this apparent phantasy was finally overthrown by numerous experiments carried out all over the world.

3. Max Planck had made another desperate move to overcome notorious misunderstandings by introducing the statistical interpretation of the second basic law. This move could work out, for it was (and most probably still is!) sheer impossible to design experiments checking the realistic background.

4. How should we estimate the significance of Max Planck's move in question? Was it a positive or a negative event?

5. On the one hand, this was definitely a theoretically correct move. Ludwig Boltzmann could ingeniously guess the correct functional relationship between the entropy and probability notions. Dr. Georg(e) Augustus Linhart could unambiguously prove the validity of the logarithmic functional relationship suggested.

6. On the other hand, Max Planck, Planck's, and Boltzmann's numerous followers have drastically over-interpreted the formula, by declaring and forcefully following through with the idea of primary importance and fundamental significance of the probability notion. Meanwhile, the latter conclusion is not correct.

7. Nonetheless this 'probabilistic approach' has caused an avalanche of further very strange speculations as for the actual physical meaning of the entropy notion.

The modern state of art is clearly described in the following nicely written and truly popular books by Prof. Dr. Arieh Ben-Naim, an outstanding specialist in theoretical biophysical chemistry [2.302–2.310].

In reading the above books we learn that all the misinterpretations of the entropy notion are being successfully overcome, whereas the most stubborn ought to be the 'probabilistic approach' stubbornly shared to the full by Prof. Ben-Naim himself, among numerous colleagues all over the world. Indeed, except for the purely mathematical parallel (logarithmic functional dependence) the thermodynamic entropy has essentially nothing to do with the notion of information. Still, any attempt to utter the latter statement ought to resemble 'a voice crying in the wilderness'.

Well, the only what remains in the current situation is just to cite the wise words by Tatyana Alexeyevna Ehrenfest-Afanassjewa she published in her paper [2.311], which might in fact be viewed as her true bequest:

"... By the preceding I am far from rejecting the application of statistical formulas, which coincide with the 'most probable from a certain point of view'. I only want to claim the impossibility of assuming the 'chance' as the ultimate rationale of any physical laws."

In fact, this is the result of her own and her spouse's life-long work on the foundations of thermodynamics and the actual interrelationship between the latter and the statistical mechanics. Noteworthy, they both, and especially Prof. Dr. Paul Ehrenfest (1880–1933), are definitely not wearing the veils of oblivion.

With this in mind we revert now to the book of Peter Boas, which has triggered all the above deliberations. We would now like to explicitly summarize what Peter had no more time for—taking into account a small volume of his first publication, the over-all political situation all around the world, plus a truly wild referee's report concerning his first attempt.

Peter has clearly demonstrated that an outstanding thermodynamic hopeful of the beginning of the XX-th century, Max Planck, had in effect no idea on how to correctly interpret the second basic law and therefore the entropy notion. In fact, his truly desperate move has resulted in over-interpreting the bold (*and, in effect, fully successful!*) guess by Boltzmann. Indeed, using the Boltzmann's famous formula, Max could still successfully solve his research problems and get the due credit. Congratulations and celebrations once again to Max!

But an interesting question as to *why* Boltzmann's formula is correct has nonetheless remained unanswered. Moreover, the over-all stance in regard to the second basic law, as Peter clearly demonstrates, was a fancy mixture of enthusiasm, respect and dissent. Themselves, the 'holy fathers' of thermodynamics could not manage to skilfully formulate this truly fundamental basic law, because their entire research activity was not connected with thermodynamics itself. Meanwhile, the main discoverer of the actual modalities of this law, N. L. S. Carnot, had already entered the ultimate business trip without arrival.

The second basic law formulations available at that time were negations/denials (perpetuum mobile is impossible!), and this way could excite no other feeling than a strong dissent. It is especially the latter feeling that is capable of creating a non-professional, purely emotional striving to look for the cases, where this evil law loses its validity. This same stance was rather widespread in regard to the first basic law as well (in fact basically the same "impossibility of perpetuum mobile" was at work!). Most probably, it is just this situation that caused even such undoubtedly competent and therefore deservedly renowned colleagues, like Rudolf Julius Emanuel Clausius, Ludwig Eduard Boltzmann, Peter Guthrie Tait, James Clerk Maxwell, to produce emotional statements of no physical value at all.

This is just how the following idea was born (here we allow ourselves to paraphrase the actual utterances): '*Well, the basic laws seem to be truly relentless, but still a non-zero probability ought to exist that in some particular cases this evil law does lose its validity*' (Boltzmann), '*There ought to be a reasonable micro-creature capable of withstanding the attacks of the evil second basic law*' (Maxwell). To our sincere regret, the latter both utterances stand for the actual '*deep fundamental physics*' behind the '*basic laws*' of the conventional '*equilibrium thermodynamics*'!

Furthermore, the fancy combination of the apparent failure to prove the 'statistical interpretation' of the first basic law and the remarkable (but not immediately explicable) success of the 'statistical interpretation' of the second basic law have caused a deep conceptual gap between the both basic laws. This gap has but nothing to do with the actual physics—it was the logical outcome of the emotional approach discussed above.

As a firm result, there remains no regular conceptual basis under the general macroscopic thermodynamics, except for the "quasi-static Carnot cycle possessed of no clear driving force to cause its actual working—and thus remaining forever in its equilibrium state"—to maintain that "perpetuum mobile does basically exist, but it must be sheer improbable, due to the basic indeterminism."

This is just why Dr. Linhart states en clair that Planck, Einstein etc. could in effect not achieve their aims, although the entirety of the mathematical formulas they have suggested and practically introduced are indeed correct and very useful.

Up to nowadays, the foundations of statistical mechanics are also still remaining at the level set up by Ludwig Boltzmann and Josiah Willard Gibbs. It is to 100% clear, that the both eminent colleagues would definitely put their seminal studies to the logical end, if they had a bit more time to stay upon earth.

Is the over-all picture really as lugubrious as depicted above?

No, of course not!

2.4 Why Linhart offers no justification for his definitions of *K* as efficiency?

The immediate answer would be: Most probably, he had postponed a more detailed discussion of this point to his later studies/ publications. Therefore, re-casting a look at Dr. Linhart's work under such an angle of view ought to be of definite interest, and we are presently working on trying to clarify this point.

2.5 Why Linhart goes in for some weird speculations that all bodies are being black at $T \to 0$ and behave like blackbody radiation?

Above we have discussed the story about the blackbody radiation and its theoretical processing. Everybody interested in the field at that time was definitely aware of the criticism by Walther Ritz and other colleagues including Max Bernhard Weinstein. A much narrower circle of professionals was aware of the results of Edgar Buckingham. Howbeit, we might assume that Dr. Linhart was attentive not only to everything being published in German and in English on the theme, but also to the destinies of the actors in the scene, in particular to their escapes from the academic circles—like those by Profs Buckingham and Strong, including George Linhart himself.

Meanwhile, the logics to become conventional—even until nowadays—was stubbornly dictating that:

(a) Everything is consistent of atoms/molecules and so on.
(b) Even in the absence of any realistic process the micro-particles mentioned above are experiencing their eternal dynamics, which is basically chaotic.
(c) This fact has been revealed experimentally already long ago, and nowadays nobody is questioning its existence.
(d) Specifically, the chaotic thermal motions of the micro-particles of any physical body produce the so-called blackbody radiation.
(e) In the absence of any realistic process any system under study ought to persist in its equilibrium state.
(f) If so, the interested theorists ought to employ the probability theory to duly describe the above-mentioned facts.
(g) The famous formula ingeniously guessed by Ludwig Boltzmann and just picked up by Max Planck (without even a minuscule conceptual analysis of the latter!) should therefore build up the true basis of the equilibrium thermodynamics/ statistical mechanics.

(h) To sum up, the entropy is nothing more and nothing less than just a *probability*.

    With the above-sketched conceptual framework in mind we ought to pursue our theoretical researches based upon the axioms listed above, by deriving our considerations from the four basic laws of the equilibrium thermodynamics (of which for the present we are aware of at least four, but who knows: There might even be much more to them?). **Important Note**: Any volitional violation of the four basic laws mentioned above (likewise attempting to be somehow skeptical in regard to them) ought to be equal to the Heresy. *Amen!*

Our immediate comment: In fact, for any active, proactive researcher it was, is and still remains to be truly very easy to violate these basic laws, especially the second one, for anybody of the 'holy fathers' did have not the faintest idea about what is the actual true physical sense of the latter. Hence, you, the 'dirty heretics', are predestined to violate them! See, e.g., the detailed contents of Refs. 34 and 35 in the publication list by Max Bernhard Weinstein (Chapter 3).

    To this end, we have seen in Chapter 4 of the present book that during his studies Dr. G. A. Linhart could definitely kill the weird, purely emotional 'red herring' of the 'probability as the only sense of entropy'.

    Specifically, George Augustus Linhart could theoretically demonstrate that

(i) There is absolutely no need to overthrow the atomistic picture of the physical matter.

(j) The above-mentioned picture is basically *fuzzy*, because of the huge numbers of the particles under study.

(k) The latter principal obstacle might be theoretically overcome, if we employ the Bayesian statistical approach.

(l) It is just in such a way that we might formally complete the missing Boltzmann–Planck inference, to properly clarify the actual physical sense of the probability function appearing under the logarithmic expression for the entropy.

(m) Indeed, this mysterious *probability/number of complexions*, or how ever we might wish to poetically denote it, is nothing more than just a handy algebraic function of the absolute temperature.

(n) It is the latter proper combination of the logarithmic function with the algebraic one that drives the entropy to its absolute zero at the absolute zero of the temperature.

Here we would not like to announce *Amen*, as we have just done above, for Dr. Linhart's achievement is nothing more and nothing less than an invitation to perform any possible kind of further works on the fundamentals and applications of thermodynamics and statistical mechanics.

Meanwhile, Dr. Linhart's professional contemporaries at the Berkeley University could definitely recognize the latter fact. Moreover, all of them could see that his unambiguous victory was by far not the final point, for the **notorious** (*no other word for it!*) '*basic third law of thermodynamics*', being therewith mathematically killed as well, was nonetheless still physically alive, as an experimentally detectable rule.

Indeed, there were young and active colleagues in the nearest proximity to Linhart who were actively working exactly on the modalities of this 'basic law' as well. The brightest example of them was Dr. William Francis Giauque (1895–1982), the winner of the Nobel Prize in 1949 for his experimentally studying the properties of matter at temperatures close to the absolute zero.

Remarkably, the fully recognized principal objective of Dr. Giauque's research was '*to demonstrate that the third law of thermodynamics is a basic natural law*', using the range of appropriate experimental tests, i.e., **just to the contrary** of what Dr. Linhart could have proven theoretically.

Now there are definitely no doubts that Linhart's result does in no way cancel the achievements of Dr. Giauque, for the results of the latter colleague could serve as a trigger for further important developments in such a generally important field as cryogenics.

With this in mind, at the first glance, in the active time of both Giauque and Linhart it is just the latter colleague that could get nothing more than the fully improper label of a 'cheap barrater'— and nothing more than this.

Still, the main point of applied/practical significance here was, is and will remain to be the cryogenics itself, if we take into account the firmly established fact that entropy is always heading to zero together with the absolute temperature. However, as entropy's heading to zero means the consequent ultimate termination for any kind of the realistic processes, it is difficult to underestimate the significance of such a field of applied science as the cryogenics based upon the fact of practical unreachability of the zero absolute temperature.

To sum up, the only valid physical question, which still remains basically unanswered—*why* the latter experimental fact ought to be true? That looks like a good poser to deal with for the younger generation of natural scientists, so here we shall stop this discussion to come back to this point once more a bit later on.

The situation around Dr. Linhart as a whole is also closely related to the materials already discussed in the present note. We might refer the interested reader to our detailed answer to the Question 2.3 asked by Prof. B. H. Lavenda. He is wondering about the 'alleged failures' of Planck, Einstein, etc. in writing down the simple expression for the heat capacity as a function of temperature.

As we could see in Chapter 4 of this book, Dr. Linhart could not only kill the 'red herrings' of the 'entropy–probability interconnection', of the 'third basic law' of thermodynamics, but also derive a handy expression for the heat capacity at the constant volume as a function of temperature.

The general functional form of the latter is just the one adopted by the Hill/Langmuir isotherms. In addition, as Dr. Linhart has carefully checked in detail, such a functional form is capable of perfectly fitting all the pertinent experimental data in the whole temperature range under study. Noteworthy, that was also the Question 2.2 by Prof. Lavenda, and above in this note we have already discussed this very important point. Dr. Linhart has not blindly mimicked the Hill/Langmuir isotherm-like functional relationships for his purposes: He has formally mathematically derived them using this same Bayesian statistical approach to the fuzzy problem of 'a huge number of atoms/molecules'.

Noteworthy, except for Linhart's result, the relevant formula by Debye is the only one that is capable of correctly fit experimental data, although its functional form is much more complicated as compared to that of Linhart, whereas the formulae by Einstein and Nernst-Lindemann are already known to cause noticeable problems when interpreting the relevant experimental data.

The reason for this might be that in deriving their formulae Einstein, Nernst, and Lindemann were using Planck's quantum-statistical approach to the heat radiation without carefully checking its actual validity for the case under study, while Debye could modify the mathematical—but not the over-all physical—standpoint to some significant extent [2.312–2.315].

Bearing all the above in mind, it ought to be psychologically clear that such a young, active, talented physical chemist as Linhart was vigorously trying to enter the field of pure physics, or at least to trigger more profound physical studies on the problem he indicated. Still, all of Linhart's attempts had turned out to be in vain, as we now know—to our sincere regret.

Further, one of the significant goals of the present Chapter is trying to answer Prof. Lavenda's question 1) (see Page 609) in connection with the gloomy picture painted by Prof. Truesdell.

Sure, the posers we face are of undoubted fundamental significance, and answering them immediately produced the difficulties of the same kind. Indeed, the posers we are dealing with ought to inevitably tend to remain rhetoric ones.

With this in mind we would humbly appreciate, at the very least, to show the possible right directions of complete answering the posers at hand. Please, note, that we have already placed an extensive discussion about this immensely important topic into Chapter 1 of the volume at hand, in connection with the book by Peter Boas Freuchen and the wild reaction in regard to that book. So here we would just like to finalize the discussion started over there.

References to Note 2

2.1 John T. Edsall (**1974**): Some notes and queries on the development of bioenergetics: Notes on some "founding fathers" of physical chemistry: J. Willard Gibbs, Wilhelm Ostwald, Walther Nernst, Gilbert Newton Lewis, *Molecular and Cellular Biochemistry*, 5(2), pp. 103–112.

2.2 William L. Jolly (**1987**): *From Retorts to Lasers: The Story of Chemistry at Berkeley*, Distributed by the College of Chemistry, University of California.

2.3 G. N. Lewis (**1926**): *Anatomy of Science*, Yale University Press, New Haven, USA; Oxford University Press, London, Great Britain.

2.4 Simón Reif-Acherman (**2008**): Otto Redlich: chemist and gentleman from the "old school," *Química Nova*, 31(7), pp. 1901–1908.

2.5 Otto Redlich (**1970**): The basis of thermodynamics, in: *A Critical Review of Thermodynamics*, E. B. Stewart, B. Gal-Or, A. J. Brainard, Editors, Mono Book Corp., Baltimore, USA (pp. 439–444).

2.6 A. V. Hill (**1910**): The possible effects of the aggregation of the molecules of hemoglobin on its dissociation curves, *J. Physiol.*, 40(Suppl.), pp. iv–vii.

2.7 Irving Langmuir (**1916**): The constitution and fundamental properties of solids and liquids. Part I. Solids, *J. Am. Chem. Soc.*, 38, pp. 2221–2295.

2.8 Irving Langmuir (**1917**): The constitution and fundamental properties of solids and liquids. Part II. Liquids, *J. Am. Chem. Soc.*, 39, pp. 1848–1906.

2.9 Irving Langmuir (**1918**): The adsorption of gases on plane surfaces of glass, mica and platinum, *J. Am. Chem. Soc.*, 40, pp. 1361–1403.

2.10 Irving Langmuir (**1932**): *Nobel Price Lecture*: Nobel Price Foundation.

2.11 Andres Ortiz (**2013**): Derivation of Hill's equation from scale invariance, *J. Uncertain Sys.*, 7, pp. 198–202.

2.12 E. B. Starikov (**2012**): George Augustus Linhart as a "widely unknown" thermodynamicist, *World Journal Condensed Matter Physics*, 2(2).

2.13 Rudolf Seising(**2009**): *Views on Fuzzy Sets and Systems from Different Perspectives: Philosophy and Logics, Criticisms and Applications*, Springer Verlag: Berlin, Heidelberg, Germany.

2.14 Barnabas Bede (**2013**): *Mathematics of Fuzzy Sets and Fuzzy Logic*. Berlin, Heidelberg, Germany.

2.15 Ludwig Boltzmann (**1871**): Zur Priorität der Auffindung der Beziehung zwischen dem zweiten Hauptsatze der mechanischen Wärmetheorie und dem Principe der kleinsten Wirkung, *Annalen der Physik*, 219, pp. 211–230.

2.16 Rudolf Clausius (**1871**): Bemerkungen zu der Prioritätsreclamation des Hrn. Boltzmann, *Annalen der Physik*, 220, pp. 265–274.

2.17 Ludwig Boltzmann (**1896**): Ein Wort der Mathematik an die Energetik, *Annalen der Physik*, 293, pp. 39–71.

2.18 Ludwig Boltzmann (**1896**): Zur Energetik, *Annalen der Physik*, 294, pp. 595–598.

2.19 Ernst Heinrich Wilhelm Schmidt (**1953**): *Die Einführung in die technische Thermodynamik und in die Grundlagen der chemischen Thermodynamik.* Fünfte berichtigte Auflage, Springer-Verlag GmbH, Berlin, Heidelberg.

2.20 Ernst Heinrich Wilhelm Schmidt and Joseph Kestin (**1949**): *Thermodynamics: Principles and Applications to Engineering.* 1st Edition, Clarendon Press, Oxford, Great Britain.

2.21 Hermann von Helmholtz (**1884**): *Studien zur Statik Monozyklischer Systeme.* Berliner Berichte am 6. März 1884.

2.22 Hermann von Helmholtz (**1895**): *Wissenschaftliche Abhandlungen,* Band 3, p. 121, Johann Ambrosius Barth, Leipzig, Germany.

2.23 E. Budde (**1892**): *Über integrierende Divisoren und Temperatur, Annalen der Physik,* 281, 751.

2.24 Max Planck (**1926**): Über die Begründung des zweiten Hauptsatzes der Thermodynamik. Berliner Berichte am 27. Dezember 1926.

2.25 J. W. Gibbs (**1906**): *The Scientific Papers: Thermodynamics.* Longmans, Green: London, New York, and Bombay.

2.26 F. Reech (**1853**): Théorie générale des effets dynamiques de la chaleur, *Journal de Mathématiques Pure et Appliquées,* 1, 357.

2.27 F. Massieu (**1869**): Sur les fonctions caractéristiques des divers fluides et sur la théorie des vapeurs, *Comptes Rendus,* 69, 858–862, 1057–1061.

2.28 F. Massieu (**1876**): *Thermodynamique: Mémoire sur les fonctions caractéristiques des divers fluides et sur la théorie des vapeurs'.* Mémoire de l'Académie des Sciences de l'Institut National de France.

2.29 Henri Poincaré (**1892**): *Thermodynamique.* Georges Carré, Éditeur, Paris, France.

2.30 P. Duhem (**1911**): *Traité d'Énergétique ou de Thermodynamique générale,* t. I et II, Éditions Jacques Gabay, Paris, France.

2.31 Ludwig Boltzmann (**1880**): Zur Theorie der Gasreibung, *Sitzungsberichte der K. u. K. Akademie der Wissenschaften zu Wien,* 81(2), pp. 117–158.

2.32 Max Planck (**1891**): *Über das Prinzip der Vermehrung der Entropie. Vierte Abhandlung, Wiedemanns Annalen der Physik und Chemie,* XLIV, pp. 385–428.

2.33 Max Planck (**1901**): Bemerkungen zu einer Abhandlung über Thermodynamik des Hrn. K. Wesendonck, *Annalen der Physik und Chemie,* 306, pp. 621–624.

2.34 Dieter Hoffmann (**2010**): *Max Planck und die moderne Physik.* Springer Verlag: Berlin, Heidelberg, Germany.

2.35 Ludwig Boltzmann (**2012**): *Wissenschaftliche Abhandlungen* (1865–1905). Friedrich Hasenöhrl (Ed), Volumes: 1–3, Cambridge University Press, Cambridge, UK.

2.36 Oskar Emil Meyer (**1899**): *Kinetic Theory of Gases: Elementary Treatise with Mathematical Appendices.* Longman, Green and Co., London, New York, Bombay. (**Our Comment**: *A clearly written and thorough treatise in the field. The original book in German has been first published in 1877, in Breslau, now Wroclaw, Poland*).

2.37 Carl Gottfried Neumann (**1875**): *Vorlesungen über die mechanische Theorie der Wärme.* Verlag und Druck: B. G. Teubner, Leipzig, Germany.

2.38 Carl Gottfried Neumann (**1891**): Bemerkungen zur mechanischen Theorie der Wärme. Berichte über die Verhandlungen der königlich sächsischen Gesellschaft der Wissenschaften zu Leipzig. *Mathematisch-physische Klasse*, 43, pp. 75–156.

2.39 Otto Wiedeburg (**1890**): *Über die Hydrodiffusion, Annalen der Physik*, 277, pp. 675–711.

2.40. Otto Wiedeburg (**1894**): *Über die Gesetze der galvanischen Polarisation und Elektrolyse, Annalen der Physik*, 287, pp. 302–345.

2.41 Otto Wiedeburg (**1894**): Das Gibbs'sche Paradoxon, *Annalen der Physik*, v. 289, pp. 684–697.

2.42 Otto Wiedeburg (**1894**): *Über die Potentialdifferenzen zwischen Metallen und Elektrolyten, Annalen der Physik*, 295, pp. 742–749.

2.43 Otto Wiedeburg (**1896**): *Über nicht umkehrbare Vorgänge. I, Annalen der Physik*, 297, pp. 705–736.

2.44 Otto Wiedeburg (**1897**): *Über nicht umkehrbare Vorgänge. II.* Gesetze der Widerstandsgrößen, *Annalen der Physik*, 297, pp. 652–679.

2.45 Otto Wiedeburg (**1897**): Ein physikalisches Entwickelungsprinzip, *Annalen der Physik*, 298, pp. 154–159.

2.46 Otto Wiedeburg (**1898**): *Über nicht umkehrbare Vorgänge. III. Die Stellung der Wärme zu den anderen Energieformen; Gesetze der spezifischen Wärme, Annalen der Physik*, 300, pp. 519–548.

2.47 Otto Wiedeburg (**1898**): Zur Frage nach der absoluten Temperatur, *Annalen der Physik*, 301, pp. 921–922.

2.48 Otto Wiedeburg (**1898**): Vergleichende Messungen der Wärmestrahlung der Metalle, *Annalen der Physik*, 302, pp. 92–110.

2.49 Otto Wiedeburg (**1899**): *Über Zustandsgleichung und Energiegleichungen, Annalen der Physik*, 305, pp. 66–82.

2.50 Otto Wiedeburg (**1900**): Energetische Theorie der Elektrizität und die Wärmeleitung der Metalle, *Annalen der Physik*, 306, pp. 758–789.

2.51 Otto Wiedeburg (**1901**): Zum zweiten Hauptsatz der Thermodynamik, *Annalen der Physik*, 310, pp. 514–547.

2.52 K. Wesendonck (**1892**): *Über Electricitätserregung bei Reibung von Gasen an Metall, Naturwissenschaftliche Rundschau*, 18, pp. 225–226.

2.53. K. Wesendonck (**1892**): *Über Electricitätserregung bei Reibung von Gasen an Metall, Annalen der Physik*, 283, pp. 529–566.

2.54 K. Wesendonck (**1895**): *Über einige Beobachtungen von Hrn. Villard den kritischen Zustand betreffend, Annalen der Physik*, 291, pp. 577–591.

2.55 K. Wesendonck (**1897**): Zur die Thermodynamik der Lumineszenz, *Annalen der Physik*, 298, pp. 706–798.

2.56 K. Wesendonck (**1899**): Zur Thermodynamik, *Annalen der Physik*, 303, pp. 444–451.

2.57 K. Wesendonck (**1899**): Zur Thermodynamik, *Annalen der Physik*, 305, pp. 809–833.

2.58 K. Wesendonck (**1900**): Weiteres zur Thermodynamik, *Annalen der Physik*, 307, pp. 746–756.

2.59 K. Wesendonck (**1902**): Einige Bemerkungen über die Arbeit des Hrn. Wiedeburg zum zweiten Hauptsatz der Thermodynamik, *Annalen der Physik*, 312, pp. 576–583.

2.60 K. Wesendonck (**1903**): Über einige Beziehungen des zweiten Hauptsatzes der Thermodynamik zur Leistung mechanischer Arbeit, *Physikalische Zeitschrift*, 4, pp. 329–333.

2.61 K. Wesendonck (**1903**): Zur Lehre von der Zerstreuung der Energie, *Physikalische Zeitschrift*, 4, pp. 589–592.

2.62 K. Wesendonck (**1904**): *Über die thermodynamische Herleitung der physikalisch-chemischen Gleichgewichtsbedingungen, Physikalische Zeitschrift*, 5, pp. 521–525.

2.63 K. Wesendonck (**1905**): *Über freie Energie, Physikalische Zeitschrift*, 6, pp. 545–548.

2.64 K. Wesendonck (**1907**): Einige Bemerkungen zu Herrn Boltzmanns Theorie der Zusammenstöße allgemeinster Art, *Physikalische Zeitschrift*, 8, pp. 179–182.

2.65 K. Wesendonck (**1913**): Zur Thermodynamik, *Verhandlungen der Deutschen Physikalischen Gesellschaft*, 15, pp. 839–856.

2.66 Giuseppe Belluzzo (**1904**): Principi di termodinamica grafica, *Il Nuovo Cimento*, 8, pp. 196–222.

2.67 Giuseppe Belluzzo (**1904**): Principi di termodinamica grafica, *Il Nuovo Cimento*, 8, pp. 241–263.

2.68 Albert Fliegner (**1901**): Thermodynamische Maschinen ohne Kreisprozess, *Vierteljahresschrift der Naturforschungsgesellschaft zu Zürich*, 46, pp. 94–-121.

2.69 Albert Fliegner (**1903**): *Über den Clausius'schen Entropiesatz*, *Vierteljahresschrift der Naturforschungsgesellschaft zu Zürich*, 48, pp. 1–48.

2.70. Prof. W. Mc F. Orr M. A. (**1904**): On Clausius' theorem for irreversible cycles, and on the increase of entropy, *Philosophical Magazine Series 6*, 8(46), pp. 509–529.

2.71 Prof. Dr. Max Planck (**1905**): On Clausius' theorem for irreversible cycles, and on the increase of entropy, *Philosophical Magazine Series 6*, 9(49), pp. 167–169.

2.72 Prof. W. McF. Orr M. A. (**1905**): On Clausius' theorem for irreversible cycles, and on the increase of entropy, *Philosophical Magazine Series 6*, 9(53), 728–730.

2.73 Roger Balian, Dirk ter Haar, and John F. Gregg (**2006**): *From Microphysics to Macrophysics*, Vol. 1. Springer-Verlag: Berlin, Heidelberg, Germany.

2.74 Roger Balian, Dirk ter Haar, and John F. Gregg (**2007**): *From Microphysics to Macrophysics*, Vol. 2. Springer-Verlag: Berlin, Heidelberg, Germany.

2.75 Anatoli Polkovnikov and Vladimir Gritsev (**2008**): Breakdown of the adiabatic limit in low-dimensional gapless systems, *Nature Physics*, 4, pp. 477–481.

2.76 Wilhelm Zwerger (**2008**): Thermodynamics: limited adiabaticity, *Nature Physics*, 4, pp. 444–446.

2.77 James Swinburne (**1904**): *Entropy, or Thermodynamics from the Engineer's Standpoint and the Reversibility of Thermodynamics*. Archibald Constable & Co., Westminster, United Kingdom.

2.78 Uwe Krey and Anthony Owen (**2007**): *Basic Theoretical Physics: A Concise Overview*. Springer-Verlag: Berlin, Heidelberg, Germany.

2.79 Jakob Johann Weyrauch (**1904**): *Über die spezifischen Wärmen des überhitzten Wasserdampfes, Zeitschrift des Vereins Deutscher Ingenieure*, 48, pp. 24–28; 50–54.

2.80 Jacobus Henricus van't Hoff (**1904**): Einfluss der Änderung der spezifischen Wärme auf die Umwandlungsarbeit, in: *Festschrift Ludwig Boltzmann, gewidmet zum sechzigsten Geburtstage*. Verlag von Johann Ambrosius Barth, Leipzig, Germany.

2.81 Arturo Giammarco (**1903**): Un caso di corrispondenza in termodinamica, *Il Nuovo Cimento (1901-1910)*, 5, pp. 377–391.

2.82 Jakob Johann Weyrauch (**1905**): *Grundriss der Wärmetheorie: Mit zahlreichen Beispielen und Anwendungen. Nach Vorträgen an der Kgl. Technischen Hochschule in Stuttgart.* Vol. 1. Stuttgart: Konrad Wittwer, Germany.

2.83. John E. J. Schmitz (**2007**): *The Second Law of Life: Energy, Technology, and the Future of Earth as We Know It.* William Andrew Publishing: Norwich, NY, U. S. A.

2.84 Max Planck (**1950**): *Scientific Autobiography and Other Papers.* Williams & Norgate: London and Edinburgh, U.K.

2.85 Albert Einstein (**1912**): *Rapport sur l'état actuel du problème des chaleurs spécifiques.* La théorie du rayonnement et les quanta, Rapports et discussions de la Réunion tenue à Bruxelles, du 30 octobre au 3 novembre 1911 Sous les auspices de M. E. Solvay. Publiés par MM. P. Langevin et M. de Broglie, Paris. Imprimerie Gauthier-Villars, France.

2.86 Max Planck (**1906**): *Vorlesungen über die Theorie der Wärmestrahlung.* Verlag von Johann Ambrosius Barth, Leipzig, Germany.

2.87 Johannes Diderik van der Waals jr. (**1911**): *Über die Erklärung der Naturgesetze auf statistisch-mechanischer Grundlage, Physikalische Zeitschrift*, XII, pp. 547–549.

2.88 M. Tribus, and E. C. McIrvine (**1971**): Energy and information, *Scientific American*, 225, pp. 179–188.

2.89 Bernhard F. Arnold (**1998**): Testing fuzzy hypotheses with crisp data, *Fuzzy Sets and Systems*, 94, pp. 323–333.

2.90 S. M. Taheri and J. Behboodian (**2001**): A Bayesian approach to fuzzy hypotheses testing. *Fuzzy Sets and Systems*, 123, pp. 39–48.

2.91 Martin Holeňa (**2004**): Fuzzy hypotheses testing in the framework of fuzzy logic, *Fuzzy Sets and Systems*, 145, pp. 229–252.

2.92 Neli R. S. Ortega, Eduardo Massad, and Cláudio José Struchiner (**2008**): A Bayesian approach to fuzzy hypotheses testing for the estimation of optimal age for vaccination against measles, *Mathematics and Computers in Simulation*, 79, pp. 1–13.

2.93 Erwin Schrödinger (**1948**): *Statistical Thermodynamics*. Cambridge University Press, Cambridge, U. K.

2.94 Evgeni B. Starikov (**2010**): Many faces of entropy or Bayesian statistical mechanics, *ChemPhysChem*, 11, pp. 3387–3394.

2.95 Evgeni B. Starikov (**2015**): *Statistical Mechanics in Bayesian Representation: How It Might Work and What Ought to be the Probability Distribution behind It*. Proceedings of the Conference: SEE-Mie2015, The city of Tsu, county of Mie, Japan; (*in Open Access, see the URL address*): https://www.researchgate.net/publication/281030679_Statistical_Mechanics_in_Bayesian_Representation_How_It_Might_Work_and_What_Ought_to_be_the_Probability_Distribution_Behind_It

2.96 Patrick Coffey (**2008**): *Cathedrals of Science: The Personalities and Rivalries That Made Modern Chemistry*. Oxford University Press: Oxford, New York, Auckland, Cape Town, Dar es Salaam, Hong Kong, Karachi, Kuala Lumpur, Madrid, Melbourne, Mexico City, Nairobi, New Delhi, Shanghai, Taipei, Toronto.

2.97 Lieut.-Col. Richard de Villamil (**1912**): *ABC of Hydrodynamics*, E. & F. N. Spon, Ltd., London and New York.

2.98 Lieut.-Col. Richard de Villamil (**1912**): *The Laws of Avanzini: Laws of Planes Moving at An Angle In Air and Water*, E. & F. N. Spon, Ltd., London and New York.

2.99 Lieut.-Col. Richard de Villamil (**1914**): *Motions of Liquids*, E. & F. N. Spon, Ltd., London and New York 1914.

2.100 Lieut.-Col. Richard de Villamil (**1917**): *Resistance of Air*, E. & F. N. Spon, Ltd., London and New York.

2.101 Lieut.-Col. Richard de Villamil (**1920**): *Soaring flight: A Simple Mechanical Solution of the Problem*, University of Michigan Library, USA.

2.102 Lieut.-Col. Richard de Villamil (**1928**): *Rational Mechanics*, E. & F. N. Spon, Ltd., London and New York 1928.

2.103 Lieut.-Col. Richard de Villamil (**1931**): *Newton: The Man*, Gordon D. Knox: London, U. K.

2.104 Daniel Berthelot (**1922**) : *La physique et la métaphysique des théories d'Einstein*, Payot et Cie, Paris, France.

2.105 Daniel Berthelot (**1924**): *La science et la vie moderne*, Payot et Cie, Paris, France.

2.106 Daniel Berthelot (**1925**): *La doctrine de la relativité et les théories d'Einstein*, Société des ingénieurs civile de France, Paris, France.

2.107 J. Wisniak (**2010**): Daniel Berthelot. Part. I. Contribution to thermodynamics, *Educ. Quim.*, 21, pp. 155–162.

2.108 J. Wisniak (**2010**): Daniel Berthelot. Part. II. Contribution to electrolytic solutions, *Educ. Quim.*, 21, pp. 238–245.

2.109 J. Wisniak (**2010**): Daniel Berthelot. Part. III. Contribution to photochemistry, *Educ. Quim.*, 21, pp. 314–323.

2.110 Moïse-Emmanuel Carvallo (**1934**): *La théorie d'Einstein démentie par l'expérience*. Chiron, Paris, France.

2.111 Moïse-Emmanuel Carvallo (**1893**): *Traité de mécanique*. Librairie Nony, Paris, France.

2.112 Moïse-Emmanuel Carvallo (**1896**): *Méthode pratique pour la résolution numérique complète des équations algébriques ou transcendantes*. Librairie Nony, Paris, France.

2.113 Spenta R. Wadia (**2006**): *The Legacy of Albert Einstein: A Collection of Essays in Celebration of the Year of Physics*. World Scientific: New Jersey, London, Singapore, Beijing, Shanghai, Hong Kong, Taipei, New Delhi,.

2.114 Alexandre Moatti (**2007**): *Einstein: Un siècle contre lui*. Odile Jacob, Paris, France.

2.115 Alexandre Moatti (**2013**): *Alterscience: Postures, dogmes, idéologies*. Odile Jacob, Paris, France.

2.116 Jeffrey Crelinsten (**2016**): *Einstein's Jury: The Race to Test Relativity*. Princeton University Press, Princeton and Oxford.

2.117 Karl Bohlin (**1887**): *Om betydelsen af lefvande kraftens Princip för frågan om dynamiska systems stabilitet*. Bihang till K. Svenska Vet.-Akad. Handlingar. Band 13. Afd. I. N:o 1. (*The meaning of the Livening Force for the issue of dynamic system's stability*).

2.118 Lars Gårding (**1998**): *Mathematics and Mathematicians: Mathematics in Sweden before 1950*. American Mathematical Society: Providence, Rhode Island, USA.

2.119 Helmut A. Abt (**2007**): Karl Petrus Teodor Bohlin, in: *The Biographical Encyclopedia of Astronomers*. Springer Science+Business Media LLC: New York, USA, pages 143–144.

2.120 K. Bohlin (**1903–1904**): Sur le choc considéré comme fondement des théories cinétiques de la pression des gaz et de la gravitation universelle, *Arkiv för matematik, astronomi och fysik, utgifvet af K. Svenska Vetenskaps-Akademin*, 1, pp. 529–540. (*The collision considered as the basis of kinetic theories of gas pressure and of universal gravitation*).

2.121 K. F. Slotte (**1904–1905**): Über die Schmelzwärme, *Öfversigt af Finska vetenskaps-societetens förhandlingar (Suomen Tiedeseura)*, 47, pp. 117–124. (On the Melting Heat)

2.122 K. F. Slotte (**1904–1905**): *Folgerungen aus einer thermodynamischen Gleichung.* Öfversigt af Finska vetenskaps-societetens förhandlingar (Suomen Tiedeseura), 47, pp. 125–127. (Implications of a Thermodynamic Equation).

2.123 Max Born (**1923**): *Atomtheorie des Festen Zustandes (Dynamik der Kristallgitter).* B. G. Teubner, Leipzig, Germany. (Atomic Theory of the Solid State: Dynamics of Crystalline Lattices).

2.124 Henrik Petrini (**1892**): Om gasers' jämvikt under invärkan af gravitationen, Öfversigt af Kongl. Vetenskaps-akademiens förhandlingar, 49, pp. 559–569. (On the equilibrium of gases under the influence of gravitation).

2.125 Henrik Petrini (**1893**): Om några grundbegrepp i den mekaniska värmeteorin, *Bihang Stockh. Akad.*, XIX(Afd. I), p. 1. (On some fundamental notions in the mechanical theory of heat).

2.126 Henrik Petrini (**1894**): Zur kinetischen Theorie der Gase, Öfversigt af Kongl. Vetenskaps-akademiens förhandlingar, 51, pp. 263–296. (On the kinetic theory of gases).

2.127 Edgar Buckingham (**1900**): *An Outline of the Theory of Thermodynamics.* The Macmillan Company: London, New York.

2.128 J. R. Nimmo and E. R. Landa (**2005**): The soil physics contributions of Edgar Buckingham, *Soil Science Society of America: Journal*, 69, pp. 328–342.

2.129 Edgar Buckingham (**1922**): *Jet Propulsion in Airplanes.* Report No. 159: National Bureau of Standards.

2.130 Edgar Buckingham (**1905**): On certain difficulties, which are encountered in the study of thermodynamics, *Philosophical Magazine Series 6*, 9, pp. 208–214.

2.131 Joseph Louis François Bertrand (**1889**): *Calcul des probabilités.* Gauthier-Villars et fils: Paris, France.

2.132 Joseph Louis François Bertrand (**1887**): *Thermodynamique.* Gauthier-Villars: Paris, France.

2.133. Antoine-Augustin Cournot (**1843**): *Exposition de la théorie des chances et des probabilités.* (*Exposition of the Theory of Chances and Probabilities*, translated into English by Oscar Sheynin), NG Verlag: Berlin, Germany, **2013**.

2.134 E.-P. Bottinelli (**1913**): *A. Cournot, métaphysicien de la connaissance.* Librairie Hachette: Paris, France.

2.135 Chester Townsend Ruddick (**1940**): Cournot's doctrine of philosophical probability, *The Philosophical Review*, 49, pp. 415–423.

2.136 Steven Weinberg (**1994**): *Dreams of a Final Theory: The Scientist's Search for the Ultimate Laws of Nature.* Vintage Books, a Division of Random House, Inc.: New York, USA.

2.137 Victor J. Stenger (**1935–2014**), James A. Lindsay, and Peter Boghossian (**2015**): Physicists are philosophers, too, *Scientific American*, May 2015: A special online publication, available at: https://www.scientificamerican.com/article/physicists-are-philosophers-too/

2.138 B. Bru (**2006**): *Les leçons de calcul des probabilités de Joseph Bertrand: Les lois du hasard.* Electronic Journal for History of Probability and Statistics. Volume 2.

2.139 Thierry Martin (**2007**): Cournot's probabilistic epistemology, in: *Augustin Cournot: Modelling Economics*, Jean-Philippe Touffut, Editor, pp. 21–40; Edward Elgar Publishing Ltd., Cheltenham, UK; Northampton, MA, USA.

2.140 *Hasard et incertitude: Le défis, qu'ils posent.* Pour La Science N° 385 - Novembre **2009**; this is a very interesting journal's issue entitled: '*Chance and uncertainty: The challenges they pose*', especially pp. 24–67, 116–123, 136–141 over there are of interest for our present discussion.

2.141 Edgar Buckingham (**1904**): Note on the deduction of Stefan's law, *Physical Review*, 17, pp. 277–280.

2.142 Ludwig Boltzmann (**1884**): Ableitung des Stefan'schen Gesetzes, betreffend die Abhängigkeit der Wärmestrahlung von der Temperatur aus der electromagnetischen Lichttheorie, *Annalen der Physik und Chemie*, 258 (6), pp. 291–294.

2.143 Max Planck; Morton Masius, transl. (**1914**): *The Theory of Heat Radiation.* P. Blakiston's Son & Company, Philadelphia, USA.

2.144 Siegfried Valentiner (**1919**): *Die Grundlagen der Quantentheorie in elementarer Darstellung.* Zweite erweiterte Auflage. Friedrich Vieweg & Sohn, Braunschweig, Germany.

2.145 Siegfried Valentiner (**1919**): *Anwendung der Quantenthypothese in der kinetischen Theorie der festen Körper und der Gase.* Zweite erweiterte Auflage. Friedrich Vieweg & Sohn, Braunschweig, Germany.

2.146 Siegfried Valentiner (**1929**): Ausdehnung fester Körper. Ausdehnung der Flüssigkeiten, in: *Handbuch der Experimentalphysik, Band VIII, II. Teil: Wärmeausdehnung, Zustandsgrößen und Theorien der Wärme*, pp. 1–75 Herausgegeben von W. Wien, F. Harms und H. Lenz. Akademische Verlagsgesellschaft M. B. H., Leipzig, Germany.

2.147 Bernard H. Lavenda (**1991**): *Statistical Physics: A Probabilistic Approach.* Dover Publications Inc., Mineola, USA.

2.148 Balfour Stewart (**1858**): An account of some experiments on radiant heat, *Transactions of the Royal Society of Edinburgh*, 22, pp. 1–20.

2.149 G. Kirchhoff (**1860**). Über die Fraunhofer'schen Linien. Monatsberichte der Königlich Preußischen Akademie der Wissenschaften zu Berlin, pp. 662–665.

2.150 G. Kirchhoff (**1860**): Über den Zusammenhang zwischen Emission und Absorption von Licht und Wärme. Monatsberichte der Königlich Preußischen Akademie der Wissenschaften zu Berlin, pp. 783–787.

2.151 G. Kirchhoff (**1860**): Über das Verhältnis zwischen dem Emissionsvermögen und dem Absorptionsvermögen der Körper für Wärme and Licht, *Annalen der Physik und Chemie*, 109, pp. 275–301.

2.152 G. Kirchhoff (**1882**): Über das Verhältnis zwischen dem Emissionsvermögen und dem Absorptionsvermögen der Körper für Wärme und Licht, *Gesammelte Abhandlungen*, pp. 571–598, Leipzig: Johann Ambrosius Barth, Germany.

2.153 Max Planck (**1901**): Über das Gesetz der Energieverteilung im Normalspectrum, *Annalen der Physik*, 309, pp. 553–563.

2.154 Jagdish Mehra and Helmut Rechenberg (**1982**): *The Historical Development of Quantum Theory.* Volume 1, Part 1. Springer-Verlag: New York, USA.

2.155 Helge Kragh (**1999**): *Quantum Generations: A History of Physics in the Twentieth Century.* Princeton University Press, Princeton, USA.

2.156 Edwin H. Hall (**1902**): Review: Theory of thermodynamics, by Edgar Buckingham, *Bulletin American Mathematical Society*, 9, pp. 173–175.

2.157 Percy Williams Bridgman (**1939**): *Edwin Herbert Hall (1855-1938).* National Academy of Sciences of the USA: Biographical Memoirs, vol. XXI, the second memoir presented to the Academy at the annual meeting on 1939.

2.158 William Walker Strong (**1908**): Die durchdringungskräftige Strahlung, *Physikalische Zeitschrift*, 9, pp. 117–119. (*The Penetrating Radiation*).

2.159 William Walker Strong (**1918**): *The New Science of the Fundamental Physics.* S. I. E. M. Co.: Mechanicsburg, Pennsylvania, USA.

2.160 William Walker Strong (**1920**): *The New Philosophy of Modern Science.* Kyle Printing Company, York, Pennsylvania, USA.

2.161 Walther Ritz (**1908**): Über die Grundlagen der Elektrodynamik und die Theorie der schwarzen Strahlung, *Physikalische Zeitschrift*, 9, pp. 903–907.

2.162 Prof. Dr. Hans-Georg Schöpf (**1978**): *Von Kirchhoff bis Planck: Theorie der Wärmestrahlung in historisch-kritischer Darstellung* , Akademie-Verlag: Berlin, Germany.

2.163 Walther Ritz (**1911**): *Gesammelte Werke. Œuvres.* Imprimerie Gauthier-Villars, Paris, France.

2.164 Walther Ritz (**1908**): On a new law of series spectra, *Astrophysical Journal*, 28, pp. 237–243.

2.165 Robert Jastrow (**1948**): On the Rydberg-Ritz formula in quantum mechanics, *Physical Review*, 73, p. 60.

2.166 Walter Ritz (**1909**): Über eine neue Methode zur Lösung gewisser Variationsprobleme der mathematischen Physik, *Journal für die Reine und Angewandte Mathematik*, 135, pp. 1–61.

2.167 J. K. MacDonald (**1933**): Successive approximations by the Rayleigh–Ritz variation method, *Physical Review*, 43, p. 830.

2.168 Walther Ritz (**1908**): Du rôle de l'éther en physique, *Rivista di scienza*, 3, pp. 260–274.

2.169 Walther Ritz (**1908**): Recherches critiques sur l'Électrodynamique Générale, *Annales de Chimie et de Physique*, 13, pp. 145–275.

2.170 Wilhelm Ostwald (**1907**): Intorno all'energetica moderna, *Rivista di scienza*, 1, pp. 16–43.

2.171 G. H. Bryan (**1908**): Diffusion and dissipation of energy, *Rivista di scienza*, 3(1), pp. 14–25; G. H. Bryan (**1908**): Diffusion and dissipation of energy, *Rivista di scienza*, v3(2), pp. 275–289.

2.172 Max Planck (**1908**): Bemerkungen zum Prinzip der Aktion und Reaktion in der allgemeinen Dynamik, *Physikalische Zeitschrift*, 9, pp. 828–830.

2.173 Hendrik Antoon Lorentz (**1906**): *Versuch einer Theorie der elektrischen und optischen Erscheinungen in bewegten Körpern.* Verlag von B. G. Teubner, Leipzig, Germany.

2.174 Max Abraham (**1909**): Zur elektromagnetischen Mechanik, *Physikalische Zeitschrift*, 10, pp. 737–741.

2.175 Max Abraham and August Föppl (**1918**): *Theorie der Elektrizität.* Erster Band: *Einführung in die Maxwell'sche Theorie der Elektrizität.* Verlag von B. G. Teubner, Leipzig, Berlin, Germany.

2.176 Max Abraham (**1920**): *Theorie der Elektrizität.* Zweiter Band: *Elektromagnetische Theorie der Strahlung.* Verlag von B. G. Teubner: Leipzig and Berlin, Germány.

2.177 Edwin Bidwell Wilson (**1905**): Review: 'The theory of electricity' by Max Abraham and August Föppl, *Bulletin American Mathematical Society*, 11, pp. 383–387.

2.178 Edwin Bidwell Wilson (**1908**): Review: 'The theory of electricity' by Max Abraham, Vol. 2, *Bulletin American Mathematical Society*, 14, pp. 230 *Bulletin American Mathematical Society* 237.

2.179 Tullio Levi-Civita (**1917**): Meccanica. Sul espressione analitica spettante al tensore gravitazionale nella teoria di Einstein, *Rendiconti della Reale Academia dei Lincei*, 26, p. 381–391. (Mechanics. On the analytic expression that must be given to the gravitational tensor in Einstein's theory.)

2.180 Tullio Levi-Civita and Ugo Amaldi (**1974**): *Lezioni di meccanica razionale.* Volumes I, II, III. Nicola Zanichelli, Bologna, Italy. (Originally published in 1923–1925 and re-published many times).

2.181 Tullio Levi-Civita (**1927**): *The Absolute Differential Calculus. Calculus of Tensors.* Blackie & Son Ltd., London & Glasgow, Great Britain.

2.182 Tullio Levi-Civita (**1964**): *The N-Body Problem in General Relativity Theory.* D. Reidel Publishing Company: Dordrecht, Holland.

2.183 Louis De Broglie (**1925**): *Recherches sur la théorie des Quanta.* Annales de Physique, 10ᵉ Série, Tome III, Janvier-Février 1925, pp. 22–128.

2.184 Gustav Mie (**1912, 1913**): Grundlagen einer Theorie der Materie, *Annalen der Physik*, 37 (1912), pp. 511–534 (I); 39 (1912), pp. 1–40 (II), 40 (1913), pp. 1–66 (III).

2.185 Gérard Gouesbet and Gérard Gréhan (**2011**): *Generalized Lorenz-Mie Theories*. Springer-Verlag: Berlin, Heidelberg, Germany.

2.186 Wolfram Hergert and Thomas Wriedt (**2012**): *The Mie Theory: Basics and Applications* Springer-Verlag: Berlin, Heidelberg, Germany.

2.187 Andrew Zangwill (**2013**): *Modern Electrodynamics*. Cambridge University Press: New York, USA.

2.188 Louis de Broglie (**1995**): *Diverses questions de mécanique et de thermodynamique classiques et relativistes Edition établie d'après un manuscrit inédit de Louis de Broglie*, édité et préfacé par Georges Lochak, Michel Karatchentzeff et Daniel Fargue, Fondation Louis de Broglie. Springer, Berlin, Heidelberg, New York, Barcelona, Budapest, Hong Kong, London, Milan, Paris, Santa Clara, Singapore, Tokyo. (*This is a publication of the unpublished sketches by Louis de Broglie on the mechanics, as well as the classical and relativistic thermodynamics*).

2.189 Albert Einstein (**1909**): Zum gegenwertigen Stand des Strahlungsproblems, *Physikalische Zeitschrift*, 10, pp. 185–193.

2.190 Walther Ritz (**1909**): Zum gegenwertigen Stand des Strahlungsproblems. (Erwiderung auf den Aufsatz des Herrn A. Einstein), *Physikalische Zeitschrift*, 10, pp. 224–225.

2.191 Walther Ritz and Albert Einstein (**1909**): Zum gegenwertigen Stand des Strahlungsproblems, *Physikalische Zeitschrift*, 10, pp. 323–324.

2.192 Mr. F. Cohen (**1909**): Zur kinetischen Gastheorie: Eine kritische Untersuchung. *Physikalische Zeitschrift*, 10, pp. 138–140; pp. 196–197.

2.193 Max Planck (**1909**): Zur kinetischen Gastheorie: Eine kritische Untersuchung, *Physikalische Zeitschrift*, 10, pp. 195–196.

2.194 David Enskog (**1911**): Über eine Verallgemeinerung der zweiten Maxwell'schen Theorie, *Physikalische Zeitschrift*, XII, pp. 56–60; *Bemerkungen zu einer Fundamentalgleichung in der kinetischen Gastheorie*, XII, pp. 533–539.

2.195 David Enskog (**1917**): *Kinetische Theorie der Vorgänge in mäßig verdünnten Gasen. I. Allgemeiner Teil.* Almqvist & Wiksells Boktryckeri A.-B., Uppsala, Sweden.

2.196 David Enskog (**1929**): Über die Entropie der Gase bei irreversiblen Prozessen, *Zeitschrift für Physik*, 54, pp. 498–504.

2.197 Sydney Chapman and T. G. Cowling (**1953**): *The Mathematical Theory of Non-uniform Gases,* Reprinted Second Edition. Cambridge University Press: Cambridge, Great Britain.

2.198 J. J. O'Connor and E. F. Robertson: *David Enskog, Biography*. http://www-history.mcs.st-andrews.ac.uk/Biographies/Enskog.html

2.199 Eberhard Zschimmer (**1909–1913**): *Das Weltwunder.* Verlag von Wilhelm Engelmann, Leipzig und Berlin, Germany, Part III, pp. 14–15.

2.200 Lord Kelvin (**1908**): Über Ätherbewegungen, hervorgerufen durch Kollision von Atomen oder Molekülen mit oder ohne Elektronen, *Physikalische Zeitschrift*, 9, pp. 2–6.

2.201 O. M. Corbino (**1908**): Das Zeeman-Phänomen und der zweite Hauptsatz der Thermodynamik, *Physikalische Zeitschrift*, 9, pp. 344–347.

2.202 Max von Laue (**1908**): Das Zeeman-Phänomen und der zweite Hauptsatz der Thermodynamik, *Physikalische Zeitschrift*, 9, pp. 617–620; p. 907.

2.203 O. M. Corbino (**1908**): Die Lichtemission seitens eines glühenden Dampfes in einem Magnetfelde unter verschiedenem Azimut, *Physikalische Zeitschrift*, 9, pp. 669–671.

2.204 H. A. Lorentz (**1908**): Zur Strahlungstheorie, *Physikalische Zeitschrift*,9, pp. 562–563.

2.205 J. H. Jeans (**1908**): Zur Strahlungstheorie, *Physikalische Zeitschrift*, 9, pp. 853–855.

2.206 Angelo Battelli (**1908**): Spezifische Wärme von Flüssigkeiten, die bei sehr tiefen Temperatur fest werden, *Physikalische Zeitschrift*, 9, pp. 671–675.

2.207 Edmund Altenkirch (**1909**): Über den Nutzeffekt der Thermosäule, *Physikalische Zeitschrift*, 10, pp. 560–568.

2.208 Josef Weiß (**1909**): Über das Planck'sche Strahlungsgesetz. (Vorläufige Mitteilung.), *Physikalische Zeitschrift*, 10, pp. 193–195.

2.209 Josef Weiß (**1910**): *Experimentelle Beiträge zur Elektronentheorie aus dem Gebiet der Thermoelektrizität*. Druck- und Verlagsgesellschaft vorm. Dölter, Emmendingen, Germany.

2.210 Johannes Königsberger and Josef Weiß (**1911**): Über die thermoelektrischen Effekte, *Annalen der Physik*, 340, pp. 1–46.

2.211 Hans Rott and Herausgeber (**1933**): *Quellen und Forschungen zur südwestdeutschen und schweizerischen Kunstgeschichte im XV. und XVI. Jahrhundert*. Band 1: *Bodenseegebiet*. Strecker und Schröder Verlag: Stuttgart, Germany.

2.212 E. B. Starikov (**2014**): 'Meyer-Neldel Rule': True history of its development and its intimate connection to classical thermodynamics, *Journal of Applied Solution Chemistry and Modeling*, 3, pp. 15–31.

2.213 N. Mehta (**2010**): Meyer-Neldel rule in chalcogenide glasses: Recent observations and their consequences, *Current Opinion in Solid State and Materials Science*, 14, pp. 95–106.

2.214 Anshuman Dalvi, N. Parvathala Reddy, and S. C. Agarwal (**2012**): The Meyer–Neldel rule and hopping conduction, *Solid State Communications*, 152, pp. 612–615.

2.215 N. Parvathala Reddy, Rajeev Gupta, and S. C. Agarwal (**2013**): Electrical conduction and Meyer-Neldel Rule in nanocrystalline silicon thin films, *Journal of Non-Crystalline Solids*, 364, pp. 69–76.

2.216 Georg Busch (**1989**): Early history of the physics and chemistry of semiconductors: From doubts to fact in a hundred years, *European Journal of Physics*, 10, pp. 254–264.

2.217 Yu. Kh. Vekilov and Ya. M. Mukovskii (**2012**): Variable range hopping conductivity in manganites, *Solid State Communications*, 152, pp. 1139–1141.

2.218 Andrew Zimmerman Jones and Daniel Robbins (**2009**): *String Theory for Dummies*. Wiley Publishing, Inc., Indianapolis, Indiana, USA.

2.219 R. Stephen Berry (**1991**): *Understanding energy: Energy, Entropy and Thermodynamics for Everyman*. World Scientific Publishing Co. Pte. Ltd.: Singapore, New Jersey, London, Hong Kong.

2.220 C. Kiefer (**1993**): Kosmologische Grundlagen der Irreversibilität, *Physikalische Blätter*, 49, pp. 1027–1029.

2.221 John S. Briggs (**2008**): A derivation of the time-energy uncertainty relation, *Journal of Physics*, Conference Series 99, 012002.

2.222 G. Wilk and Z. Włodarczyk (**2011**): Generalized thermodynamic uncertainty relations, *Physica A (Statistical Mechanics and its Applications)*, 390, pp. 3566–3572.

2.223 Guo Lina and Du Jiulin (**2011**): Thermodynamic potentials and thermodynamic relations in non-extensive thermodynamics, *Physica A (Statistical Mechanics and its Applications)*, 390, pp. 183–188.

2.224 K. Urmossy, G. G. Barnaföldi, and T. S. Biró (**2011**): Generalized Tsallis statistics in electron–positron collisions, *Physics Letters B*, 701, pp. 111–116.

2.225 Ira Mark Egdall (**2016**): *Einstein Relatively Simple: Our Universe Revealed in Everyday Language.* World Scientific Publishing Company, Pte. Ltd., New Jersey, London, Singapore, Bei Jing, Shanghai, Hong Kong, Taipei, Chennai.

2.226 Joachim G. Leithäuser (**1957**): *Köpfe des XX. Jahrhunderts: Werner Heisenberg.* Colloquium Verlag Otto H. Hess, Berlin-Dahlem, Germany.

2.227 Werner Gent (**1962**): *Die Philosophie des Raumes und der Zeit: Historische, kritische und analytische Untersuchungen.* Bände I und II. Georg Olms Verlagsbuchhandlung, Hildesheim, Germany.

2.228 Herausgegeben und kommentiert von Armin Hermann (**1968**): *Albert Einstein/Arnold Sommerfeld: Briefwechsel. Sechzig Briefe aus dem goldenen Zeitalter der modernen Physik.* Schwabe & Co., Basel/ Stuttgart.

2.229 Armin Hermann (**1969**): *Frühgeschichte der Quantentheorie (1899 – 1913).* Physik Verlag, Mosbach in Baden, Germany.

2.230 Ulrich Röseberg (**1978**): *Quantenmechanik und Philosophie.* Akademie-Verlag: Berlin, Germany.

2.231 Prof. Dr. Hans-Georg Schöpf (**1978**): *Von Kirchhoff bis Planck. Theorie der Wärmestrahlung in historisch-kritischer Darstellung.* Springer Fachmedien: Wiesbaden, Germany.

2.232 Gereon Wolters (**1987**): *Mach I, Mach II, Einstein und die Relativitätstheorie.* Walter de Gruyter: Berlin, New York.

2.233 Mary Jo Nye, Joan L. Richards, and Roger H. Stuewer (**1992**): *The Invention of Physical Science. Intersections of Mathematics, Theology and Natural Philosophy since the Seventeenth Century Essays in Honor of Erwin N. Herbert.* Kluwer Academic Publishers: Dordrecht, Holland.

2.234 Diana Kormos Barkan (**1999**): *Walter Nernst and the Transition to Modern Physical Science.* Cambridge University Press: Cambridge, United Kingdom.

2.235 Yong Zhu (**2003**): *Large-Scale Inhomogeneous Thermodynamics and Application for Atmospheric Energetics.* Cambridge International Science Publishing: Cambridge, U. K.

2.236 Yoshikata Koga (**2007**): *Solution Thermodynamics and Its Application to Aqueous Solutions: A Differential Approach.* Elsevier: Amsterdam, Holland.

2.237 Jeremy Dunning-Davies (**2010**): *Concise Thermodynamics: Principles and Applications in Physical Science and Engineering.* Woodhead Publishing: Oxford, Cambridge, Philadelphia, New Delhi.

2.238 Alexander Komech (**2012**): *Quantum Mechanics: Genesis and Achievements*. Springer Science+Business Media: Dordrecht, Holland.

2.239 Helge Kragh (**2012**): *Niels Bohr and the Quantum Atom: The Bohr Model of Atomic Structure, 1913–1925*. Oxford University Press, Oxford, U. K.

2.240 Massimiliano Badino and Bretislav Friedrich (**2013**): Much polyphony but little harmony: Otto Sackur's groping for a quantum theory of gases, *Physics in Perspective*, 15, pp. 295–319.

2.241 Yoshikata Koga (**2016**): Spectroscopy vs. thermodynamics, or G vs. GE in studies of aqueous solutions, *Journal of Molecular Liquids*, 219, pp. 1006–1009.

2.242 N. Bohr, H. A. Kramers, and J. C. Slater (**1924**): Über die Quantentheorie der Strahlung, *Zeitschrift für Physik*, XXIV, pp. 69–87.

2.243 W. Bothe and H. Geiger (**1925**): *Über das Wesen des Compton-Effekts; ein experimenteller Beitrag zur Theorie der Strahlung*, *Zeitschrift für Physik*, XXXII, 639–663.

2.244 C. D. Ellis and W. A. Wooster (**1927**): The average energy of disintegration of radium E, *Proc. Roy. Soc. London A*, 117, pp. 109–112.

2.245 Lise Meitner and W. Orthmann (**1930**): *Über eine absolute Bestimmung der Energie der primären Strahlen von Radium E*, *Zeitschrift für Physik*, LX, 143–155.

2.246 B. W. Sargent (**1933**): The maximum energy of the β-Rays from uranium X and other bodies, *Proc. Roy. Soc. London A*, 132, pp. 659–673.

2.247 C. D. Ellis and N. F. Mott (**1933**): Energy relations in the β-ray type of radioactive disintegration, *Proc. Roy. Soc. London A*, 141, pp. 109–123.

2.248 R. S. Shankland (**1936**): An apparent failure of the photon theory of scattering, *Phys. Rev.*, 49, pp. 8–13.

2.249 C. Jacobsen (**1936**): Correlation between scattering and recoil in the Compton effect, *Nature*, 138, p. 25.

2.250 W. Bothe and H. Maier-Leibnitz (**1936**): Eine neue experimentelle Prüfung der Photonenvorstellung, *Zeitschrift für Physik*, CII, pp. 143–155.

2.260 R. S. Shankland (**1936**): The scattering of gamma rays, *Phys. Rev.*, 50, p. 571.

2.261 Theo A. F. Kuipers, Dov M. Gabbay, Paul Thagard, and John Woods (**2007**): *General Philosophy of Science: Focal Issues.* Elsevier: North Holland Editorial Company, Holland.

2.262 Max Planck (**1926**): Physikalische Gesetzlichkeit, *Naturwissenschaft*, 14, pp. 249–261.

2.263 Max Planck (**1926**): Über die Begründung des zweiten Hauptsatzes der Thermodynamik, Berichte der Berliner Akademie der Wissenschaft, am 27. Dezember 1926.

2.264 M. W. Zemansky (**1957**): Fashions in thermodynamics, *Am. J. Phys.*, 25, p. 349.

2.265 Peter Fong (**1963**): *Foundations of Thermodynamics.* Oxford University Press: Oxford, UK.

2.266 C. Carathéodory (**1909**): Untersuchungen über die Grundlagen der Thermodynamik, *Math. Ann.*, 67, pp. 355–386.

2.267 C. Carathéodory (**1925**): Über die Bestimmung der Energie und der absoluten Temperatur mit Hilfe von reversiblen Prozessen, Berichte der Berliner Akademie der Wissenschaft, am 8. Januar 1925.

2.268 Max Born (**1921**): Kritische Betrachtungen zur traditionellen Darstellung der Thermodynamik, *Physikalische Zeitschrift*, 22, pp. 218–224; 249–254; 282–286.

2.269 A. E. Ruark (**1925**): The proof of the corollary of the Carnot's theorem, *Phil. Mag.*, 49, pp. 584–585.

2.270 Tatiana Ehrenfest-Afanassjewa (**1925, 1926**): Zur Axiomatisierung des zweiten Hauptsatzes der Thermodynamik, *Zeitschrift für Physik*, 33, pp. 933–945, 1925; 34, p. 638, 1926.

2.271 A. Landé (**1926**): *Axiomatische Begründung der Thermodynamik durch Carathéodory*, in: *Handbuch der Physik. Band IX. Theorien der Wärme*; F. Henning, Ed., Verlag von Julius Springer, Berlin, Germany.

2.272 J. Farkas (**1895**): Vereinfachte Ableitung des Carnot-Clausius'schen Satzes, *Mathematische und Physikalische Blätter aus Ungarn*, IV, pp. 7–11.

2.273 N. Schiller (**1874**): Einige experimentelle Untersuchungen über elektrische Schwingungen, *Annalen der Physik*, 228, pp. 535–565.

2.274 N. Schiller and R. Colley (**1875**): Ein Versuch über die elektrodynamische Wirkung des Polarisationsstroms, *Annalen der Physik*, 231, pp. 467–469.

2.275 N. Schiller (**1876**): Elektromagnetische Eigenschaften ungeschlossener elektrischer Ströme, *Annalen der Physik*, 235, pp. 456–473.

2.276 N. Schiller (**1876**): Elektromagnetische Eigenschaften ungeschlossener elektrischer Ströme, *Annalen der Physik*, 235, pp. 537–553.

2.277 N. Schiller (**1877**): Berichtigung zum Aufsatze über die elektromagnetischen Eigenschaften ungeschlossener Ströme, *Annalen der Physik*, 236, pp. 333–335.

2.278 Н. Шиллер (**1879**): Некоторые приложения механической теории тепла к изменениям состояния упругого тела, Журнал Русского Физико-Химического Общества, 11, pp. 55–77.

2.279 N. Schiller (**1890**): Über eine mögliche, aus den Joule-Thomson'schen Abkühlungsversuchen herzuleitende Form der Zustandsgleichung für Gase, *Annalen der Physik*, 276, pp. 149–156.

2.280 N. Schiller (**1894**): Über die durch einen äußeren Druck verursachte isothermische Änderung der Spannkraft gesättigten Dampfes, *Annalen der Physik*, 289, pp. 396–400.

2.281 N. Schiller (**1894**): Über die von der Variation elektrostatischer Energie abgeleiteten elektrischen ponderomotorischen Kräfte, *Annalen der Physik*, 289, pp. 432–446.

2.282 N. Schiller (**1897**): Einige Versuche über Verdampfung von Flüssigkeiten durch einen hohen Gasdruck, *Annalen der Physik*, 296, pp. 755–759.

2.283 N. Schiller (**1899**): Die Bedeutung des osmotischen Druckes in der Thermodynamik der Lösungen, *Annalen der Physik*, 303, pp. 291–306.

2.284 N. Schiller (**1900**): Einige thermodynamisch abzuleitende Beziehungen zwischen den Größen, die den physikalischen Zustand einer Lösung charakterisieren, *Archives Néerlandaises des Sciences Exactes et Naturelles, sér. 2*, 5, pp. 118–147.

2.285 N. Schiller (**1901**): Der Begriff des thermischen Verkehrs als Grundlage des zweiten thermodynamischen Hauptsatzes, *Annalen der Physik*, 310, pp. 313–325.

2.286 N. Schiller (**1901**): Zur Thermodynamik gesättigter Lösungen, *Annalen der Physik*, 310, pp. 326–348.

2.287 N. Schiller (**1901**): Zur Thermodynamik ungesättigter Lösungen, *Archives Néerlandaises des Sciences Exactes et Naturelles, sér. 2*, 6, pp. 497–549.

2.288 N. Schiller (**1902**): Das Gesetz der Partialdichtigkeitsänderung eines Lösungsmittels mit der Concentration der Lösung, *Annalen der Physik*, 313, pp. 588–599.

2.289 Проф. Николай Николаевич Шиллер (**1902**): Основные законы термодинамики, Типография Императорского университета Св. Владимира. Акционерное О-во печатного и издательского дела Н. Т. Корчак-Новицкого, Киев, Российская Империя.

2.290 N. Schiller (**1904**): Einige Bedenken betreffend die Theorie der Entropievermehrung durch Diffusion der Gase bei einander gleichen Anfangsspannungen der letzteren, in: *Festschrift Ludwig Boltzmann, gewidmet zum sechzigsten Geburtstage*, Meyer, Editor, pp. 350–366.

2.291 N. Schiller (**1906**): Die Bedeutung der Unstetigkeit der ersten Derivierten des Druckes nach der Temperatur bei der Feststellung der Phasenregel, *Zeitschrift für physikalische Chemie*, 54, pp. 451–462.

2.292 N. Schiller (**1907**): Eine Bemerkung über die Beziehung zwischen der absoluten Temperatur und der kinetischen Energie eines thermodynamischen Systems, *Annalen der Physik*, 327, pp. 573–578.

2.293 Н. Шиллеръ (**1908**): Замѣчания объ аналитическомъ представленіи второго закона термодинамики, Журнал Русскаго Физико-Химическаго Общества, 40, pp. 85–111.

2.294 Н. Шиллеръ (**1910**): Замѣчания объ аналитическомъ представленіи второго закона термодинамики, Журнал Русскаго Физико-Химическаго Общества, 42, pp. 117–128.

2.295 William Thomson and Lord Kelvin (**1857**): On the thermoelastic, thermomagnetic and pyro-electric properties of matter, *Quarterly Journal of Pure and Applied Mathematics*, 1, pp. 57–77.

2.296 William Thomson and Lord Kelvin (**1878**): On the thermoelastic, thermomagnetic and pyro-electric properties of matters, *Philosophical Magazine*, 5, pp. 4–27.

2.297 Woldemar Voigt (**1889**): Über adiabatische Elasticitätsconstanten, *Annalen der Physik*, 272, pp. 743–759.

2.298 Woldemar Voigt (**1903**): *Thermodynamik*. Volume 1, Leipzig, Verlag von G. J. Göschen, Leipzig, Germany.

2.299 K. Martinas and I. Brodszki (**2000**): Thermodynamics of Gyula Farkas—A new (old) approach to entropy, *Per. Polytech. Ser. Chem. Eng.*, 44, pp. 17–27.

2.300 Т. А. Афанасьева-Эренфест (**1928**): Необратимость, односторонность и второе начало термодинамики, Журнал Прикладной Физики, 5, pp. 3–30. (Irreversibility, lopsidedness and the second basic law of thermodynamics.)

2.301 E. H. Lieb and J. Yngvason (**1999**): The physics and mathematics of the second law of thermodynamics, *Phys. Rep.*, 310, pp. 1–96, as well as the references therein.

2.302 A. Ben-Naim (**1992**): *Statistical Thermodynamics for Chemists and Biochemists.* Springer Science+Business Media: New York, USA.

2.303 A. Ben-Naim (**2007**): *Entropy Demystified, the Second Law of Thermodynamics Reduced to Plain Common Sense*, World Scientific, Singapore.

2.304 A. Ben-Naim (**2008**): *A Farewell to Entropy: Statistical Thermodynamics Based on Information*, World Scientific, Singapore.

2.305 A. Ben-Naim (**2010**): *Discover Entropy and the Second Law*, World Scientific, Singapore.

2.306 A. Ben-Naim (**2012**): *Entropy and the Second Law: Interpretation and Misss-Interpretationsss*, World Scientific, Singapore.

2.307 A. Ben-Naim (**2014**): *Statistical Thermodynamics, with Applications to Life Sciences*, World Scientific, Singapore.

2.308 A. Ben-Naim (**2014**): *Discover Probability: How to Use It, how to Avoid Misusing It and how It Affects Every Aspect of Your Life*, World Scientific, Singapore.

2.309 A. Ben-Naim (**2015**): *Information, Entropy, Life and the Universe: What We Know and what We Do Not Know*, World Scientific, Singapore.

2.310 A. Ben-Naim (**2016**): *The Briefest History of Times*, World Scientific, Singapore.

2.311 T. Ehrenfest-Afanassjewa (**1958**): On the use of the notion "probability" in physics, *American Journal of Physics*, 26, pp. 388–392.

2.312 Robert Lowrie and A. M. Gonas (**1965**): Dynamic elastic properties of polycrystalline tungsten, 24–1800°C, *J. Appl. Phys.*, 36, pp. 2189–2192.

2.313 Naohiro Soga (**1966**): Comparison of measured and predicted bulk moduli of tantalum and tungsten at high temperatures, *J. Appl. Phys.*, 37, pp. 3416–3420, 1966.

2.314 R. P. Singh and G. S. Verma (**1968**): Validity of Nernst-Lindemann equation in the range 24–1800°C in tungsten, *Solid State Communications*, 6, pp. 113–114.

2.315 Yiannis N. Kaznessis (**2012**): '*Statistical Thermodynamics and Stochastic Kinetics: An Introduction for Engineers*', pages 167-170. Cambridge University Press: Cambridge, New York, Melbourne, Madrid, Cape Town, Singapore, São Paulo, Delhi, Tokyo, Mexico City.

[**Note 3**]: To sum up, the entire set of the chapters in the present monograph has definitely been the 'long of it', while now comes the 'short'.

Thermodynamics is a unique branch of our knowledge, for unlike most of the other ones it is truly inhomogeneous. What do we mean?

If to consider the thermodynamics history in detail, just what we have attempted to perform in the volume at hand, we clearly recognize one important point. Sure, there might be much more points to recognize, or the sense of the point in question might even be quite different. Howbeit, the final judge should anyway be the readership!

Meanwhile, here we dare to point out the following important (to our mind!) feature:

Basically, there were and are two clearly recognizable flavors/tastes/branches in thermodynamics: the irrational and the rational ones.

3.1 Irrational vs. rational thermodynamics: what is the zest?

The irrational one is tightly related to such outstanding names as Ludwig Boltzmann and Max Planck. Other names conventionally mentioned in this connection were just the enthusiastic followers of the former both colleagues. For example, Albert Einstein's actual role in and the final contributions to the field we are discussing here seem to be truly ambiguous.

Two posers immediately arise in this connection:

1. *Why, at all, this entire thermodynamics' branch is to be dubbed 'irrational'?*

Our answer is immediate: First of all, we just follow here the opinion of a lot of different clearly noticeable colleagues all around the world in different time periods.

To sum, up what we have discussed in Chapters 1 to 4 and also above here, shows definite logical gaps in Boltzmann and Planck's theoretical constructions. Does this mean that the both colleagues were completely in error?

Surely not, not at all!

The both of them were heading for the solution of the problem, they have suggested their own proper solutions, but were duly following their own way. Either Boltzmann or Planck should really have nothing to blame for, except for somewhat much too aggressive presentation of their results. Of the both Planck was rather a definite champion in the '*aggressive marketing*' (sorry, but no other appropriate term ought to come to mind when reading Boltzmann's and Planck's papers published right at the borderline of the XIX-th and XX-th centuries).

Howbeit, no professional questions are remaining to any of the above both colleagues. Still, the entire raw of their followers did and does still follow some peculiar way, to our mind.

This ought to be connected with the so-called '*operational approach in the scientific research*' and is mainly concentrated on the theoretical research branch. To put it shortly, at the first glance the '*operationists*' (from here on, let us dub the adepts of the operational approach this way) were and are driving a very clear and indisputable train of thoughts:

'*...Theorists have provided us with some handy formulae. So let us try to explain our experiments. ...*

... Aha, we see that mostly we get the proper fits. Well, perhaps this and that regions of the data might be approximated somewhat better way – So what? We just modify the mathematics a bit by ourselves, or we might ask some other theoretician to do this at some leisure time. Still, our main point has been clearly reached. All those dreadful theories and maths ought to look like so unattractively strange – that we surely feel a giant respect for those who have the mood and leisure for dealing with this – but we ourselves ought to deal with the realistic things instead...'

The above is a truly widespread stance. The only pertinent criticism in this regard might be as follows. Even if the mathematical formulas at hand are fully correct, and even if we do not need to anyhow modify them—this is *by far not the very proof* that all this toolbox at hand is perfectly describing the actual situation as a whole—once and forever.

Sometimes somebody manages to find new undoubtful phenomena, effects—fortunately, this happens every now and then in all the research fields. These novel findings ought to be properly described and explained in that we might gain access to their intrinsic mechanisms, to their interrelationships to something already well known. Only this way it is possible for us to make a reasonable use of the good novelties involved.

This is why, immense ought to be the role of those who sacrifice their '*mood and leisure to dig for treasures, but finding the earthworms*', to paraphrase Goethe's Faust. These are just the adepts of the rational branch, who study everything—what has come, what comes, and even what might be expected to come—in their respective fields—study thoroughly and in detail.

The above is just the main idea of the present volume: The irrational branch is *never false/bad/negative*; it is *throughout positive*, **but not quite sufficient**.

We should never forget the adepts of the rational branch, for otherwise we cannot perceive the entire picture. This is how we arrive at the second and the very last poser of the volume and hand:

2. *Who ought to belong to the 'rational thermodynamics' branch'—and why?*

We would not like to restrict our investigation scope to the number of individuals we are highlighting here in Chapters 1–4.

Sure, several lines do exist in the direction 'rational development of thermodynamics'. Indeed, the main line, which is long known and well recognized, ought to be the research line followed by Josiah Willard Gibbs in the United States. A very important addition to be made here: J. W. Gibbs has borrowed from August Friedrich Horstmann the seminal rational idea he was actively developing. Fortunately, these both colleagues could be in the direct personal contact.

In turn, August Friedrich Horstmann was thoughtfully reacting to the criticism expressed by Friedrich Mohr, as concerns Rudolf Clausius' way of developing the ideas of N. L. S. Carnot. As a result, Horstmann could have success in deciphering the actual rational sense of the second basic law, of the entropy. Further, he could even share these key ideas with Gibbs. Still, his health state had not allow him to pursue going this seminal way. Meanwhile, Prof. Gibbs had also to untimely enter his very last 'business trip'.

It is also important to mention here the British thermodynamics' school, where a renowned physical chemist George Downing Liveing, who was duly representing the 'rational branch' by adequately and rationally developing the ideas of Hon. William Thomson, Lord Kelvin—but to our sincere regret, only up to his tragic road accident. Meanwhile, other outstanding British thermodynamicists, like e.g., James Clerk Maxwell and Peter Guthrie Tait, were clearly sharing the 'irrational branch', with Maxwell's giving life to some peculiar 'demons'—and Tait's turning the then unclear notion of entropy into the ultimate driving force (entropy production as an entropic driving force).

While Maxwell could be successful in a number of important physical fields, so that every school pupil all over the world knows Maxwell's name, Tait's name is instead practically forgotten nowadays—but his strange idea is nonetheless stubbornly underlying the modern physics till nowadays. Consequently, Maxwell's demons are still populating the pages of scientific periodicals.

Noteworthy, the both phantoms named are steadily decorating such valid and fruitful modern scientific fields as quantum computing and synergetics, respectively. Sure, these mascots are very nice, especially if we take into account their unique antiquary values, but the both have nothing to do with any true physical basics.

Howbeit, at the beginning of the XX-th century everything had looked like that the 'operationists' could completely pick up the baton. The pertinent rationalists were no more active/proactive—or even among us at all. On the one hand, such competent detractors/censors of the 'irrationality' as Ernst Mach, Wilhelm Ostwald and other notable specialists in the field were either slowly growing old and ill, entering their last 'business trips'—or just not primarily interested in the field of thermodynamics, like, for example Felix Auerbach. On the other hand, many renowned operationists, like the renowned experimental physicists Percy William Bridgman, Walter Nernst, the theoretical physicist Max Planck, the theoretical and practical engineer Albert Einstein, etc. have become the Nobel Prize winners in physics.

Meanwhile, in response to the latter good news, the revolutionary, passionate thermodynamics was doing but nothing more than intensively 'producing' the 'new basic natural laws'.

At the very same time, the engineering school of thermodynamics was developing the true, working theory of energetics, which is nothing more than just the more general term for thermodynamics. To be mentioned here is the Zeuner–Mollier–Bošnjaković thermodynamic school (Technical University of Dresden, Germany). Further, definitely mentioned ought to be much less known, but still tremendously important, Austrian school founded by an outstanding, worldly known metallurgist Prof. Dr. Hans Jüptner, Freiherr von Jonstorf (1853–1941). In particular, his apprentice, Dr. Oskar Nagel, represents this school. He was born in Austria in 1874 and studying over there. Then, Dr. Nagel had moved to the USA in looking for his working position, although his further destiny is practically unclear. We would like to list important works and translations by the Austrian colleagues just mentioned [3.1–3.23] plus about a dozen of patents in metallurgy by Dr. Oskar Nagel. In this Chapter we have already mentioned the Russian school of engineering and chemical thermodynamics.

Of extreme interest for our present discussion would be to point out the involvement of the engineers into the methodological studies (see, e.g., the relevant works by Prof. Jüptner and Dr. Oskar Nagel [3.14, 3.15–3.18, 3.20]).

This is just to show that thermodynamics wasn't killed, it was live and active, and it was producing sensible, important, realistic results. The question was and is just about thermodynamics' actual foundations.

In the USA one of the well-known adepts of the rational engineering thermodynamics was Prof. Dr. Joseph Henry Keenan (1900–1977), who was systematically following Prof. Gibbs' approach. The first edition of his thermodynamics handbook (1941) gives a very clear picture of the thermodynamics foundations [3.24].

However, much less conceptual clarity could be detected in the later edition (1965), where he and his followers have introduced a 'reformulation of thermodynamics' [3.25]. Most probably, they had this way succumbed to the '*aggressive marketing*' still performed by the adepts of the '*irrational*' branch at their time, and, to our mind, the very foundations of thermodynamics were completely smashed as a result.

In this connection, a good question sounds as follows: *Were there at least some physicists, who did not follow the 'irrational' line in thermodynamics?*

Yes, of course!

In this volume we have just pointed out the Scandinavian physicists: Peter Boas Freuchen and Nils Engelbrektsson. Peter Boas was trying to attract attention to the inconsistencies in the foundations of thermodynamics at his time and show the way to overcome them. Nils Engelbrektsson had successfully rebuilt the macroscopic thermodynamics in the fully logical, fully rational way—and could fortunately come across Karl Alexius Franzén, who could successfully verify the theory by Nils, by adding the definite physical–chemical fleur to the whole story.

Max Bernhard Weinstein was not only trying to find the ways to building-up the different thermodynamics and kinetics; he was also duly struggling against the 'quaint approaches to scientific research'.

The last, but not the least, in this chapter we have also mentioned the authors of the '*Karlsruher Course of Physics*'.

Furthermore, we should also pay tribute to the numerous colleagues in different countries all around the world, so we have tried to mention here at least a small number of them, who were adepts of the 'rational branch', or, being the adepts of the 'irrational wave' were delivering valid research results.

The good poser is: Why could not they build a unique 'rational front' to cope with the 'irrational wave'?

Well, the commonplace sagacity would dictate that: *when a man is alone on the battlefield he would never be a soldier. But poor, sad, invisible little people, when they are well and truly knit together, are stronger than any giant.*

One might further reason that the '*operationists*' had more chances to be 'knit together', for they in their entirety were following one and the same direction, and had finally managed to penetrate lower organizational levels of the matter beyond the conventional macroscopic one. Hence, the conventional macroscopic part of thermodynamics, with all of its successes, with all of its 'irks and quirks', could be quietly turned in to the engineers (whether they are mechanical or chemical engineers, does not matter!). Instead, the modern physics ought to be busy with atoms, molecules, etc.

Meanwhile, it is just at this point that the world of probability comes, which is governing *basically everything*—because we, together with all of our surroundings, are consistent of the molecules, atoms, which in turn are consistent of nuclei, and electrons, with the former and latter being electrically charged—and the most important point—there are always *huge numbers* of them.

How to cope with this fact without the proper statistical consideration?

This is but again a purely emotional statement having still something to do with the actual state of affairs, but showing no constructive way to any correct and detailed study of the situation.

In fact, there are a lot of fundamental, truly philosophic problems [18–21], which seemingly turn our broad endless pavement nicely depicted above into a narrow blind alley.

So, how should we proceed in such a case?

No problem at all, the '*operationists*' tell us, we could hire clever physicists who would properly play with this entire philosophy to wipe out all the 'imaginary obstacles' (such actors are always in the nearest vicinity, like, e.g., Prof. Dr. Hans Reichenbach (1891–1953)).

Well, the whole story, as we depict it here, seems to be truly forlorn—the physicists are properly doing their job, whereas chemists, biologists, engineers etc. theirs.

So what?

This might truly be a sad story, if there would never be guys, like Dr. Georg(e) Augustus Linhart, who could in fact manage to find how to skillfully reconcile the gap between the conventional classical macroscopic thermodynamics and the novel microscopic theories.

Linhart's first and foremost step was to kill the 'red herring' of some imaginary 'fundamental role' to be allegedly played by the probability notion. The consequent Bayesian approach to the mathematical statistics could be throughout helpful in proving the notoriously famous Boltzmann–Planck formula. The final answer is: Yes, sure, the thermodynamic entropy does indeed represent a function of probability—*but this is by far not the final chapter of the story*!

The next Linhart's step was to demonstrate that this notorious mystic probability is simply a mathematical function of the absolute temperature. Moreover, owing to this peculiar functional relationship, the entropy should properly come to zero at the zero absolute temperature.

Meanwhile, our experimental experience clearly shows us that no realistic process might bring the system to the zero absolute temperature.

Hurray! We have found some basic natural law! Then, we would just need to properly formulate it.

If we remain stick at the emotional interpretation of the Boltzmann–Planck formula, we should then immediately arrive at the third basic law of thermodynamics—together with Prof. Dr. Walter Nernst (1864–1941) and his associates.

However, if we adopt the standpoint of A. F. Horstmann, J. W. Gibbs, P. B. Freuchen, N. Engelbrektsson, K. A. Franzén, G. A. Linhart, M. B. Weinstein, and other colleagues from the 'rational' branch, we would arrive at a seemingly trivial, but in principle very important point:

The absolute zero temperature can **physically** never be reachable because the entropy can **physically** never come to zero. Mathematically, the entropy might well come to zero, but **never** physically! The energetics teaches us that entropy is always correspondent to the ubiquitous obstacles/hindrances/resistances. If there is a non-zero driving force of whatever nature, there will be the pertinent resistances to it. This is how the the active, useful energy behind the driving force, will **always** end up as the waste, useless energy.

Prof. Dr. Franz Eugen Simon (1893–1956), the apprentice of W. Nernst, was in effect absolutely right, when exclaiming, that '*it is impossible to deprive any system of its entropy*'!

Meanwhile, our over-all experimental experience (the physics here is just a nice particular example, while the story is truly general) teaches us that nothing comes at zero prices; to accomplish something we have to exert some efforts to achieve our goals—by spending some available resources.

Specifically, to exert efforts we have to have some resources, which would be enough to allow us exerting any desirable effort.

On the other hand, there is the following problem: What should be the proper method to spend the resources in this case? It is but not difficult to see the true direction. Indeed, usually we cannot reach our goals just so; we always have to overcome some hindrances, some obstacles—this is our life in general, whatever the level of our consideration: macroscopic, mesoscopic, nanoscopic, microscopic, ultramicroscopic, etc.—or, otherwise—physical, chemical, biological, sociological, economical, or even our personal everyday life.

Hence, the answer to the poser is just before your eyes: We spend our resources *first* to compensate the Hindrances—and *thereafter* to achieve our actual goals.

The hindrances are truly ubiquitous, this is why—it is the above statement that truly corresponds to the basic law, so that all the other laws and rules are simply some logical consequences from the latter one.

To put the whole story in the energy terms, we write as follows:

The total energy is always constant. On the other hand, all the forms of the matter are consistent of atoms, molecules, supramolecules, etc. In turn, we have to note that supramolecules are consistent of molecules; the latter ones are consistent of atoms; the latter ones are consistent of nuclei and electrons, etc. *At all the levels mentioned above* the lower-level components composing the object at the current consideration level are somehow connected to each other, they are interacting with each other.

The energy of this interaction is just the potential energy—or, as one might purposefully denote it, this is just the '*energy stock*' to be used as a universal currency. This interaction potential energy in any particular case is the only measure of what energy resources are truly available to us.

Nonetheless, the latter story ought to be of truly little avail, if we do not know how to efficiently convert the available potential energy into the kinetic energy or—putting it in the ancient terms—into the vis viva, i.e., into the livening force.

In effect, it is just the vis viva that serves as a sole source of the driving force for any realistic process.

Further, the driving force thus available ought to solve two basic problems:

(A) First, to overcome the hindrances by fully equilibrating them;

(B) Second, to reach the actual goal without any resistance from elsewhere.

From the energetics standpoint it is then throughout logical to relate the entropy notion to the waste energy, so that the only physically reasonable assignment of entropy ought to be the ubiquitous hindrances/obstacles.

To sum up, we might formulate the *only valid* basic law of thermodynamics as the law of conservation and transformation of energy as follows:

To be capable of achieving any specific goal, the system ought to get the vis viva from the energy stock somehow (interactions between the system's components), and then spend the latter to achieve the complete equilibrium with the hindrances.

The intensity of the obstacles/hindrances ought to grow with the increase in the driving force's intensity from its zero, whereas the latter means the zero Hindrances as well.

The hindrance should then always reach its maximum, whereas we have to properly ensure that the driving force is enough to equilibrate the maximum hindrance.

Hence, any realistic process might run, if the corresponding system is possessed of the energy stock rich enough to provide the proper amounts of the vis viva. In other words, the driving force should be capable of duly following the hindrance till its maximum, to finally compensate the latter. To sum up, the livening/driving force intensity ought to be versatile to the level high enough to be capable of compensating the hindrance at the maximum of the latter.

Thus, the permissible maximum of the vis viva ought to perfectly correspond to the maximum of the hindrance, because the total energy must be constant.

The last but not the least point in the whole story is that driving the vis viva to its maximum would drive the energy stock, i.e., the potential energy, to its minimum, due to the fact that the total energy must be constant.

With all this in mind, we note that, on the one hand, the minimum of the potential energy corresponds to the equilibrium state of the system. On the other hand, this same state corresponds to the maximum kinetic energy and the maximum entropy. The latter both compensate each other, so that at last we arrive at the true saddle point on the *energy hypersurface* of our system.

Meanwhile, looking at the story purely mathematically, our process could bring us to the unambiguous saddle point, that is, to the *minimax/maximin* state of the system under study.

This is why the proper mathematics to describe the entire above story is nothing more and nothing less than just the game theory we have discussed in this chapter.

Interestingly, the above-sketched story is in fact tightly related to the theory of Prof. Dr. Kenichi Fukui (1918–1998). Moreover, what we suggest ought to be viewed as the further development of Prof. Fukui's theory.

3.2   Thermodynamics and the theory of intrinsic reaction coordinate by Kenichi Fukui

Indeed, Prof. Fukui had suggested the fundamental microscopic theory of chemical reactions and could win the Nobel Prize for his contribution in 1981.

He introduces his ideas as follows (we have highlighted the most important fragments, to our mind, for more details see the works by Prof. Fukui and his colleagues [3.26–3.38]):

"...It has already been pointed out that the detailed mechanism of a chemical reaction along the reaction path can be discussed on the basis of the orbital interaction argument. For this purpose, however, it is necessary that the problem of how the chemical reaction path is determined be solved.

The evaluation of the route of a chemical reaction and its rate in terms of the potential energy surface is performed using a statistical–mechanical formulation established by Eyring. Many other papers appeared in which the rate expression was derived wave-mechanically using the potential energy function...

The center line of the reaction path, so to speak, the idealized reaction coordinate, which I called the "intrinsic reaction coordinate" (IRC), seemed to have been, rather strangely, not specifically defined until then. For that reason, I began with the general equation which determines the line of force mathematically. Although my papers themselves were possibly not very original, they later turned out to develop in a very interesting direction and opened up the method of calculating the quasi-static change of nuclear configuration of the reacting system starting from the transition state and proceeding to a stable equilibrium point. I termed the method of automatic determination of the molecular deformation accompanying a chemical reaction, 'reaction ergodography'.

In a reacting system with no angular momentum it is possible to obtain the IRC by the use of a space-fixed Cartesian coordinate system. All of the calculated examples mentioned previously belong to this case. However, in a reaction in which rotational motion exists, it is necessary to discuss the IRC after separating the nuclear configuration space from the rotational motion."

Fukui's theory was based upon the conventional approach to the chemical reactions kinetics introduced in the book [3.39].

In connection with this, it is important to mention here that both Prof. Dr. Samuel Glasstone (1897–1969) and Prof. Dr. Henry B. Eyring (1901–1981) were the clear adepts of the 'irrational' thermodynamics' branch, as can be clearly seen from their well-known publications [3.40–3.42].

Most probably, this is just why Prof. Fukui was considering solely the '*potential energy hypersurface*' in his theory, thereby completely throwing away the entropy contributions.

Specifically, the very idea of the '*saddle point on the energy hypersurface*' as the very '*transition state of a reaction*' ought to be a true theoretical milestone on the way to understand the microscopic mechanisms of chemical reactions. By stubbornly restricting the further consideration to the potential energy hypersurface is leading to the conceptual blind alley. This is not 'a stone to Prof. Fukui's garden'—he was just the colleague who could go one tremendously important step to the correct solution of the problem.

There is but one more step to go, which could be accomplished by the young enthusiastic colleagues.

The volume of the present monograph does not allow describing all the important details of the step to go, so here we shall just solely indicate the true direction, to our mind.

Our suggestion here would be to consider the '*full energy hypersurface*', instead of the '*potential energy hypersurface*', because it is just this way that we do take into account the notoriously overlooked entropic effects as well.

Indeed, as we have already seen above here, the potential energy ought to be just the energy stock necessary to generate the true kinetic energy, the true livening force. It is the livening force that is spent to equilibrate the hindrances/obstacles (that is, the entropy effects). Due to the energy conservation law, it is but the kinetic energy available that would be to some extent spent for equilibrating the entropy (that is, the hindrances on the way to our goal).

The attentive reader would interrupt us here by exclaiming: What should then be your breaking news then? We do remain here at the Fukui's 'potential energy hypersurface'—just look at the line you draw! potential energy → kinetic energy → hindrances to overcome. So what?

Sure, we respond—here you are but right to 90%. The resting 10% is the following—*nowadays but still missing*—realization that among all the wealth of the atomic movements composing the over-all trajectory of any chemical reaction involved there is a certain group of them **helping** the mentioned reaction to proceed, as well as some other group **interfering** the reaction in question.

Remarkably, the former group could be considered '*enthalpic*', whereas the latter ought to be 'entropic'. The wealth of the '*enthalpic*' dynamics ought to be such as not only to equilibrate the available '*entropic*' dynamics, but also to achieve the actual goal of the reaction under study.

The problem of (enzymatic) catalysis should also be considered here—just in short: Indeed, the enzyme, the catalyst ought to somehow *enlarge the wealth of the 'enthalpic' dynamics and/or hinder the 'entropic' one.*

Thus, the very important step still to go ought to be the proper classification of the molecular/atomic dynamics of relevance to the process under study. It is this classification that ought to extend the 'potential energy hypersurface' to the 'total energy hypersurface', by making the grounded difference between the 'enthalpic' and the 'entropic' effects.

To sum up the entire discussion taken place in the volume at hand, one of the important, but still unsolved problems ought to be the detailed interrelationship among the Linhart's statistical mechanics, the macroscopic theory by Nils Engelbrektsson and the work by Max Bernhard Weinstein to bridge the gap between the macroscopic and microscopic levels of consideration. To our mind, this ought to help conceptually enabling the classification mentioned above.

References to Note 3

3.1 Albert Sauveur, Übersetzt von Hans Jüptner, and Freiherr von Jonstorff (**1898**): *Die Mikrostructur des Stahles und die currenten Härtungstheorien.*

(***Our comment***: Albert Sauveur (1863–1939) was a Belgian-born USA metallurgist, who published his works in the USA, see, e.g., his own work: Albert Sauveur (**1912**): *Metallography and Heat Treatment of Iron and Steel.* 1)

3.2 Hans Jüptner and Freiherr von Jonstorff. Traduit par Ernest Vlasto (**1891**): *Traité pratique de chimie métallurgique.*

3.3 Hans Jüptner and Freiherr von Jonstorff. Translated by Charles Salter (**1902**): *Siderology: The Science of Iron.*

3.4 Hans Jüptner and Freiherr von Jonstorff. Translated by Oskar Nagel (**1908**): *Heat Energy and Fuels.*

3.5 Hans Jüptner and Freiherr von Jonstorff (**1894**): *Heizversuche an Kesselfeuerungen.*

3.6 Hans Jüptner and Freiherr von Jonstorff (**1894**): *Die Untersuchung von Feuerungs-Anlagen.*

3.7 Hans Jüptner and Freiherr von Jonstorff (**1896**): *Die Einführung Einheitlicher Analysenmethoden.*

3.8 Hans Jüptner and Freiherr von Jonstorff (**1900**): *Grundzüge der Siderologie: Erster Theil.*

3.9 Hans Jüptner and Freiherr von Jonstorff (**1905**): *Lehrbuch der chemischen Technologie der Energien: Erster Band: Erster Theil.*

3.10 Hans Jüptner and Freiherr von Jonstorff (**1906**): *Lehrbuch der chemischen Technologie der Energien: Lehrbuch der chemischen Technologie der Energien: Erster Band: Zweiter Theil.*

3.11 Hans Jüptner and Freiherr von Jonstorff (**1906**): *Lehrbuch der chemischen Technologie der Energien: Zweiter Band.*

3.12 Hans Jüptner and Freiherr von Jonstorff (**1907**): *Beiträge zur Theorie der Eisenhüttenprozesse.*

3.13 Hans Jüptner and Freiherr von Jonstorff (**1908**): *Lehrbuch der chemischen Technologie der Energien: Dritter Band.*

3.14 Hans Jüptner and Freiherr von Jonstorff (**1910**): *Das chemische Gleichgewicht auf Grund mechanischer Vorstellungen.*

3.15 Oskar Nagel (**1908**): Evolution und Energie. *Annalen der Philosophie*, 7, pp. 251–256.

3.16 Oskar Nagel (**1908**): Versuch einer energetischen Geschichtsauffassung. *Annalen der Philosophie*, 7, pp. 257–276.

3.17 Oskar Nagel (**1908**): Zur Entstehung der Arten. *Annalen der Philosophie*, 7, pp. 387–392.

3.18 Oskar Nagel (**1908**): Politische Ökonomie und Energetik. *Annalen der Philosophie*, 7, pp. 417–428. (***Our Comment:*** *This journal was edited by Wilhelm Ostwald*).

3.19 Oskar Nagel (**1908**): *The Mechanical Appliances of the Chemical and Metallurgical Industries.*

3.20 Oskar Nagel (**1908, 1914**): *Die Romantik der Chemie.*

3.21 Oskar Nagel (**1909**): *Producer Gas Fired Furnaces.*

3.22 Oskar Nagel (**1909**): *The Transportation of Gases, Liquids and Solids by Means of Steam, Compressed Air and Pressure Water.*

3.23 Oskar Nagel (**1911**): *The Layout, Design and Construction of Chemical and Metallurgical Plants.*

3.24 Joseph H. Keenan (**1941**): *Thermodynamics.* Wiley & Sons, Inc.: New York, USA.

3.25 George N. Hatsopoulos, Joseph H. Keenan (**1965**): *Principles of General Thermodynamics.* Wiley & Sons, Inc.: New York, London, Sydney.

3.26 K. Fukui (**1970**): *J. Phys. Chem.*, 74, p. 4161.

3.27 K. Fukui, in R. Daudel, and B. Pullman, Editors (**1974**): *The World of Quantum Chemistry.* Reidel, Dordrecht, p. 113.

3.28 K. Fukui, S. Kato, and H. Fujimoto (**1975**): *J. Am. Chem. Soc.*, 97, p. 1.

3.29 S. Kato and K. Fukui (**1976**): *J. Am. Chem. Soc.*, 98, p. 6395.

3.30 K. Ishida, K. Morokuma, and A. Komornicki (**1977**): *J. Chem. Phys.*, 66, p. 2153.

3.31 B. D. Joshi and K. Morokuma (**1977**): *J. Chem. Phys.*, 67, p. 4880.

3.32 A. Tachibana and K. Fukui (**1978**): *Theor. Chim. Acta*, 49, p. 321.

3.33 A. Tachibana and K. Fukui (**1979**): *Theor. Chim. Am,* 51(189), p. 275.

3.34 K. Fukui (**1979**): *Recl. Trav. Chim. Pays-Bas*, 98, p. 75.

3.35 A. Tachibana and K. Fukui (**1980**): *Theor. Chim. Acta*, 57, p. 81.

3.36 K. Fukui, A. Tachibana, and K. Yamashita (**1981**): *Int. J. Quantum Chem. Symp.*, 15, p. 621.

3.37 Kenichi Fukui (**1981**): The path of chemical reactions—the IRC approach, *Accounts of Chemical Research*, 14, pp. 363–368.

3.38 Kenichi Fukui (**1982**): The Role of Frontier Orbitals in Chemical Reactions (Nobel Lecture), *Angewandte Chemie International Edition*, 21, pp. 801–809.

3.39 Samuel Glasstone, K. J. Laider, and Henry Eyring (**1941**): *The Theory of Rate Processes.* McGraw-Hill, New York, USA.

3.40 Samuel Glasstone (**1942**): *Introduction to Electrochemistry*. D. Van Nostrand Company: Princeton, Toronto, London, New York.

3.41 H. Eyring, J. Walter, and G. E. Kimball (**1946**): *Quantum Chemistry*. John Wiley & Sons: New York, USA.

3.42 Samuel Glasstone (**1947**): *Thermodynamics for Chemists*. D. Van Nostrand Company: Princeton, Toronto, London, New York.

[Note 4]: This is why we would greatly appreciate pursuing here the discussion started in the previous chapters and supplementary notes.

4.1 Institutionalizing scientific mafia—or—just doing normal scientific research

Concerning the contributions by Prof. Tait to thermodynamics, we had already started to discuss his role in Chapter 1 of the volume at hand, where we learned that in effect Tait was an active proponent of Prof. Mohr's constructive criticism against R. Clausius' treatment of thermodynamics. From Chapter 1 we also learned that, to wit, there had been a true constructive answer to Prof. Mohr's criticism, delivered not by Prof. Tait, but by Prof. Dr. August Friedrich Horstmann. Nonetheless, Horstmann's name had somehow gone lost for the academic community, although Josiah Willard Gibbs could manage to borrow and employ Horstmann's thermodynamic ideas. Prof. Gibbs was, is not and will definitely not be wearing the veil of oblivion. Howbeit, solely Dr. George Augustus Linhart was systematically continuing to develop those Horstmann–Gibbs ideas. What happened but to Dr. Linhart? For the answer to this poser see Chapter 4 of the volume at hand. Was Linhart indeed the only colleague busy with Horstmann–Gibbs ideas seriously and systematically?

Likewise, we have lost the name of Peter Boas Freuchen, who was basically the only colleague upon the world, who dared to systematically analyze Horstmann's contribution in detail, see Chapter 1 of the volume at hand.

On the other hand, Prof. Weinstein has correctly underlined a strange role of Albert Einstein in the then scientific environment, and nowadays we are reading the numerous modern publications by the professional physicists about basically the same topic.

Above we have already cited the striking reference to some 'scientific mafia' from the preface to Prof. Lavenda's thermodynamics book (see Page 11). In this connection, right in Chapter 1 of the present volume we had noticed Albert Einstein's strange activities in parallel to the fierce struggle by Ludwig Boltzmann and Max Planck against any manifestation of energetics, and we continued analyzing them in the Chapter 3, as well as in the Supplementary Notes throughout the volume at hand.

With all this in mind, the question would be: How to bring all these facts together, to work out the correct standpoint in regard to thermodynamics. This way we would definitely pay tribute to George Augustus Linhart, Max Bernhard Weinstein, Nils Engelbrektsson, Karl Alexius Franzén, Peter Boas Freuchen, August Friedrich Horstmann, and Max Theodor Trautz. We start here with Albert Einstein.

In the most recent literature there are reports sounding truly paradoxically [4.1–4.3]. The abstract of the work [4.2] sounds like as follows:

Abstract: (*The English translation*) *Mid-1920s Einstein and his Hungarian colleague Szilard have developed a refrigerator that did not involve any moving parts. This device has been reconstructed at the University of Oldenburg.*

Reading the relevant report leads to a truly serious conclusion: Albert Einstein, a genius physicist of all the times and all the peoples appeared to be a genius of engineering as well!

Indeed, a careful thinking over does stimulate us to look at Albert Einstein's story somewhat more attentively. After thoughtfully digging in detail through the old literature we encounter a very interesting work by Albert Einstein [4.4]. This paper of him describes an interesting engineering solution in the field of detecting and measuring 'small electricity amounts'. The paper contains a glaring conclusion the English translation of which we would like to re-publish here:

… Since the increase in the sensitivity of methods for electrostatic measurements is important for the study of radioactivity, I hope that a physicist could be interested in this matter. I would greatly appreciate to share with the latter my further thoughts on the subject. To the plan described above I could have come during my reflections on how to establish and measure the spontaneous charges of conductors.* Such

charges are required for building up the molecular theory of heat and analogous to the Brownian motion. With my plan described above I hope to have brought the latter problem one step closer to its solution.

Albert Einstein did not consider himself a physicist. Was he an engineer indeed? Well, a truly good poser! The most recent publication [4.3] is a review-essay analyzing the recent literature about Albert Einstein's engineering activity, especially one of the recent monographs [4.5]. The reviewer of the latter book writes, we cite [4.3]:

"For decades, Albert Einstein was portrayed as absorbed in theories, aloof of experiments and inventions. But in recent years, several books have tracked his activities in technological matters. ...

... Illy lucidly discusses Einstein's later technical labors and collaborations. Among his best-known innovative designs are refrigerators with Leo Szilard (from 1926 to 1930 they submitted 37 applications for patents, 28 of which were approved).

...In some ways, the book cannot stand by itself. Illy does not explain why Einstein even knew some of his main collaborators, or more importantly, why they chose to work on specific inventions. Why did Einstein and Nernst start designing refrigerators? Why did Einstein know Szilard? And why did he too work on cooling systems? How did Einstein meet Gustav Bucky in Germany? Why did Bucky move to the US? Illy neatly explains how Einstein met some lesser-known individuals, such as Rudolf Goldschmidt, but apparently he presupposes that readers will already know some of Einstein's collaborators. ...

... As an inventor, Einstein was often foiled by ignorance, by not seeking the published literature: when planning a thermocouple experiment ca. 1898 without knowing about Michelson's experiment of 1887; when designing the "Maschinen" to measure tiny electrical charges though knowing little about contemporary electrometers; when designing a uranium torsion balance in 1912 without knowing about the Eötvös experiment of 1889; when musing about meandering rivers; when writing that flight literature lacked discussions of the carrying capacity of airplane wings; and in 1942 when helping Bucky (re)invent an airplane airspeed indicator, which had already been patented by someone else two years earlier. ..."

The picture we actually see from the above: A truly talented engineer-theorist readily working on physical problems in vogue is inspired by his results and looking for the opportunity to excite interest among the physicists. Sure, in his full-time activities he had anything to do with theoretical physics, not to speak of the natural philosophy. The imperfect working style was not only Einstein's personal feature—his collaborator, Leo Szilárd was working the same way, see the careful analysis by Prof. Dr. Stephen Jay Kline (1922–1997) [4.6].

Prof. Kline has persuasively shown the negative impact of the Leo Szilárd's immense fascination by the 'Maxwell demon', which caused and still causes rather strange activities in different fields. In his essay, Alberto A. Martinez is also expressing a fully vindicated wish to look forward to more detailed, more thorough investigations concerning the actual activity of Albert Einstein.

Howbeit, for us here is important, that Max Bernhard Weinstein (see Chapter 3) knew very well, who was Albert Einstein in effect, for, as we know, Max Bernhard, among his numerous working activities, was also successful as a patent referee. Hence, Weinstein's last popular publication about the relativity theory is a clear and authoritative demonstration that Einstein was '*often foiled by ignorance*' not only in the purely engineering invention field.

4.2   Einstein's contribution to thermodynamics, statistical mechanics, and philosophy

In the Supplementary Note 21 to Chapter 1 as well as in the Supplementary Note 2 to the present chapter, we were monitoring Albert Einstein's refereeing activity in the *Beiblätter von Annalen der Physik*. So, why was he performing this work? Most probably, he would like to review the current thermodynamic literature of his time, to initiate his own working in the field. Actually, a very good move!

Nonetheless, what A. Einstein was producing in the field of thermodynamics did not look like a careful analysis of the fundamental basics, but rather like just a trivial manipulation with known mathematical formulae to reach more or less predestined results.

At the same time, other colleagues were indeed performing true investigations of the basics in the field, as we know now: Dr. Nils Engelbrektsson, Dr. Georg(e) Augustus Linhart (see Chapters 2 and 4 of the present book). Meanwhile, much more serious attention deserves Einstein's contribution to statistical mechanics—we shall not consider it here in detail, by sending the reader to the competent literature sources (see, for example, the serious handbook [4.7] in the field). A propos, the latter book contains a very interesting dedication, which is in effect also tightly connected with the achievements of Albert Einstein:

> *This book is dedicated to Ilya Prigogine for his encouragement and support and because he has changed our view of the world.*

Later on in this Supplementary Note we shall come back to the seminal contribution by Prof. Dr. Prigogine in more detail.

Concerning Einstein's further contributions, as we already know from Chapter 3 here, Max Planck was actively driving Albert Einstein into the natural philosophy. What was the result?

Einstein was a definite and *leal* adept of the "new revolutionary standpoint" in physics. He had summarized these ideas in a separate book [4.8]. We would now like to cite some fragments, for they appear to be of extreme importance for our present discussion. We shall highlight the points being in our opinion central to our present discussion, and share our thoughts on the theme.

We summarize:
Again, the rich variety of facts in the realm of atomic phenomena forces us to invent new physical concepts. Matter has a granular structure; it is composed of elementary particles, the elementary quanta of matter. Thus, the electric charge has a granular structure and, most important from the point of view of the quantum theory, so has energy.

Photons are the energy quanta of which light is composed. Is light a wave or a shower of photons? Is a beam of electrons a shower of elementary particles or a wave?

*These fundamental questions are forced upon physics by experiment. In seeking to answer them **we have to abandon the description of atomic events as happenings in space and time, we have to retreat still further from the old mechanical view. Quantum physics formulates laws governing crowds and not*** *individuals. Not properties, but probabilities are described, not laws disclosing the future of systems are formulated, but laws governing the changes in time of the probabilities and relating to great congregations of individuals.*

Of extreme interest and importance would be the question, how a radical standpoint of the above kind could be compatible with the reality?

The authors of the above book readily answer the poser as follows:

Physics and Reality

What are the general conclusions, which can be drawn from, the development of physics indicated here in a broad outline representing only the most fundamental ideas?

Science is not just a collection of laws, a catalogue of unrelated facts. It is a creation of the human mind, with its freely invented ideas and concepts. Physical theories try to form a picture of reality and to establish its connection with the wide world of sense impressions. Thus, the only justification for our mental structures is whether and in what way our theories form such a link.

We have seen new realities created by the advance of physics. However, this chain of creation can be traced back far beyond the starting point of physics. One of the most primitive concepts is that of an object.

The concepts of a tree, a horse, any material body, are creations gained based on experience, though the impressions from which they arise are primitive in comparison with the world of physical phenomena. A cat teasing a mouse also creates, by thought, its own primitive reality. The fact that the cat reacts in a similar way toward any mouse it meets shows that it forms concepts and theories, which are its guide through its own world of sense impressions.

"Three trees" is something different from "two trees." Again, "two trees" is different from "two stones." The concepts of the pure numbers 2, 3, 4 ... freed from the objects from which they arose, are creations of the thinking mind which describe the reality of our world.

The psychological subjective feeling of time enables us to order our impressions, to state that one event precedes another. Nevertheless, to connect every instant of time with a number, by the use of a clock, to regard time as a one-dimensional continuum, is already an invention. This same way we understand the concepts of Euclidean, non-Euclidean geometry, and the space surrounding us, as a three-dimensional continuum.

Physics really began with the invention of mass, force, and an inertial system. These concepts are all free inventions. They led to the formulation of the mechanical point of view. For the physicist of the early nineteenth century, the reality of our outer world consisted of particles with simple forces acting between them and depending only on the distance. He tried to retain as long as possible his belief that he would succeed in explaining all events in nature by these fundamental concepts of reality.

The difficulties connected with the deflection of the magnetic needle, the difficulties connected with the structure of the ether, induced us to create a more subtle reality.

The important invention of the electromagnetic field appears. A courageous scientific imagination was necessary to realize fully that not the behavior of bodies, but the behavior of something between them, that is, the field, may be essential for ordering and understanding events. Later developments both destroyed old concepts and created new ones.

Absolute time and the inertial coordinate system were abandoned by the relativity theory.

The background for all events was no longer the one-dimensional time and the three-dimensional space continuum, but the four-dimensional time-space continuum, another free invention, with new transformation properties. The inertial coordinate system was no longer necessary. Every coordinate system is equally suited for the description of events in nature. The quantum theory again created new and essential features of our reality.

Discontinuity replaced continuity. Instead of laws governing individuals, probability laws appeared.

Indeed, the reality created by modern physics is far removed from the reality of the early days. However, the aim of every physical theory remains mainly the same. With the help of physical theories, we try to find our way through the maze of observed facts, to order and understand the world of our sense impressions. We want the observed facts to follow logically from our concept of reality.

Without the belief that it is possible to grasp the reality with our theoretical constructions, without the belief in the inner harmony of our world, there could be no science. This belief is and always will remain the fundamental motive for all scientific creation.

Throughout all our efforts, in every dramatic struggle between old and new views, we recognize the eternal longing for understanding, the ever-firm belief in the harmony of our world, continually strengthened by the increasing obstacles to comprehension.

What about the thermodynamics, dear colleagues? Aren't we missing something very important here?

The answer is no, *not at all!* In a separate chapter of the cited book we read:

Probability Waves

*If, according to classical mechanics, we know the position and velocity of a given material point, and which external forces are acting, then we can predict using the mechanical laws, the whole of its future path. The sentence: "The material point has such-and-such position and velocity at such-and-such an instant," has a definite meaning in classical mechanics. If this statement were to lose its sense, our argument (**p. 32**) about foretelling the future path would fail.*

***Page 32**: "... Knowing the initial velocity and its change, we can find the velocity and position of the planet at the end of the time interval. By a continued repetition of this process, the whole path of the motion may be traced without further recourse to observational data. This is, in principle, the way mechanics predicts the course of a body in motion, but the method used here is hardly practical.*

In practice, such a systematic procedure would be extremely tedious as well as inaccurate. Fortunately, it is unnecessary; mathematics furnishes a short cut, and makes possible precise description of the motion in much less ink than we use for a single sentence.

The conclusions reached in this way can be proved or disproved by observation. The same kind of external force is recognized in the motion of a stone falling through the air and in the revolution of the moon in its orbit, namely, that of the earth's attraction for material bodies. Newton recognized that the motions of falling stones, of the moon, and of planets are only very special manifestations of the universal gravitational force acting between any two bodies. In simple cases, the motion may be described and predicted by the aid of mathematics.

In remote and extremely complicated cases, involving the action of many bodies on each other, a mathematical description is not so simple, but the fundamental principles are the same."

"In the early nineteenth century, scientists wanted to reduce all physics to simple forces acting on material particles that have definite positions and velocities at any instant. Let us recall how we described motion when discussing mechanics at the beginning of our journey through the realm of physical problems. We drew points along a definite path showing the exact positions of the body at certain instants and then tangent vectors showing the direction and magnitude of the velocities. This was both simple and convincing. ...

... We become indifferent to the fate of the individual gas particles. Our problem is of a different nature. For example, we do not ask, "What is the speed of every particle at this moment?" However, we may ask, "How many particles have a speed between 1000 and 1100 feet per second?" We care nothing for individuals. What we seek to determine do average values typify the whole aggregation. It is clear that there can be some point in a statistical method of reasoning only when the system consists of a large number of individuals.

By applying the statistical method, we cannot foretell the behavior of an individual in a crowd. We can only foretell the chance, the probability that it will behave in some particular manner. If our statistical laws tell us that one-third of the particles have a speed between 1000 and 1100 feet per second, it means that by repeating our observations for many particles, we shall really obtain this average, or in other words, that the probability of finding a particle within this limit is equal to one-third.

Similarly, to know the birth rate of a great community does not mean knowing whether any particular family is blessed with a child. It means a knowledge of statistical results in which the contributing personalities play no role.

Our first step away from classical physics was abandoning the description of individual cases as objective events in space and time. We were forced to apply the statistical method provided by the probability waves.

Once having chosen this way, we are obliged to go further toward abstraction. Probability waves in many dimensions corresponding to the many-particle problems must be introduced. Let us, for the sake of briefness, call everything except quantum physics, classical physics. Classical and quantum physics differ radically.

Classical physics aims at a description of objects existing in space, and the formulation of laws governing their changes in time. Nevertheless, the phenomena revealing the particle and wave nature of matter and radiation, the apparently statistical character of elementary events such as radioactive disintegration, diffraction, emission of spectral lines, and many others, forced us to give up this view. Quantum physics does not aim at the description of individual objects in space and their changes in time. There is no place in quantum physics for statements such as: "This object is so-and-so, has this-and-this property." Instead, we have statements of this kind: "There is such-and-such a probability that the individual object is so-and-so and has this-and-this property." There is no place in quantum physics for laws governing the changes in time of the individual object. Instead, we have laws governing the changes in time of the probability."

In connection with the above citations, please note that there exists a truly rich literature about the theme, in practically all the languages around the world—just to mention a few bright examples, see the references [4.9–4.12]. To sum up, the logical sequence of what Albert Einstein—and all the following authors—would like to teach us might be cast as follows:

First, we adopt the idea about the atomic structure of the matter. Indeed, there is really nothing to demur. Sure, everything consists of atoms and/or of molecules, which in turn contain a number of atoms somehow bound together.

Second, in speaking of individual atoms/molecules, we should definitely bid farewell to the conventional physics, for there is a novel (at its respective time) quantum theory which urges us to employ the language of the probability theory. Moreover, the same obviously holds, when we start considering a group of atoms/molecules.

The latter two steps imply that we are not speaking any more about some particles with definite coordinates in the three-dimensional space to be changed in the course of the one-dimensional time. From now on, we speak about some "probability waves" to be defined by some fancy functions on an abstract and truly multi-dimensional continuous spatial-temporal space. What is but still remaining discrete, quantized—ought to be the energy—first of all, itself as well as any carrier of the latter (some quasi-particles like photons, i.e., the quanta of light, phonons, i.e., the vibrational quanta, and so on, so forth).

As to the question of how should we now employ the probability theory—there is a very important mathematical question of *what should then be the correct probability distribution*. No problem ought to be the correct answer. There is a solid physical mathematics. For the thorough analysis of the latter we would like to send the interested readership to the detailed and truly ingenious treatise by Prof. Dr. Lavenda [4.13].

Importantly, in addition, we do not need to somehow infer or prove the validity of the Boltzmann–Planck's formula—it is *just obvious* that the *probability* notion duly underlies everything in the whole universe.

From such a viewpoint, Prof. Dr. Onsager's theory (see Supplementary Note 2, the present chapter, for the details of our discussion on the latter theme) is indeed marvelous. Still, only one recalcitrant notion is remaining, namely '*the rate of entropy growth*'. Indeed, this means that there remains some intrinsic time to define and follow this rate.

And it was Prof. Dr. Ilya Romanovich Prigogine (1917–2003), who could finally achieve the actual breakthrough, in perfectly completing Albert Einstein's efforts by 'killing the time'—to wit, ultimately and forever—and arriving at the true irreversible thermodynamics.

Indeed, it ought to be of tremendous importance for our present discussion to put here the exact formulation of Prof. Prigogine's bright idea [4.14]. We shall cite here the fragment from the latter report by him, which appears to be the most important for our present discussion:

"*Caractéristiques du temps thermodynamique*

Nous pouvons résumer comme suit les principales caractéristiques du temps thermodynamique :

Le temps thermodynamique est non-métrique, c'est-à-dire qu'il ne se réduit pas à la mesure des longueurs. Il est au contraire arithmétique car la source d'entropie introduisant des vitesses réactionnelles chimiques, sa détermination exige le dénombrement de particules.

Issu du second principe, le temps thermodynamique apparait nécessairement comme une notion statistique. Il perd son sens simple à l'échelle des processus élémentaires. Enfin, le temps thermodynamique est essentiellement local. Il est engendré par les processus irréversibles qui se passent a un endroit bien déterminé de l'espace.

Notons, que dans notre théorie élémentaire, non relativiste, le temps astronomique et le temps thermodynamique sont lies par la formule de transformation simple (13.4). Il n'en sera plus de même dans une théorie plus générale invariantive ou apparaitra une différence fondamentale entre le temps thermodynamique défini à partir d'un invariant, l'entropie, et le temps ordinaire défini comme quatrième composant d'un vecteur (1).

> *(1) Cf. E. A. Milne [4.15–4.17], qui distingue très clairement le temps 'newtonien' du 'temps de la probabilité' ou 'temps de la radioactivité'. Cette dernière conception est très proche du temps thermodynamique."*

"The characteristics of the thermodynamic time

We can now summarize the main features of the thermodynamic time:

The thermodynamic time is not metric, that is, it cannot be reduced to a length measurement. Instead, it is algebraic, for the determination of the entropy source during the introduction of chemical reaction rates requires enumeration of the particles.

According to the Second Basic Law, the thermodynamic time is necessarily introduced as a statistical concept. However, it loses its simple meaning at the scale of elementary processes. So finally, the thermodynamic time ought to be essentially local. It occurs owing to irreversible processes and has a well-defined point in the space.

Note that in our elementary non-relativistic theory the conventional astronomical time can always be transformed into the thermodynamic time by the simple formula (13.4). Even in a more general theory there ought to be substantial difference between the thermodynamic time, which is defined by an invariant, the entropy, and the ordinary time, which should be displayed as a fourth component of a vector (1).

> *(1) See E. A. Milne [4.15–4.17], who is very clearly distinguishing among the "Newtonian" time, "probability time" or*

"radioactivity time." It is the latter concept that is very close to the thermodynamic time."

To sum up, in fact we ought to have no real time, no real space (sure, there is but nothing apart from the spatial-temporal continuum), while the only basic and intrinsic relationships ought to be the pairwise relationships among the entropy, irreversibility, and probability.

The final trustworthy refinement of the entire conceptual gallimaufry we have seen above here had in effect come with the work by Dr. George Augustus Linhart; see Chapter 4 of the book at hand.

But what ought to be wrong with interpreting the interconnection between the entropy and irreversibility?

Such an interconnection does indeed exist, that is not the point—both entropy and irreversibility are definitely related to each other—we might well agree with the holy fathers of thermodynamics that there is even a rather tight correlation between the both notions.

There should be but somewhat more to the story in that to assume a direct cause-and-effect relationship between the both in this case ought to be expressly vague.

Indeed, thanks to the highly professional efforts of such renowned colleagues as Ernest Rutherford, Niels Bohr, Hans Geiger, Otto Hahn, Pjotr Leonidowitsch Kapiza, as well as Ludwig Boltzmann, Erwin Schrödinger, Werner Heisenberg, Albert Einstein, Louis de Broglie, Max Planck—and many-many other more or less known researchers every schoolchild nowadays knows:

There is not only the macroscopic world we perceive with our senses, but many more levels of the matter as well, which underlie all of our conventional pictures, starting from mesoscopic, via nanoscopic and down to the world of intrinsically microscopic systems.

In addition, thanks to a brilliant thought experiment by Erwin Schrödinger we are now well aware that 'a **microscopic** cat might be **both** dead **and** alive, which is contrary to our conventional **macroscopic** experiences' [4.18].

Sure, it is truly difficult to directly observe, what exactly happens at the levels beyond our direct imagination and experimentation.

In such a situation, as soon as we might reliably observe correlations, the only reasonable move ought to be dropping the direct cause-and-effect relationships among the observables, if we would like to interpret our observations skillfully. It is just the point, where we ought to adopt the standpoint of an outstanding British psychologist, Charles Edward Spearman (1863–1945), who had suggested how to significantly broaden the interpretational scope of the observed correlations.

Specifically, Prof. Spearman means, we cite [4.19]: *"But another— theoretically far more valuable—property may conceivably attach to one among the possible systems of values expressing the correlation, that is, that a measure might be afforded of the hidden underlying cause of the variations."*

In fact, what we have read above is just the conceptual basis for the nowadays tried-and-true method of the multidimensional statistics: the factor analysis of correlations. The details of the approach, the references to the important literature sources could be found in the book [4.20].

Thus, the ostensible mathematical 'blind corner' of the 'mysterious interconnection' between the entropy and irreversibility should not be a serious hurdle for the physical deliberations. The main point here ought to be our clear understanding of the entropy's true physical sense.

In effect, we should expect that the processes contributing to the livening/driving force and those contributing to entropy (useful energy devaluation) belong to quite different factors competing with each other. Specifically, by carrying out systematical experimental studies and by interpreting the set of observed correlations using the factor analysis, we might surely learn a lot about all the possible states of all the 'Schrödinger's cats' ever available, as well as about all of their eternal partnerships and competitions.

A propos, the idea behind the factor analysis might also serve as a methodological basis to resolve the well-known Einstein– Podolski–Rosen (EPR) paradox. The idea suggested by EPR is that quantum mechanics, despite its undoubted success in a wide variety of experimental scenarios, is actually an incomplete theory.

In other words, there is some yet undiscovered general theory of nature to which quantum mechanics relates as a kind of statistical approximation (albeit an exceedingly successful one). Unlike quantum mechanics, the more complete theory contains variables corresponding to all the "elements of reality."

There must be some unknown mechanism acting on these variables to give rise to the observed effects of "non-commuting quantum observables," i.e., the Heisenberg uncertainty principle. Such a theory is called a **hidden variable theory**—for more details see, e.g., the most recent textbooks and the references therein [4.21, 4.22].

Sure, the quantum theories, as well as everything in relation to them, were, are and would undoubtedly remain a great achievement of the physics in the XX-th century. Still, there are serious posers to be solved and resolved, so that the story is by far not finalized. One characteristic and interesting example of the latter statement (from the standpoint of our present discussion) ought to be studies by Yuri Kornyushin as to the interrelationship between the quantum mechanics and energetics [4.23, 4.24].

At this very point we would like to stop this interesting discussion, because any detailed consideration of the interconnection between the factor analysis and quantum theory, as well as that between the former and the natural-scientific field as a whole ought to be far beyond our present topic.

Meanwhile, the true, meaningful, insightful research work on the macroscopic thermodynamics, statistical mechanics, as well as on the quantum theory could not be properly finalized. The reasons ought to be obvious—finiteness of our lifetimes, methodological errors—plus purely political reasons worldwide. Therefore, research in the directions listed remains even nowadays an interesting, hot, *attractive* topic—*to attract* young, diligent, and positively ambitious colleagues.

With this in mind, the only point we would like to mention here would be the true relationship between the statistical mechanics and thermodynamics at the beginning of the XX-th century, that is, at the time of the physical revolution we mentioned already. We have already published several aspects of our present story and would like to refer to that work [4.25–4.31] (All the listed papers by Starikov are in open access, except for the 1st one):

Furthermore, to clearly show what kind of 'surf' was and still is 'ahead' of us, what Max Bernhard Weinstein would like to caution us against (see Chapter 3), we would like to present here the story by a truly forgotten American physicist, William Walker Strong, whom we have encountered in the previous note to this chapter and learned about his picture of the modern physics. From Prof. Strong's citation we might hear many voices (see Note 2)—it is a true polyphony!

We have cited above the book by Charles Ruhla, who praises Blaise Pascal and Niels Bohr as the two holy fathers of the 'physics of chance'.

Remarkably in this connection, Niels Bohr has considered the 'disappearance of energy' just as 'a probable event. Ludwig Boltzmann was not excluding such a possibility as well. We have already discussed the activities of the both earlier in this book.

Thus, we have to be sincerely indebted to Prof. Dr. William Walker Strong for his clearly showing us the 'surf', which is 'ahead'—and which we have to carefully avoid by any means, if we are not the allegiant adepts of the 'operational approach'.

To summarize, we have to tell solely a couple of words concerning the actual role of Blaise Pascal (1623–1662). He was indeed one of the holy fathers of the modern probability theory, and, moreover, he was an active physicist. Along with this he was a fine philosopher.

Due to his swiftly deteriorating health state he had to gradually stop his research work and concentrate on thinking over the philosophic backgrounds of what he was steadily working on at his happier time. As a result he could publish at least some part of his very important and valuable conclusions. The collection of his last writings is entitled '*Pensées*' [4.32]. Concerning the probability notion we find the following interesting and important notes in Pascal's '*Pensées*':

"*Mais est-il probable que la probabilité assure?—Différence entre repos et sureté de conscience. Rien ne donne l'assurance que la vérité. Rien ne donne le repos que la recherche sincère de la vérité.*

Probabilité – Chacun peut mettre, nul ne peut ôter."

"*But is it likely that the likelihood ensures?—As to a difference between the rest and the safety awareness. Nothing gives the assurance but the truth. Nothing gives the rest but the sincere search for the truth.*

Probability—Everyone can put, no one can take away."

After attentively reading these lines the immediate poser arises: What ought to be the 'physics of chance' according to Blaise Pascal?

To our mind, Pascal's second statement comes to the point of some intrinsically fundamental interrelationship between the probability and irreversibility. Meanwhile, his first statement does not encourage us to rely upon the 'sole probabilistic reasoning' and stop at this point. Instead, he would probably like to encourage us to stubbornly look for the very truth.

To sum up, if we are not just blindly following the 'operational approach', the interconnection between the chance and irreversibility should never be the final point. To arrive at some reasonable result of our research, we ought to try finding the true answer to the poser:

> **Why**, in any particular case, 'everyone can put', but 'no one can take away'?

Instead, what Prof. Dr. Strong describes in his book ought to be a kind of fluffy carnival parade, but definitely not a serious research activity. Most probably, Prof. Dr. Strong was trying to call his contemporary colleagues to action by showing them the strange picture under construction. But who knows the very truth now, whom should we ask?

4.3   Statistical thermodynamics

Let us now try to make an unbiased summary of what was happening in the field of thermodynamics and statistical mechanics at the end of XIX-th and beginning of the XX-th centuries. We might clearly see that after the untimely departures of such outstanding colleagues as Ludwig Boltzmann and Josiah Willard Gibbs, the main research activity seems to be directed to the 'operational digestion' of the results by the both. Had the story to stop?

Not at all! Sure, the 'surf' was 'ahead', but George Linhart was not alone. Indeed, in the Russian literature, there was an attempt to carefully infer the Boltzmann–Planck formula and analyze in detail the physical sense of the inference as well as of its pertinent result.

Prof. Dr. Vladimir Alexandrovich Michelson (1860–1927) undertook this attempt and could manage to publish his results in the Russian mathematical journal in 1887 [4.33]:

This portrayal of Prof. Dr. Michelson has been taken from
http://physiclib.ru/books/item/f00/s00/z0000052/st081.shtml

Here we shall translate some interesting fragments of the above paper into English and comment them.

"Among all the attempts to infer the second law of thermodynamics from the general equations of mechanics, which I am aware of, several are definitely erroneous and arrive at the Carnot's law (known in advance in the Clausius' form) only due to several errors in the inference course, which cancel each other (see 1 and 3); Then, several other ones aren't exact and rigorous enough, but they still seem to be much too general owing to insufficiently clear formulation of the initial conditions (see, e.g., 3); There is also another group of works based on such particular and voluntary conditions that it might be truly difficult to imagine any realistic situation, where the second law ought to be applicable (see, e.g., works 4–6). Finally, the best analytical inferences available (see the further works by Clausius and Boltzmann, 7) are extremely complicated in both their assumptions and their conclusions, and apart from this "it's truly a trick and a half" to grasp the difference among the assumptions voluntarily adopted and the conclusions rigorously inferred over there.

Bearing all this in mind I have tried to separate here the most evident and necessary assumptions as to the thermal motion among the huge number of all the available ones. Moreover, I would greatly appreciate to simplify the very mathematical inference, without any loss of the necessary rigor and generality.

[1] *Szily. Pogg. Ann. 145, pp. 295–302 к Erg. 7, pp. 74.*

[2] *Oppenheim, Wied. Ann. 15, pp. 495.*

[3] *J. J. Müller. Pogg. Ann. 152, p. 105–131.*

[4] *A. Ledieu. Comptes Rendus, T. 77 & 78.*

[5] *L. Boltzmann. Sitzber. d. Wien. Ak. Bd. 53. Abth. II. pp. 195–220.*

[6] *Szily. Pogg. Ann. 160 pp. 435–454.*

[7] *Clausius. Pogg. Ann. 142, pp. 433–461 & 150, pp. 106–130;*

L. Boltzmann. Sitzber. d. W. Ak. Bd. 63 Abth. II pp. 712–732 & Bd. 76 Abth. II pp. 373–435."

Then, the detailed formal inference follows, starting from the equations of motion for some extended system of material points, A, consisting of a body, K, which is of interest for us—together with all the bodies somehow interacting with the latter. The fact of the interaction requires us to consider the corresponding energy, that is, the potential energy dependent on the positions of all the bodies/points in the space.

The physical logic dictates then that the displacements of the bodies/points would change the potential energy and therefore produce the forces capable of performing some (useful) work.

In Chapter 1 here we had already learned about the ultimate caution expressed by Prof. Dr. Mohr concerning the attempts to infer the formal mathematical expressions for the second basic law assuming a reversible process of some general kind. That was the research forcefully introduced by Clausius and followed by Boltzmann and other colleagues just mentioned above. Prof. Mohr suggested not to be fascinated by purely mathematical inferences but to follow the actual physics of the phenomena instead. Above in this volume we were also discussing the risk of being fully absorbed by the 'reversibility' notion, which has nothing to do with the actual physics, but solely with the cyclic gadget suggested by N. L. S. Carnot.

Prof. Michelson is not referring to Prof. Dr. Mohr, but he duly follows his caution. Indeed, like all of his predecessors, he is interested in finding some definite analytic expression for the reversible process. Still, unlike the vast majority of his predecessors, including Clausius and Boltzmann themselves, he is not considering the separate dynamics of all the particular material points, which is sheer unknown in effect. Instead, he suggests taking into account the system as a whole, by assuming that the system K consists of some noticeable number of the material points. At the first glance, the difference among the standpoints by Michelson and those by his predecessors seems to be truly negligible. Nonetheless, it is essential.

Specifically, Michelson's way of consideration ought to provide us with much more conceptual space, because any assumption clearly improbable for a separate point can be rendered probable for the system as a whole by the law of the large numbers.

With this in mind Prof. Michelson declares the conventional thermal motion to be stationary, to point out that the total state of the system as a whole won't be changing, if a number of identical points would be involved into some kind of mutual exchange of their positions and/or velocities.

Mechanically, Prof. Michelson defines a stationary many-particle dynamics of such a kind, if in the course of the latter:

1. The very type of the functional dependence of the potential energy on the particles' coordinates remains to be unchanged.

$$U \equiv U(x_1, y_1, z_1, ..., x_N, y_N, z_N); \qquad \text{(M1)}$$

2. The value of the total kinetic energy value, being the symmetrical function of the particles' velocities, is experiencing oscillations of the regular—or even irregular—kind, so that the average value of the kinetic energy over some finite, although, possibly, rather short time period, i, isn't dependent upon choice and duration of the latter (please note that here i doesn't correspond to the imaginary unit of the complex numbers).

$$T = \sum \frac{m}{2} v^2. \qquad \text{(M2)}$$

$$\frac{1}{i} \int_0^i T \, dt = \bar{T} = \text{const.} \qquad \text{(M3)}$$

3. The total energy of the system as a whole is equal to:

$$T + U = E = \text{const.} \qquad \text{(M4)}$$

That is, the latter ought to remain constant all the time.

The temperature of the body under study in such a physical state ought to be proportional to the average *vis viva*, $\bar{T}$, of the stationary movement defined above, whereas the coefficient of proportionality, according to Clausius*, ought to be chosen as the expression Σmc, to

---

*Pogg. Ann. 142. p. 458.

ensure that various particles composing the body under study might in fact be possessed of different heat capacities, *c*.

Hence, with this in mind, we might cast the body's temperature as follows:

$$T = \frac{\sum \frac{m}{2} \overline{v^2}}{\sum mc}. \tag{M5}$$

Most probably, such a temperature definition could be justified by the Clausius' virial theorem,* in conjunction with one of the Boltzmann's[†] theorems stating that there is an equilibrium between the kinetic energies of the neighboring particles for two bodies in contact with each other.

Still, to apply such theorems consistently we have to assume that our thermometers are the bodies, where internal forces are infinitesimally small, or proportional to the external ones. In such a case, we use the normal thermometer and are dealing with the absolute scale of temperature.

Furthermore, Prof. Michelson points out that there are in effect no strictly stationary dynamical regimens, for there are no realistic systems, wherein no forces could be exerted. Hence, contrariwise, any (at the first glance) stationary thermal state of any body would be somehow varying all the time—for, in effect, the parameters to be constant in order to guarantee the onset of a truly stationary state ought to be changing in reality.

But if the variations in question are taking place in such a way that it is just ceasing at any arbitrary time point of any changes in the parameters involved (i.e., physically, ceasing of any changes in the environmental conditions) that might render the body's dynamics really stationary, then we might consider all these variations **reversible**.

Indeed, for the process to be reversible it should be both necessary and sufficient that the internal state of the body under study (the disposition of the particles and the distribution of their velocities over the magnitudes and directions) would every moment be correspondent to the environmental conditions, so that the latter

---

Pogg. Ann. 141. p. 124 and Rühlmann: *Handbuch der mechanischen Wärmetheorie,* v. I, p. 437.

[†]Sitzungsberichte der Wiener Akademie, *Band* 53, pp. 195–201 and Rühlmann: v. I, pp. 453–456.

correspondence might mimic the situations leading to the stationary dynamics. In other words, any change in the body's internal state ought to match the variations in the environmental conditions: This is why, the latter have to be extremely slow, or the corresponding parameter changes must be infinitesimally small.

Further on, Prof. Michelson's paper contains no more detailed deliberations in the field of natural science, so that the formal mathematical inference begins: The Hamiltonian equations of motions are being cast and then systematically solved.

Taking into account the fact that during the integration of our equations the total energy, that is, E = T + U remains constant, as well as the variation of the latter, plus the variations of the environmental conditions, we arrive at Eq. (M17),

$$\delta(i \cdot 2\overline{T}) = i\left\{ \delta E - \sum \left(\overline{\frac{\partial U}{\partial a}}\right)\delta a + \left(\overline{\frac{\partial U}{\partial b}}\right)\delta b + \left(\overline{\frac{\partial U}{\partial c}}\right)\delta c \right\} \quad \text{(M17)}$$

*where the horizontal lines above the expressions denote their average values during the **time-lapse i = t₁ – t₂**.*

Please note that the 'c' here is not the heat capacity appearing in Eq. M5, but one of the three Cartesian axes (a, b, c). Here we have just taken over the designations of the Michelson's original work.

*It is evident that the right-hand part of Eq. (17), which is denoted Eq. (M17a) below, stands for **the work performed by the average values of the external forces during the time-lapse** δt,*

$$-\sum \left(\overline{\frac{\partial U}{\partial a}}\right)\delta a + \left(\overline{\frac{\partial U}{\partial b}}\right)\delta b + \left(\overline{\frac{\partial U}{\partial c}}\right)\delta c \quad \text{(M17a)}$$

that is, during the transition from one of the stationary dynamics regimens to another one any system's state ought to be infinitesimally near to its previous counterpart. In other words, this is nothing more than the work transferred from the body K under study to all of its surrounding, and we denote this whole term by δL_e.

Thus, we might arrive at Eq. (M18).

$$\delta(i \cdot 2\overline{T}) = i(\delta E + \delta L_e). \quad \text{(M18)}$$

But if the body K under study would get from its surrounding some amount of heat, dQ, then the energy conservation law dictates Eq. (M19),

$$J \cdot \delta Q = \delta E + \delta L_e, \quad \text{(M19)}$$

where J stands for the mechanical equivalent of calorie. Eq. (M18) might then be recast as Eq. (M19a)

$$\delta(i \cdot 2\bar{T}) = i \cdot J \cdot \delta Q \qquad \text{(M19a)}$$

and leads to Eq. (M20).

$$\delta Q = \frac{1}{J} \cdot \frac{\delta(i \cdot 2\bar{T})}{i} = \frac{\bar{T}}{J} \cdot 2\delta \log(i \cdot \bar{T}) \qquad \text{(M20)}$$

In dividing the latter expression term-wise by our definition of absolute temperature, Eq. (M5), we might finally arrive at Eq. (M21), or, taking into account that all of c's in Eq. (M5)

$$\frac{\delta Q}{T} = \frac{\sum mc}{J} \cdot \delta \log(i^2 \cdot \bar{T}^2) \qquad \text{(M21)}$$

are assumed to be constant, we get Eq. (22),

$$\frac{dQ}{T} = d\frac{\sum mc}{J} \cdot \log(i^2 \cdot \bar{T}^2), \qquad \text{(M22)}$$

that is, we arrive at the notations using the full differential. By integrating the latter expression over the closed cycle we get Eq. (M23)

$$\oint \frac{dQ}{T} = 0, \qquad \text{(M23)}$$

that is the very expression of the second law of thermodynamics in the form of Clausius.

$$S = \int \frac{dQ}{T} = \frac{\sum mc}{J} \cdot \log(i^2 \cdot \bar{T}^2) + \text{Const}, \qquad \text{(M24)}$$

Obviously, the analytical expression of the entropy S might be cast in the form of Eq. (M24), whereas the detailed analytic expression for the "free energy" notion introduced by Helmholtz* would then be re-cast in the form shown by Eq. (M25).

$$F = E - J \cdot \bar{T} \cdot \text{Const},$$
$$F = E - \bar{T} \cdot \log(i^2 \cdot \bar{T}^2) - J \cdot \bar{T} \cdot \text{Const}. \qquad \text{(M25)}$$

To sum up, we would greatly appreciate to justify our above-sketched approach by noticing the following facts.

Clausius guesses that the convectional approach to the variables' variation wouldn't be applicable to the thermal motion, because the

---

**H. von Helmholtz: Sitzungsberichte der Berliner Akademie, der 2 Februar 1882; Wissenschaftliche Abhandlungen, II, pp. 958–978.*

approach in question doesn't take into account that coordinates of the separate particles might in the course of time experience not only just oscillations within some rather narrow limits, but also become infinitely growing, or simply finite, instead of being infinitesimally small. To get rid of such an inconvenience Clausius* introduces his own, specific variation method based upon the generalized notion of phase as a function, whose values are determining the respective time points and dispositions of the particle in its initial and modified states.

But I guess that if during our variations we would strictly follow the individuality of the separate particles constituting the system, the aim formulated by Clausius can't always be achievable, for any infinitesimally small variation of the environmental parameters might cause any finite dodging and/or change in the trajectory of one and the same separate particle due to the extreme improperness and disorder[†] of the thermal motions. Hence in fact, no choice of the phase might properly guarantee limiting the changes of the particles' disposition by the vibrations within some narrow limits during any arbitrarily long time-lapse.

To really achieve his aim Clausius ought to go one more step in the direction of his generalizations and agree to view the correspondence between different dynamical regimens not only at some different moments of time from the standpoint of different particles, but also taking into account the particles themselves without assuming that they would be moving identically both in the initial and in the modified regimen.

*To put this in more detail, let us re-consider the actual sense of the difference between the particles constituting our system in various dynamical regimens of the latter. Indeed, we might note that in the initial dynamic regimen some arbitrary point 1, with the coordinates x_1, y_1, z_1, might in the modified regimen be equivalent not only to itself, but also to some other point h of {**Our note: chemically**} the same type and with the coordinates x_h, y_h, z_h. Bearing this in mind we shall consider the variations dx_1, dy_1, dz_1 to be **not** the differences between the respective coordinates of the particle 1 in **different** dynamical regimens and at the corresponding time points‡, **but** the differences between the coordinates x_1, y_1, z_1 in the initial regimen and the coordinates x_h, y_h, z_h in the modified regimen, provided the resulting differences are infinitesimally small.*

**Pogg. Ann. 142, pp. 433–461 and 150, pp. 106-130;* Rühlmann, I, p. 440.

[†]*Unordnung, as Helmholtz puts this, l. c. pp. 972 (15).*

‡*As Clausius indeed does this.*

The main point is that even if during the variation of stationary dynamics a trajectory of some particular point 1 would be significantly shifted and finitely deformed, there will always be some other (but chemically identical) point h which in the modified regimen would take place of the point 1—if not to 100% exactly, then just rather approximately. If in the course of time the point h would not be obeying these conditions anymore, we might find the next pertinent point, and so on, and so forth.

This way, it ought to be always possible to 'synthesize' a 'combined' trajectory in the modified regimen, using different (but chemically identical) particles, to infinitesimally closely mimic the trajectory of the initially chosen particle in the initial regimen.

And, to consistently choose the infinitesimally small elements of the second order for these artificial trajectories, which in fact describe the transitions among different parts of the actual trajectory, we ought to assume all these elements to be situated on the equipotential surfaces, in order to ensure that no work is required to perform such imaginary transitions.

Still, even such a choice does not seem to be mandatory, since from the beginning on these artificial trajectories are representing some imaginary way to characterize the realistic variations of the system's actual state, but not the actual trajectories of the particles, which constitute the system under study, whatever the actual dynamical regimen is present at this time point. This is why the artificial trajectories might even remain discontinuous. Now let us apply the Clausius' approach just in the pairwise way: To the actual trajectory in the initial regimen and to the artificial one in the modified regimen. Then his 'phase variations' would as a result be oscillating within really narrow limits.

In summing up, when speaking about the stationary dynamics, I guess that although the above-sketched approach wouldn't be much less exact as compared to the conventional one, it is quite possible to make use of the latter by viewing the variable values at the time points t and t + dt to be correspondent to each other, even without giving a hostage to fortune that the variations in question would experience infinite growth.

That the general equations of motion would be applicable even with such an approach to the variations ought to be clear from the following physical consideration as well: Indeed, any exchange of positions between the identical particles of the system under study does not require any additional spending of work, for any positive work to perform one of the displacements would be equal to the negative work to carry out another one.

Prof. Michelson was 24 years old, when working on the above topic; he was an apprentice of an outstanding Russian physicist of that time, Prof. Dr. Alexander Grigorievich Stoletov (Александр Григорьевич Столетов, 1839–1896). Howbeit, it took him some 7 years to publish this very important work. Why this work was so important?

Michelson could manage not only rigorously proving the famous Boltzmann–Planck formula, but also demonstrating that the notorious W in the expression $S = k * \ln(W)$ ought to be the function of the temperature and time. This was indeed a serious thrust against the 'mystic interrelationship between the entropy and probability', for Michelson's work had clearly shown what the notorious '*magic thermodynamic probability, W*' ought to be possessed of a clear-cut physical sense, whatever handy identifier might be chosen for this function of the measurable temperature and time.

Interestingly, after publishing the above-mentioned paper Prof. V. A. Michelson was continuing to use statistical physics. In the years 1887–1890 he started studying one of the most fashionable problems of that time, namely the problem of blackbody radiation. In using classical statistics to describe radiating atoms he could finally arrive at some theoretical conclusions concerning the actual functional form of the Kirchhoff's emission–absorption law, which had turned out to be wrong.

Meanwhile, it is well known that it was Max Planck, who could correctly derive the formula for this law. Remarkably, Planck's formula is presently viewed as a result of profound thermodynamic and electro-magnetic theoretical studies. According to Planck's own confession, it was but nothing more than just a '*fortunately guessed law having only a formal meaning*'. This '*fortunate guess*' could definitely help him to arrive at his nowadays-famous conclusion about the '*energy quanta*'. In effect that was a result of Max Planck's careful analysis of the actual physical sense of the Kirchhoff's law. Both Ludwig Boltzmann and Albert Einstein were also very actively participating in the relevant research work.

Still, the only poser was and is remaining unanswered: What should then be the true physical rationale for the Boltzmann–Planck's formula?

All the details of that entire story might be found in the book [4.34]. *Meanwhile, to our mind, it is at this very point that thermodynamics started to perish by its fragmentation. Indeed, it was definitely not necessary to struggle for answering the above-mentioned poser for arriving at the quantum theory of the blackbody radiation. And the clear successes of the resulting quantum physics are presently undoubted.*

Meanwhile, the latter successes did and do not cancel the validity and importance of the poser formulated, whereas the quantum physics/kinetic gas theory themselves are by far not helpful in properly answering it.

It is clear then, why Prof. Michelson had at last quit this research field—and noticed the following interesting 'logical gap' in his *'physics course'* (originally published in 1913, but we reference here its 9th edition [4.35]) (our English translation of this citation follows):

It is possible to derive some very important consequences from the Carnot's theorem in regard to the practical problems.

For all the realistic engines any change in their efficiency coefficient starting from the ideal one can only follow in the direction of its reduction. Consequently, in any heat engine all the heat borrowed from the heater cannot be fully converted into useful work. And still no machine has been built as yet, that could be capable of converting even at least a half of the heat developed by the fuel into the work.

The reason for such a limited conversion of the heat into work doesn't lie in the imperfection of our machines—instead, it is owing to the very laws of nature: i.e., to the discordancy of molecular thermal motion.

Remarkably, Prof. Michelson had addressed his above-cited book to the students of technical universities—and the book contains *no information* about *how exactly* the intrinsic *discordance* of the thermal motion would finally lead to the *intrinsically low* efficiency coefficients of the real machines. To our mind, that was just a desperate move by Prof. Michelson to recognize his above-mentioned failure in considering the physics of the thermal phenomena. Meanwhile, his very first work in the field we have analyzed above had in effect presented *the first formal inference* of the famous Boltzmann–Planck logarithmic formula for the Clausius entropy notion [*sic*], shown how to deal with the 'reversibility'—but this in fact seminal paper is never mentioned in the available biographical sketches about Prof. Michelson's work!

Nonetheless, this is, to the best of our knowledge, the *first publication* all over the world, in which the *logarithmic function* for the Clausius' entropy notion *is mathematically formally* derived, starting from the general equations of motion for the multi-particle systems. Remarkably, Prof. Michelson *duly mentions* the possibility to consider the latter systems from the standpoint of the probability theory, *but never goes this way*! Instead, he is in fact trying to find *the truly physical basis* for the *discordance* of the notorious thermal motions. And he succeeds in solving this problem: The expression under the sign of logarithm turns out to be dependent **both** on the *temperature* **and** *time* [*sic*], but is never delivering the expression for some '*magic probability*' possessed of some unclear physical sense.

Actually, Prof. Michelson's work in question was the very first step to clarifying the physical basis of the second basic law of thermodynamics, together with the actual physical sense of the entropy notion. This is clearly seen in the final mathematical part of his paper, where the analytic expression for the Helmholtz' Free Energy is also being derived—The readership might immediately recognize that entropy in effect stands for nothing more than the *lost energy*, but Prof. Michelson wasn't dwelling on this important point. Instead, he is deliberating in detail about how one might mathematically consistently deal with the stationary discordant thermal motion of a huge number of (chemically) identical particles, *without departing from the actual physics of the phenomenon*. This theme is definitely of extreme importance, no doubt, but his deliberations could not lead him to the recognition of the quantum nature of the matter—although his work ought to remain the guiding star showing that, although there should be some interrelationship between the entropy and the probability notions, the latter is by far not a 'magic natural law' governing everything starting from the (ultra-) micro- till the macroscopic events. Meanwhile, that was Dr. Georg(e) Augustus Linhart who could show the true way to the ultimate solution of the problem (see Chapter 4 of the book at hand). We have discussed the consequences in the work [4.25].

In this connection, there were several other colleagues who were dealing in detail with the notion of *free energy*, but they mostly remain in the shadow somehow. One of them in Russian Empire was Prince Boris Borisovich Gallitzin (1862–1916).

This portrayal of Prof. Dr. Gallitzin has been taken from
https://ru.wikipedia.org/wiki/Голицын,_Борис_Борисович

Prof. Dr. Gallitzin was from his young years on dealing with thermodynamics and could publish a number of interesting papers in the field, in one of the then highly reputable journals, namely, in the *Proceedings of the Imperial Academy of Sciences of Russia* [4.36–4.38]. Prince B. B. Gallitzin started his education as a naval officer, first graduating from the Naval Cadet Corps and then from the Nikolaev Marine Academy. During his youth time he was fond of physics, chemistry, astronomy, and other sciences. In 1887 he had decided to leave the army and continue his education at the University of Strasbourg. In 1890 he had returned to Russia after acquiring the university diploma in Strasbourg. In 1892 Prince B. B. Gallitzin began teaching at the University of Moscow and had published his work entitled: "Research on mathematical physics: Part I. General properties of dielectrics in terms of the mechanical theory of heat. Part II. About radiant energy" in the journal entitled 'Mathematical Transactions of Moscow'. Then, in early 1893 he had presented this same work to the faculty as his master's thesis. Meanwhile, Gallitzin's thesis had met a very harsh negative assessment by reviewers A. G. Stoletov (the mentor of Prof. A. V. Michelson) and A. P. Sokolov. As a consequence, B. B. Gallitzin had to ultimately leave the University of Moscow, while starting to hold his lectures at the University of Yuryev (in the earlier time: Dorpat/ Dörpt, in Swedish/German, now—Tartu in Estonian), where he became a professor of physics. 'Moreover, Prof. Gallitzin had to drop his research in the field of thermodynamics and started working as a geophysicist, seismologist, he was really successful in these both important fields as well.

Meanwhile, Gallitzin's work on the thermodynamic notion of the free energy is but of our primary interest here.

Among those colleagues, who were also active in the field was the protagonist of this monograph, Prof. Dr. Max Bernhard Weinstein (Ref. 33 in Weinstein's publication list, see Chapter 3 in the book at hand). Weinstein's work was an outstanding critical review of the relevant efforts by different authors at the time, which had somehow remained unnoticed, most probably due to the sad fact that this was actually the last publication of him before his untimely departure.

Remarkably, at the same time another very important contribution to the theme of the free energy had been delivered by one of the outstanding Polish physicists, Prof. Dr. Władysław (Ladislaus) Natanson (1864–1937) [4.39], who was one of the leading personalities in the Polish Academy of Science, professor of physics and the rector of the Jagiellonian University of Krakow (the then name of this city was Krakau), Poland.

This portrayal of Prof. Dr. Natanson has been taken from
http://www.archiwum-nauki.krakow.pl/pl/wystawy/natanson.html

In their comprehensive treatises about the free energy, both Gallitzin and Natanson would like to show that both the Helmholtz function and the Gibbs function are of very general validity and applicability. Moreover, there are no doubts that the actual physical sense of the thermodynamic entropy ought to be correspondent to the 'lost energy', which cannot be used anymore for carrying out useful work. In effect, many more colleagues all over the world were studying the actual sense of the free energy notion. As we could also see from the literature review, in the field of chemistry and engineering science, they could achieve much more appreciable success in this field than the majority of the physicists, who had just stuck at the very physical sense of the entropy notion. To sum up, the free energy of Helmholtz and that of Gibbs are intrinsically related to each other (for all the details cf. the recent work [4.31]).

4.4 Thermodynamic works were/are steady, world-wide, but not fragmented conceptually

That we are recalling here Max Bernhard Weinstein's really harsh and highly professional criticism in regard to the relativity theories by Albert Einstein (see Chapter 3)—and then even daring ourselves to critically analyze in detail his actual professional activity in detail—might cause allegations to our address.

Indeed, the allegations might concern in two serious points:

(a) Squeezing out a dignified distinguished colleague;
(b) Ignorance of the Holocaust, whose victim was the colleague in question;

To the point (a) we would like to note that Einstein's cult still existing already since decades ought to require a critical reassessment of his actual professional activities.

Nobody would like to humiliate his truly unique, outstanding stature—but, in general, we ought to refrain from ascribing to anyone the credits definitely not belonging to him/her.

Noteworthy, just opposite situations are very frequent, when truly dignified and distinguished colleagues are longer time wearing the veils of oblivion. The present volume aims at reconciling a number of such gaps in the field of thermodynamics. The volume of the present monograph would surely not allow simultaneous closing all the existing gaps, but here we would greatly appreciate mentioning a couple of the colleagues, who were fully squeezed-out for absolutely unclear reasons. The first colleague in this row ought to be an outstanding, prominent Swiss theoretical physicist: Prof. Dr. Ernst Carl Gerlach Stueckelberg von Breidenbach (1905–1984).

Prof. Dr. Stueckelberg is depicted here during his visit to the Princeton University, 1927, and is here the third man from the left. The photo has been borrowed from http://www.sps.ch/en/articles/physicsanecdotes/ aportraitofstueckelbergasayoungman9/

The information about him might be found in the following sources [4.40–4.42]:

The main work by Prof. Stueckelberg was his monograph, co-authored with one of his pupils, Paul B. Scheurer [4.43].

This book is just a systematic attempt to construct the reliable physical basis for the 'statistical mechanics', based upon the fundamental mechanics plus thorough natural-philosophical analysis, and thus aiming at the true physical basis for thermodynamics, irrespective of the actual level of the matter organization—from truly macroscopic till truly microscopic ones.

That was a unique attempt to systematically bridge the conceptual gap between thermodynamics and 'statistical mechanics'. In fact, it should be considered a continuation of Michelson's foundational work we have discussed above. Moreover, Max Bernhard Weinstein envisaged such a train of thoughts in his 1914 paper criticizing Einstein's theories, but he had no more 'vis viva' and, consequently, lifetime left to thoroughly dwell on this very important theme.

The result of that Prof. Dr. Stueckelberg's effort was truly sad: Nobody had taken this interesting and important baton! Meanwhile, his co-author Paul B. Scheurer had ultimately changed to the field of the philosophy of science [4.44].

Another interesting colleague, who had practically disappeared from our horizon, ought to be an outstanding theoretical biophysicist from Japan, Prof. Dr. Motoyosi Sugita (1905–1990).

Prof. Dr. Motoyosi Sugita in 1975.

In the Internet there is only a short note in English about him—to be found under the following URL address: http://www.eoht.info/page/Motoyosi+Sugita

Here we learn that Prof. Sugita was vividly interested in and productively working on thermodynamics and was one of the founding members of the Society for Studies on Entropy in Japan: http://entropy.ac/

Born in 1905, Motoyosi Sugita had graduated from the Physics Department of the Tokyo Imperial University's Faculty of Science in 1929. Thereafter he worked at the Kobayashi Institute of Physical Research (Kobayashi Riken) and the Naval Engineering School as a researcher in physics. After the end of the Second World War in 1945 he joined the teaching staff at the Tokyo Industrial University, which was later on re-organized into the Hitotsubashi University. Motoyosi

Sugita was actively involved into the teaching and research activity in the field of physics up to his retirement in 1969. Then, he was an active member of the Society for Studies on Entropy, which had been founded in 1983, until his final departure in 1990.

About this colleague and his work I have learned from Dr. Kazumoto Iguchi, who has most recently published a review paper about Prof. Sugita's life and work [4.45]. Moreover, he has re-published Prof. Sugita's thermodynamics monograph with his comments [4.46]. The bibliography of Prof. Sugita is in fact very rich. Here we would just like to refer to a number of his papers published in English and German on the theme we are now discussing [4.47–4.52].

Prof. Sugita and his collaborators have initiated a thorough study of biochemical processes based upon a skillful combination of thermodynamics and the systems theory.

The last, but not the least—concerning the problem of the Holocaust and its over-all impact—we would like to remind here about the tragic story of an outstanding German biophysicist, a definite victim of Holocaust, Dr. Kurt Stern.

Information about him is truly scarce in the Internet. Noteworthy, he should not be confused with the following two colleagues:

Curt Stern (1902–1981), a German–American geneticist.

Kurt Günter Stern (1904–1956), a German–American biochemist.

They both had indeed emigrated from Germany after Hitler came to power and were continuing their professional activities in American Universities.

But there was also a namesake of them who was living in Frankfurt/Main, Deutschordenstraße, 78 till the year of 1933. It is possible to fetch the following information about him in the German-speaking Internet (we present here the pertinent English translation):

Stern, Kurt

Kurt Stern was born in 1892 as the son of a wealthy mill owner Oskar Stern and his wife Eugenie, née Rosenthal, in Wroclaw. He was the younger brother of the physicist and Nobel laureate of the year 1943, Otto Stern (17.02.1888–08.17.1969) and also Berta Kamm, née Stern, born on *10.16.1889, who emigrated to the US after 1933, like her brother Otto.*

Dr. Phil. Kurt Stern was working in Frankfurt as a botanist, whereas his chief was Friedrich Dessauer, at the "Institute of Physical Foundations of Medicine" and lived since the mid-20s in the Deutschordenstraße, 78. Since the optimum conditions for the experiments at the Institute were not met, he pursued his research on the physical clock of plants in the basement of his house.

Kurt Stern had been suspended from his working position in 1933, immediately after the transfer of power to the Nazis and had obviously been caused to sell his house under duress. The Gestapo confiscated his scientific equipment, including very valuable microscopes. The entire home furnishing was auctioned.

Immediately thereafter, Kurt Stern escaped to Paris. From over there, he had moved to the US in 1934/1935.

Dr. Karsten Krakow, the current owner and occupant of the house in the Deutschordenstraße, 78, has initiated the installation of the following stumbling block (a small memorial plate on the pavement near the house entrance, in German: 'Stolperstein'):

Kurt Stern
Date of Birth: 8/7/1892
Deportation: To Paris in 1933 and New York
Date of death: 12/19/1938 (suicide)

As Kurt Stern ought to be a colleague of Motoyosi Sugita, according to the above information, we have carefully looked for the available references to his works, and might now present his publications listing [4.53–4.58].

This listing is definitely not complete, but we might immediately recognize that Dr. Stern was studying a truly interesting and important topic—the electrophysiology of plants. And Dr. Krakow might definitely be right—'physical clocks of plants' is a particular topic of their physiology.

The whole topic was long time considered to be of huge interest and even nowadays seems to still remain being expressly interesting. The information about how the story was developing after the untimely departure of Dr. Kurt Stern could be borrowed from a number of sources and the references therein [4.59–4.61].

The main achievement of Dr. Kurt Stern was his considering thermodynamics as the pertinent tool for the physically correct analysis of electrophysiological events. His untimely departure was a serious negative thrust to the scientific community. What serious reasons could drive Kurt to a suicidal decision is a good poser for the historians of science. At the first glance, Holocaust, being definitely the cornerstone factor does not exceed all the possible reasons.

Howbeit, the main point to be stressed: **Scientific research is eternal, it will never stop**.

References to Note 4

4.1 K. W. Graff (**2005**): *Albert Einstein als Erfinder in den Jahren 1907 bis 1933*. Dissertation an der Universität Stuttgart, Germany. (*Albert Einstein as an Inventor in the years 1907 to 1933*).

4.2 Wolfgang Engels (**2006**): *Der Volkskühlschrank von Albert Einstein und Leo Szilard. Physik in unserer Zeit*, v. 37, p. 144.

4.3 Alberto A. Martinez (**2014**): The questionable inventions of the clever Dr. Einstein. *Metascience*, v. 23, pp. 49–55.

4.4 Albert Einstein (**1908**): Eine neue elektrostatische Methode zur Messung kleiner Elektrizitätsmengen. *Physikalische Zeitschrift*, v. 9, pp. 216–217.

4.5 József Illy (**2012**): *The Practical Einstein: Experiments, Patents, Inventions*. Johns Hopkins University Press, Baltimore, USA.

4.6 Stephen J. Kline (**1989**): *The Low-Down on Entropy and Interpretive Thermodynamics*. DCW Industries: Lake Arrowhead, CA, USA.

4.7 Linda E. Reichl (**2009**): *A Modern Course in Statistical Physics*. Wiley-VCH, 3rd Edition, a Wiley-Interscience Publication, John Wiley & Sons, INC.: New York, Chichester, Weinheim, Brisbane, Singapore, and Toronto.

4.8 Albert Einstein, Leopold Infeld (**1967**): '*Evolution of Physics*'. Cambridge University Press: London, U. K.

4.9 *Ordnung aus dem Chaos,* Bernd-Olaf Küppers (Hrsg.) (**1987**) —Piper, München, Zürich: With the contributions by such eminent specialists as A. Dress, H. Haken, B. Hess, B.-O. Küppers, Ch. v. d. Malsburg, M. Markus, H. Meinhardt, D. Pörschke, P. H. Richter, as well as H.-J. Scholz, K. Schulten and P. Schuster.

4.10 Charles Ruhla (**1989**): *La physique du hasard: de Blaise Pascal à Niels Bohr*. Hachette: Paris, France. Cambridge University Press. The English translation of this book: Charles Ruhla (**1992**): *The Physics of Chance*: *From Blaise Pascal to Niels Bohr*. Oxford University Press: Oxford, New York, Tokyo.

4.11 Vinay Ambegaokar (**1996**): *Reasoning about Luck: Probability and Its Uses in Physics*. Cambridge University Press: Cambridge, New York, Melbourne.

4.12 Richard Morris (**1999**): *The Universe, the Eleventh Dimension, and Everything – What We Know and How We Know It*. Four Walls, Eight Windows: New York, USA. The German translation of this book: Richard Morris (**2001**): *Gott würfelt nicht. Universum, Materie und kreative Intelligenz*. Europa Verlag: Hamburg, Wien.

4.13 Bernard H. Lavenda (**1991**): *Statistical Physics: A Probabilistic Approach*, John Wiley & Sons, INC.: New York, Chichester, Weinheim, Brisbane, Singapore, and Toronto.

4.14 I. Prigogine (**1947**): Étude thermodynamique des phénomènes irréversibles, Éditions Desoer, Liège, Belgique.

4.15 E. A. Milne (**1935**): *Relativity, Gravitation and the World-Structure*. Oxford University Press: Oxford, U.K.

4.16 E. A. Milne (**1940**): Cosmological theories. *Astrophysics Journal*, v. 90, pp. 129–158.

4.17 E. A. Milne (**1952**): *Modern Cosmology and the Christian Idea of God*. Clarendon Press: Oxford, U.K.

4.18 Erwin Schrödinger (**1935**): Die gegenwärtige Situation in der Quantenmechanik. *Naturwissenschaften*, v. 48, p. 807; v. 49, p. 823; v. 50, p. 844.

4.19 C. Spearman (**1904**): The proof and measurement of association between two things. *The American Journal of Psychology*, v. 15, pp. 72–101.

4.20 S. James Press (**2005**): *Applied Multivariate Analysis: Using Bayesian and Frequentist Methods of Inference*. Dover Publications, Inc.: Mineola, New York, USA.

4.21 Manjit Kumar (**2010**): *Quantum: Einstein, Bohr, and the Great Debate about the Nature of Reality*. W. W. Norton, New York, USA 2010 (pp. 305–306, among others).

4.22 John Archibald Wheeler, Wojciech Hubert Zurek (**2014**): *Quantum Theory and Measurement*. Princeton University Press, Princeton, USA 14 July 2014 (pp. 357–358, among others).

4.23 Yuri Kornyushin (**2008**): *Understanding Kinetic Energy Paradox in Quantum Mechanics*, arxiv.0806.3232.

4.24 Yuri Kornyushin (**2015**): *Studying Quantum Mechanics. Selected Topics*. LAP Lambert Academic Publishing, Saarbrücken, Germany.

4.25 E. B. Starikov (**2010**): Many faces of entropy or Bayesian statistical mechanics. *ChemPhysChem*, v. 11, pp. 3387–3394.

4.26 E. B. Starikov (**2012**): George Augustus Linhart as a "widely unknown" thermodynamicist. *World Journal of Condensed Matter Physics*, v. 2, pp. 101–116.

4.27 E. B. Starikov (**2013**): Entropy is 'anthropomorphic': Does this lead to interpretational devalorisation of entropy-enthalpy compensation? *Monatsh. Chem.*, v. 144, pp. 97–102.

4.28 E. B. Starikov (**2013**): Valid entropy-enthalpy compensation: Its true physical-chemical meaning. *J. Appl. Sol. Chem. Model.* v. 2, pp. 240–245.

4.29 E. B. Starikov (**2014**): 'Meyer-Neldel rule': True history of its development and its intimate connection to classical thermodynamics. *J. Appl. Sol. Chem. Model.* v. 3, pp. 15–31.

4.30 E. B. Starikov (**2014**): What Nicolas Leonard Sadi Carnot would like to tell us in fact. *Pensée Journal*, v. 76, pp. 171–214 (see the fully proofread version of this paper here: https://www.researchgate.net/publication/263163306_Starikov_Carnot_proof-read_version)

4.31 E. B. Starikov (**2015**): The interrelationship between thermodynamics and energetics: The true sense of equilibrium thermodynamics. *J. Appl. Sol. Chem. Model.* v. 4, pp. 19–47.

4.32 *Pensées de Blaise Pascal* (**1946**): Nelson, Éditeurs: Paris, Londres, Edimbourg et New York, pp. 443–444.

4.33 В. А. Михельсонъ (**1887**): *Простѣйшій выводъ второго закона термодинамики изъ началъ аналитической механики. Матем. сб. том* 13, номер 2, pp. 229–244. (*The simplest inference of the thermodynamics' second basic law from the basics of the analytical mechanics*)

4.34 Prof. Dr. Hans-Georg Schöpf (**1978**): *Von Kirchhoff bis Planck: Theorie der Wärmestrahlung in historisch-kritischer Darstellung.* Springer-Verlag, Berlin, Heidelberg, Germany.

4.35 В. А. Михельсон (**1938**): Физика. Т. I. Механика. Молекулярная физика. Термодинамика. (Physics. Volume 1. Mechanics, Molecular Physics, Thermodynamics)

4.36 Кн. Б. Голицынъ (**1894**): О *свободной энергіи, Извѣстія Императорской Академіи Наукъ, томъ* 1, *выпускъ* 4, 387–394. (*About the free energy*).

4.37 Fürst B. Galitzine (**1895**): *Über die Molecularkräfte und die Elasticität der Molecüle, Извѣстія Императорской Академіи Наукъ,* томъ 3, выпускъ 1, 1–53. (*About the molecular forces and the elasticity of molecules*)

4.38 Кн. Б. Голицынъ (**1896**): О *свойствахъ мельчайшихъ частицъ матеріи, Извѣстія Императорской Академіи Наукъ,* томъ 4, выпускъ 3, 293–314. (*On the properties of the tiniest particles of the matter*).

4.39 Ladislaus Natanson (**1892**): *Über thermodynamische Potentiale, Zeitschrift für Physikalische Chemie,* v. X, pp. 734–747. (*About thermodynamic potentials*).

4.40 Robert P. Crease and Charles C. Mann (**1985**): The physicist that physics forgot: Baron Stueckelberg's brilliantly obscure career. *The Sciences*, v. 25, pp. 18–23.

4.41 Gérard Wanders (**1985**): Ernst Carl Gerlach Stueckelberg von Breidenbach (1905–1984). *Physikalische Blätter*, v. 41, pp. 22–23.

4.42 Jan Lacki, Henri Ruegg, and Gérard Wanders (**2009**): *E. C. G. Stueckelberg, An Unconventional Figure of Twentieth Century Physics. Selected Scientific Papers with Commentaries.* Birkhäuser: Basel, Boston, and Berlin.

4.43 E. C. G. Stueckelberg de Breidenbach et P. B. Scheurer (**1974**): *Thermocinétique phénoménologique galiléenne.* Birkhäuser: Basel and Stuttgart.

4.44 Paul Scheurer (**1979**): *Révolutions de la science et permanence du réel.* Presses Universitaires de France: Croisées, Paris, France.

4.45 Kazumoto Iguchi (**2016**): Motoyosi Sugita: A "widely unknown" Japanese thermodynamicist, who explored the 4th law of thermodynamics for creation of the theory of life. *Open Journal of Biophysics*, v. 6, pp. 125–232.

4.46 Motoyosi Sugita (**2016**): 過渡的現象の熱力学： 生物体の熱力学の構築に向けて． 新版. (In Japanese: *Thermodynamics of Transient Phenomena: Toward Constructing Thermodynamics of Organisms. A Novel Edition by Dr. Kazumoto Iguchi*). http://www.taiyo-g.com/shousai182.html

4.47 Motoyosi Sugita (**1933**): Bemerkung über den Planck'schen Beweis des Zweiten Hauptsatzes. *Proceedings of the Physical-Mathematical Society of Japan*, v. 15, pp. 108–113.

4.48 Motoyosi Sugita (**1953**): Thermodynamical analysis of life I: Thermodynamics of transient phenomena. *Journal of Physical Society of Japan*, v. 8, pp. 697–703.

4.49 Motoyosi Sugita (**1953**): Thermodynamical analysis of life II: On the maximum principle of transient phenomena. *Journal of Physical Society of Japan*, v. 8, pp. 704–709.

4.50 Motoyosi Sugita (**1953**): Thermodynamical analysis of life III: Mathematical analysis of metabolism. *Journal of Physical Society of Japan*, v. 8, pp. 709–713.

4.51 Motoyosi Sugita (**1961**): Functional analysis of chemical systems in vivo using a logical circuit equivalent. *Journal of Theoretical Biology*, v. 1, pp. 415–430.

4.52 Nobuo Fukuda, Motoyosi Sugita (**1961**): Mathematical analysis of metabolism using an analogue computer: I. Isotope kinetics of iodine metabolism in the thyroid gland. *Journal of Theoretical Biology*, v. 1, pp. 440–459.

4.53 Kurt Stern (**1922**): Zur Elektrophysiologie der Berberisblüte. *Zeitschrift für Botanik*, v. 14, p. 234–248.

4.54 Kurt Stern (**1922**): Über polare elektronastische Erscheinungen. *Berichte der Deutschen Botanischen Gesellschaft*, v. 40, pp. 43–51.

4.55 Kurt Stern (**1922**): Über polare elektronastische Erscheinungen. *Berichte der Deutschen Botanischen Gesellschaft*, v. 40, pp. 52–59.

4.56 Kurt Stern (**1924**): *Elektrophysiologie der Pflanzen*. Springer Verlag: Berlin, Germany.

4.57 Kurt Stern (**1925**): Bewegungen kontraktiler Organe an Pflanzen, in A. Bethe, G. v. Bergmann, G. Embden, A. Ellinger (Eds.), *Handbuch der normalen und pathologischen Physiologie mit Berücksichtigung der* *experimentellen Pharmakologie*, pp. 94–107, Springer Verlag: Berlin, Germany.

4.58 Kurt Stern (**1933**): *Pflanzenthermodynamik*. Springer Verlag, Berlin: Germany.

4.59 Georges Ungar (**1963**): *Excitation (American Lecture Series, Number 524)*. Published by Charles C. Thomas: Springfield, Illinois, USA.

4.60 L. George Lawrence (**1969**): Electronics and the living plant. *Electronics World*, v. 10 (October), pp. 25–28.

4.61 B. Setlow (**1997**): Georges Ungar and Memory Transfer. *Journal of the History of Neuroscience*, v. 6, pp. 181–192.

[**Note 5**]: Below we have just in part re-formulated the English text presented in the Internet.

The Zero-th Basic Law

Subject:

"If each of some two systems is in thermal equilibrium with a third one, all of them are in thermal equilibrium with each other." This statement is called the zero-th law of thermodynamics.

Deficiencies:

If two systems are in thermal equilibrium, their temperatures are equal, and if their temperatures are equal they are in thermal equilibrium. From this fact follows the zero-th law. There is no doubt that the statement is correct.

However, the statement represents such a simple conclusion that it is truly hard to understand, how it could reach the status of a "law of thermodynamics."

Who reckons that there is a profound meaning hidden behind the words, should remember that several other statements about other equilibria could be formulated which nobody would call a "law" of anything, since the content of these statements is obvious.

Since the zero-th law is often cited in the context of statistical thermodynamics, we shall, for comparison, consider the chemical equilibrium. In the conventional statistical thermodynamics, the chemical potential plays a role that is rather similar to that of the temperature: Together with temperature it is one of the two parameters in the probability distribution of the energy. Thus, in addition to the zero-th law we could formulate an analogous "law" for chemical equilibria.

"If each of some two systems is in chemical equilibrium with a third one, all of them are in chemical equilibrium with each other."

Phenomenological thermodynamics shows us that we can formulate yet various other "zero-th laws": A separate one for each of the terms in Gibbs fundamental equation:

$$dE = TdS – pdV + \mu\, dn + vdp + Fds + \omega dL + \psi dm + \varphi dQ + Id\varphi...$$

Here:

T = absolute temperature,
S = entropy,
p = pressure,
V = volume,
μ = chemical potential,
n = amount of substance,
v = velocity,
p = momentum,
F = momentum flow,
s = displacement, ω = angular velocity,
L = angular momentum,
ψ = gravitational potential,
m = mass,
φ = electric potential,
Q = electric change,
I = electric current,
φ = magnetic flux.

So, we could formulate for three bodies, which by means of inelastic collisions attain the same velocities:

"If each of some two systems are in velocity equilibrium with a third one, all of them are in velocity equilibrium with each other."

Origin:

The need for the formulation of the zero-th law seems to arise, when temperature and chemical potential are introduced in statistical mechanics. Then it has to be shown that one of the parameters in the probability distribution has the property of that quantity which is familiar to us and which we call temperature. However, even in this context the zero-th law is nothing more than the expression of the transitivity for the relevant physical quantity.

Ultimate Disposal:

> *We suggest **not** to treat the zero-th law at school. What should be the actual relevance of the subject for school physics? Already this sole particular example helps us to understand why thermodynamics is so unpopular at universities and at schools. With no other intensive quantity we make such a great play as with the temperature, and with no other extensive quantity we make such a big fuss as with the entropy. Sometimes, thermodynamics may remind the emperor's new clothes.*

> *Regarding the disposal of the zero-th law in particular, we have to apply to our colleagues at the universities. We recommend our students: Don't allow persuading you that there is a problem, where there is none.*

> *Friedrich Herrmann, Karlsruhe Institute of Technology*

Meanwhile, no English translation of the chapter about the third basic law has been published [91, 92], so here we shall place our own translation of the original German version. We remember in this connection that neither Peter Boas Freuchen (Chapter 1), nor Nils Engelbrektsson (Chapter 2) did make any fuss about "some basic law" in this regard, with Max Bernhard Weinstein (Chapter 3) just following them, while George Augustus Linhart (Chapter 4) has formally mathematically proven a complete absence of any kind of "the third basic law."

Subject:

"There is no process running in the finite dimensions, during which a system can be cooled down to absolute zero." This is just one of the many permissible formulations of the third law of thermodynamics.

Deficiency:

So, why exactly this law at all appears to us worth mentioning?

The impossibility statements of this kind could be immediately available in a large heap. Indeed, there is no process running in the finite dimensions, during which an air-filled canister could be pumped out to be completely empty. Further, it is impossible to fully exhaust a water-filled bathtub with a handy bucket. We rightfully perceive the statements of this kind as trivial ones—and certainly not as the basic natural laws.

But when it comes to the entropy notion, it looks like pretty different. Indeed, the latter notion is mostly delivered to us in such an esoteric packaging that we find any uninhibited handling of this notion to be extremely difficult. Sure, all the statements about entropy acquire usually such a tremendous value, that finally it appears to be in no perceptible relation to the simple physical properties delivered by this actual notion.

To sum up, we encounter the entropy with so much awe and put so much metaphysics in this rather intelligible term that the comparison of the third law of thermodynamics with some unencumbered pipe or tub seems to be quite a disrespectful behavior. Yet all of the above statements are of the same kind. And the simple analogies ought to describe the situations in a much clearer way than any of the current formulations of the third basic law.

Origin:

The pioneering formulation of this basic law goes back to W. Nernst. And from his student F. Simon comes the wording: "It is impossible to completely deprive any substance of its entropy." Meanwhile, in fact this law just fills the mathematical gap left by the 2nd law, for it simply allows an exact determination of the integration constant value, when calculating the entropy according to the Clausius formulation.

Ultimate Disposal:

The high respect for the inventors and investigators of this law should not prevent us from viewing the things somewhat more soberly. Indeed, the rule in question does in no way belong to the historical altar, but straight into our conventional mathematical toolbox.

Georg Job, University of Hamburg

5.14 Acknowledgement

I would greatly appreciate to express my sincere thanks to my daughter Amalia for her careful proof-reading.

5.15 References

1. Clifford Ambrose Truesdell (**1966**): *Six Lectures on Modern Natural Philosophy*, Springer-Verlag, Berlin, Heidelberg, Germany.

2. Clifford Ambrose Truesdell (**1973**): *Tragicomedy of Classical Thermodynamics*, Springer-Verlag, Wien, New York.

3. Clifford Ambrose Truesdell, Subramanyam Bharatha (**1977**): *The Concepts and Logic of Classical Thermodynamics as a Theory of Heat Engines Rigorously Constructed upon the Foundation Laid by S. Carnot and F. Reech*, Springer-Verlag, New York, Berlin, Heidelberg, London, Paris, Tokyo.

4. Clifford A. Truesdell (**1980**): *The Tragicomical History of Thermodynamics, 1822–1854*, Springer-Verlag, New York, Berlin, Heidelberg, London, Paris, Tokyo.

5. Clifford Ambrose Truesdell (**1984**): *An Idiot's Fugitive Essays on Science: Methods, Criticism, Training, Circumstances*. Springer-Verlag: New York, Berlin, Heidelberg, London, Paris, and Tokyo.

6. Clifford Ambrose Truesdell (**1988**): *The Tragicomical History of Thermodynamics, 1822–1854*, (Studies in the history of mathematics and the physical sciences), Springer-Verlag: New York, Heidelberg, Berlin.

7. Peter Atkins (**2007**): *The Four Laws that Drive the Universe*. Oxford University Press, Inc.: Oxford, New York, Auckland, Cape Town, Dar-es-Salaam, Hong Kong, Karachi, Kuala Lumpur, Madrid, Melbourne, Mexico City, Nairobi, New Delhi, Shanghai, Taipei, Toronto.

8. Edward Eugene Daub (**1981**): Rudolf Clausius. In: *Dictionary of Scientific Biography*, Charles Coulston Gillispie, Editor-in-Chief, Vol. 03, pp. 303–311.

9. Edward Eugene Daub (**1966**): *Rudolf Clausius and the Nineteenth Century Theory of Heat*. University of Wisconsin Publishing, Madison, USA.

10. Edward Eugene Daub (**1967**): Atomism and thermodynamics. *ISIS*, Vol. 58, pp. 292–303.

11. Edward Eugene Daub (**1969**): *Probability and Thermodynamics: Reduction of the Second Law*. ISIS, Vol. 60, pp. 318–330.

12. Edward Eugene Daub (**1970**): *Entropy and Dissipation*. Historical studies in physical sciences, Vol. 2, pp. 321–354.

13. Stephen G. Brush (**1976**): *The Kind of Motion We Call Heat. Books 1, 2*, North-Holland Publishing Company, Amsterdam, New York, Oxford.

14. Stephen G. Brush, Nancy S. Hall (**2003**): *Kinetic Theory of Gases. An Anthology of Classic Papers with Historical Commentary*. World Scientific Publishing, Singapore, London.

15. Gerald Holton, Stephen G. Brush (**2010**): *Physics, the Human Adventure: From Copernicus to Einstein and Beyond*. Rutgers University Press: New Brunswick, New Jersey, USA; London, U. K.

16. Theodor A. Wand (**1868**): *Kritische Darstellung des zweiten Satzes der mechanischen Wärmetheorie*. Repertorium für Experimentalphysik und für physikalische Technik, Band 4, pp. 281–322; pp. 369–405.

17. F. Alexander Bais, J. Doyne Farmer (**2007**): *Physics of Information*, SFI Working Paper 2007-08-029, (a publication of Santa Fe Institute for studying complex systems).

18. Linda Claire Burns (**1991**): *Vagueness. An Investigation into Natural Languages and the Sorites Paradox*. Science+Business Media, Dordrecht, the Netherlands.

19. Oswald Hanfling (**2000**): *Philosophy and Ordinary Language. The Bent and Genius of our Tongue*. Routledge: Tailor & Francis Group, London and New York.

20. Oswald Hanfling (**2001**): What is wrong with sorites arguments? *Analysis*, Vol 61, pp. 29–35.

21. Dominic Hyde (**2011**): Sorites paradox. In: *Stanford Encyclopedia for Philosophy*: http://plato.stanford.edu/entries/sorites-paradox/

22. Mark W. Zemansky (**1957**): Fashions in thermodynamics, *American Journal of Physics*, Vol. 25(6), pp. 349–351.

23. *Emotions: How we change what others think, feel, believe and do*. http://changingminds.org/explanations/emotions/emotions.htm

24. Проф. А. Щукарёвъ (**1912**): Введенiе въ курсъ физики. Ученiе объ энергiи и энтропiи въ элементарномъ изложенiи. Типографiя Т-ва И. Д. Сытина, Москва, Russia. ('*Introduction to the Course of Physics. The Doctrines of Energy and Entropy in an Elementary Exposition*').

25. '*Die Kultur der Gegenwart: Ihre Entwicklung und Ziele*. Herausgegeben von Paul Hinneberg. Dritte Abteilung: *Anorganische Wissenschaften*, unter Leitung von E. Lecher. Erster Band: *Physik*, unter Redaktion von E. Warburg; Verlag von B. G. Teubner in Leipzig und Berlin, Germany, **1915**. (*The Modern Culture: Its Development and Aims*, edited by Paul Hinneberg. The Third Part: *Inorganic Sciences*, under the leadership of E. Lecher. The First Volume: *Physics*, edited by E. Warburg).

26. '*Fünf Wiener Verträge*'. Verlag Franz Deuticke, Wien, Austria – especially in the following volumes: 1st Cycle, '*Krise und Neuaufbau in den exakten*

Wissenschaften', 1933, 2nd Cycle, '*Alte Probleme – Neue Lösungen in den exakten Wissenschaften*', 3rd Cycle, '*Neuere Fortschritte in den exakten Wissenschaften*', 1936, 4th Cycle, '*Neue Wege exakter Naturerkenntnis*', 1939. ('*Five Lectures from Vienna*', '*Crisis and Reconstruction in the Exact Natural Sciences*', '*New Solutions for the Exact Natural Sciences*', '*A Newer Progress in the Exact Sciences*', '*New Directions in the Exact Natural Sciences*').

27. P. T. Landsberg (**1984**): Can entropy and "order" increase together? *Phys. Lett.* Vol. 102A, pp. 171–173.

28. Friedrich Stadler (**2015**): *Der Wiener Kreis: Ursprung, Entwicklung und Wirkung des Logischen Empirismus im Kontext*. Springer International Publishing, Switzerland.

29. András Máté, Miklós Rédei, Friedrich Stadler (**2011**): *Der Wiener Kreis in Ungarn/The Vienna Circle in Hungary,* Springer-Verlag: Vienna, Austria.

30. Stephen J. Kline (**1995**): *Conceptual Foundations for Multidisciplinary Thinking*, Stanford University Press: Redwood City, CA, USA.

31. Peter A. Corning and Stephen J. Kline (**1998**): *Thermodynamics, Information, and Life Revisited, Part I: 'To Be or Entropy'* , Systems Research and Behavior Science, Vol. 15(4), pp. 273–295.

32. Peter A. Corning and Stephen J. Kline (**1998**): *Thermodynamics, Information, and Life Revisited, Part II: 'Thermoeconomics' and 'Control Information'*, Systems Research and Behavior Science, Vol. 15(6), pp. 453–482.

33. Stephen J. Kline (**1999**): *The Low-Down on Entropy and Interpretive Thermodynamics*, DCW Industries.

34. M. Tribus, E. C. McIrvine (**1971**): Energy and information, *Scientific American*, Vol. 224 (September 1971), pp. 178–184.

35. Johannes Zernike (**1972**): *Entropy: The Devil on the Pillion: A Popular Exposition*. Kluwer: Deventer, the Netherlands.

36. Arno Höpfner (**1974**): A Review of the book by J. Zernike, *Angewandte Chemie*, Vol. 86, pp. 748–749 (in German).

37. K. G. Denbigh, J. S. Denbigh (**1985**): *Entropy in Relation to Incomplete Knowledge*. Cambridge University Press: Cambridge, U. K.

38. Laurence Sklar (**1987**): A review of the book by K. G. Denbigh and J. S. Denbigh, *Philosophy in Review*, Vol. 7, pp. 54–55.

39. Dénes Petz (**2001**): Entropy, von Neumann and the von Neumann Entropy. In: *John von Neumann and the Foundations of Quantum Physics*, M. Rédei and M. Stöltzner, Editors, Kluwer, Dordrecht, the Netherlands.

40. Craig Callender (**2001**): Taking thermodynamics too seriously, *Studies in History and Philosophy of Science Part B. Studies in History and Philosophy of Modern Physics*, Vol. 32, pp. 539–553.

41. А. А. Брандт: Основания термодинамики. Ч. 1: Основные законы. Газы. Пг., 1915; Ч. 2. Пары. Жидкости. Пг., 1918. (A book on the foundations of thermodynamics published in two parts).

42. Н. А. Быков (**1916**): Термодинамика, Петроград: Типо-литогр. И. Трофимова, Russia. (A fundamental handbook on thermodynamics and its foundations).

43. В. В. Шарвин (**1922**): ‚Энергия: сохранение и вырождение'. Москва, "Мир," Russia. (Energy, its Conservation and Devaluation).

44. А. А. Саткевич (**1906**): ‚О формулировке 1-го закона термодинамики', Журнал Русского Физико-Химического Общества, Номер 7 за 1906 год. (*As to the Formulation of the 1st Law of Thermodynamics*, a publication in the Journal of the Russian Physical-Chemical Society, 1906).

45. Bohdan von Szyszkowski (**1908**): ‘*Experimentelle Studien über Kapillare Eigenschaften der wässerigen* Lösungen von Fettsäuren', Zeitschrift für Physikalische Chemie, v. 64, pp. 385–414.

46. В. В. Кузнецов, В. К. Усть-Качинцев (**1976**): Физическая и коллоидная химия. Москва: Высшая школа, USSR. (A Course of Physical and Colloidal Chemistry).

47. H. P. Meissner and A. S. Michaels (**1949**): Surface tensions of pure liquids and liquid mixtures, *Ind. Eng. Chem.*, Vol. 41(12), p. 2782.

48. D. K. Chattoraj and K. S. Birdi (**1984**): *Adsorption and the Gibbs Surface Excess*. Plenum Press: New York and London.

49. Zbigniew Adamczyk (**1987**): Non-equilibrium surface tension for mixed adsorption kinetics, *Journal of Colloids and Interface Science*, Vol. 120, pp. 477–485.

50. Erwin A. Vogler (**1989**): A simple mathematical model of time-dependent interfacial tension, *Journal of Colloids and Interface Science*, Vol. 133, pp. 228–236.

51. Ludger O. Figura and Arthur A. Teixeira (**2007**): *Food Physics: Physical Properties – Measurement and Application*. Springer-Verlag: Berlin, Heidelberg, New York.

52. Felipe Suárez and Carmen M. Romero (**2011**): Apparent molar volume and surface tension of dilute aqueous solutions of carboxylic acids, *J. Chem. Eng. Data*, Vol. 56, pp. 1778–1786.

53. M. Kuncheva, I. Panchev, M. Kamburova, N. Radchenkova, and I. Boyadzhieva (**2015**): *Surface Active Properties of a Newly Synthesized*

Biopolymer from Halophytic Microorganisms, Scientific Works of the University of Food Technologies, Plovdiv, Bulgaria, Vol. LXII, pp. 487–489.

54. Б. А. Шишковскій (**1909**): Энергія и Энтропія. Вступительная лекція. Киев: Университетъ Св. Владимира, Russian Empire. (Energy and entropy: An inaugural lecture).

55. Franz-Serafin Exner (**1919**): *Vorlesungen über die physikalischen Grundlagen der Naturwissenschaften*. Deuticke, Wien, Austria.

56. Franz-Serafin Exner (**1923**): *Vom Chaos zur Gegenwart. Eine kulturhistorische Studie*. Eigenverlag, Wien, Austria.

57. Michael Stöltzner (**1999**): *Vienna Indeterminism: Mach, Boltzmann, Exner*. Synthese, Vol. 119, pp. 85–111.

58. Michael Stöltzner (**2002**): *Vienna Indeterminism II: From Exner's Synthesis to Frank and von Mises*. [Preprint]: http://philsci-archive. pitt.edu/624/

59. Michael Stöltzner (**2003**): *Causality, Realism and the Two Strands of Boltzmann's Legacy (1896-1936)*. PhD Thesis, University of Bielefeld, Germany: https://pub.uni-bielefeld.de/download/2304524/2304527

60. Heinrich Zankl (**2006**): *Fälscher, Schwindler, Scharlatane. Betrug in Forschung und Wissenschaft*. Wiley-VCH Verlag GmbH &Co. KGaA, Weinheim, Germany.

61. W. B. Jensen (**2009**): August Horstmann and the origins of chemical thermodynamics, *Bull. Hist. Chem.*, Vol. 34, pp. 83–91.

62. W. B. Jensen (**2013**): George Downing Liveing and the early history of chemical thermodynamics, *Bull. Hist. Chem.*, Vol. 38, pp. 37–51.

63. Alexander Y. Kipnis: Early chemical thermodynamics: Its duality embodied in van't Hoff and Gibbs. In: *Van't Hoff and the Emergence of Chemical Thermodynamics*, Willem J. Hornix, S. H. W. M. Mannaerts, Editors, DUP Science, 2001, pp. 212–242.

64. Lynde Phelps Wheeler (**1970**): *Josiah Willard Gibbs: The History of a Great Mind*, Archon Books: Hamden, CT, USA.

65. *Selected Works of Mao Tse-tung*, Vol. 3, 1965, pp. 61–62.

66. P. Perrot (**1998**): *A to Z of Thermodynamics*, Oxford University Press, Oxford Great Britain.

67. J. W. Gibbs (**1873**): A method of geometrical representation of the thermodynamic properties of substances by means of surfaces, *Transactions of the Connecticut Academy of Arts and Sciences* 2, Dec. 1873, pp. 382–404 (the quotation on the p. 400).

68. R. Baierlein (**2003**): *Thermal Physics,* Cambridge University Press: Cambridge, Great Britain.

69. Howard Reiss (**1965**): *Methods of Thermodynamics,* Dover Publications: New York, USA.

70. Haywood (**1974**): A critical review of the theorems of thermodynamic availability, with concise formulations, *Journal of Mechanical Engineering Science,* Vol. 16, pp. 160–173; pp. 258–267.

71. Robert Pauli (**1896**): *Der erste und zweite Hauptsatz der mechanischen Wärme-Theorie und der Vorgang der Lösung: eine energetische Theorie des chemischen Molecüls,* Fischers Technologischer Verlag, M. Krayn, Berlin.

72. Max Rudolphi-Darmstadt (**1897**): *Wärmelehre. Zum Buche vom Robert Pauli,* Beiblätter zu den Annalen der Physik und Chemie Vol. 21, pp. 714–715.

73. J. E. Trevor (**1897**): As to the book by Robert Pauli. New Books, *J. Phys. Chem.,* Vol. 1, pp. 499–501.

74. Wolfgang Pauli (**1981**): *Theory of Relativity,* Dover Publications: New York, USA.

75. Wolfgang Pauli and Charles P. Enz (**1973**): *Pauli Lectures on Physics. Volume 3: Thermodynamics and the Kinetic Theory of Gases.* MIT Press, Cambridge, Massachusetts, London, Great Britain.

76. E. B. Starikov, B. Nordén (**2007**): Enthalpy-entropy compensation: a phantom or something useful? *J. Phys. Chem.,* Vol. 111 B, pp. 14431–14435.

77. E. B. Starikov, I. Panas, B. Nordén (**2008**): Chemical-to-mechanical energy conversion in biomacromolecular machines: A plasmon and optimum control theory for directional work. 1. General considerations. *J. Phys. Chem. B,* Vol. 112, pp. 8319–8329.

78. E. B. Starikov, D. Hennig, B. Nordén (**2008**): Protein folding as a result of 'self-regulated stochastic resonance': A new paradigm? *Biophys. Rev. Lett.* Vol. 3, p. 343–363.

79. E. B. Starikov, B. Nordén (**2009**): Physical rationale behind the nonlinear enthalpy-entropy compensation in DNA duplex stability, *J. Phys. Chem. B,* Vol. 113, pp 4698–4707.

80. E. B. Starikov, B. Nordén (**2009**): DNA duplex length and salt concentration dependence of enthalpy-entropy compensation parameters for DNA melting, *J. Phys. Chem. B,* Vol. 113, pp. 11375–11377.

81. E. B. Starikov, D. Hennig, H. Yamada, R. Gutierrez, B. Nordén, and G. Cuniberti **(2009)**: Screw motion of DNA duplex during translocation through pore. I. Introduction of the model, *Biophys. Rev. Lett.* Vol. 4, pp. 209–230.

82. B. Nordén, E. B. Starikov **(2012)**: Entropy-enthalpy compensation may be a useful interpretation tool for complex systems like protein-DNA complexes: An appeal to experimentalists, *Appl. Phys. Lett.* Vol. 100, p. 193701.

83. B. Nordén, E. B. Starikov **(2012)**: Entropy–enthalpy compensation as a fundamental concept and analysis tool for systematical experimental data, *Chem. Phys. Lett.* Vol. 538, pp. 118–120.

84. B. Nordén, E. B. Starikov **(2012)**: Entropy-enthalpy compensation: Is there an underlying microscopic contribution to mechanism? (Invited chapter) In: *Current Microscopy Advances in Science and Technology,* A. Mendez-Vilas (Ed.), Vol. 2, Formatex Research Center, Spain, pp. 1492–1503.

85. E. B. Starikov **(2013)**: 'Entropy is anthropomorphic': does this lead to interpretational devalorisation of entropy-enthalpy compensation? *Monatsh. Chem.,* Vol. 144, pp. 97–102.

86. E. B. Starikov **(2013)**: Valid entropy–enthalpy compensation: Fine mechanisms at microscopic level, *Chem. Phys. Lett.* Vol. 564, p. 88–92.

87. E. B. Starikov **(2013)**: Entropy–enthalpy compensation and its significance—in particular for nanoscale events, *J. Appl. Sol. Chem. Model.* Vol. 2, pp. 126–135.

88. E. B. Starikov **(2013)**: Valid entropy-enthalpy compensation: it's true physical-chemical meaning, *J. Appl. Sol. Chem. Model.* Vol. 2, pp. 240–245.

89. E. B. Starikov **(2014)**: 'Meyer-Neldel rule': True history of its development and its intimate connection to classical thermodynamics, *J. Appl. Sol. Chem. Model.* Vol. 3, pp. 15–31.

90. E. B. Starikov **(2015)**: The interrelationship between thermodynamics and energetics: the true sense of equilibrium thermodynamics, *J. Appl. Sol. Chem. Model.* Vol. 4, pp. 19–47.

91. F. Herrmann und G. Job **(2002)**: *Altlasten der Physik*, AULIS Verlag in der STARK Verlag GmbH: Hallbergmoos, Germany.

92. http://www.physikdidaktik.uni-karlsruhe.de/publication/Historical_burdens/

93. Frederick William Lanchester (**1914**): Aircraft in the warfare: The dawn of the fourth arm. communications number V and VI; *Engineering*, Vol. 98, pp. 422–423; pp. 452–454.

94. Frederick William Lanchester (**1916**): *Aircraft in Warfare, the Dawn of the Fourth Arm.* Constable and Company, Limited, London, Great Britain.

95. Rufus Isaacs (**1965**): *Differential Games: A Mathematical Theory with Applications to Warfare and Pursuit, Control and Optimization*, John Wiley & Sons, Inc.: New York, USA.

96. Zhen Dong Dai, Qun Ji Xue (**2009**): Progress and development in thermodynamic theory of friction and wear, *Science in China, Series E, Technological Sciences*, Vol. 5, pp. 844–849.

97. Mehdi Amiri, Michael M. Khonsari (**2010**): On the thermodynamics of friction and wear—A review, *Entropy*, Vol. 12, pp. 1021–1049.

98. Michael M. Khonsari, Mehdi Amiri (**2013**): *Introduction to Thermodynamics of Mechanical Fatigue.* CRC Press, Taylor & Francis Group: Boca Raton, London, New York.

99. M. Banjac, A. Vencl, S. Otović (**2014**): Friction and wear processes: Thermodynamic approach, *Tribology in Industry*, Vol. 36, pp. 341–347.

Index